I0131977

Membrane Technologies for Heavy Metal Removal from Water

This book offers lucid treatment of fundamental concepts related to potential applications and prospects of different membranes for wastewater decontamination by removing heavy metals. Divided into two sections, it provides an overview of different sources of water contamination, their impacts on human health and the environment, and compares traditional methods used to nullify these impacts. Further, it covers different mature membrane technologies such as polymeric membranes, poly-ceramic membranes, carbon-based membranes and many more, followed by pertinent case studies.

Features:

- Focuses on the removal of heavy metals using membrane-based technologies.
- Discusses pertinent criteria to select suitable membranes.
- Includes feasibility studies and applications of different mature and emerging membranes.
- Describes heavy metals' occurrence and transport in an aqueous system with an overview of the adverse effects.
- Reviews challenges and opportunities associated with using different membranes.

This book is aimed at graduate students and researchers in materials science, water engineering and wastewater treatment.

Membrane Technologies for Heavy Metal Removal from Water

Edited by
Juhana Jaafar, Asad A. Zaidi
and Muhammad Nihal Naseer

CRC Press
Taylor & Francis Group
Boca Raton London New York

CRC Press is an imprint of the
Taylor & Francis Group, an **informa** business

Designed cover image: © Juhana Jaafar

First edition published 2024
by CRC Press
2385 NW Executive Center Drive, Suite 320, Boca Raton FL 33431

and by CRC Press
4 Park Square, Milton Park, Abingdon, Oxon, OX14 4RN

CRC Press is an imprint of Taylor & Francis Group, LLC

© 2024 selection and editorial matter, Juhana Jaafar, Asad A. Zaidi and Muhammad Nihal Naseer; individual chapters, the contributors

Reasonable efforts have been made to publish reliable data and information, but the author and publisher cannot assume responsibility for the validity of all materials or the consequences of their use. The authors and publishers have attempted to trace the copyright holders of all material reproduced in this publication and apologize to copyright holders if permission to publish in this form has not been obtained. If any copyright material has not been acknowledged please write and let us know so we may rectify in any future reprint.

Except as permitted under U.S. Copyright Law, no part of this book may be reprinted, reproduced, transmitted, or utilized in any form by any electronic, mechanical, or other means, now known or hereafter invented, including photocopying, microfilming, and recording, or in any information storage or retrieval system, without written permission from the publishers.

For permission to photocopy or use material electronically from this work, access www.copyright.com or contact the Copyright Clearance Center, Inc. (CCC), 222 Rosewood Drive, Danvers, MA 01923, 978-750-8400. For works that are not available on CCC please contact mpkbookspermissions@tandf.co.uk

Trademark notice: Product or corporate names may be trademarks or registered trademarks and are used only for identification and explanation without intent to infringe.

ISBN: 9781032353050 (hbk)
ISBN: 9781032353067 (pbk)
ISBN: 9781003326281 (ebk)

DOI: 10.1201/9781003326281

Typeset in Times
by codeMantra

Contents

Preface

"Water is the driving force of all nature." – Leonardo da Vinci

A beautiful proverb which portrays the essentiality of water for life throughout the universe. Scientists searching for life beyond Earth and beyond our solar system search for planets that are in the "Goldilocks zone," not too close or too far from their Sun so that water could exist in a liquid state. This shows how much of a part water takes to make us live. The reduction in the availability of clean water resources is heart wrenching. People from certain parts of the Earth with no proper access to fix this issue have no other choice rather than being exposed to contaminated water every day. Heavy metals are released into water bodies as products from industrialization. These non-biodegradable and carcinogenic pollutants find their way into the environment, threatening human health and the ecosystem. Various water purification technologies have been invented and are being practiced for years to reduce the severity of having heavy metal components in the water. In that case, the technological advancement in membrane development has led to an increased use of membranes for filtration and extraction of heavy metals from wastewater. Accordingly, this book focuses on heavy metals as the prime contaminant of water and membrane technologies as the saviors in making sure clean water is available for users.

There are two sections that make up this book whereby Section 1 is an introductory part with five general topics with an overall view about heavy metals and membrane technology. On the other hand, Section 2 covers 18 topics where each of the topics is highlighted specifically according to the type of membrane technology, type of heavy metals or different adsorbents used for heavy metal remediation.

Chapter 1 in Section 1 is about the origin of water pollution because of heavy metals. Sources of different heavy metals, their characteristics and the impacts of heavy metals are covered in this chapter. Also, the indexing approach of heavy metal pollution and available analytical methods for the detection of heavy metals have also been elucidated.

In Chapter 2, the potential of upcoming membrane technologies in removing heavy metals has been justified. Different membrane applications such as pressure-driven membranes, forward osmosis, electrified membranes, liquid membranes and hybrid membranes are highlighted. Also, the membrane modules and the usage of correct modules associated with the application have also been emphasized. Further information about membrane fouling and pre-treatment strategies have also been covered in this part.

Chapter 3 is about the health risks associated with heavy metals in water. The toxicity mechanisms of different types of heavy metals have been focused on.

Adding on, Chapter 4 specifically explains about metal oxides and their toxicity in the form of nanoparticles toward individual and the environment.

The last chapter of this section, which is Chapter 5, revolves around the theoretical details about advanced and commonly used membrane technologies. Comparisons have been made on the basis of heavy metal removal, water recovery and reclamation. The purpose of this chapter is to amalgamate relevant details about membrane technology by gathering up required literatures to further identify the research gaps.

Chapter 6 as the opening of Section 2 starts with biopolymeric membranes for heavy metal removal. Being an eco-friendly way to eliminate heavy metals, this type of membrane secured its place due to its biodegradable nature.

As in Chapter 7, bioremediation of heavy metals using plants and bacteria has been explained thoroughly.

Moving on with Chapter 8, a common heavy metal which is boron and different technologies related with its removal from seawater have been discussed and compared. Seawater reverse osmosis and the future of this desalination treatment is elaborated on in this segment.

Chapter 9 deals with an interesting study involving fabrication and characterization of chelating membranes using semi-interpenetrating polymer networks. Apart from being porous and permeable, this membrane is proven to provide a better complexation reaction toward heavy metals.

Chapter 10 focuses on graphene and graphene-bonded nanomaterials in alleviating heavy metals from water and wastewater. The process of synthesizing and modifying nanomaterials has been included in this portion. Apart from explaining several graphene-based nanocomposites, their adsorption process and the adsorptive mechanisms have been considered too in this chapter.

Chapter 11 provides an in-depth analysis about the integration of photocatalysts in membrane technologies to remove heavy metals. Numerous photocatalysts, mechanisms of photocatalytic reaction and the factors affecting the process is covered in this. Rather than highlighting the fabrication and characterization techniques, this chapter also offers membrane fouling control measures and underlines the future prospects of this method.

Chapter 12 reviews the utilization of ionic liquid membranes for heavy metal expulsion. The selectivity and adsorption capacity of ionic liquids and the role in promoting the extraction of heavy metals has been debated. The practical application of combining ionic liquids in the membrane matrix and its ability to perform on a large scale have also been analyzed.

Chapter 13 comprehensively discusses the investigation of different membrane technologies using life cycle assessment. This approach is applied in order to know the impact of the membrane toward the environment. The integration of membrane technology with other separation technique is stressed and analyzed on the basis of economical and marketing capability.

Chapter 14 offers an overall view about the usage of nanofiber membranes in removing chromium, which is a heavy metal. The drawbacks of readily available chromium removal technologies is studied and the significance of adopting nanoparticles incorporating nanofiber membranes and the mechanisms involved are presented.

Chapter 15 is committed to providing a comprehensive review about available nano-adsorbents for heavy metal removal. The feasibility and adsorptive capability of various nanoparticles to play the role of heavy metals' captivators are discussed.

Chapter 16 expounds about the employment of polymeric membranes in removing heavy metals. Common polymers and nanomaterials that make up polymeric membranes and the physicochemical properties they offer to withstand the operation of water and wastewater treatment are also reviewed in this chapter.

Chapter 17 is devoted to discussing about the part that the bacterial cell transmembrane plays as a biopolymer in eliminating mercury, an extremely hazardous heavy metal from industrial wastewater. Apart from detoxifying heavy metals by not harming nature, the bacteria applied through this method are said to be able to provide some additional benefits.

Chapter 18 discusses the influence of ionic/zwitterionic materials in polymer membranes for heavy metal rejection. The improvement in terms of membrane performance provided by this material has been scrutinized and the challenges associated with its adsorptive performance have been studied too.

Equipped with critical reviews and skillful suggestions from several well-known researchers around the world, this book is likely to provide the readers with most extensive and trustworthy literature that has ever been published in this field and will undoubtedly serve as a potent source of information for those interested in this field. Finally, we would like to express our thanks to each of the contributors to this book for their dedication and willingness to share knowledge, expertise, passion and time.

About the editors

Juhana Jaafar is an Associate Professor in the School of Chemical and Energy Engineering, Universiti Teknologi Malaysia. She is also the Deputy Director of Advanced Membrane Technology Research Centre (AMTEC), a Higher Institute Centre of Excellent (HICOE) rewarded by the Ministry of Education, Malaysia. Her outstanding outputs in research are evident from her having received distinguished awards at the national and international level including the World Class Professor award 2021 under the Ministry of Education, Culture, Research and Technology of Indonesia, Grand Awards in Seoul International Invention Fair (SIIF 2019) and many more. She is also active in writing for scientific publications in high-impact international and national journals. To date, she has published more than 350 papers in ISI-indexed journals with an H-index of 34.

Muhammad Nihal Naseer is a mechanical engineer from the National University of Sciences & Technology (NUST)–Pakistan. He started his research career in 2018 from the Laboratory of Applied Sciences at NUST–Pakistan. His research interests lie in the area of thermal sciences, energy-water-waste-food nexus and nanotechnology. In 2019, he was research assistant in the NANOCAT Research Centre of the University of Malaya–Malaysia. He has more than ten publications to his credit published in peer reviewed journals and conferences. He has published articles in reputed journals of the field such as *Energy Reports*. In addition, he has filed two innovations for patent and edited a book *Utilization of Thermal Potential of Abandoned Wells*.

Asad A. Zaidi is Associate Professor at the Department of Mechanical Engineering, Faculty of Engineering Sciences and Technology, Hamdard University–Pakistan. He has also served as head of undergraduate program, assistant professor of thermal sciences and research supervisor at the National University of Sciences and Technology–Pakistan. He is an experienced researcher with a demonstrated history of working in the higher education industry. He is skilled in materials, renewable energy and low-cost water production technologies. His research interest spans thermal modeling, energy system analysis and auditing, renewable energy resources assessment, energy-water-waste nexus, and water-biomass-wind and waves energies technology. He is an author or co-author of more than 65 papers in international refereed journals and conferences. He has also given several invited/plenary talks at international conferences.

Contributors

Robina Khan
Department Microbiology, Hazara University, Maneshra, Pakistan

Zohaib Abbas
Department of Environmental Sciences and Engineering, Government College University, Allama Iqbal Road, 38000 Faisalabad, Pakistan

Hafiz Affan Abid
Electrospun Materials & Polymeric Membranes Research Group, National Textile University, Faisalabad, Pakistan

Sharjeel Abid
DGM Processing, Beacon Impex Pvt. Ltd., Faisalabad, Pakistan

Zainul Abideen
Dr. Muhammad Ajmal Khan Institute of Sustainable Halophyte Utilization, University of Karachi, 75270, Pakistan

Augustine Chioma Affam
Civil Engineering Department, School of Engineering and Technology, University of Technology Sarawak, Persiaran Brooke, 96000 Sibu, Sarawak, Malaysia Centre of Research for Innovation and Sustainable Development (CRISD), University of Technology Sarawak, Sibu, Malaysia

Ouafa Tahiri Alaoui
Faculté des Sciences et Techniques, Département de chimie, Laboratoire Chimie Physique, Matériaux et Environnement, Université Moulay Ismail, BP 509 Boutalamine, Errachidia, Maroc

Shafaqat Ali
Department of Environmental Sciences and Engineering, Government College University, Allama Iqbal Road, 38000 Faisalabad, Pakistan.

Meryem Amar
Faculté des Sciences et Techniques, Département de chimie, Laboratoire Chimie Physique, Matériaux et Environnement, Université Moulay Ismail, BP 509 Boutalamine, Errachidia, Maroc

Aatif Amin
Department of Microbiology, Faculty of Life Sciences, University of Central Punjab, Lahore-54000, Pakistan

Sorth Ansari
US Pakistan Center for Advanced Studies in Water, Mehran University of Engineering and Technology, Jamshoro, 76062, Pakistan

Irfan Aziz
Dr. Muhammad Ajmal Khan Institute of Sustainable Halophyte Utilization, University of Karachi, 75270, Pakistan

Hamid Barkouch
Faculté des Sciences et Techniques, Département de chimie, Laboratoire Chimie Physique, Matériaux et Environnement, Université Moulay Ismail, BP 509 Boutalamine, Errachidia, Maroc

Marion Bellier
School of Sustainable Engineering and the Built Environment, Arizona State University, Tempe, Arizona 85287-3005, United States

Haad Bessbousse
Ecole Supérieure de Technologie de Sidi Bennour, Laboratoire de Management de l'Agriculture Durable, Université Chouaib Doukkali, Avenue Jabran Khalil Jabran, B.P. 299-24000, El Jadida, Maroc

Abdullahi Haruna Birniwa
Department of Chemistry, Sule Lamido University, PMB 048 Kafin-Hausa, Nigeria.

Sayed Muhammad Ata Ullah Shah Bukhari
Departrment of Microbiology, Quaid-i-Azam
    University, Islamabad-45320, Pakistan.

Adil Denizli
Chemistry Department, Faculty of Science,
    Hacettepe University.

Sisem Ektirici
Chemistry Department, Faculty of Science,
    Hacettepe University.

Muhammad Faisal
University of Punjab, Pakistan

Shahnaz Ghasemi
Sharif Energy, Water and Environment
    Institute, Sharif University of Technology,
    Azadi Avenue, P.O. Box 11365-9465,
    Tehran, Iran

C. Godinez-Seoane
Department of Chemical and Environmental
    Engineering, Technical University of
    Cartagena (UPCT), Campus Muralla del
    Mar, E-30202, Cartagena, Spain

Bouchaib Gourich
Laboratory of Process and Environmental
    Engineering, Higher School of Technology,
    Hassan II University, 20000, Casablanca,
    Morocco

Maria Hanif
Department of Biotechnology, Lahore College
    for Women University, Lahore, Pakistan

Maria Hasnaian
Department of Biotechnology, Lahore College
    for Women University, Lahore, Pakistan

Gaohong He
R&D Center of Membrane Science and
    Technology, School of Chemical
    Engineering, Dalian University of
    Technology, Dalian, 116024 China

Tanveer Hussain
Department of Textile Engineering, National
    Textile University, Faisalabad, Pakistan
    Electrospun Materials & Polymeric

Membranes Research Group, National
    Textile University, Faisalabad, Pakistan

Abdulmalik Hussaini
Department of Civil Engineering, Federal
    University Dutsin-Ma, Dutsin-Ma P.M.B.
    5001, Katsina State, Nigeria

Arun M. Isloor
Membrane Technology Laboratory,
    Department of Chemistry, National Institute
    of Technology Karnataka, Surathkal,
    Mangalore, India

Ahmad Hussaini Jagaba
Department of Civil Engineering, Abubakar
    Tafawa Balewa University, Bauchi, Nigeria

Mitra Jalilzade
Chemistry Department, Faculty of Science,
    Hacettepe University.

Muhsin Jamal
Department of Microbiology, Abdul Wali Khan
    University, Mardan-23200, Pakistan

Muhammad Awais Khalid
Department of Environmental Sciences, Faculty
    of Bio Sciences, University of Veterinary and
    Animal Sciences, Lahore, Pakistan.

Aamir Khan
Institute of Environmental Sciences and
    Engineering (IESE), School of Civil and
    Environmental Engineering (SCEE),
    National University of Sciences and
    Technology (NUST), Islamabad, Pakistan

Hera Naheed Khan
Department of Microbiology & Molecular
    Genetics, University of the Punjab, Lahore
    54590, Pakistan

Rafi Ullah Khan
Institute of Polymer and Textile Engineering,
    University of the Punjab, Lahore, 54590
    Pakistan.

Sher Jamal Khan
Institute of Environmental Sciences and
    Engineering (IESE), School of Civil and

Environmental Engineering (SCEE),
National University of Sciences and
Technology (NUST), Islamabad, Pakistan

Nura Shehu Aliyu Yaro
Department of Civil Engineering, Ahmadu
Bello University, Zaria 810107, Kaduna
State, Nigeria

Ibrahim Mohammed Lawal
Department of Civil Engineering, Abubakar
Tafawa Balewa University, Bauchi, Nigeria.
Department of Civil and Environmental
Engineering, University of Strathclyde,
Glasgow, UK

Laurent Lebrun
Laboratoire Polymères, Biopolymères,
Surfaces, Université de Rouen, 76821 Mont-
Saint-Aignan Cedex, France

L.J. Lozano-Blanco
Department of Chemical and Environmental
Engineering, Technical University of
Cartagena (UPCT), Campus Muralla del
Mar, E-30202, Cartagena, Spain

Rasool Bux Mahar
US Pakistan Center for Advanced
Studies in Water, Mehran University of
Engineering and Technology, Jamshoro,
76062, Pakistan

Urwa Mahmood
Department of Textile Engineering,
National Textile University,
Faisalabad, Pakistan

Sheeraz Ahmed Memon
Institute of Environmental Engineering
and Management, Mehran University of
Engineering and Technology, Jamshoro,
Pakistan

Syed Ibrahim Gnani Peer Mohamed
Membrane Technology Laboratory,
Department of Chemistry,
National Institute of Technology
Karnataka, Surathkal, Mangalore, India

Neelma Munir
Department of Biotechnology, Lahore College
for Women University, Lahore, Pakistan

Noura Najid
Laboratory of Process and Environmental
Engineering, Higher School of Technology,
Hassan II University, 20000, Casablanca,
Morocco

Muhammad Nihal Naseer
Department of Mechanical Engineering,
National University of Science and
Technology Islamabad, Pakistan.

Muhammad Asif Nawaz
Department of Biotechnology, Shaheed Benazir
Bhutto University, Sheringal, Dir (Upper),
Pakistan

Ahsan Nazir
Department of Textile Engineering,
National Textile University, Faisalabad,
Pakistan Electrospun Materials &
Polymeric Membranes Research Group,
National Textile University, Faisalabad,
Pakistan

Azmatullah Noor
Department of Civil and Environmental
Engineering, Universiti Teknologi
PETRONAS, Bandar Seri Iskandar 32610,
Perak Darul Ridzuan, Malaysia.

V.M. Ortiz-Martínez
Department of Chemical and Environmental
Engineering, Technical University of
Cartagena (UPCT), Campus Muralla del
Mar, E-30202, Cartagena, Spain

François Perreault
School of Sustainable Engineering and the Built
Environment, Arizona State University,
Tempe, Arizona 85287-3005, United States

Abdul Majeed Pirzada
Department of Environmental Sciences, Sindh
Maderasatul Islam University, Karachi,
Pakistan

Naveed Ahmed Qambrani
US Pakistan Center for Advanced Studies in Water, Mehran University of Engineering and Technology, Jamshoro, 76062, Pakistan

Sunbul Rasheed
Department of Microbiology, Faculty of Life Sciences, University of Central Punjab, Lahore-54000, Pakistan

Sana Raza
Departrment of Microbiology, Abdul Wali Khan University, Mardan-23200, Pakistan.

Muhammad Rizwan
US Pakistan Center for Advanced Studies in Water, Mehran University of Engineering and Technology, Jamshoro, 76062, Pakistan

Umm E Ruman
Department of Chemistry, University of Gujrat, Gujrat, 57200, Pakistan

Aneela Sabir
Institute of Polymer and Textile Engineering, University of the Punjab, Lahore, 54590 Pakistan.

María José Salar García
Department of Chemical and Environmental Engineering, Technical University of Cartagena (UPCT), Campus Alfonso XIII, Aulario C, E-30203, Cartagena, Spain

Dalhatu Saleh
Department of Civil Engineering, Faculty of Engineering Technology, Nigerian Army University, Biu, PMB 1500, Borno State Nigeria.

S. Sánchez-Segado
Department of Chemical and Environmental Engineering, Technical University of Cartagena (UPCT), Campus Muralla del Mar, E-30202, Cartagena, Spain

Zirwa Sarwa
Department of Biotechnology, Lahore College for Women University, Lahore, Pakistan

Muhammad Shafiq
Institute of Polymer and Textile Engineering, University of the Punjab, Lahore, 54590 Pakistan.

Liloma Shah
Departrment of Microbiology, Abdul Wali Khan University, Mardan-23200, Pakistan.

Iqra Shahid
Department of Environmental Sciences, Faculty of Bio Sciences, University of Veterinary and Animal Sciences, Lahore, Pakistan.

Amna Siddique
Department of Textile Technology, National Textile University, Faisalabad, Pakistan

Usman Bala Soja
Department of Civil Engineering, Federal University Dutsin-Ma, Dutsin-Ma P.M.B. 5001, Katsina State, Nigeria

Shamas Tabraiz
School of Natural and Applied Sciences, Canterbury Christ Church University, Canterbury, CT1 1QU, United Kingdom.

Izza Taufiq
Institute of Microbiology and Molecular Genetics, University of the Punjab, Lahore, Pakistan

Deniz Türkmen
Chemistry Department, Faculty of Science, Hacettepe University, Türkiye

Fatima Tu Zahra
Institute of Microbiology and Molecular Genetics, University of the Punjab, Lahore, Pakistan

Abdullahi Kilaco Usman
Civil Engineering Department, University of Hafr Al-Batin, Hafr Al-Batin, Saudi Arabia.

Jean-François Verchère
Laboratoire Polymères, Biopolymères, Surfaces, Université de Rouen, 76821 Mont-Saint-Aignan Cedex, France

Maria Wasim
Institute of Polymer and Textile Engineering,
University of the Punjab, Lahore, 54590,
Pakistan

Fatna Zaakour
Ecole Supérieure de Technologie de Sidi
Bennour, Laboratoire de Management de
l'Agriculture Durable, Université Chouaib
Doukkali, Avenue Jabran Khalil Jabran, B.P
299-24000, El Jadida, Maroc.

Ihsan Elahi Zaheer
Department of Environmental Sciences
and Engineering, Government College
University, Allama Iqbal Road, 38000
Faisalabad, Pakistan.

Muhammad Zeeshan
German Environment Agency, Section II 3.3,
Schichauweg 58, 12307, Berlin, Germany.
Technische Universitat Berlin, Water
Treatment, KF4, Str. des 17. Juni 135, 10623,
Berlin, Germany.

Asma Zouitine
Ecole Supérieure de Technologie de Sidi
Bennour, Laboratoire de Management de
l'Agriculture Durable, Université Chouaib
Doukkali, Avenue Jabran Khalil Jabran, B.P
299-24000, El Jadida, Maroc.

Muhammad Zubair
Department of Chemistry, University of Gujrat,
Gujrat, 57200, Pakistan

Section 1

Introduction

1 Sources of Water Contamination by Heavy Metals

Ahmad Hussaini Jagaba, Ibrahim Mohammed Lawal,
Abdullahi Haruna Birniwa, Augustine Chioma Affam,
Abdullahi Kilaco Usman, Usman Bala Soja,
Dalhatu Saleh, Abdulmalik Hussaini,
Azmatullah Noor and Nura Shehu Aliyu Yaro

1.1 INTRODUCTION

Water is the most important commodity on the planet, and it is essential for all living organisms in its purest form because it makes up the bulk of living tissues, and life cannot exist without it [1]. Heavy metals have been released into the environment because of poor industrial waste disposal on land and in bodies of water because of intense industry and unplanned urbanization in the eco-system, as well as other natural and man-made activities. This has been a big source of concern in terms of safe drinking water [2]. With increasing population and industrialization, the demand for pure water continues to rise. Drinking water is scarce in many parts of the world, making water contamination a recent source of concern for the entire world. The quality of water resources has been compromised by a variety of contaminants discharged from various sources. To be more specific, heavy metal discharge contributes greatly to water pollution [3]. Heavy metals are a collection of naturally occurring metals with high *densities* (3.5–7 g/cm^3) and *atomic weights* > 23 or *atomic numbers* > 20 that have high densities [2]. Metals and metalloids have been extracted from minerals for millennia and have been used by mankind. Heavy metals are classified as such because of their large atomic weight or density. They are harmful to both the environment and people. Some heavy metals, on the other hand, are rarely hazardous [4]. Heavy metal concentrations (see Figure 1.1) can be naturally high in some areas due to geogenic factors; some heavy metals are micronutrients, while others are non-essential heavy metals [5]. Heavy metals are a unique class of contaminants found in water reservoirs. Because they do not filter themselves out of water but rather concentrate in reservoirs and penetrate the food chain, they have a strong ecological impact. Metal levels are frequently indicated by a surge in metal concentrations in a reservoir's bottom sediment [6, 7]. Heavy metal contamination can be found in soil, water, and the atmosphere, among other places. Because of their substantial inhibitory effects on biodegradation activities, heavy metals are directly linked to environmental pollution and biological toxicity issues [8]. The presence of heavy metals in wastewater released from laboratories and industry is a major source of environmental concern. They are persistent environmental pollutants that cannot be completely biodegraded and must instead be converted into harmless forms. Streams become unsuitable when these metal ions are present in excess amounts in an aqueous discharge due to the negative effects associated with ingestion [9].

DOI: 10.1201/9781003326281-2

1.1.1 PROPERTIES OF HEAVY METALS

Most of the metals have a lower level of reactivity than lighter metals and usually have a higher density. In the Earth's crust, they are comparatively scarce/less plentiful. They are incredibly tough and difficult to cut. Heavy metals have a low thermal expansivity and a high tensile strength, and their melting points range from low to high. In hydroxides they are normally insoluble and in sulphides they are exceedingly insoluble. Heavy metal complexes are mostly colourful. As a result, they create coloured water solutions. Heavy metals are micronutrients that must be consumed in small amounts, whereas light metals must be consumed in higher amounts [10]. They are one of the longest-lasting contaminants in wastewater. Because they are non-biodegradable and poisonous, they persist in wastewater [11]. Unlike many organic pollutants, which disintegrate into H_2O and CO_2, heavy metals begin to rack up in the environment, particularly in lake, estuary, and marine sediments [12].

Heavy metals form water-soluble complexes when they mix with anions. Because of their high solubility, these complexes can be transported and can bioaccumulate in a range of animals, particularly in the aquatic biome, transported to other areas, thus contributing to their toxicity [13]. Heavy metals should be handled with care because some are less reactive than others. Heavy metals pollute the air, water, and soil, causing health issues in plants, animals, and humans. Heavy metals are non-biodegradable and are unable to degrade. They have a long lifespan in the soil. Heavy metals bioaccumulate in the system when they are eaten or inhaled by humans. As a result, they are considered dangerous. Some heavy metals are considered essential elements because they are necessary for survival. The necessary elements are required for the creation of skeletal structures, the regulation of acid-base equilibrium, and the preservation of the colloidal system. They also play a role in the production of essential enzymes, structural proteins, and hormones. They can, however, be poisonous in excessive quantities, as seen in Figure 1.2. They have infiltrated our atmosphere, waterways, and soils [14].

The principal elements required by the body, macrominerals, and trace elements are divided into three categories:

- The basic blocks of most biological stuff require four important main components. According to their atomic number, these are nitrogen, hydrogen, oxygen, and carbon.
- The macrominerals are a group of seven key elements that are necessary for the ionic balance of structural molecules, nucleic acids, and amino acids to be maintained. These elements are sodium, chlorine, magnesium, sulphur, potassium, phosphorous, and calcium based on their atomic mass.
- The trace elements are made up of 13 elements including chromium, copper, iodine, iron, silicon, arsenic, nickel, vanadium, molybdenum, manganese, cobalt, selenium, and zinc [15].

Although non-essential metals have no role in the body, they can cause toxicity by lowering the body's level of a vital element. Nuclear proteins and DNA have been shown to interact with heavy metals, causing damage at specific sites. They have also been discovered to activate signalling pathways [15].

The following is a list of heavy metals in order of their toxicity to living species: silver > copper > zinc > nickel > lead > cadmium > chromium > stannum > iron > manganese > aluminium. They stop biodegradation from happening. Each heavy metal has a different composition, physical qualities, and chemical structure. Arsenic, manganese, nickel, titanium, molybdenum, vanadium, chromium, iron, zinc, silver, copper, cadmium, tin, cobalt, platinum, gold, mercury, and lead is a group of heavy metals that have densities of more than 5 g/cm^3 and are more commonly found in our daily lives [15].

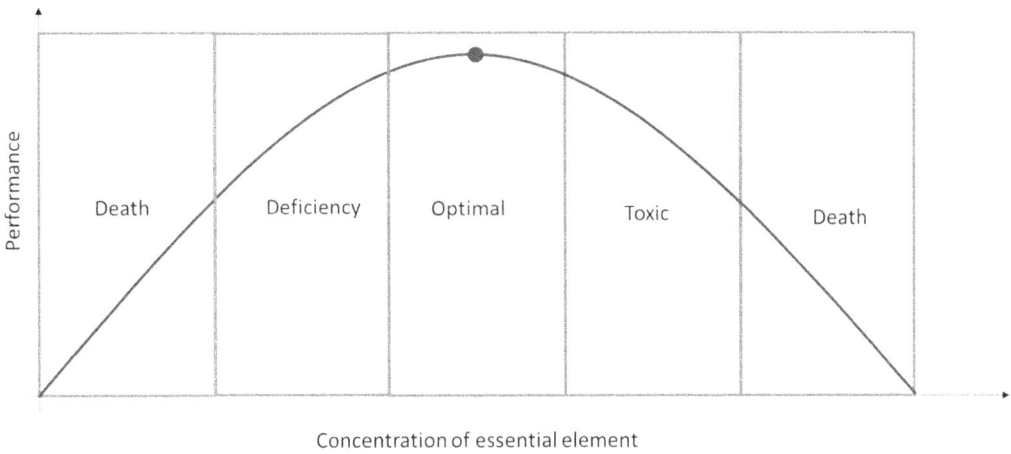

FIGURE 1.1 The correlation between an individual's performance and the concentration of a heavy metal [15].

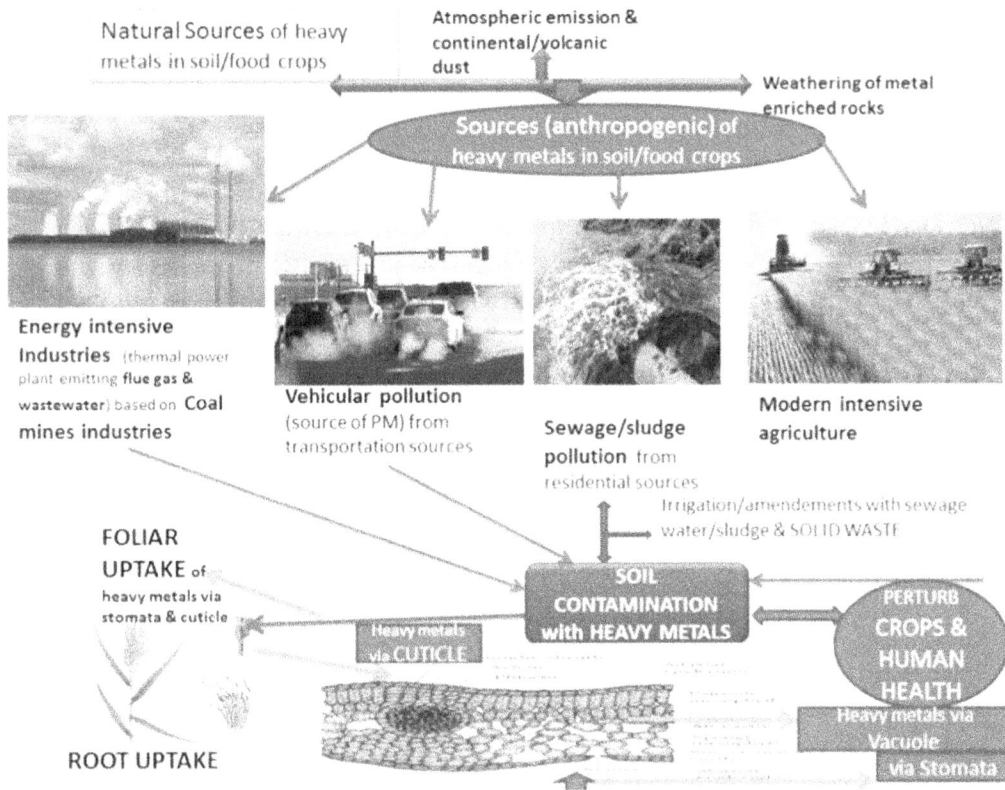

FIGURE 1.2 Natural and anthropogenic sources of heavy metal contamination and mechanisms of their entrance with resulting impacts on biota and humans [20].

1.1.2 USES OF HEAVY METALS

Heavy metals are employed in a range of industrial, residential, and agricultural processes in modern cultures, and their production and consumption are on the rise [4]. The typical usage of heavy metals is determined by general features such as electrical conductivity and reflection. Other common properties include density, strength, and durability. Life and ecosystems require a wide range of heavy metals; they are required to keep the body's metabolism running smoothly [16]. Other applications are dependent on the element's quality, such as its biological role as nutrients or toxins, or other atomic qualities. Sport, mechanical engineering, military ordnance, and nuclear physics are some of the other applications that take advantage of their comparatively high density. When maximum weight in a compact space is necessary, such as in boats, heavy metals are employed; they are also employed as ballast and in aeroplanes and other motor vehicles, as well as for balancing weights on wheels, gyroscopes, propellers, and centrifugal clutches. Tungsten and uranium are employed in military ordnance to make armour-plating, armour-penetrating missiles, and nuclear weapons more effective. Heavy metals are used in linear accelerators and radiotherapy applications for radiation shielding and beam focusing. Different heavy metals have recently been used as the core atom in artificially constructed "bioinorganic" catalysts for specific chemical reactions. Heavy metals are important in the hydrosphere because they interact with geologically produced soil/sediment samples, which might affect biological processes.

1.1.3 IMPACTS OF HEAVY METALS

Water pollution is proportional to the number of pollutants present. As a result, the quality of both subsurface and surface water must be monitored frequently. The potential risks of heavy metals have sparked worldwide concern due to their stability, accumulation, and biological amplification. Heavy metal contamination is a severe environmental problem in many regions, especially rich ones, despite increasingly stringent environmental protection regulations in recent decades [17]. According to a recent study conducted by the European Environment Agency (EEA), it is estimated that heavy metals are still present in high concentrations in 75%–96% of European waters. Heavy metal toxicity in drinking water causes fatigue, affects the nervous system, and has a severe impact on mental health. Heavy metal toxicity can also cause changes in blood composition and damage to key organs such as the liver and kidneys. Long-term exposure to these heavy metals can cause muscle, musculoskeletal, and neurological damage, as well as diseases like muscular dystrophy (skeletal muscle weakening), Alzheimer's disease (brain problem), Parkinson's disease, and multiple sclerosis [18]. Any heavy metal that is present more than the allowed limit in drinking water might cause poisoning. Therefore, heavy metal toxicity in water should be investigated, and various solutions for dealing with contaminated water should be applied to limit the risk of health problems [19]. There are around 50 elements classified as heavy metals, with 17 being the most dangerous. The level of toxicity is determined by the metal, its functions, and the species that are exposed to the heavy metal. Metal poisoning causes DNA damage, sulphydryl homeostasis disturbance, and lipid peroxidation by releasing free radicals. In farm, residential, pharmaceutical, and manufacturing settings, heavy metals can enter the body through the swallowing of contaminated food, by inhaling contaminated air, by drinking contaminated water, and by skin contact [15].

1.2 SOURCES OF HEAVY METALS

Heavy metals can be split into two categories based on their natural and human sources. They primarily end up in the soil, water, air, and the contact between them.

1.2.1 Natural Sources

Several investigations have identified various natural heavy metal sources. Natural heavy metal emissions occur under a variety of different and unique environmental conditions. This could be due to a variety of natural geographical phenomena, such as rock weathering, mineral degradation, forest fires, volcanic eruptions, biogenic sources, sea-salt sprays, evaporation from soil and water surfaces, and wind-borne soil particles. Natural heavy metal emissions are a local and worldwide environmental hazard that must be addressed. Every year, about 1.72107 mg/kg As is released as volcanic ash, while the Earth's crust contains 4.011016 mg/kg As and another 4.87106 mg/kg is released by undersea volcanoes. Metals are harvested from natural sources by the industrial sector, and these metals are then released into the atmosphere. Metals can be liberated from their endemic spheres by natural weathering processes and can end up in a variety of environments; for example, the resultant particles can end up in aquatic systems such as seas and rivers, contaminating water [19]. Complexes of oxides, sulphides, silicates, sulphates, hydroxides, organic compounds, and phosphates are also considered salts of heavy metals. Lead, arsenic, copper, zinc, cadmium, nickel, mercury, and chromium are the most prevalent heavy metals [21].

1.2.2 Anthropogenic Sources

Anthropogenic sources are often regarded as the primary contributors to rising levels of heavy metal contamination in many environments. Heavy metals from anthropogenic sources have been proven to exceed natural flows in a few circumstances. Heavy metals are produced because of a variety of human activities. The daily manufacture of products to suit the demands of a huge population has been proven to contribute more to environmental pollution [21]. Wastewater, industries, agriculture, mining, and runoffs all release pollutants into various environments. Heavy metal contamination can be caused by a variety of industrial activities and their effluents. In comparison to others, some sectors may constitute a significant hazard to various ecosystems. Because of anthropogenic activities like smelting or treating metal ores, extraction, electroplating production, mining practices, coal-fired power and heat production, leather production, burning of fossil fuels, metallurgic activities, automobile exhaust that releases lead, insecticides, waste incineration, landfills, heavy metals can enter the environment as agricultural, residential, and industrial waste.

1.2.3 Diverse Sources

Heavy metals are abundant in the Earth's crust, and runoff can carry them into the ecosystem. Furthermore, rushing water causes heavy metal build-up in water reservoirs. Heavy metals may also be released into the environment via waste. Construction of ports, power generation and transmission, railways, drainage systems, dredging, housing and recreational systems, and waste treatment systems are also potential sources. Heavy metal residues could be present in some of the materials utilized in these activities [22]. Heavy metal impurities can be present in a wide range of items, which include detergents, fertilizers, and refined petroleum products. As a result, the various sources of heavy metals can be categorized as follows: waste material and industrial processes.

1.2.3.1 Sources of Waste Materials

1.2.3.1.1 Agriculture Wastes

Agricultural systems have expanded and developed in response to increased food demand [23]. Increased pollution loads in rivers, lakes, aquifers, and coastal waters have resulted from the overuse and misuse of agrochemicals (insecticides, pesticides, herbicides, fungicides, and chemical fertilizers), sediments, water, sludge, organic matter, animal feeds, and drugs on farms designed to increase productivity. Water contamination with carcinogens and other harmful compounds, as a result, poses known threats to aquatic ecosystems, human health, and economic activity. Agricultural waste dumped in aquatic ecosystems has had a number of negative effects on aquatic animals, including fish, by concentrating toxins directly from dirty water and passing them through the food chain.

1.2.3.1.2 Biomedical Waste

Body parts, organs, tissues, blood, and biological fluids, as well as dirty linen, cotton, bandages, and plaster casts from infected and contaminated areas, and spent needles, syringes, and other sharps, can all be found in medical waste. It contains pathogens in their visible and bulk forms. Biological, chemical, and radioactive contaminants can pollute the water supply. Pathogens leaching into the water can pollute it and cause sickness. Water pollution can be caused by heavy metals found in chemical waste. They can enter biological systems via the biological magnification process. The inappropriate disposal of biomedical waste may have a negative influence on water quality because various chemicals may leak into groundwater from waste dumping sites. Heavy metals and polycyclic aromatic hydrocarbons are abundant in biomedical waste, resulting in undesirable levels of dangerous elements that can damage surface and groundwater [24].

1.2.3.1.3 Industrial Effluents

Some of the biggest sources of contamination are municipal trash, home sewage, and industrial waste released directly into natural water systems. Water contamination results from untreated waste discharge. Manufacturing effluents discharged untreated into water bodies are a major source of surface and groundwater contamination. Wastewater containing bacteria, heavy metals, fertilizers, radionuclides, medications, and personal care products makes its way into surface water resources, causing irreversible harm to the aquatic ecosystem and humans. These contaminants lower the amount of usable water available, increase purifying costs, harm aquatic resources, and have an influence on the food supply. Pollutants such as dangerous heavy metals, acids, agrochemicals, dyes, and other untreated waste discharged by industry contribute to water contamination. Discharged items, which frequently create pollution, also cause a loss of biodiversity in the aquatic ecosystem, as well as human health hazards such as cholera, diarrhoea, and other illnesses [25].

1.2.3.1.4 Electronic Waste/(E-Waste)

The manufacture of electronic items is one of the most significant sources of e-waste. E-waste is high in heavy metals, hazardous chemicals, and cancer-causing substances. E-waste contains heavy metals such as lead, cadmium, mercury, arsenic, and nickel, as well as persistent organic chemicals such as brominated flame retardants, phthalates, polychlorinated biphenyls, nonylphenol, triphenyl phosphate, and others. Heavy metals are one of the most harmful compounds found in e-waste due to their toxicity, mobility, and non-biodegradability. Heavy metal contamination is extremely dangerous to water supplies, as well as to soil. Heavy metals in soil, for example, can be washed away by rain and can end up in surrounding ponds; heavy metals can also contaminate groundwater through leaching, which is particularly problematic in acidic environments. The uncontrolled dumping and improper recycling of e-waste endangers human health and the environment in general. They cause immunological illnesses, neurological system disorders, malignancies, and skin ailments, among other things. The proper treatment and disposal of e-waste using appropriate beneficial practices helps avoid certain diseases of the skin, respiratory, digestive, immunological, endocrine, and nervous systems, including cancer. The exponential development in the use of electrical and electronic equipment (EEE) to bridge the digital gap has a negative impact on the environment and human health when telecommunication wastes are not disposed of scientifically. Adopting existing legislation and guidelines following international norms and practices is critical for a healthy e-waste management system [26].

1.2.3.2 Sources from Industrial Processes

1.2.3.2.1 Mining

Heavy metal contamination is most visible in mining districts and abandoned mine sites, and pollution levels drop as one walks away from the mines. Mining is the process of extracting minerals and other geological materials from underground deposits. The mining sector is responsible for extracting the metals and minerals that our society needs for survival. Mining activity is a major source of water pollution. The potential for these compounds to contaminate groundwater and surface water is increased by large amounts of water produced throughout various stages of mining. Surface water is mostly contaminated by numerous sources such as toxic chemical spills, waste material erosion, and mine contamination through water discharge [27]. During the mining and extraction of various elements from their ores, several metals are released. Because of mining procedures, heavy metals are leached and reach aquatic environments. Acid mine damage may develop as those mined regions encounter air and water.

1.2.3.2.2 Mineral Extraction

Mineral processing releases heavy metals into the environment, either directly or indirectly: initially, during the extraction process, when the necessary size is reduced or the surface area is increased to create more effluents, and, later, when heavy metals are leached into the soil from ores.

1.2.3.3.3 Electroplating

Electroplating is a plating technique that uses electrical flow to remove cations of a desired substance from a solution and coat a conductive device with a thin coating of the desired substance, such as metal. It's mostly used to apply a layer of metal beneath a desired component to a surface that doesn't have that property otherwise. Effluents from the electroplating industry pollute the air, water, and soil. Electroplating is a major polluter since it releases harmful compounds and heavy metals into the environment via water, air emissions, and solid waste. In several electroplating processing sectors, heavy metals such as nickel, iron, lead, zinc, chromium, cadmium, and copper have been detected in significant amounts.

1.2.3.3.4 Smelting

Nonferrous metal smelting is one of the most significant human-caused causes of heavy metal pollution in the environment. Metals generated during smelting are released into the environment, including water, soil, and plants, and can eventually enter human bodies via food chains or by direct ingestion, posing a health risk. In numerous studies high levels of lead and cadmium have been found in the blood and urine of persons, particularly children, who live near nonferrous metal smelters. Huge amounts of exhaust fumes containing a variety of heavy metals can be emitted into the air during long-term zinc smelting processes, as well as considerable amounts of smelting wastes produced in piles and spoil heaps. According to studies, zinc smelting activities in Hezhang released over 50 tonnes of mercury and 450 tonnes of cadmium into the environment between 1989 and 2001. Heavy metal contamination was also found in the local surface water, air, and soil compartments, according to the findings.

1.2.3.3.5 Automobile Exhaust Fumes

Motor vehicle exhaust emissions may contain heavy metals. Vehicle wear items like tyres, brakes, and catalysts are also linked to their release. Vehicle movement-induced re-suspension of metal-enriched road dust could be a significant source of pollution, particularly along routes with large traffic volumes and a high number of heavy vehicles [28]. Motor vehicle exhaust gases are the primary source of lead contamination in soils, plants, and the air. Lead tetraethyl was added to gasoline for many years and released during mechanical engine operation. Another cause of pollution with this element along highways could be grease from motor vehicles. The effect of automotive exhaust fumes on lead concentration levels in bread has been demonstrated. There was a link between the concentration of lead in bus terminals and the volume of vehicles counted. Anthropogenic sources of metals in roadway dust are possible. Humans and animals living near road construction sites are exposed to heavy metal contamination of edible plants [29].

1.2.3.3.6 Geothermal Sources

Trace heavy metal pollution from geothermal sources is very likely [30]. Heavy metal poisoning of the environment is a severe problem in liquid-dominated hydrothermal reservoirs with salty fluids at high temperatures and pressures. Natural sources include volcanic eruptions, geysers, seismic activity, fumaroles, and hot springs, which are commonly found nearby [31]. Volcanic eruptions produce volcanic ash, which settles in the water system, creating turbidity, acidity, and low pH. Because of the influence of aerosols containing the strong mineral acids H_2SO_4, HCl, and HF in the plumes, the surface coatings on fresh volcanic ash are extremely acidic. As a result, when ash from a recent eruption comes into touch with water, it has the potential to reduce the pH below levels safe for aquatic life.

1.2.3.3.7 Power Plants

Power producing plants are another source of heavy metals. Thermal pollution from nuclear and fossil fuel facilities is substantial in bodies of water. The Environmental Protection Agency (EPA) estimates that thermoelectric power stations generate 50%–60% of all harmful pollutants emitted into surface waters by all industrial categories under the Clean Water Act (CWA). Power plants operated by using activated coal are the largest source of hazardous pollution among the numerous types of thermoelectric producing units. They can contaminate water bodies by releasing mercury into the atmosphere through boiler flues. Furthermore, these businesses produce ash, which may contain heavy metal particles such as uranium. About half of the 1,100 steam-electric power plants in operation in the United States are coal-fired power plants. Every year, these facilities discharge millions of tonnes of dangerous heavy metals into waterways around the country, including arsenic, selenium, lead, mercury, boron, and cadmium [32].

1.2.3.3.8 Urban Surfaces

Deposits from numerous sources sink to the ground on urban surfaces. The urban surface road and urban soil are the two main surfaces. Heavy metals in city surface roadways come from a variety of internal and external sources. Intrinsic sources include domestic emissions, building facade weathering, pavement surface weathering, and precipitation on previously vulnerable surfaces (atmospheric aerosols) [33].

1.2.3.3.9 Plumbing and Fittings

Chemical compounds are present in water treatment techniques and materials used in plumbing and fittings [34]. Water stagnation in plumbing pipes and distribution system may raise cadmium, copper, lead, iron, or zinc concentrations [13]. Aluminium, sulphate, and sodium, as well as lead, cadmium, copper, nickel, antimony, zinc, or iron, are found in pipes and fittings, whereas manganese, chromium, barium, and arsenic are prevalent in water bodies [35].

1.2.3.3.10 Irrigation Water

The heavy metal concentrations in irrigation water from diverse sources, such as canal water, tube-well water, and sewage water, were significantly varied. The highest quantities of heavy metals were found in sewage water [36].

1.2.3.3.11 Tannery Activities

The tannery business, which uses a range of chemicals in the tanning process, is widely acknowledged as a significant source of heavy metals in the environment, posing considerable environmental dangers around the world. Hazardous chemicals and other heavy toxic trace metals, organic debris, lime, and sulphide, as well as a vast volume of liquid and solid wastes, are found in tannery effluents. Several businesses that are typically associated with tanneries, such as footwear, animal glue, and paint, produce harmful compounds as a result of their processes [37].

1.3 SOURCES, CHARACTERISTICS, USES, AND IMPACTS OF INDIVIDUAL HEAVY METALS

There are a plethora of sources of drinking water such as rivers, lakes, ponds, etc. The most important driver for rising levels of trace metals in water bodies, particularly heavy metals, through water runoff is the expansion of industries and urbanization [38]. Heavy metals are actively the major cause in the rise of numerous waters-borne diseases, including diabetes, Alzheimer's disease, and different forms of cancer. Extended exposure of humans to trace metals may succeed in gradually advancing muscular dystrophy, Parkinson's disease, and multiple sclerosis. Trace metals, for example, arsenic, lead, mercury, and cadmium, are among the ten elements of public concern according to the WHO [39].

1.3.1 Arsenic

1.3.1.1 Characteristics of Arsenic

Arsenic is the world's 21st most common element. It is found in the environment as an odourless and tasteless hazardous metallic element [40]. It can be found as an oxide, sulphide, or metal salt in a variety of substances and is frequently observed at extremely low concentrations in practically all situations. Arsenic exists in trace concentrations and in four oxidation states including As(V), As(0), As(−III), and As(III), The two most common in nature, however, are As(V) and As(−III), both within the pH range of 6–9. Inorganic salts, organic salts, and gaseous salts are the most common forms [2]. Arsenic has an *atomic mass* of 74.92–75 amu, an atomic number of 33, a *density* of 5.72 g/cm^3, a *melting point* of 817°C, *heat of fusion* of 370.3 kJ/kg, *boiling point* of 613°C, *heat of vaporization* of 426.77 kJ/kg, *linear coefficient of thermal expansion* of 5.6×10^{-6}, *specific heat* of 328 J/kg K, and *electrical resistivity* of 26×10^{-6} Ω. An estimated 1.5 mg/L of this metalloid is on the Earth's surface, 3 ng/m^3 in the air, and 10 mg/L in freshwater [2]. It exhibits a complex chemistry. Significant concentrations > 1,000 ng/cm^3 have been measured near industrial sources [29].

1.3.1.2 Sources of Arsenic

Arsenic can have either anthropogenic or natural origins. Arsenic is discharged to the environment through the smelting of copper, zinc, and lead and can also be generated from the manufacture of other chemicals and glasses [41]. Anthropogenic activity, geochemical reactions, and biological actions, among other things, may be linked to the presence of arsenic underground. It is, however, a source of contamination of drinking water sources due to numerous industrial and agricultural operations. Arsenic is found in the environment through natural processes, especially, geochemical reactions, biological activities, volcanic eruptions, and weathering of rocks [42]. It is classified as geogenic, biogenic, and anthropogenic based on the source of occurrence. Generally, thermal power plants, mining, smelting operations, fuels, geothermal activity, mineral dissolution, (pyrite oxidation), discharge of industrial and metallic waste, herbicides, fossil fuel burning, wood preservatives, reductive dissolution, and desorption are the main arsenic sources for groundwater contamination. However, the weathering of rocks and minerals and volcanoes are geogenic sources. It is estimated that the environment is exposed to 12,000 tonnes of arsenic annually and European Union's 27 member states are responsible for the generation of 3 billion metrics tonnes globally [2].

1.3.1.3   Uses of Arsenic

Arsenic has been considered to have huge agricultural, medicinal, and commercial value worldwide. It is commercially used in the production and manufacture of industrial devices such as semiconductors, transistors, wood, paper, pigments, dyes, and glass, and in pharmaceuticals and hide tanning processes. It is also utilized in the manufacture of agro-based products such as crop desiccants, pesticides, herbicides, fungicides, and other valuable additives used in livestock feeds in agriculture [43]. Arsenic has also been used to treat infectious parasites in dogs and poultry, such as filariasis and blackheads, as well as to eliminate tapeworms in dairy animals. Furthermore, arsenic-based drugs are found useful and have been applied in public for the treatment of diseases in the tropics and serve as an anticancer agent in public health.

1.3.1.4   Impact of Arsenic

Because a substantial number of people worldwide are exposed to high arsenic concentrations, 100 g/L or higher, which is greater than the standard, arsenic has been categorized by the WHO as one of the ten substances of public health concern. Arsenic has a negative impact on the health of a variety of organisms. It negatively influences the endocrine, hepatic, reproductive, and nervous systems. It undermines muscles and coagulates protein. Drinking water with arsenic concentrations above 50 µg/L has an adverse effect on human health. Lung, skin, bladder, and kidney cancers, as well as eye, liver, central nervous system, heart, neurological, and respiratory problems, can all be caused by arsenic exposure [44]. Neurobehavioural impacts, respiratory disorders, and hormonal and reproductive system effects are all non-carcinogenic consequences of arsenic. Inorganic arsenic is extremely dangerous, especially its trivalent species because it is a teratogen and carcinogen when compared to its organic counterpart. Arsenic species that are inorganic are known to be more dangerous than those that are organic. Furthermore, As(III) has been reported to be more mobile, soluble, genotoxic, and cytotoxic, increasing the risk of developing arsenic-related illnesses. Arsenic poisoning can be produced via eating, inhalation, parenteral route, and skin contact to some extent. In rural areas away from human activity, arsenic concentrations in the air can range from 1 to 3 ng/m^3 and in urban areas from 20 to 100 ng/m^3. The concentration of arsenic in water is normally less than 10 g/L, which is the WHO's maximum contamination limit unless the sources are near mining sites and mineral resources [2].

1.3.2   NICKEL

1.3.2.1   Sources of Nickel

Thermal power facilities and waste incinerators discharge nickel into the atmosphere. Silver refineries, electroplating, zinc-based casting, and battery industries produce detergents and effluents that end up in water bodies. Wind-blown dust, weathered rocks and soils, volcanic emissions, forest fires, vegetation, specks of dust from volcanic emissions, and the weathering of rocks and soils are all natural sources of atmospheric nickel. Nickel can enter the body through breathing, drinking, eating, and smoking cigarettes. The burning of residual and fuel oils, incineration of municipal waste, and nickel mining and processing are the main anthropogenic sources of nickel emissions to the atmosphere. In nickel and nickel alloy production plants, as well as electroplating, smelting operations, thermal power plants, plumbing and fittings, welding, grinding, cutting operations, vehicle manufacture, battery production, and many other metal-involved production activities.

1.3.2.2   Characteristics of Nickel

Nickel is a heavy metal that is frequently available in metropolitan areas. It is a silvery-white hard, flexible, and elastic metal that is found in the environment in very low amounts. Nickel comes in two commercial forms: sulphide ores and silicate oxide.

1.3.2.3 Uses of Nickel

Living beings require a modest amount of nickel to block the activity of some enzymes and hormones. Steel manufacturing, alloy manufacture, and electroplating of nickel sulphides account for around 42%, 36%, and 18% of nickel's environmental usage, respectively [33].

1.3.2.4 Impacts of Nickel

Nickel is a potentially carcinogenic metal that has been linked to several pathologic consequences in humans. Allergic dermatitis can be caused by skin contact with metallic or soluble nickel compounds. Nickel consumption and overdose can cause significant weight loss, dizziness, birth complications, asthma, hair loss, lung inflammation, fibrosis, emphysema, allergic reactions (dermatitis), heart disorders, chronic bronchitis, reduced lung function, respiratory problems, nasal sinus, and cancer of the nose, larynx, and lung [39].

1.3.3 ZINC

1.3.3.1 Sources of Zinc

Zinc waters can come from human activities or geological rock weathering, such as industrial and domestic wastewater outflows, as well as from animals, where it is crucial for cytoplasmic integrity. The released materials from mine and metallurgic operations, smelting, electroplating, plumbing and fittings, vehicle emissions, and all man-made sources of zinc in the build-up environment are commercial products containing zinc, such as fertilizers [33].

1.3.3.2 Uses of Zinc

Zinc is considered vital for immune responses.

1.3.3.3 Impacts of Zinc

Acrodermatitis enteropathica, diabetic mellitus, high myopia, schizophrenia, and other diseases are linked to excessive zinc intake in humans [44]. Zinc increases the risk of cardiovascular disease and psychological problems. Neural abnormalities, hypertension, nausea, and stomach injury are all possible side effects. It is also responsible for neurotoxic effects on humans [23].

1.3.4 LEAD

1.3.4.1 Sources of Lead

Lead comes from both natural and man-made sources. Lead output has surged by 232%, reaching 11.3 Mt/year. Lead enters the body mostly through breathing and food, and it circulates in the bloodstream as soluble salts, protein complexes, or ions. It mostly builds up in the bones [33]. Corrosion of leaded pipelines in a water transportation system and corrosion of leaded paints can both introduce lead into water. Lead sources include different industrial (mining, fossil fuel burning, manufacturing), agricultural, and household processes. Paints, automotive exhaust/traffic emissions, lead-acid batteries, e-waste, smelting operations, material weathering, plumbing and fittings, coal-fired thermal power stations, ceramics, and the bangle industry are all potential sources. Lead is commonly found in wastewater from industries such as electroplating, electrical, steel, and explosive makers, among others. Electronic trash, lead-acid battery effluents, and metal industry discharge all contribute to its presence in aquatic systems.

1.3.4.2 Characteristics of Lead

Lead is a glossy, bluish-white metal that is extremely soft, malleable, and ductile. Sulphide, cerussite ($PbCl_2$), and galena are all examples of heavy and soft metals. It is a metal with atomic number

82, atomic mass 207.2, density of 11.4 g/cm³, melting and boiling temperatures of 327.4°C and 1,725°C, respectively, that belongs to group IV and period 6 of the periodic table. Blood lead levels in children below 10 µg/dl have been considered safe [29].

1.3.4.3 Uses of Lead
Metal items, cables, pipelines, paints, and pesticides all contain lead.

1.3.4.4 Impacts of Lead
Lead is one of the four most toxic non-essential heavy metals, and inorganic forms are absorbed through food, drink, and breathing. It's a dangerous metal that accumulates quickly in the human body. It causes metabolic toxicity and enzyme inhibition in most people. It is a potentially carcinogenic substance with neurotoxic properties. It causes haemoglobin biosynthesis disruption, miscarriages, blood pressure elevation, and behavioural problems in youngsters such as hostility, impulsive conduct, and hyperactivity [39]. It harms the liver, kidneys, and the urinary system, as well as the nervous system, the brain, and the excretory system. It also disrupts the basic physiological processes of cells, resulting in mental lapses and memory loss [13]. It also causes hearing problems and intestinal issues in humans [24]. Mild symptoms such as sleeplessness, weariness, and weight loss have been recorded, as well as more serious side effects such as neurotoxicity and nephrotoxicity [44].

1.3.5 MERCURY

1.3.5.1 Sources of Mercury
Mercury is a persistent, poisonous, and bio-accumulative chemical that is typically introduced in an inorganic form into aquatic systems by businesses. Its contamination is linked to the manufacturing of electrical equipment, pressure gauges, and thermometers, barometers, among other things. It can also come from industrial sources such as coal combustion, acid rain-induced soil leaching, fluorescent lamps, thermal power plants, chlor-alkali plants, hospital waste, and electrical appliances, etc. [41]. The main natural sources of mercury include degassing from the Earth's crust, volcanic eruptions, and evaporation from natural water bodies [19].

1.3.5.2 Characteristics of Mercury
At ordinary temperature and pressure, mercury is the sole liquid metal. It has an atomic number of 80, and an atomic weight 200.6, mass density of 13.6 g/cm³, melting point of −13.6°C, and boiling point of 357°C, and is mainly recovered as a by-product of ore processing [29].

1.3.5.3 Uses of Mercury
Thermometers, sphygmomanometers, barometers, and blood pressure monitors all contain metallic mercury.

1.3.5.4 Impacts of Mercury
Mercury has been linked to cancer. Its short-term mercury exposure can harm the lungs and kidneys. It can also harm the neurological system, resulting in haemorrhagic gastritis and colitis. When mercury is ingested in large quantities, it causes neurotoxicity. It causes involuntary abortion in women who are pregnant. There has been a wide range of cognitive, psychological, sensory, cardiovascular, gastrointestinal, and motor abnormalities declared [44]. Long-term exposure to elemental mercury vapour has been linked to unsteady walking, chronic cough, poor focus, hearing loss, ataxia, general debilitation, tremulous speech, clouded vision, and poor psychomotor function. Inorganic mercury has the potential to be embryotoxic and teratogenic [17].

1.3.6 COPPER

1.3.6.1 Sources of Copper

Copper is released into the atmosphere by both natural and human-made sources. The most prevalent natural causes include wind-blown dust, sea spray, volcanoes, rotting vegetation, and forest fires [33]. Vehicle emissions from nonferrous metal production, mechanical abrasion, iron and steel production, coal combustion, wood production, industrial applications (mining, electroplating, smelting operations, plumbing, and fittings), and phosphate fertilizer production are all examples of anthropogenic copper emissions [44].

1.3.6.2 Uses of Copper

In humans, copper is required to produce haemoglobin and certain enzymes.

1.3.6.3 Impacts of Copper

High copper consumption can harm the liver and kidneys, as well as produce gastrointestinal problems.

1.3.7 CHROMIUM

1.3.7.1 Sources of Chromium

Natural chromium concentrations in the atmosphere and aquatic habitats are minimal. It is unable to dissolve in water and hence penetrates deeper into the soil to reach subsurface water. As a result, chromium levels in polluted wells may be dangerously high. The largest sources of chromium emission are metallurgical, chemical, and refractory industries [39]. Chrome-electroplating, metal processing, chromium salts manufacturing, chrome pigment production, leather tannery operation, mining, photography, industrial coolants application, textile manufacturing, painting, chrome alloys, and stainless-steel welding all cause releases through air and wastewater. Chromium also enters the atmosphere from coal combustion and is eventually passed on to soil and water bodies. High chromium concentrations can also be caused by heavy traffic. Chrome plating of several motor vehicle parts can be linked to chromium in the urban environment [33].

1.3.7.2 Characteristics of Chromium

Chromium is a silver-grey metal with a glossy sheen. The global production of chromium (Cr) has expanded by 514% during the last five decades, reaching 11.3 Mt/year [4]. Anthropogenic acts may cause chromium contamination in natural water.

1.3.7.3 Uses of Chromium

The human body requires chromium for a variety of processes, including fat and carbohydrate catabolism, as well as glucose and blood pressure management. Chromium affects the metabolism of lipids and sugars.

1.3.7.4 Impacts of Chromium

Chromium is known to cause allergies, skin rashes, irritations, nosebleeds, stomach ulcers, respiratory problems, immune system damage, genetic material changes, lung cancer, kidney damage, and death in extreme cases [39]. Chromium(VI) is a human carcinogen [44]. Chromium composites can increase the risk of lung cancer, damage the circulatory system, and cause nerve tissue to disintegrate. Chromium's coexistence with other metals may increase glycogen levels in organs that are under stress from metal exposure [24]. Consuming too much Cr through food or dirty water can result in kidney damage, intestinal haemorrhage, and gastrointestinal stromal tumours [45]

1.3.8 CADMIUM

1.3.8.1   Sources of Cadmium

Cadmium (Cd) is a predetermined by-product of lead or zinc refining processes, but it is also found in nature as a raw ore. Cd can be recycled after it has been removed from the ores. Cadmium is found in the Earth's crust at a quantity of about 0.1 mg/kg and occurs in a mixture with zinc. It's found naturally in zinc, lead, and copper ores. Cadmium can also be found in food [29]. Cadmium is naturally released in huge amounts into the environment. The weathering of rocks releases about half of the cadmium into rivers, whereas the rest is released by anthropogenic activity. Cadmium comes from a variety of sources, including industry, mining, electroplating, sewage, waste batteries, paint sludge, incinerations, zinc smelting, e-waste, and fuel combustion [41], and plumbing and fittings sources [13]. Other sources of cadmium include cigarette smoking and synthetic phosphate fertilizer manufacture. Cd levels in urban roads are also high due to the wear and tear of rubber backing on carpets and fragments of car tyres. Cadmium-containing products are rarely recycled, but are regularly thrown away with household waste, polluting the environment, especially when the waste is disposed.

1.3.8.2   Characteristics of Cadmium

Cadmium is found in natural sediments, which also contain other elements. It's used in plating, cadmium-nickel batteries, phosphate fertilizers, stabilizers, and alloys, among other things. However, it lacks the requisite beneficial characteristics to sustain life [40]. Cadmium has an atomic number of 48, an atomic weight of 112.4, a density of $8.65g/cm^3$, and melting and boiling temperatures of $320.9°C$ and $765°C$, respectively. A urine creatinine excretion of 10 mol/mol was deemed by the WHO to be a "critical limit" below which renal damage would not occur [29].

1.3.8.3   Uses of Cadmium

Cadmium is used extensively in industries that deal with electroplating, Ni-Cd batteries, pigments, alloys, coatings, plastic stabilizers, and cigarettes. Cd coatings provide excellent corrosion resistance, particularly in high-stress settings where protection is required. Cadmium compounds are employed as stabilizers in PVC products, colour pigments, a variety of alloys, and, most recently, rechargeable nickel-cadmium batteries. Cadmium metal has mostly been utilized as an anticorrosion agent (cadmiation) [39].

1.3.8.4   Impacts of Cadmium

Cadmium is the sixth element, and it is a significant toxin that harms humans. It's the most dangerous heavy metal found in industrial waste. It can build up in the bodies of living beings as well as in water reservoirs, causing calcium shortage and bone fractures by disrupting calcium metabolism [33]. In humans, it has a long half-life of roughly 1,033 years and exhibits persistent behaviour. Cadmium compounds are very toxic to living beings and accumulate in the ecosystem even at low concentrations of 1 mg/kg. When consumed in large doses, it causes damage to the kidneys, liver, reproductive organs, skeleton, blood, DNA, and immunological and nervous systems. It causes pulmonary oedema, bone softening, and prostate and lung cancer. This could possibly lead to death [17]. It is possibly carcinogenic and affects the gastrointestinal tract [44]. Thus, many governments and international organizations consider cadmium to be a harmful contaminant.

1.3.9 MOLYBDENUM, SILVER, AND VANADIUM

1.3.9.1   Sources of Molybdenum

Spent catalyst, naturally occurring, agricultural, industrial waste.

1.3.9.2   Sources of Silver

Naturally occurring/water treatment sources.

1.3.9.3 Sources of Vanadium

Spent catalyst, sulphuric acid production plants.

1.4 INDEXING APPROACH FOR HEAVY METAL POLLUTION

Established indexing approaches such as HPI, CI, HEI, and the recently developed "heavy metal contamination index (HCI)" have been used to determine the level of heavy metal contamination in water and groundwater [46, 47].

1.4.1 HEAVY METAL POLLUTION INDEX (HPI)

The HPI distributes weights based on the inverse proportionality of each component's recommended standard values, which is widely accepted around the world. Mohan's HPI model describes the combined impact of heavy metals on water quality [48]. For each metal parameter, the first step is to calculate the relative weight (W_i). W_i is the inverse of the maximum/upper permissible concentrations (MAC). This index assigns a rating or weightage (W_i) between 0 and 1 and is inversely proportional to the parameter's standard permitted value (S_i). In this indexing method, the critical pollution index value for drinking water is set at 100 [49]. HPI is calculated using the formulae:

$$HPI = \frac{\sum_{i=1}^{n} W_i Q_i}{\sum_{i=1}^{n} W_i} \tag{1.1}$$

where Q_i and W_i denote the sub-index and unit weight assigned to the ith parameter, respectively, and n denotes the number of parameters considered. In the second step, the sub-index (Q_i) for the ith parameter is calculated for each heavy metal using equation (1.2):

$$Q_i = \frac{|M_i - I_i|}{|S_i - I_i|} \times 100 \tag{1.2}$$

where M_i is the measured metal value in the ith sample, I_i is the ideal value or desirable limit of the ith parameter, and S_i is the standard permitted limit. The numerator represents the numerical difference between the two values; however, the algebraic sign (−) is omitted. According to WHO (2011) guidelines, the required values for water quality are given in Table 1.1. Surface water quality standards for drinking water are provided in Table 1.2.

TABLE 1.1
Standard values for computation of pollution indices

Parameter	Weightage (K/MAC)	Standard acceptable limit (g/L)	Highest permissible limit (g/L)	Maximum admissible limit (g/L)
As	0.02	50	10	50
Cadmium	0.37	5	3	3
Chromium	0.02	50	50	50
Copper	0.001	1,000	2,000	1,000
Iron	0.006	300	200	200
Manganese	0.02	100	500	50
Nickel	0.05	20	20	20
Lead	0.75	100	10	1.5
Zinc	0.0002	5,000	3,000	5,000

TABLE 1.2
Surface water quality standards for drinking water [46]

Characteristics	WHO drinking water guidelines (µg/L)
Cu	2000
Mg	–
Mn	400
Fe	300
Cr	50
Cd	3
Pb	10

1.4.2 CONTAMINATION INDEX (CI)

CI measures the extent of pollution or the cumulative effects of several quality indicators that are harmful to drinking water. It assesses the total level of water quality contamination and is calculated separately for each parameter. The total number of contamination factors that exceed the upper permissible limits is referred to as CI [50]. It was estimated using the following method to add up the cumulative effects of several water quality criteria considered dangerous for domestic water:

$$CI = \sum_{i=1}^{n} \left\{ \frac{C_{ai}}{C_{si}} - 1 \right\} \quad (1.3)$$

where C_{ai} and C_{si} are the lowest and upper allowed values of the ith component, respectively.

Heavy metal concentrations beyond the permitted limits were not considered while calculating CI values. The degree of contamination (CI) is commonly used as a metric for assessing the extent of metal pollution in water [51]. Based on the CI values, the monitoring sites were divided into three categories, as shown in Table 1.3. Depending on the extent of contamination, CI values are divided into three categories: low (CI < 1), medium (CI = 1–3), and high (CI > 3).

1.4.3 EVALUATION INDEX OF HEAVY METALS (HEI)

To determine the water quality in terms of heavy metal content, a heavy metal analysis is mostly used. The HEI assigns a score to each heavy metal's accumulation rate in the overall water quality, which is determined using equation (1.4):

$$HEI = \sum_{i=1}^{n} \frac{H_i}{H_{imax}} \quad (1.4)$$

where H_i and H_{imax} are the ith parameters monitored and permitted values, respectively.

TABLE 1.3
Degree of contamination based on CI values [46]

CI values	Category
CI < 1	Low
1 ≤ CI < 3	Medium
CI > 3	High

TABLE 1.4

Water quality classifications based on HMI values [46]

Range of HMI values	Category
HMI < 50	Excellent
50 ≤ HMI < 100	Good
100 ≤ HMI < 200	Poor
200 ≤ HMI < 300	Very poor
HMI ≥ 300	Unsuitable for drinking

The HEI, like the HPI, determines the overall water quality in terms of heavy metal content. This index is mostly used for simple computation steps and is beneficial for interpreting pollution levels [52]. The index value, which ranges from 0 to 1, represents the relative/subjective importance of each quality concern and is inversely proportional to each metal's standard. Water samples are classified for HEI using multiples of the mean value obtained.

1.4.4 HEAVY METAL INDEX (HMI)

The HMI values can be computed using the formula:

$$\text{HMI} = \sum_{i=1}^{n}\left[p_i \times \frac{M_i}{S_i} \right] \times 100 \tag{1.5}$$

where p_i specifies the weight of each parameter that is given. The actual heavy metal concentrations in water samples are denoted by M_i and S_i, whereas the acceptable limit values for the ith parameter are set by WHO drinking water recommendations. Table 1.4 shows how the HPI values are categorized for water samples.

1.5 ANALYTICAL METHODS FOR HEAVY METALS DETECTION

A metal ion detector is a device or instrument that detects and quantifies metal ions in the environment. Before we can remove these dangerous metal ions from water samples, we need a technology that can detect their presence and provide a quantitative assessment of pollution levels so that the best removal strategy can be selected. To this end, time and cost-effective detection processes that are also environmentally friendly should be developed. A detection method must also be sensitive enough to accurately identify even minute levels of metal ions. A significant number of analytical methods for identifying heavy metals in environmental and biological samples have been developed and implemented, according to the literature [53]. Heavy metals can be identified by chromatographic analysis, electrical analysis, instrumental neutron activation analysis, spectrometry analysis, and hyphenated methods. Hybrid approaches have been widely applied in speciation analysis. Inductively coupled plasma mass spectrometry (ICP-MS) and inductively coupled plasma atomic emission spectrometry (ICP-AES) are becoming increasingly popular in this field owing to their merits of concurrent detection of multiple elements, shorter processing time, maximum throughput, and far less sample intake. High sensitivity, a huge linear range, and powerful anti-interference capabilities are only a few of the benefits of the ICP-MS. Heavy metal biomonitoring is an exciting new technology for determining heavy metal pollution and toxicity in the environment and biosphere. The simplest way for evaluating the extent of heavy metal contamination using environmental matrices is chemical analysis; however, it offers insufficient information on the cumulative impact and potential toxicity on organisms and ecosystems. Biomonitoring, which is based on collecting and analysing the tissues and fluids of an individual organism, can be used to enhance chemical tests.

1.5.1 BROAD CATEGORIES OF HEAVY METAL ION DETECTION TECHNIQUES

Heavy metal ion detection techniques can be broadly divided into three main categories:

1.5.1.1 Spectroscopic Detection

Highly sensitive techniques such as graphite furnace atomic absorption spectrometry, atomic fluorescence spectrometry (AFS), X-ray fluorescence spectrometry, atomic absorption spectroscopy (AAS), neutron activation analysis, inductively coupled plasma mass spectroscopy, and inductively coupled plasma-optical emission spectrometry have been used to detect heavy metal ions [54]. With very low detection limits, they can concurrently determine the concentration of heavy metal ions for several elements. These methods, however, are quite expensive, and they require the use of skilled personnel to run the sophisticated equipment.

1.5.1.1.1 Atomic Absorption Spectroscopy

AAS, as shown in Figure 1.3, is a qualitative and quantitative analytical approach that is used for elements that can be assessed based on spectral lines and varied spectral line attenuation degrees [55]. An atomization system, a detecting system, a spectroscopic system, a sharpline light source, and a synchronous modulation system of power supply are the five key components of the AAS. The atomization system is primarily used to provide energy for sample solution atomization, evaporation, drying, and atomization, as well as to generate the fundamental atomic vapour. The detecting system is used to reduce background emission interference, increase the signal-to-noise ratio, and extend bulb life. Exceptional selectivity, excellent precision, a wide analytical range, and strong anti-interference capacity are some of the advantages of AAS in determining elements. Its steady spectral line can be utilized to determine trace and ultra-trace elements. However, the disadvantages of AAS include the need to change the light source bulb while identifying the elements and the inability to examine many elements at once. During this point, using AAS for soil environment monitoring should be considered a necessary component of ecological and environmental conservation [56].

1.5.1.1.2 Graphitic Furnace

A graphitic furnace consists of an open-ended cylindrical graphitic tube with a sample hole input. The sample is deposited into a graphite tube, which is heated before being atomized. The atomized material is subsequently evaluated. Atomization happens in a graphitic furnace AAS setting where temperature does not change rapidly because the sample is not on the furnace wall, resulting in more repeatable results [58]. Longer analysis times than flame sampling and a lower number of elements that may be evaluated by this method are some of the drawbacks of a heavy metal ion detection device.

1.5.1.1.3 Atomic Fluorescence Spectrometry

AFS depicted in Figure 1.4 is a fast-evolving technology for detecting heavy metals that relies on the excitation of atoms in the vapour state to be examined by radiation of a certain wavelength, resulting in atomic fluorescence. The intensity of atomic fluorescence is related to the concentration. The elements in the soil can be effectively assessed using fluorescence intensity. There are two forms of AFS: dispersive and nondispersive. AFS and AAS are quite similar, except for the fact that the light source and other components are not aligned in a straight line, but rather at a 90° right angle, to prevent the excitation light source's radiation from influencing the atomic fluorescence detection

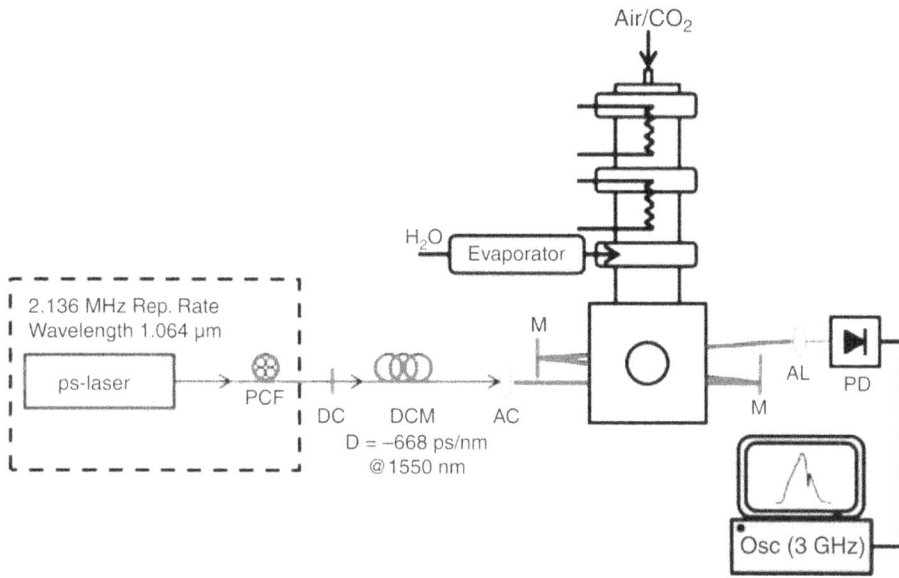

FIGURE 1.3 Supercontinuum absorption spectroscopy set-up for temperature and multi-species measurements [57].

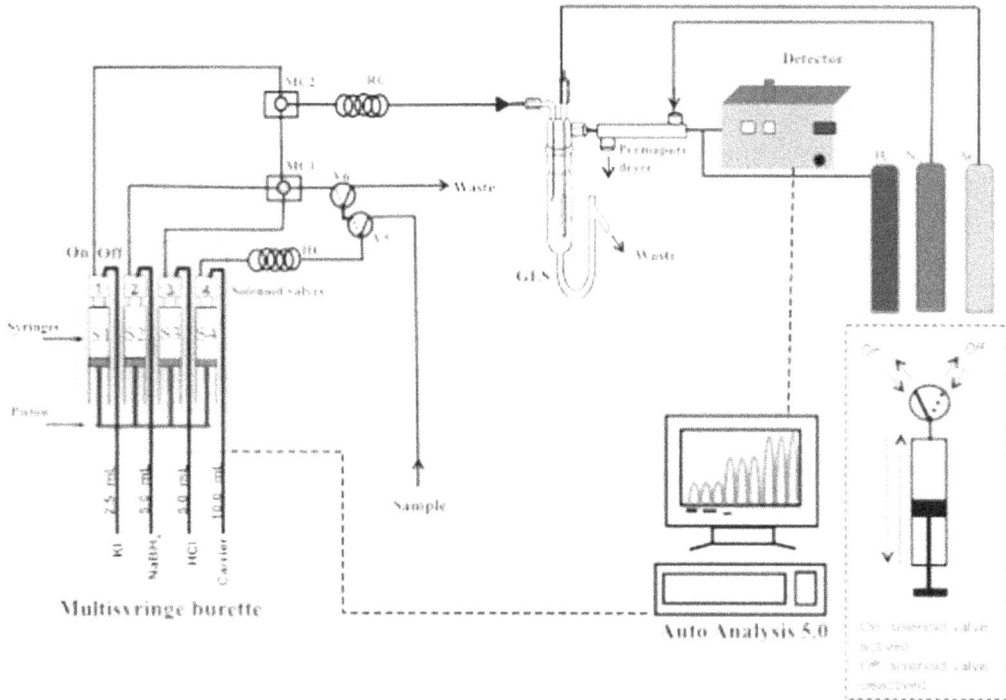

FIGURE 1.4 Scheme of the MSFIA-HG-AFS for As, Sb, and Se determination [59].

signal. It offers atomic emission an absorption benefit. Nevertheless, its drawbacks include limited applicability and the utilization of only few elements. As a result, analysts continue to be interested in promoting and using AFS for heavy metal detection studies. The identification of trace elements in stream sediments, soil, coal, diverse minerals, and rock was the first application field of AFS. Scientists have conducted more in-depth research on AFS as this technology has progressed. There have been some situations where AFS has been combined with other methodologies since its inception [56].

1.5.1.2 Electrochemical Detection Methods

Electrochemical methods are inexpensive, simple to use, and dependable, with straightforward protocols for monitoring contaminated samples. When compared to other spectroscopic techniques, electrochemical methods have the benefit of having a comparatively short analytical time [60, 61]. These electrochemical techniques, however, have substantial limitations as compared to spectroscopic and optical approaches, such as lower sensitivity and higher detection limits (LOD). Furthermore, these techniques may require design tweaks and upgrades to improve their efficacy in detecting heavy metal ions in specific cases. Minimal limit of detection, great sensitivity, huge surface area, superior repeatability, higher signal-to-noise ratio, and selective sensing of several metal ions are all advantages of electrochemical technologies. It's a tool that turns chemical information from a sample into an analytical signal. Chemical information derived from the system's physical properties or the reaction of species present in the analyte is used in the receptor unit, which is then converted into an electrical signal by the transducer system, a potentiostat/galvanostat-based electrochemical workstation. Various signal amplification techniques can be used to deduce the final analytical usable results. Among the many electrochemical techniques accessible are potentiometric, amperometric, voltammetric, coulometric, impedance measurement, and electrochemiluminescent approaches.

1.5.1.2.1 Potentiometry

The electromotive force (EMF) is measured without utilizing any electric current in this method. Quantitative investigation of ions in solutions is usually done using potentiometric techniques [62]. It is, nevertheless, particularly effective for detecting heavy metal ions due to characteristics such as minimal rates, fast reaction time, great specificity, and a broad response range. This approach has some disadvantages, such as high detection limits and low sensitivity. For heavy metal ions, such modified electrodes have been shown to have higher sensitivity and lower detection limits [63, 64]

1.5.1.2.2 Amperometry and Voltammetry

Heavy metals detection utilizing voltammetric techniques is a very sensitive approach capable of detecting concentrations in the nanomolar and even picomolar ranges. In electrochemical-enhanced sensing techniques, the electrochemical transduction substance is crucial. Nanostructured platforms, spanning from carbon nanomaterials to metallic nanoparticles, and other natural nanostructured adsorbents are being studied extensively in this area [65]. Heavy metal ions are routinely detected using this approach. To obtain a current-voltage curve, current is measured at various applied potentials. Because of its excellent accuracy, reduced detection limits, and great sensitivity, voltammetry is an extensively used technique. Voltammetry can take many forms, but the essential technique, which involves measuring current by adjusting the potential, is the same for all. Cyclic voltammetry, pulse voltammetry, and stripping voltammetry are some of the voltammetry modes used to analyse heavy metal ions [66].

1.5.1.2.3 Linear Sweep Voltammetry

This entails performing a linear potential sweep with conventional electrodes to obtain an I–V curve and polarography, with one electrode being an electrochemically active electrode.

1.5.1.2.4 Combining Stripping Procedures with Voltammetric Approaches

Stripping approaches can be coupled with a variety of pulse voltammetric techniques to produce new identification techniques like square-wave anodic stripping voltammetry, differential pulse anodic stripping voltammetry, linear sweep anodic stripping voltammetry, etc., all of which are very good at detecting trace levels of heavy metal ions with very limited limit of detection.

1.5.1.2.5 Galvanostatic Techniques

The electric potential is measured while an electric current is applied in galvanostatic procedures. A current source (galvanostat) controls the current between the working and counter electrode, and the associated potential is observed throughout the reference and working electrodes [67]. Galvanostatic techniques use less complicated equipment than potentiostatic techniques, but they have the drawback of causing large double-layer charging impacts. Galvanostatic stripping chrono-potentiometry (SCP) is said to be less sensitive when organic material is present.

1.5.1.3 Optical Methods of Detection

In the optical detection methods, optical fibres, particular indicator dyes, integrated optics, iono-phores, capillary-type devices, and several other technologies are commonly utilized.

1.5.1.3.1 Indicator Dye-Based Sensors

This type of sensor works by combining a heavy metal ion binding reaction with an indicator dye, which results in a change in the binding reagents' absorbance or fluorescence. The indicator func-tions as a transducer for the heavy metal ion in such heavy metal ion sensors because direct visual detection is difficult. Heavy metal ions operate as "quenchers" when they interact with another set of indicators, causing both static and dynamic quenching of indicator dye luminescence [68]. Many indicators are non-selective, meaning they will bind to a variety of metal ions [69].

1.5.1.3.2 Ionophore-Based Sensors

Due to the limitations of indicator reagents, ionophores (ion-complexing organic compounds) have been used to detect heavy metal ions using optical techniques. Because of their ion car-rier, complexing capabilities, and ability to interact with ions, these ionophores are ideal for heavy metal ion specification. The extraction of ions into membranes employing ion carriers is another way for detecting heavy metal ions. A proton-selective chromo-ionophore generates the optical signal, and heavy metal ions are detected via selective binding of heavy metals with ionophores [70, 71].

1.6 CONCLUSION

Heavy metals are a unique class of contaminants found in water reservoirs. They are a collection of naturally occurring metals with high densities and atomic weights. Heavy metal contamination can be found in soil, water, and the atmosphere, among other places. Because of their substantial inhibitory effects on biodegradation activities, heavy metals are directly linked to environmental pollution and biological toxicity issues. This can be largely attributed to the primary and second-ary sources of heavy metals. Thus, this chapter defined heavy metals, their properties, uses, and impacts on the environment. The chapter went further to discuss the various sources of metals. These include natural, anthropogenic, and diverse sources. It was discovered that natural heavy metal emissions occur under a variety of different and unique environmental conditions. This could be due to a variety of natural geographical phenomena. Anthropogenic sources are often regarded as the primary contributors to rising levels of heavy metal contamination in many environments. Heavy metal contamination can also be caused by a variety of industrial activities and their efflu-ents. The diverse source was classified into (i) waste sources: agricultural, biomedical, industrial, and electronic waste, (ii) industrial processes: mining, mineral extraction, smelting, electroplat-ing, automobile exhaust fumes, geothermal sources, plumbing and fittings, power plants, urban surfaces, irrigation water source, tannery activities. The sources, characteristics, uses, and impacts of individual heavy metals were extensively discussed in the chapter. Various established indexing approaches used for heavy metal contamination level assessment in water and groundwater were also highlighted in this chapter. A significant number of analytical methods for identifying heavy metals in environmental samples have been discussed. A detection method must also be sensitive to accurately identify even minute levels of metal ions. In this chapter, heavy metal ion monitoring techniques were split into three types: (i) spectroscopic detection: graphitic furnace, AFS, and AAS, (ii) electrochemical methods of detection: potentiometry, amperometry, voltammetry, linear sweep voltammetry, stripping methods in combination with voltametric techniques, galvanostatic techniques, (iii) optical methods of detection: indicator dye-based sensors and ionophore-based sensors. According to several sources, it was discovered that the most generally used approach for heavy metal detection is AAS, which, while being a destructive technology, has a high level of dependability and precision.

REFERENCES

1. Krishna, A.K., M. Satyanarayanan, and P.K. Govil, Assessment of heavy metal pollution in water using multivariate statistical techniques in an industrial area: a case study from Patancheru, Medak District, Andhra Pradesh, India. *Journal of hazardous materials*, 2009. 167(1–3): pp. 366–373.
2. Ahmed, S.F., et al., Heavy metal toxicity, sources, and remediation techniques for contaminated water and soil. *Environmental Technology & Innovation*, 2022. 25: p. 102114.
3. Jagaba, A.H., et al., Combined treatment of domestic and pulp and paper industry wastewater in a rice straw embedded activated sludge bioreactor to achieve sustainable development goals. *Case Studies in Chemical and Environmental Engineering*, 2022: p. 100261.
4. Jagaba, A.H., et al., Synthesis, characterization, and performance evaluation of hybrid waste sludge biochar for COD and color removal from agro-industrial effluent. *Separations*, 2022. 9(9): p. 258.
5. Jagaba, A.H., et al., Circular economy potential and contributions of petroleum industry sludge utilization to environmental sustainability through engineered processes-a review. *Cleaner and Circular Bioeconomy*, 2022: p. 100029.
6. Yaro, S.N.A., et al., The influence of waste rice straw ash as surrogate filler for asphalt concrete mixtures. *Construction*, 2022. 2(1): pp. 118–125.
7. Abdullahi, S.S.a., et al., Comparative study and dyeing performance of as-synthesized azo heterocyclic monomeric, polymeric, and commercial disperse dyes. *Turkish Journal of Chemistry*, 2022. 46: pp. 1–12.
8. Birniwa, A.H., et al., Polymer-based nano-adsorbent for the removal of lead ions: kinetics studies and optimization by response surface methodology. *Separations*, 2022. 9(11): p. 356.
9. Chaemiso, T.D. and T. Nefo, Removal methods of heavy metals from laboratory wastewater. *Journal of Natural Sciences Research*, 2019. 9(2): pp. 36–42.
10. Birniwa, A.H., et al., Synthesis of gum arabic magnetic nanoparticles for adsorptive removal of ciprofloxacin: equilibrium, kinetic, thermodynamics studies, and optimization by response surface methodology. *Separations*, 2022. 9(10): p. 322.
11. Noor, A., et al., Parametric optimization of additive manufactured biocarrier submerged in sequencing batch reactor for domestic wastewater treatment. *Heliyon*, 2023. 9(4): p. e14840.
12. Birniwa, A.H., et al., Recent trends in treatment and fabrication of plant-based fiber-reinforced epoxy composite: a review. *Journal of Composites Science*, 2023. 7(3): p. 120.
13. Cipriani-Avila, I., et al., Heavy metal assessment in drinking waters of Ecuador: Quito, Ibarra and Guayaquil. *Journal of Water and Health*, 2020. 18(6): pp. 1050–1064.
14. Birniwa, A.H., et al., Polypyrrole-polyethyleneimine (PPy-PEI) nanocomposite: an effective adsorbent for nickel ion adsorption from aqueous solution. *Journal of Macromolecular Science, Part A*, 2021. 58(3): pp. 206–217.
15. Briffa, J., E. Sinagra, and R. Blundell, Heavy metal pollution in the environment and their toxicological effects on humans. *Heliyon*, 2020. 6(9): p. e04691.
16. Jagaba, A.H., et al., Parametric optimization and kinetic modelling for organic matter removal from agro-waste derived paper packaging biorefinery wastewater. *Biomass Conversion and Biorefinery*, 2022: p. 1–18.
17. Dökmeci, A.H., Environmental impacts of heavy metals and their bioremediation, in *Heavy Metals-Their Environmental Impacts and Mitigation*. 2020, IntechOpen. pp. 1–16. http://dx.doi.org/10.5772/intechopen.95103
18. Birniwa, A.H., et al., Innovative and eco-friendly technologies for the upgradation of pharmaceutical wastewater treatment processes, in Afzal Husain Khan, Nadeem A Khan, Mu. Naushad, Hamidi Abdul Aziz (Eds.), *The Treatment of Pharmaceutical Wastewater*. 2023, Elsevier. pp. 367–398.

19. Kapoor, D. and M.P. Singh, Heavy metal contamination in water and its possible sources, in Daniel Junqueira Dorta, Danielle Palma de Oliveira (Eds.), *Heavy Metals in the Environment*. 2021, Elsevier. pp. 179–189.

20. Rai, P.K., et al., Heavy metals in food crops: health risks, fate, mechanisms, and management. *Environment International*, 2019. 125: pp. 365–385.

21. Masindi, V. and K.L. Muedi, Environmental contamination by heavy metals. *Heavy Metals*, 2018. 10: pp. 115–132.

22. Izah, S.C. and T.C. Angaye, Heavy metal concentration in fishes from surface water in Nigeria: potential sources of pollutants and mitigation measures. *Sky Journal of Biochemistry Research*, 2016. 5(4): pp. 31–47.

23. Birniwa, A.H., et al., Application of agricultural wastes for cationic dyes removal from wastewater, in Subramanian Senthilkannan Muthu, Ali Khadir (Eds.), *Textile Wastewater Treatment*. 2022, Springer. pp. 239–274.

24. Sonone, S.S., et al., Water contamination by heavy metals and their toxic effect on aquaculture and human health through food chain. *Letters in applied NanoBioScience*, 2020. 10(2): pp. 2148–2166.

25. Tracy, J.W., et al., Sources of and solutions to toxic metal and metalloid contamination in small rural drinking water systems: a rapid review. *International Journal of Environmental Research and Public Health*, 2020. 17(19): p. 7076.

26. Jagaba, A., et al., Evaluation of the physical chemical bacteriological and trace metals concentrations in different brands of packaged drinking water. *Engineering Letters*, 2021. 29(4): pp. 1552–1560.

27. Islam, M., et al., Heavy metals pollution sources of the surface water of the Tunggak and Balok river in the Gebeng industrial area, Pahang, Malaysia. *International Journal of Energy and Water Resources*, 2022: pp. 1–8.

28. Noor, A., et al., Treatment innovation using biological methods in combination with physical treatment methods, in Afzal Husain Khan, Nadeem A. Khan, Mu. Naushad, Hamidi Abdul Aziz (Eds.), *The Treatment of Pharmaceutical Wastewater*. 2023, Elsevier. pp. 217–245.

29. Ojo, A.A., et al., Review on heavy metals contamination in the environment. *European Journal of Earth and Environment*, 2017. 4(1): pp. 1–6. ISSN: 2056–5860.

30. Jagaba, A.H., et al., A systematic literature review of biocarriers: central elements for biofilm formation, organic and nutrients removal in sequencing batch biofilm reactor. *Journal of Water Process Engineering*, 2021. 42: p. 102178.

31. Sabadell, J.E. and R.C. Axtmann, Heavy metal contamination from geothermal sources. *Environmental Health Perspectives*, 1975. 12: pp. 1–7.

32. Zhaoyong, Z., J. Abuduwaili, and J. Fengqing, Heavy metal contamination, sources, and pollution assessment of surface water in the Tianshan Mountains of China. *Environmental Monitoring and Assessment*, 2015. 187(2): pp. 1–13.

33. Hanfi, M.Y., M.Y. Mostafa, and M.V. Zhukovsky, Heavy metal contamination in urban surface sediments: sources, distribution, contamination control, and remediation. *Environmental Monitoring and Assessment*, 2020. 192(1): pp. 1–21.

34. Lokeshwari, H. and G. Chandrappa, Effects of heavy metal contamination from anthropogenic sources on Dasarahalli tank, India. *Lakes & Reservoirs: Research & Management*, 2007. 12(3): pp. 121–128.

35. Jagaba, A.H., et al., A systematic literature review on waste-to-resource potential of palm oil clinker for sustainable engineering and environmental applications. *Materials*, 2021. 14(16): p. 4456.

36. Hussain, S., et al., Different sources of irrigation water affect heavy metal accumulation in soils and some properties of guava fruits. *Environmental Science and Pollution Research*, 2022: pp. 1–10.

37. Bhuiyan, M.A.H., et al., Investigation of the possible sources of heavy metal contamination in lagoon and canal water in the tannery industrial area in Dhaka, Bangladesh. *Environmental Monitoring and Assessment*, 2011. 175(1): pp. 633–649.

38. Ahmad, M., et al., A review of environmental contamination and remediation strategies for heavy metals at shooting range soils. *Environmental Protection Strategies for Sustainable Development*, 2012: pp. 437–451.

39. Singh, A., D. Singh, and H. Yadav, Impact and assessment of heavy metal toxicity on water quality, edible fishes and sediments in lakes: a review. *Trends in Biosciences*, 2017. 10(8): pp. 1551–1560.

40. Järup, L., Hazards of heavy metal contamination. *British Medical Bulletin*, 2003. 68(1): pp. 167–182.

41. Verma, R. and P. Dwivedi, Heavy metal water pollution-A case study. *Recent Research in Science and Technology*, 2013. 5(5): pp. 98–99.

42. Sardar, K., et al., Heavy metals contamination and what are the impacts on living organisms. *Greener Journal of Environmental Management and Public Safety*, 2013. 2(4): pp. 172–179.

43. Micó, C., et al., Assessing heavy metal sources in agricultural soils of an European Mediterranean area by multivariate analysis. *Chemosphere*, 2006. 65(5): pp. 863–872.

44. Vu, C.T., et al., Bioaccumulation and potential sources of heavy metal contamination in fish species in Taiwan: assessment and possible human health implications. *Environmental Science and Pollution Research*, 2017. 24(23): pp. 19422–19434.

45. Vijayakumar, C.R., D.P. Balasubramani, and H.M. Azamathulla, Assessment of groundwater quality and human health risk associated with chromium exposure in the industrial area of Ranipet, Tamil Nadu, India. *Journal of Water, Sanitation and Hygiene for Development*, 2022. 12(1): pp. 58–67.

46. Dash, S., S.S. Borah, and A. Kalamdhad, A modified indexing approach for assessment of heavy metal contamination in Deepor Beel, India. *Ecological Indicators*, 2019. 106: p. 105444.

47. Lawal, I.M., et al., Multi-criteria performance evaluation of gridded precipitation and temperature products in data-sparse regions. *Atmosphere*, 2021. 12(12): p. 1597.

48. Jagaba, A., et al., Water quality hazard assessment for hand dug wells in Rafin Zurfi, Bauchi State, Nigeria. *Ain Shams Engineering Journal*, 2020. 11(4): pp. 983–999.

49. Herath, I.K., et al., Heavy metal toxicity, ecological risk assessment, and pollution sources in a hydropower reservoir. *Environmental Science and Pollution Research*, 2022: pp. 1–18.

50. Rajkumar, H., P.K. Naik, and M.S. Rishi, A new indexing approach for evaluating heavy metal contamination in groundwater. *Chemosphere*, 2020. 245: p. 125598.

51. Reza, R. and G. Singh, Heavy metal contamination and its indexing approach for river water. *International Journal of Environmental Science & Technology*, 2010. 7(4): pp. 785–792.

52. Ogunlaja, A., et al., Risk assessment and source identification of heavy metal contamination by multivariate and hazard index analyses of a pipeline vandalised area in Lagos State, Nigeria. *Science of the Total Environment*, 2019. 651: pp. 2943–2952.

53. Jagaba, A., et al., Degradation of Cd, Cu, Fe, Mn, Pb and Zn by Moringa-oleifera, zeolite, ferricchloride, chitosan and alum in an industrial effluent. *Ain Shams Engineering Journal*, 2021. 12(1): pp. 57–64.

54. Jagaba, A., et al., Derived hybrid biosorbent for zinc(II) removal from aqueous solution by continuous-flow activated sludge system. *Journal of Water Process Engineering*, 2020. 34: p. 101152.

55. Saeed, A.A.H., et al., Modeling and optimization of biochar based adsorbent derived from Kenaf using response surface methodology on adsorption of Cd^{2+}. *Water*, 2021. 13(7): p. 999.

56. Jin, M., et al., Review of the distribution and detection methods of heavy metals in the environment. *Analytical Methods*, 2020. 12(48): pp. 5747–5766.

57. Braeuer, A., Absorption spectroscopy, in Andreas Braeuer (Ed.), *Supercritical Fluid Science and Technology*. 2015, Elsevier. pp. 347–366.

58. Alloway, B.J., Sources of heavy metals and metalloids in soils, in Brain J. Alloway, Jack T. Trevors (Eds.), *Heavy Metals in Soils*. 2013, Springer. pp. 11–50.

59. Ferreira, S.L.C., et al., Speciation analysis of antimony in environmental samples employing atomic fluorescence spectrometry–review. *TrAC Trends in Analytical Chemistry*, 2019. 110: pp. 335–343.

60. Jagaba, A.H., et al., Kinetics of pulp and paper wastewater treatment by high sludge retention time activated sludge process. *Journal of Hunan University Natural Sciences*, 2022. 49(2): pp. 242–251.

61. Jagaba, A.H., et al., Effect of hydraulic retention time on the treatment of pulp and paper industry wastewater by extended aeration activated sludge system, in *2021 Third International Sustainability and Resilience Conference: Climate Change*. 2021. IEEE.

62. Al-dhawi, B.N.S., et al., Treatment of synthetic wastewater by using submerged attached growth media in continuous activated sludge reactor system. *International Journal of Sustainable Building Technology and Urban Development*, 2022. 13(1): pp. 2–10.

63. Jagaba, A.H., et al., Toxic effects of xenobiotic compounds on the microbial community of activated sludge. *ChemBioEng Reviews*, 2022. 9(5): pp. 497–535.

64. Birniwa, A.H., et al., Adsorption behavior of methylene blue cationic dye in aqueous solution using polypyrrole-polyethylenimine nano-adsorbent. *Polymers*, 2022. 14(16): p. 3362.

65. Yaro, N., et al., Geopolymer utilization in the pavement industry-an overview, in *IOP Conference Series: Earth and Environmental Science*. 2022. IOP Publishing.

66. Yaro, N.S.A., et al., Comparison of response surface methodology and artificial neural network approach in predicting the performance and properties of palm oil clinker fine modified asphalt mixtures. *Construction and Building Materials*, 2022. 324: p. 126618.

67. Mishra, S., et al., Heavy metal contamination: an alarming threat to environment and human health, in Ranbir Chander Sobti, Naveen Kumar Arora, Richa Kothari (Eds.), *Environmental Biotechnology: For Sustainable Future*. 2019, Springer. pp. 103–125.

68. Jagaba, A.H., et al., Effect of environmental and operational parameters on sequential batch reactor systems in dye degradation, in Subramanian Senthilkannan Muthu, Ali Khadir (Eds.), *Dye Biodegradation, Mechanisms and Techniques*. 2022, Springer. pp. 193–225.

69. Abdullahi, S.S.a., et al., Facile synthesis and dyeing performance of some disperse monomeric and polymeric dyes on nylon and polyester fabrics. *Bulletin of the Chemical Society of Ethiopia*, 2021. 35(3): pp. 485–497.

70. Jagaba, A.H., et al., Waste derived biocomposite for simultaneous biosorption of organic matter and nutrients from green straw biorefinery effluent in continuous mode activated sludge systems. *Processes*, 2022. 10(11): p. 2262.

71. Jagaba, A.H., et al., Diverse sustainable materials for the treatment of petroleum sludge and remediation of contaminated sites: a review. *Cleaner Waste Systems*, 2022: p. 100010.

2 Recent Trends of Promising Membrane Technologies for Heavy Metal Removal from Water and Wastewater

Zainul Abideen, Maria Hanif,
Zirwa Sarwa, Irfan Aziz, Neelma Munir and
Maria Hasnaian

2.1 INTRODUCTION

2.1.1 Water Crises

Modern society assumes water as a conventional source of disposing of waste. The results of waste disposal are more diverse and reveal the disturbing levels of pollution, heavy metals, synthetic chemicals, and human wastes in the ecosystem. The whole world is facing several problems due to the lack of freshwater. Approximately 1.2 billion people have insufficient access to freshwater. Freshwater pollution, urbanization, and industrialization are the primary roots of the water crisis (Abdullah *et al.*, 2019). Water pollution is a global issue that needs to be addressed. Industrial wastes as well as disposal water from coal stations and other pollutant areas present a challenging water treatment scenario due to the presence of higher concentrations of inorganic compounds like heavy metals, which may cause damage to the delicate aquatic ecosystems (Khilji *et al.*, 2022; Mujeeb *et al.*, 2023; Zainab et al., 2023). To overcome the dilemma of water shortage, the world seeks to explore all the available options in reducing the overexploitation of freshwater resources (Abideen *et al.*, 2020, 2022; Hasnain *et al.*, 2023). The human population and associated water degradation increases the scarcity in clean water supply and increases salinity (Obotey Ezugbe and Rathilal 2020; Munir *et al.*, 2022; Abideen *et al.*, 2023).

2.1.2 WASTEWATER GENERATION

With the passage of time, conventional wastewater treatment processes have succeeded to some extent in treating overflows for discharge purposes. Improvements in the treatment of wastewater are necessary to make it reusable for industrial and agricultural purposes. Membrane technology appears a preferred choice for reclaiming water from different wastewater streams and ensuring its reuse (El Batouti *et al.*, 2021). In developing and industrialized countries, a high amount of contaminants enter the municipal water supply systems through human activities and increase public health and environmental concerns along with excessive amount of wastewater generation (Tlili and Alkanhal, 2019).

DOI: 10.1201/9781003326281-3

2.1.3 WASTEWATER TREATMENT TECHNOLOGIES

The evolving contaminants that occur in water and soil are persistent in the environment and include pharmaceutical as well as personal care products. Three emerging treatment technologies, including membrane filtration, advanced oxidation processes, and UV irradiation, provide alternatives for better protection of freshwater discharges (Rodriguez-Narvaez et al., 2017). Environmental experts have also focused on several biological, physical, and thermal technologies to minimize water pollution. Electrokinetic and advanced phytoremediation of wastewater by using catalysts hold great potential and are still at a developmental stage. Several treatment technologies of wastewater have been adopted for the management of aquatic systems. While these technologies have their own limitations besides cost-effectiveness, the generation of secondary pollutants needs to be emphasized. Unlike adsorption, which is extracellular, 'accumulation' is an intracellular process to remove pollutants from water (Adeola and Forbes, 2021). The technologies shown in Figure 2.1 are commonly used in the treatment of wastewater.

2.1.4 HEAVY METALS AND COMMON TECHNIQUES USED FOR THEIR REMOVAL

Heavy metals are the main pollutants in industrial wastewater. Five heavy metals, namely, arsenic (As), cadmium (Cd), chromium (Cr), lead (Pb), and mercury (Hg), are carcinogenic and show toxicity even in trace amounts, posing a threat to both the environment and human health. Industrial production processes are a major contributor of heavy metals in the aquatic environment. In order to separate heavy metals from wastewater, membranes are used as primary separation elements (Castro-Muñoz et al., 2021). Several membranous materials were examined for separation of heavy metals from industrial effluents. Nanocomposite membranes exhibit better removal efficiency than polymeric membranes and are able to offer the complete removal of heavy metals (Alshahrani et al., 2021).

2.2 MEMBRANE TECHNOLOGY FOR WASTEWATER TREATMENT

Membrane technology is a more efficient method than conventional separation methods. The membrane acts as a barrier between two phases, which helps in separating and limiting the transport of many chemical compounds based on their structure. The efficiency of the membrane depends upon the membrane itself as it behaves as a semi-permeable layer between two phases and also transports between two phases. Possible optimizations (such as the pyrolysis temperature, solution pH) allow the increase of the adsorption capabilities of membrane structures, leading to the removal of organic contaminants (Aliyu et al., 2018). This technology has more efficient rates for removing heavy metals and oil droplets from industrial waste (Figure 2.2). The integration of membranous surfaces with photocatalysts is helpful in characterizing heavy metals according to their compositions and hydrophilic nature. Also, the coupling of membranes with photocatalysts reduces the fouling of membranes. The physical and chemical characterizations of membranes are necessary to identify their basic structure and properties and to predict their potential in numerous environmental applications (Yalcinkaya et al., 2020).

2.2.1 PRESSURE-DRIVEN MEMBRANE PROCESSES

Several types of materials are used for the fabrication of membranous surfaces. Fabrication of the membrane is done by combining the membrane with metal oxides and nanotubes. These combinations enhance the membrane characteristics by modifying them and pressure is enhanced, which induces the osmotically driven membrane process for detoxification of wastewater. Reverse osmosis (RO) and forward osmosis (FO) of wastewater stimulate pressure through the membrane to initiate the process of desalination (Chollom, 2014). Wastewater pollutants such as reactive dyes and oxidizing agents can be removed by applying FO and RO techniques (Al-Najar et al., 2020). Pressure-driven membrane processes mostly rely on hydraulic pressure, which helps in achieving heavy metal separation. The most effective types of pressure-driven membrane processes are

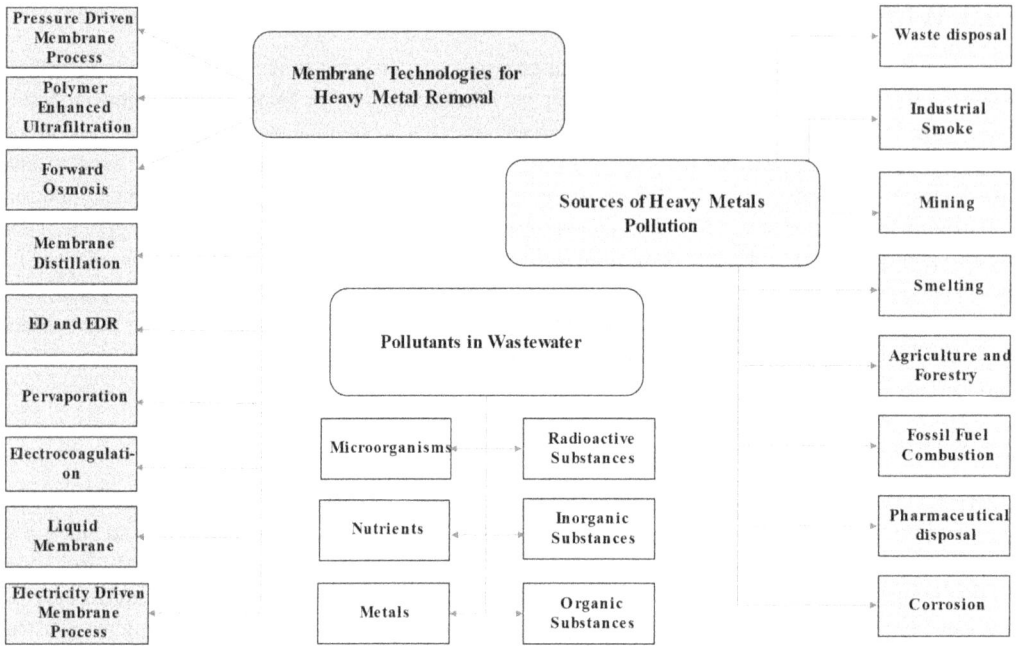

FIGURE2.1. Sources of the heavy metal and waste water pollutions and their removal efficiency by using membrane technologies

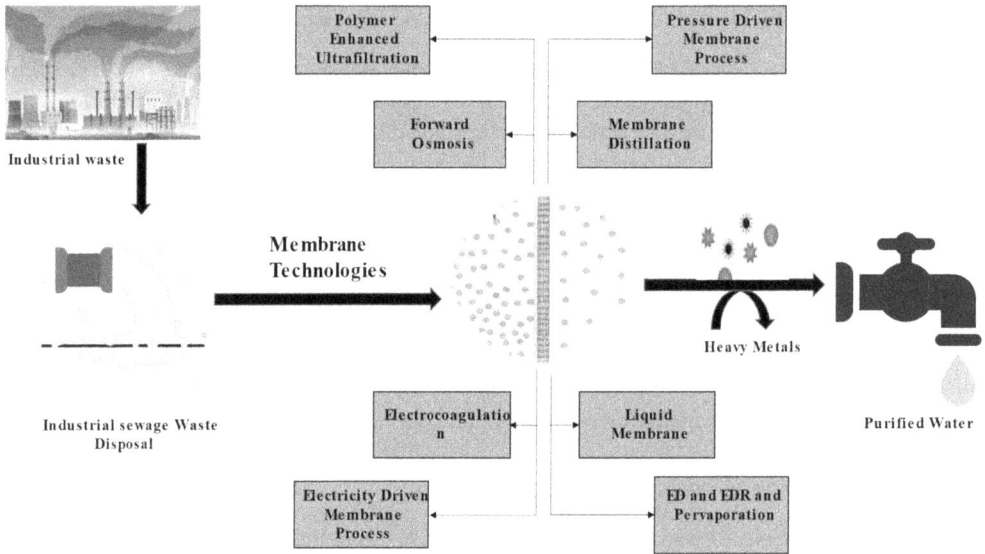

FIGURE 2.2. Membrane technologies for removing trace metals from different industrial waste

ultrafiltration, RO, and nanofiltration (NF). The main difference in these types lies in the pore sizes of their membrane. The process of RO exhibits a strong hydrostatic pressure against natural osmosis to absorb the water molecules onto the membrane surface (Zhang *et al.*, 2021).

The membranes' ultrafiltration phenomena are undertaken by fouling with iron and manganese. Ultrafiltration is undertaken by gravity-driven membrane filtration, which is ultimately used to treat manganese-contaminated surface water at high concentrations. Extremely short and stable ripening period of iron and manganese favors their removal, effectively inhibiting active catalytic manganese oxides and rapid colonization of iron and manganese-oxidizing bacteria within the biofilm.

A highly rough, porous, and heterogeneous biofilm on the membrane is responsible for stabilization. Additionally, pre-coating the membrane with manganese oxides effectively enhances the removal of manganese and iron. Several developments determine the treatments of manganese-contaminated water resources and encourage the extensive applications of membrane technologies in pressure-driven processes for detoxification of wastewater (Tang *et al.*, 2020).

2.2.2 POLYMER-ENHANCED ULTRAFILTRATION

The antifouling property of the membrane surface is usually enhanced by accomplishing the zwitterionic layer on the hydrophilic membrane surface. The strong hydrophilic properties and the zwitterionic effect of the membrane surface can effectively treat wastewater. Moreover, the monomer of sulfo-betaine methacrylate is treated with UV light through polymerization on the hydrophilic membrane proliferate membrane features. The modified membrane therefore has a higher flux rate, which results in enhanced hydrophilicity and permeability (Chiao *et al.*, 2020). The absorption equilibrium indicates the potential of effective separation in removing sulfate from the different effluents (Lin *et al.*, 2021). Table 2.1 presents different polymers used in the fabrication method.

2.2.3 FORWARD OSMOSIS

The prevalence of the heavy metal antimony in textile waste is highly toxic to human health. FO technology is used to remove the metal antimony from industrial and printing wastewater. FO with different pH and NaCl concentrations indicate that the water and the reverse salt flux of the membrane surface are proportional to the concentration of pH and NaCl (Wang and Liu, 2021). The removal of antimony may be further achieved by the addition of chromium as a co-existing ion in the FO process. The efficiency rate of FO in membrane technologies for the removal of heavy metals from wastewater is higher than 99.7% and this process also emphasizes the importance of

TABLE 2.1

Fabrication methods and polymers used for the preparation of polymeric membranes widely used in water and wastewater treatments

Membrane	Fabrication methods	Polymers in fabrication process	Solutes retained	References
Microfiltration	Phase inversion, stretching, and track etching	Polyvinylidene fluoride (PVDF), polytetra flourethylene (PTFE), polypropylene (PP), polyethylene (PE), and polyetheretherketone (PEEK)	Bacteria, fat, oil, grease, colloids, organics, micro-particles	Lalia *et al.* (2013)
Ultrafiltration	Phase inversion, solution wet spinning	Polyacrylonitrile (PAN), polysulfone (PS), polyether sulfone (PES), poly(phthazine ether sulfone ketone) (PPESK), poly(vinyl butyral), PVDF	Proteins, pigments, oils, sugar, organics, microplastics	Rohini Singh and Dutta (2020)
Nanofiltration	Interfacial, polymerization, layer-by-layer deposition, phase inversion	Polyamide (PA), PS, polyols, & polyphenols	Pigments, sulfates, divalent cations, divalent anions, lactose, sucrose, sodium chloride	Rajindar Singh and Hankins (2016)
Reverse osmosis	Phase inversion, solution casting	CA/triacetate, aromatic PA, polypiperzine, polybenziimidazoline	All contaminants including monovalent ions	Muro *et al.* (2012)
Membrane distillation	Phase inversion, stretching, electrospinning	PTFE, PVDF	Water vapors	Muro *et al.* (2012)

water quality. Furthermore, the substantial accumulation of antimony metal in FO forward osmosis was examined to understand its impact on membrane fouling (Meng *et al.*, 2020). FO is helpful in recovering water from oily wastewater. FO follows the principles of natural osmosis to provide a concentration gradient to draw water molecules from the feed solution (Haupt and Lerch, 2018).

2.2.4 Electro-Dialysis (ED) and Electro-Dialysis Reversal (EDR)

Exponential industrial development has resulted in the generation of huge quantities of organic and inorganic pollutants in wastewater. There is an urgent need to shift toward the application of advanced treatment technologies of wastewater by adopting more efficient and sustainable alternatives. The process of ED and EDR bonds the electricity and ion-permeable membranes to separate dissolved ions from water. This process generates an electric potential difference through the ion-permeable membrane, which facilitates ion transmission from the dilute solution to the concentrated solution (Campione *et al.*, 2018). There are two types of ion exchange membranes used during ED. Type one ion exchange membrane is anion permeable, and the other is cation permeable. The diluted stream faces depletion of ions while the concentrated stream becomes rich in ions due to the concentration difference (Garg *et al.*, 2022).

EDR consists of reversal of electrodes of the membrane, which results in reversing the movement of ions though the membrane. Both ED and EDR are effective in treating wastewater in order to remove various dissolved solids and ionized particles from the wastewater (Singh and Hankins, 2016). The rate of water recovery is very high in ED and EDR. EDR is more effective for wastewater streams with high salinities and this process is quite expensive to operate. Non-ionized compounds like viruses and bacteria may not be removed by EDR, which implies the need for further post-treatment of wastewater to make it virus and bacteria free (Galama *et al.*, 2014). Table 2.2 presents the applications of FO, ED, and EDR.

TABLE 2.2
Applications of FO, ED, and EDR

FO application	Result	ED, EDR application	Results	References
Raw municipal waste	Up to 70% water recovery	Treatment of almond industry wastewater	94% recovery of water	Zhao *et al.* (2017)
Coke-oven wastewater	96%–98% removal of cyanide, phenols, and Chemical Oxygen Demand	Treatment of university sewage	Removal of inorganic carbon, cations, & anions. 23%–52% removal of Chemical Oxygen Demand, Biological Oxygen Demand, color, turbidity, & Total Organic Carbon	
Reduction in gas volume field produced water	50% of volume reduced	Municipal wastewater treatment	100% effectiveness in treatment to meet discharge standards & removal of Cl^-, Mg^{2+}, Ca^{2+}	Haupt and Lerch (2018)
Coal mine wastewater desalination	More than 80% of volume of mine water recovered	Treatment of drainage wastewater for agricultural purposes	Removal of heavy metals and Na^+ up to 99%	Abou-Shady (2017)
Sewage (primary effluent)	Low water recovery due to internal concentration polarization and fouling	Tannery wastewater treatment	92%–100% removal of COD, color, NH_3-H, Cr	Deghles and Kurt (2016)
Domestic wastewater	Over 90% contaminant removal	Heavy metals (cadmium and tin) removal from the electroplating industry wastewater	Successful removal of Cd (74.8%) & Sn (64.5%)	Sivakumar *et al.* (2014)
		Wastewater treatment from China Steel Corporation	92% desalination rate, 98% Cl^- removal, 80% SO_4 removal, & 51% COD removal	Chao and Liang (2008)

2.2.5 Pervaporation

The pervaporation separation technique is based on the combination of membrane permeation and evaporation, for the removal of liquid mixtures that are based on a preference. The liquid mixture stays on one side of the membrane while evaporation occurs on the other side. The vapors are then condensed to the liquid phase, which facilitates the mass transport of components across the membrane. This mode of mass transport has been mostly applied for ethanol-water separation. The process of pervaporation is effective in treating groundwater or wastewater for micro-irrigation purposes (Cui *et al.*, 2020). The structure and chemical nature of the membrane are such that there is higher affinity for the components needed for separation. Pervaporation involves partial pressure, the flow rate of feed, and temperature and it is an ecofriendly technology. It separates liquid mixtures under highly sensitive operating conditions (Mei *et al.*, 2020).

2.2.6 Electrically Driven Membrane Processes for Chemical-Free Heavy Metal Ion Removal

The process of development and designing of heavy metal ion separation systems from wastewater plays a dynamic role in cleaning the environment. The capabilities of nanoporous membranes, made up of silicon carbide with nitrogen and fluorine under hydrostatic pressure, play an important role in interaction with water molecules and metal ions (Karimzadeh *et al.*, 2021). Conventional technologies for the recovery of ammonia from wastewater streams require an adequate amount of energy input. The combination of flow electrode capacitates deionization with the monovalent cation exchange membrane and predicts ammonia recovery from wastewater. During the charging of the electrode, the sodium and ammonium ions are transported toward the cathode due to the potential gradient. Conversion of ammonium ions into uncharged NH_3 takes place. When the electric polarity of the electrode is reversed, the charged species of sodium travel back to the spacer region and uncharged NH_3 remains trapped in the cathode layer, which results in the formation of an ammonia-rich solution. Under optimal operating conditions, a relatively low electrical energy consumption ensures the efficient recovery of uncharged ammonia from wastewater (Chen *et al.*, 2021).

2.2.7 Heavy Metal Detoxification by Electrified Membranes

Electrified membranes have the potential to overcome the limitations of conventional membrane technologies. Electrified membranes exhibit enhanced functions for separation as the sustainability of these membranes stimulates new applications in wastewater treatment. Approximately 60% of electrified membranes are prepared using metals, metal oxides, carbonaceous materials, and polymers (Liu *et al.*, 2020). The conductivity of carbonaceous materials is more than that of other membrane types. Most carbonaceous membranes may act as either a cathode or an anode. The anodes that exhibit the highest activity in oxygen evolution reactions displays strong interactions with electrogenerated hydroxyl radicals. This strong interaction of radicals helps in the formation of a higher anodic oxide, which partially oxidizes the organic compounds. A voltage below the evolution potential of oxygen of the carbonaceous membranes was applied to initiate the process of electrophoresis and direct oxidation and reduction reactions. Carbon nanotubes are used as conductive materials for electrified membranes due to their high electric conductivity and porosity. Electrified membranes with high electric conduction and surface area can be synthesized by deposition of membranes with graphene-based materials (Sun *et al.*, 2021).

2.2.8 Electrocoagulation

Electrocoagulation is playing an emerging role in removing contaminants from wastewater that are difficult to remove by using chemical treatment systems and filtration. Heavy metals and petroleum hydrocarbons may be successfully removed by implementing the technique of electrocoagulation

(El-Ashtoukhy *et al.*, 2020). Among the techniques of electrochemical processes, the most efficient one is electrocoagulation. Electrochemical production of destabilization agents from anodes is promising in removing pollutants and pathogens from wastewater. The electrocoagulation process of chromium depends on the current density and pH of membranes. The removal efficiency of chromium increases with the increase in current density and pH ~ 8, whereas the removal efficiency of chromium drops by decreasing the current density. During the electrocoagulation experiments, iron (Fe) electrodes were employed (Peng and Guo, 2020).

2.2.9 Heavy Metal Recovery by Membrane Distillation (MD)

MD technology involves the separation of heavy metals and organic ions from wastewater, which is based on the volatility of metal ions by using heat. The passage of water vapors depends on the vapor pressure gradient across the hydrophobic microporous membrane. Heat-driven processes are used to separate feed solutions with high water content (Wang *et al.*, 2018a). Tubes integrating the crystallization method ensure the production of pure water and salt crystals from wastewater using different sources. MD proves to be a promising technology for treating highly concentrated heavy metal solutions. Different hollow fiber membranes are treated with saturated NaCl solution, which has a smaller membrane pore size that makes the membrane surface more compact for MD (Ryu *et al.*, 2020). Membrane fouling has excellent resistance for concentrated water supply. The critical operating parameters in MD integrated with crystallization are saturation levels and temperature. Small filament diameter in the hollow membrane crystallizes the impurities at the inlet. Concentrated zinc and nickel solutions are used in MD to get pure crystals and distilled water. The efficiency of the distillation process relies on the distillate purity and temperature polarization (Lou *et al.*, 2020).

2.2.10 Removal of Heavy Metals by the Liquid Membrane

The extraction process of heavy metals using the emulsion liquid membrane is an effective technique for treating industrial liquid wastes. An optimal level of stability in the emulsified liquid membrane is necessary to make this technique applicable at an industrial scale. A small diameter of emulsion provides a stable larger mass transfer area. The liquid membrane separation of heavy metals is a kinetic process, which explains the increasing difference between the permeation rates of components. Emulsified oils and viruses can be removed by using the ultrafiltration membrane technique (Kazemi *et al.*, 2013). Electro-spun liquid membranes formed through electrospinning exhibit efficient microfiltration (MF) of heavy metals from industrial wastes. Ionic liquids in these modified membranes are used as an extractant in the separation of heavy metals ions. The main advantage of using ionic liquids is for dissolution of biopolymers, which are insoluble in most organic solvents. Adsorption and pore size of about 0.1–5 μm of MF membrane reject the transmission of microorganisms like bacteria and protozoa (Gong *et al.*, 2020). The removal of heavy metal ions is based on their interaction at functional sites on the nanofiber surface, which increases their removal efficiency. This extraction separation technique is ecofriendly, which removes metal ions by using an organic phase and an aqueous phase. Ionic liquids in the membranes can be used as an ideal substitute due to their stability and non-volatility (Foong *et al.*, 2020).

2.3 HYBRID MEMBRANE PROCESSES

The combination of one or more membrane techniques to make one process is known as the hybrid membrane process. These processes are coagulation, ion exchange, adsorption (or other membrane processes), which enhance performance as compared to a single process. The hybrid process is considered to be more significant as it can improve the quality of treated water (Ang *et al.*, 2015).

In order to reduce the fouling of membranes, other membrane processes are also used such as coagulation, flocculation, and sedimentation. Usually, hybrid membrane processes are used to yield water for drinking purposes as well as for other applications such as irrigation, janitorial services, and as a coolant at the industrial level. For the purification of drinking water, low-pressure membrane processes such as MF and ultrafiltration (UF) are used in combination with activated carbon. This combination leads to the removal of particulate matter and protozoa including Giardia, pathogenic bacteria, Cryptosporidium, and other organic matters (Stoquart *et al.*, 2012).

2.3.1 FO-RO HYBRID SYSTEMS

FO-RO hybrid systems are used both for the enhanced treatment of wastewater and for desalination of seawater simultaneously. In the FO-RO hybrid process, seawater is diluted by using low-salinity wastewater in order to reduce the pressure required by RO for desalination. The chemical potential of seawater provides a gradient concentration, which helps to diffuse water from wastewater (Cath *et al.*, 2010). The wastewater stream and the saline stream are present at the opposite side of the semi-permeable membrane. In the FO-RO hybrid process, water molecules transfer from the wastewater stream to the seawater stream; thus the seawater stream becomes diluted for RO. In order to enhance the concentration gradient, brine from RO is recycled into the seawater feed tank. Thus, desalination of seawater contributes to wastewater treatment. The FO-RO hybrid system has various advantages as it uses low external energy for solvent-solute separation and it also helps in the removal of the contaminant (Cath *et al.*, 2006).

2.3.2 MEMBRANE BIOREACTORS

One of the other significant hybrid processes is the use of membrane bioreactors (MBRs). This process requires a combination of biological processes such as activated sludge and membrane processes, e.g., ultrafiltration (UF), NF, and MF. This method is also used for the treatment of wastewater (Singh and Hankins, 2016). For over two decades, MBRs have often been used for their efficiency compared to conventional activated sludge processes. Conventional activated sludge processes don't have enough potential to cope with the variations in effluent flow rates and their composition as well. The MBR system occupies enough space as compared to conventional treatment systems (Judd, 2016). Nowadays, two systems of MBRs are mostly in use. These are side stream MBRs and immersed MBRs. In side stream MBRs, the membrane or filtration element is placed outside of the bioreactor, which requires a pumping system for the transfer of biomass during the filtration process. This system is significant as the cleaning of the membrane module is convenient. However, the application of this system is limited due to the requirement for high energy and pressure (Yang *et al.*, 2006). In immersed or submerged MBRs, membranes are submerged into the tank having the biological sludge. This system is commonly used for its simplicity and the low usage of energy compared to side stream MBRs. It also has the same disadvantage of cleaning membrane units as in side stream MBRs (Obotey Ezugbe and Rathilal, 2020).

2.3.3 MEMBRANE DISTILLATION

MD process has been used for more than 50 years. This technique is also explored and modified according to modern applications for commercial use (Singh and Hankins, 2016). MD is used for the separation of volatile substances using heat. The water vapor from the sample is transported using a hydrophobic microporous membrane (Nagy, 2018). Thus, this heat-based process is highly beneficial for the separation of samples having higher water content. MD uses low-grade thermal energy to provide the required pressure. Different pressures are required at the feed side and at the product side of the membrane (Wang *et al.*, 2018b). In one experiment Alkhudhiri *et al.* (2013) used MD for the treatment of water from the Arabian Gulf. Kiai *et al.* (2014) treated table olive wastewater with high phenolic content. In order to evaluate the effects of phenolic concentration, membranes of different pore sizes were used. Calabrò *et al.* (1991) treated textile wastewater and observed the energy efficiency of the MD system with respect to distillate flux, distillate purity, and temperature polarization. The improvement in the driving force on membranes may enhance the energy efficiency of MD. It is a useful treatment method for textile effluents and the recovery of water for reuse. In order to maintain heat in the system, the membrane material should be of low thermal conductivity. In the MD system, the membrane has low affinity for water to guard against wetting of the membrane. In MD, the pore size ranges between 0.1 μm and 1 μm. MD systems have many advantages, for example, they can be easily driven by the use of wind or solar energy. They use low hydrostatic pressure as compared to an RO system. MD uses pressure near atmospheric pressure. Due to the large pore size, the chances of membrane fouling are minimized. The separation of non-volatile materials from volatile materials leads to 100% separation of the feed product (Deshmukh and Elimelech, 2017). On the other hand, there are major drawbacks of MD. The non-availability of specifically designed MD membranes may enhance risks when other membranes are used. This may cause the wetting of the membrane and in turn organic components can accumulate as a result, which makes the pretreatment process more costly (Sanmartino *et al.*, 2016).

2.4 MEMBRANE MODULES AND SELECTION

For commercial applications like at the industrial level, large membrane areas are required. Modules are composed of these large membrane areas. Membrane modules are of four types, which are discussed below.

2.4.1 PLATE-AND-FRAME MODULE

The plate-and-frame module is one of the oldest modules and it contains membranes, feed spacers, and product spacers. The metallic frame covers all of these components (Gu *et al.*, 2011). The main purpose of these spacers is to prevent the sticking of membranes with each other such that the free flow of feed and product is permissible. This type of module is used for the treatment of wastewater having a high content of solid particles such as landfill leachates.

2.4.2 TUBULAR MODULE

This module is packed into an outer covering tubular-shaped shell. A fiberglass or porous stainless-steel pipe is inserted into the shell having another semi-permeable membrane. The untreated fluid is pressed into the tube under pressure. This fluid passes through the perforated pipe and is then collected through a permeate outlet (Obotey Ezugbe and Rathilal, 2020). This tubular module is designed to treat fluids having a higher amount of suspended solids.

2.4.3 SPIRAL WOUND MODULE

The spiral wound module is commonly used in RO and NF. This module is largely dependent on the high membrane surface area. This module consists of membranes, permeate spacers, and feed spacers constrained around a central collection tube. All of its components are then packed around a tubular pressure vessel. The untreated fluid enters into this module through the tangent position. The fluid travels through the membrane surface and permeate spacers and then finally into the collection tube (Kucera, 2019). This module has been successfully used at a commercial level or for large-scale operations (Lee *et al.*, 2011).

2.4.4 Hollow Fiber Module

This module consists of a bundle of hollow tubes in a pressure vessel. Hollow fibers are made up of 200-μm-thick porous nonselective support layers (Van der Bruggen and Isotherm, 2015). The design of the hollow fiber module varies according to its use. It is either made of shell side feed type or bore side feed type. Shell side feed type is used for high-pressure-purpose applications while bore side feed type is used under low pressure. One of the significant advantages of this module is that it contains a large membrane in a single module (Obotey Ezugbe and Rathilal, 2020).

2.4.5 Concentration Polarization or Polymerization

Concentration polymerization is common to all membrane filtration processes. In this technology, the concentration of the particle near the membrane surface is higher than that of the fluid and the accumulation of solute particles takes place on the membrane. This accumulation on the surface of the membrane creates a permeate flow through the membrane. This process creates a huge difference in the concentration of particles on both sides of the membrane (Klyuchnikov *et al.*, 2020). The concentration difference causes the movement of solvent molecules backward until the equilibrium phase is reached. FO also has concentration polymerization that prevents the easy movement of particles across the membrane. The formation of the boundary layer occurs due to the accumulation of particles at the membrane. Methods to reduce concentration polymerization are pretreatments, fluid management, and effective cleaning of wastewater reducing particles that contribute to the concentration polarization (Chi *et al.*, 2017).

In order to deal with internal concentration polarization, membrane modifications are applied in FO. The application of cellulose acetate in the permeable membrane reduces concentration polarization and improves the fundamental ability of the membrane to draw water (Su *et al.*, 2018; Bhattacharjee *et al.*, 2020). Flow dynamics and vibrations in membrane modules are efficient in controlling concentration polarization. The addition of spacers in the membrane filtration systems induces hydrodynamic conditions, which may reduce the polarization effect. Spacers like cylindrical rods and thin wire increase turbulence in membrane filtration systems to enhance the permeation process. By applying chemical and physical cleaning procedures with backwashing and back flushing, the effect of concentration polarization can be inhibited (Bhattacharjee *et al.*, 2020). Table 2.3 presents the properties of various membrane modules and polymers used for firm formation.

2.5 MEMBRANE FOULING AND PRETREATMENT STRATEGIES

The deposition of solid particulates, microbes, as well as organic materials onto the membrane surface or within the membrane pores may cause the fouling of membranes, which may decrease the permeate flux. The deposition of foulants on the membrane forms a layer of cake which resists

---

TABLE 2.3

Basic properties of various membrane modules and polymers used for firm formation

Property	Plate and frame	Tubular	Spiral wound	Hollow fiber	Polymer used
Packing density ft^2/ft^3 (m^2/m^3)	45–150 (148–492)	60–120 (20–374)	150–380 (492–1,247)	150–1,500 (492–4,924)	CA
Fouling potential	Moderate	Low	High	Very high	PS and PES
Ease of cleaning	Good	Excellent	Poor	Poor	PVDF
Relative manufacturing cost	High	High	Moderate	Low	Polyamide (PA)

permeate movement (Speth *et al.*, 1998). The performance ability of the membrane is greatly affected due to membrane fouling while this also hinders the movement of permeates. More pressure is required for a higher rate of membrane fouling. Consequently, higher pressure is required for the permeate movement from the membrane (Kucera, 2015). Membrane fouling affects membrane performance such as high energy consumption and reduction in the membrane filtration area. Foulants are of different forms, e.g., colloidal fouling, bio-fouling, organic and inorganic fouling, etc. (Amy, 2008). Colloids may be organic, inorganic, or in composite forms. These colloids consist of microorganisms, biological debris of polysaccharides, lipoproteins, oils, clay, silt, manganese oxide, and iron. All these suspended particles accumulate on the membrane with the passage of time (Burn and Gray, 2016). The formation of biofilms on membranes is commonly referred as bio-fouling. These biofilms consist of microbial cells and extracellular polymeric substance. All these microbes are fed on the assimilated nutrients on the membrane and grow consequently, thus blocking the pores of the membrane and increasing permeate flow (Matin *et al.*, 2011). Inorganic fouling comprises inorganic salts that are deposited on the membrane surface. Calcium sulfate, calcium carbonate, and silicon dioxide are some of the inorganic salts that are deposited on the membrane surface (Van de Lisdonk *et al.*, 2000). On the contrary, organic fouling occurs by adsorption of organic compounds onto the membrane surface. The accumulation of organic matter could affect permeate movement by the membrane (Amy, 2008). Membrane fouling is also affected by various other factors such as pH, ionic strength, and many other membrane characteristics like roughness and hydrophobicity (Li and Elimelech, 2004).

2.5.1 METHODS OF FOULING CONTROL: MEMBRANE CLEANING

The mechanism of membrane separation depends on size exclusion. There are several other techniques which are designed to reduce fouling in membranes. Some of these techniques are boundary layer velocity control, turbulence inducers, and membrane material modification (Jagannadh and Muralidhara, 1996). Other techniques are also recommended by Williams and Wakeman (2000), such as feed pretreatment, flow manipulation, rotating membranes, and gas sparging. Membrane cleaning is used for the restoration of the permeation flux of a membrane. This membrane may be destroyed due to fouling. Cleaning of membranes is classified into three processes such as physical, chemical, biochemical, and physico-chemical.

Physical cleaning involves the treatment of the membrane by using mechanical methods for the removal of foulants from the membrane (Zhao *et al.*, 2000). Physical cleaning is further classified into periodic back flushing, pneumatic cleaning, ultrasonic cleaning, sponge ball cleaning, etc. In periodic back flushing, pressure is applied on the permeate side of the membrane. The pressure applied should be higher than the filtration pressure (Zhao *et al.*, 2000). For industrial purposes, backwashing is commonly used as a fouling reversal technique, though this technique is not sufficient for the removal of irreversible fouling. Irreversible fouling is caused by clogging of membrane pores having suspended material (Yigit *et al.*, 2010). Pneumatic cleaning involves air sparging, air lifting, and air scouring. This method uses high pressure for the cleaning of the membrane. Shear force from air is usually applied for the destabilization of foulants from the membrane surface. One of the advantages of this cleaning method is that no chemical is used; however, the high cost of pumping air makes this method less usable (An *et al.*, 2010).

In ultrasonic cleaning, ultrasonic waves are used for cleaning by introducing agitation in the liquid medium. In this method, foulants are cleaned by the transmission of energy during the formation, growth, and collapse of bubbles (Li *et al.*, 2002). Sponge ball cleaning method is used for cleaning the surface of a membrane which is made up of polyurethane material. Foulants are removed as the sponge ball moves through the permeate (Zhao *et al.*, 2000). Bioactive agents such as enzymes, enzyme mixtures, and signaling molecules are used for the removal of foulants in a process known as biochemical cleaning (Maartens *et al.*, 1996). Since in the biochemical cleaning method chemicals are not used for membrane cleaning, it is a more sustainable method which does not damage the membranes as is the case in physical and chemical methods. Another frequently used method is the physico-chemical cleaning method which involves both a chemical and a physical method for the cleaning of foulants from the membrane surface. Chemical agents are used with the physical method to enhance the efficiency of the cleaning process. Chemically enhanced backwashing (CEB) is an example of the physico-chemical cleaning method (Popović et al., 2010).

2.5.2 Pretreatment Strategies for Membrane Processes

Pretreatment is one of the initial steps for the treatment of wastewater. The successful membrane process depends on feed pretreatment. Pretreatment helps in the utilization of energy besides reducing the chances of membrane fouling. It alters the physical, biological, as well as chemical properties of wastewater, which leads to efficient membrane separation (Huang et al., 2009). Pretreatment of wastewater can be carried out by using different methods. Physico-chemical methods include coagulation, adsorption, and softening techniques (Tong et al., 2019). Sardari et al. (2018) performed an experiment by using electrocoagulation as a pretreatment to direct contact membrane distillation (DCMD). As per the findings of his experiments, 57% of the water was recovered from the dissolved solids. Physico-chemical pretreatment methods are sufficient for the removal of solid content and organic particles (Hube et al., 2020). Another method for pretreatment in the membrane process is pre-filtration. It includes packed bed filters, strainers, filter cloths, and a low-pressure membrane process (Huang et al., 2009).

2.6 CONCLUSIONS

There are various membrane techniques for the treatment of wastewater. Some of the techniques have been discussed along with their applications and advantages and a few examples have also been provided to understand the mechanism. Some of the disadvantages of these membrane technologies are also listed. Since there is a global water crisis, there is a need for finding ways to obtain clean water for daily life activities. In this context, use of membrane technologies and further advancements could serve the purpose of producing clean water. Therefore, further research on this aspect is required with innovative amendments in all of the above-mentioned technologies.

2.7 RECOMMENDATIONS FOR FURTHER RESEARCH

The technology of treatment of water and wastewater is getting better and revolutionized day by day by using different membrane technology methods. Extensive research has been carried out in the recent past; still there is a lot of room for improvement that needs to be addressed to cope with the current requirement of clean water for drinking and other purposes. In the MD process, major research is needed to be done to understand the concept of temperature polarization. The development of suitable membranes according to the desired application will make the process more viable. To make FO more cost-effective, solute recovery methods need substantial improvement. Salt-based draw solutes recovery methods are also suitable for the treatment of wastewater.

REFERENCES

Abideen, Z., Ansari, R., Hasnain, M., Flowers, T. J., Koyro, H. W., El-Keblawy,... Ajmal Khan, M. (2023). Potential use of saline resources for the biofuel production using halophytes and marine algae: Prospects and pitfalls. *Frontiers in Plant Science, 14,* 1172.

Abideen, Z., Cardinale, M., Zulfiqar, F., Koyro, H. W., Rasool, S. G.,... Siddique, K. H. (2022). Seed Endophyte bacteria enhance drought stress tolerance in Hordeum vulgare by regulating, physiological characteristics, antioxidants and minerals uptake. *Frontiers in Plant Science, 13,* 980046.

Abideen, Z., Koyro, H. W., Huchzermeyer, B., Ansari, R., Zulfiqar, F., & Gul, B. (2020). Ameliorating effects of biochar on photosynthetic efficiency and antioxidant defence of Phragmites karka under drought stress. *Plant Biology, 22*(2), 259–266.

Abdullah, N., Yusof, N., Lau, W., Jaafar, J., & Ismail, A. (2019). Recent trends of heavy metal removal from water/wastewater by membrane technologies. *Journal of Industrial and Engineering Chemistry, 76,* 17–38.

Adeola, A. O., & Forbes, P. B. (2021). Advances in water treatment technologies for removal of polycyclic aromatic hydrocarbons: Existing concepts, emerging trends, and future prospects. *Water Environment Research, 93*(3), 343–359.

Al-Najar, B., Peters, C. D., Albuflasa, H., & Hankins, N. P. (2020). Pressure and osmotically driven membrane processes: A review of the benefits and production of nano-enhanced membranes for desalination. *Desalination, 479,* 114323.

Aliyu, U. M., Rathilal, S., & Isa, Y. M. (2018). Membrane desalination technologies in water treatment: A review. *Water Practice & Technology, 13*(4), 738–752.

Alkhudhiri, A., Darwish, N., & Hilal, N. (2013). Produced water treatment: application of air gap membrane distillation. *Desalination, 309*(15), 46–51.

Alshahrani, A., Alharbi, A., Alnasser, S., Almihdar, M., Alsuhybani, M., & AlOtaibi, B. (2021). Enhanced heavy metals removal by a novel carbon nanotubes buckypaper membrane containing a mixture of two biopolymers: Chitosan and i-carrageenan. *Separation and Purification Technology, 276,* 119300.

Amy, G. (2008). Fundamental understanding of organic matter fouling of membranes. *Desalination, 231*(1–3), 44–51.

An, Y., Wu, B., Wong, F. S., & Yang, F. (2010). Post-treatment of upflow anaerobic sludge blanket effluent by combining the membrane filtration process: fouling control by intermittent permeation and air sparging. *Water and Environment Journal, 24*(1), 32–38.

Ang, W. L., Mohammad, A. W., Hilal, N., & Leo, C. P. (2015). A review on the applicability of integrated/hybrid membrane processes in water treatment and desalination plants. *Desalination, 363,* 2–18.

Bhattacharjee, C., Saxena, V., & Dutta, S. (2020). Static turbulence promoters in cross-flow membrane filtration: A review. *Chemical Engineering Communications, 207*(3), 413–433.

Burn, S., & Gray, S. (2016). *Efficient Desalination by Reverse Osmosis: A Guide to RO Practice.* IWA Publishing, London.

Calabrò, V., Drioli, E., & Matera, F. (1991). Membrane distillation in the textile wastewater treatment. *Desalination, 83*(1–3), 209–224.

Campione, A., Gurreri, L., Ciofalo, M., Micale, G., Tamburini, A., & Cipollina, A. (2018). Electrodialysis for water desalination: A critical assessment of recent developments on process fundamentals, models and applications. *Desalination, 434,* 121–160.

Castro-Muñoz, R., González-Melgoza, L. L., & García-Depraect, O. (2021). Ongoing progress on novel nanocomposite membranes for the separation of heavy metals from contaminated water. *Chemosphere, 270,* 129421.

Cath, T. Y., Childress, A. E., & Elimelech, M. (2006). Forward osmosis: principles, applications, and recent developments. *Journal of Membrane Science, 281*(1–2), 70–87.

Cath, T. Y., Hancock, N. T., Lundin, C. D., Hoppe-Jones, C., & Drewes, J. E. (2010). A multi-barrier osmotic dilution process for simultaneous desalination and purification of impaired water. *Journal of Membrane Science, 362*(1–2), 417–426.

Chen, T., Xu, L., Wei, S., Fan, Z., Qian, R., Ren, X.,.... Chen, H. (2021). Ammonia-rich solution production from coal gasification gray water using chemical-free flow-electrode capacitive deionization coupled with a monovalent cation exchange membrane. *Chemical Engineering Journal, 433,* 133780.

Chi, X. Y., Zhang, P. Y., Guo, X. J., & Xu, Z. L. (2017). Interforce initiated by magnetic nanoparticles for reducing internal concentration polarization in CTA forward osmosis membrane. *Journal of Applied Polymer Science, 134*(25), 44852.

Chiao, Y.-H., Chen, S.-T., Sivakumar, M., Ang, M. B. M. Y., Patra, T., Almodovar, J.,.... Lai, J.-Y. (2020). Zwitterionic polymer brush grafted on polyvinylidene difluoride membrane promoting enhanced ultrafiltration performance with augmented antifouling property. *Polymers, 12*(6), 1303.

Chollom, M. N. (2014). *Treatment and Reuse of Reactive Dye Effluent from Textile Industry Using Membrane Technology.*

Cui, K., Li, P., Zhang, R., & Cao, B. (2020). Preparation of pervaporation membranes by interfacial polymerization for acid wastewater purification. *Chemical Engineering Research and Design, 156,* 171–179.

Deshmukh, A., & Elimelech, M. (2017). Understanding the impact of membrane properties and transport phenomena on the energetic performance of membrane distillation desalination. *Journal of Membrane Science, 539,* 458–474.

El-Ashtoukhy, E. Z., Amin, N., Fouad, Y., & Hamad, H. (2020). Intensification of a new electrocoagulation system characterized by minimum energy consumption and maximum removal efficiency of heavy metals from simulated wastewater. *Chemical Engineering and Processing-Process Intensification, 154,* 108026.

El Batouti, M., Al-Harby, N. F., & Elewa, M. M. (2021). A review on promising membrane technology approaches for heavy metal removal from water and wastewater to solve water crisis. *Water, 13*(22), 3241.

Foong, C. Y., Wirzal, M. D. H., & Bustam, M. A. (2020). A review on nanofibers membrane with amino-based ionic liquid for heavy metal removal. *Journal of Molecular Liquids, 297*, 111793.

Galama, A., Saakes, M., Bruning, H., Rijnaarts, H., & Post, J. (2014). Seawater predesalination with electrodialysis. *Desalination, 342*, 61–69.

Garg, A., Gautamb, P., Salunkeb, D. (2022). Advanced treatment technologies for industrial wastewater. *Advanced Industrial Wastewater Treatment and Reclamation of Water*, P. Perona (Ed.). (pp. 25–44). Springer.

Gong, X.-Y., Huang, Z.-H., Zhang, H., Liu, W.-L., Ma, X.-H., Xu, Z.-L., & Tang, C. Y. (2020). Novel high-flux positively charged composite membrane incorporating titanium-based MOFs for heavy metal removal. *Chemical Engineering Journal, 398*, 125706.

Gu, B., Kim, D., Kim, J., & Yang, D. R. (2011). Mathematical model of flat sheet membrane modules for FO process: Plate-and-frame module and spiral-wound module. *Journal of Membrane Science, 379*(1–2), 403–415.

Hasnain, M., Munir, N., Abideen, Z., Zulfiqar, F., Koyro, H. W., El-Naggar, A.,... Yong, J. W. H. (2023). Biochar-plant interaction and detoxification strategies under abiotic stresses for achieving agricultural resilience: A critical review. *Ecotoxicology and Environmental Safety, 249*, 114408.

Haupt, A., & Lerch, A. (2018). Forward osmosis application in manufacturing industries: A short review. *Membranes, 8*(3), 47.

Huang, H., Schwab, K., & Jacangelo, J. G. (2009). Pretreatment for low pressure membranes in water treatment: A review. *Environmental Science & Technology, 43*(9), 3011–3019.

Hube, S., Eskafi, M., Hrafnkelsdóttir, K. F., Bjarnadóttir, B., Bjarnadóttir, M. Á., Axelsdóttir, S., & Wu, B. (2020). Direct membrane filtration for wastewater treatment and resource recovery: A review. *Science of the Total Environment, 710*, 136375.

Jagannadh, S. N., & Muralidhara, H. (1996). Electrokinetics methods to control membrane fouling. *Industrial & Engineering Chemistry Research, 35*(4), 1133–1140.

Judd, S. J. (2016). The status of industrial and municipal effluent treatment with membrane bioreactor technology. *Chemical Engineering Journal, 305*, 37–45.

Karimzadeh, S., Safaei, B., Jen, T.-C., & Oviroh, P. O. (2021). Enhanced removal efficiency of heavy metal ions from wastewater through functionalized silicon carbide membrane: A theoretical study. *Journal of Water Process Engineering, 44*, 102413.

Kazemi, S. Y., Hamidi, A. S., & Chaichi, M. J. (2013). Kinetics study of selective removal of lead (II) in an aqueous solution containing lead (II), copper (II) and cadmium (II) across bulk liquid membrane. *Journal of the Iranian Chemical Society, 10*(2), 283–288.

Khilji, S. A., Munir, N., Aziz, I., Anwar, B., Hasnain, M., Jakhar,... Yang, H. H. (2022). Application of algal nanotechnology for leather wastewater treatment and heavy metal removal efficiency. *Sustainability, 14*(21), 13940.

Kiai, H., García-Payo, M., Hafidi, A., & Khayet, M. (2014). Application of membrane distillation technology in the treatment of table olive wastewaters for phenolic compounds concentration and high quality water production. *Chemical Engineering and Processing: Process Intensification, 86*, 153–161.

Klyuchnikov, A., Ovsyannikov, V. Y., Lobacheva, N., Berestovoy, A., & Klyuchnikova, D. (2020). Hydrodynamic methods for reducing concentration polarization during beer processing by membranes. *Paper presented at the IOP Conference Series: Earth and Environmental Science*, Bristol.

Kucera, J. (2015). *Reverse Osmosis: Industrial Processes and Applications*. John Wiley & Sons.

Kucera, J. (2019). *Desalination: Water from Water*. John Wiley & Sons.

Lee, K. P., Arnot, T. C., & Mattia, D. (2011). A review of reverse osmosis membrane materials for desalination— Development to date and future potential. *Journal of Membrane Science, 370*(1–2), 1–22.

Li, Q., & Elimelech, M. (2004). Organic fouling and chemical cleaning of nanofiltration membranes: Measurements and mechanisms. *Environmental Science & Technology, 38*(17), 4683–4693.

Li, J., Sanderson, R., & Jacobs, E. (2002). Ultrasonic cleaning of nylon microfiltration membranes fouled by Kraft paper mill effluent. *Journal of Membrane Science, 205*(1–2), 247–257.

Lin, W., Zhang, B., Ye, X., & Hawboldt, K. (2021). Sulfate removal using colloid-enhanced ultrafiltration: Performance evaluation and adsorption studies. *Environmental Science and Pollution Research, 28*(5), 5609–5624.

Liu, Y., Gao, G., & Vecitis, C. D. (2020). Prospects of an electroactive carbon nanotube membrane toward environmental applications. *Accounts of Chemical Research, 53*(12), 2892–2902.

Lou, X.-Y., Xu, Z., Bai, A.-P., Resina-Gallego, M., & Ji, Z.-G. (2020). Separation and recycling of concentrated heavy metal wastewater by tube membrane distillation integrated with crystallization. *Membranes, 10*(1), 19.

Maartens, A., Swart, P., & Jacobs, E. (1996). An enzymatic approach to the cleaning of ultrafiltration membranes fouled in abattoir effluent. *Journal of Membrane Science, 119*(1), 9–16.

Matin, A., Khan, Z., Zaidi, S., & Boyce, M. (2011). Biofouling in reverse osmosis membranes for seawater desalination: Phenomena and prevention. *Desalination, 281*, 1–16.

Mei, X., Ding, Y., Li, P., Xu, L., Wang, Y., Guo, Z.,... Xiao, Y. (2020). A novel system for zero-discharge treatment of high-salinity acetonitrile-containing wastewater: Combination of pervaporation with a membrane-aerated bioreactor. *Chemical Engineering Journal, 384*, 123338.

Meng, L., Wu, M., Chen, H., Xi, Y., Huang, M., & Luo, X. (2020). Rejection of antimony in dyeing and printing wastewater by forward osmosis. *Science of the Total Environment, 745*, 141015.

Mujeeb, A., Abideen, Z., Aziz, I., Sharif, N., Hussain, M. I., Qureshi, A. S., & Yang, H. H. (2023). Phytoremediation of potentially toxic elements from contaminated saline soils using Salvadora persica L.: Seasonal evaluation. *Plants, 12*(3), 598.

Munir, N., Hasnain, M., Roessner, U., & Abideen, Z. (2022). Strategies in improving plant salinity resistance and use of salinity resistant plants for economic sustainability. *Critical Reviews in Environmental Science and Technology, 52*(12), 2150–2196.

Nagy, E. (2018). *Basic Equations of Mass Transport through a Membrane Layer*. Elsevier.

Obotey Ezugbe, E., & Rathilal, S. (2020). Membrane technologies in wastewater treatment: A review. *Membranes, 10*(5), 89.

Peng, H., & Guo, J. (2020). Removal of chromium from wastewater by membrane filtration, chemical precipitation, ion exchange, adsorption electrocoagulation, electrochemical reduction, electrodialysis, electrodeionization, photocatalysis and nanotechnology: A review. *Environmental Chemistry Letters, 18*, 1–14.

Popović, S., Djurić, M., Milanović, S., Tekić, M. N., & Lukić, N. (2010). Application of an ultrasound field in chemical cleaning of ceramic tubular membrane fouled with whey proteins. *Journal of Food Engineering, 101*(3), 296–302.

Rodriguez-Narvaez, O. M., Peralta-Hernandez, J. M., Goonetilleke, A., & Bandala, E. R. (2017). Treatment technologies for emerging contaminants in water: A review. *Chemical Engineering Journal, 323*, 361–380.

Ryu, S., Naidu, G., Moon, H., & Vigneswaran, S. (2020). Selective copper recovery by membrane distillation and adsorption system from synthetic acid mine drainage. *Chemosphere, 260*, 127528.

Sanmartino, J. A., Khayet, M., García-Payo, M., Hankins, N., & Singh, R. (2016). *Desalination by Membrane Distillation*. Elsevier.

Sardari, K., Fyfe, P., Lincicome, D., & Wickramasinghe, S. R. (2018). Combined electrocoagulation and membrane distillation for treating high salinity produced waters. *Journal of Membrane Science, 564*, 82–96.

Singh, R., & Hankins, N. (2016). *Emerging Membrane Technology for Sustainable Water Treatment*. Elsevier.

Speth, T. F., Summers, R. S., & Gusses, A. M. (1998). Nanofiltration foulants from a treated surface water. *Environmental Science & Technology, 32*(22), 3612–3617.

Stoquart, C., Servais, P., Bérubé, P. R., & Barbeau, B. (2012). Hybrid membrane processes using activated carbon treatment for drinking water: A review. *Journal of Membrane Science, 411*, 1–12.

Su, X., Li, W., Palazzolo, A., & Ahmed, S. (2018). Concentration polarization and permeate flux variation in a vibration enhanced reverse osmosis membrane module. *Desalination, 433*, 75–88.

Sun, M., Wang, X., Winter, L. R., Zhao, Y., Ma, W., Hedtke, T.,... Elimelech, M. (2021). Electrified membranes for water treatment applications. *ACS ES&T Engineering, 1*(4), 725–752.

Tang, X., Xie, B., Chen, R., Wang, J., Huang, K., Zhu, X.,... Liang, H. (2020). Gravity-driven membrane filtration treating manganese-contaminated surface water: flux stabilization and removal performance. *Chemical Engineering Journal, 397*, 125248.

Tlili, I., & Alkanhal, T. A. (2019). Nanotechnology for water purification: Electrospun nanofibrous membrane in water and wastewater treatment. *Journal of Water Reuse and Desalination, 9*(3), 232–248.

Tong, T., Carlson, K. H., Robbins, C. A., Zhang, Z., & Du, X. (2019). Membrane-based treatment of shale oil and gas wastewater: The current state of knowledge. *Frontiers of Environmental Science & Engineering, 13*(4), 1–17.

Van de Lisdonk, C., Van Paassen, J., & Schippers, J. (2000). Monitoring scaling in nanofiltration and reverse osmosis membrane systems. *Desalination, 132*(1–3), 101–108.

Van der Bruggen, B., & Isotherm, F. (2015). Freundlich isotherm. *Encyclopedia of Membranes*, E. Drioli and L. Giorno (Eds.). (pp. 24–60). Springer Berlin Heidelberg.

Wang, Y.-N., Goh, K., Li, X., Setiawan, L., & Wang, R. (2018a). Membranes and processes for forward osmosis-based desalination: Recent advances and future prospects. *Desalination, 434*, 81–99.

Wang, Y.-N., Goh, K., Li, X., Setiawan, L., & Wang, R. (2018b). Membranes and processes for forward osmosis-based desalination: Recent advances and future prospects. *Desalination, 434*(15), 81–99.

Wang, J., & Liu, X. 2021. Forward osmosis technology for water treatment: Recent advances and future perspectives. *Journal of Cleaner Production, 280*, 124354.

Williams, C., & Wakeman, R. (2000). Membrane fouling and alternative techniques for its alleviation. *Membrane Technology, 2000*(124), 4–10.

Yalcinkaya, F., Boyraz, E., Maryska, J., & Kucerova, K. (2020). A review on membrane technology and chemical surface modification for the oily wastewater treatment. *Materials, 13*(2), 493.

Yang, W., Cicek, N., & Ilg, J. (2006). State-of-the-art of membrane bioreactors: Worldwide research and commercial applications in North America. *Journal of Membrane Science, 270*(1–2), 201–211.

Yigit, N., Civelekoglu, G., Harman, I., Koseoglu, H., & Kitis, M. (2010). Effects of various backwash scenarios on membrane fouling in a membrane bioreactor. Gökçekus, H., Türker, U., LaMoreaux, J. (Eds). *Survival and Sustainability* (pp. 917–929). Springer.

Zhang, Z., Wu, Y., Luo, L., Li, G., Li, Y., & Hu, H. (2021). Application of disk tube reverse osmosis in wastewater treatment: A review. *Science of the Total Environment, 792*, 148291.

Zhao, Y.-j., Wu, K.-f., Wang, Z.-j., Zhao, L., & Li, S.-s. (2000). Fouling and cleaning of membrane-A literature review. *Journal of Environmental Sciences, 12*(2), 241–251.

Zainab, R., Hasnain, M., Ali, F., Dias, D.A., El-Keblawy, A., & Abideen, Z. (2023). Exploring the bioremediation capability of petroleum-contaminated soils for enhanced environmental sustainability and minimization of ecotoxicological concerns. *Environmental Science and Pollution Research, 30*, 104933–104957.

3 Health Risks Associated with the Presence of Heavy Metals in Water

Sayed Muhammad Ata Ullah Shah Bukhari, Liloma Shah,
Sana Raza, Muhammad Asif Nawaz and Muhsin Jamal

3.1 INTRODUCTION

Freshwater covers about 3% of the surface of the earth; however, something which is clear is that different factors have limited the availability of water consumption for humans. Actually, just 61% of entire water resources are groundwater resources, which comprise approximately 20% of freshwater within the ecosphere (Hussain *et al.*, 2019). In 2015, it was reported by the World Health Organization (WHO) that approximately 71% of the world population utilized a drinkable water service at a minimum and it was also reported that by 2025 part of the population of the world would be living in regions of water stress (Alidadi *et al.*, 2019). Anyhow, the quality of water is a significant factor for societies engaged in all type of activities and it must be controlled and checked particularly for human health (Egbueri and Unigwe, 2020). Currently, heavy metals are one of the most significant factors which affect the quality of water (Egbueri and Unigwe, 2020). Presently, inhalation (breathing) and oral (eating/drinking) are the major routes for exposure of human beings to such metals (Martin and Griswold, 2009). Unlike organic contaminants which are biodegradable, ions of heavy metals are unable to be degraded to a safer form; hence, they stay in the environment for a longer period. Likewise, they show toxicity for living beings and within tissues of humans, animals and plants if they are accumulated, and numerous illnesses are caused by them due to biomagnification and bioaccumulation within the food web (Bhaskar *et al.*, 2010). From a nutritional viewpoint, essential metals such as selenium (Se), chromium (Cr), manganese (Mn), iron (Fe) and copper (Cu) are needed by humans in very minute amounts for metabolism, but in higher concentrates, they are generally lethal (Shah *et al.*, 2020). Numerous metals are toxic when their amounts are higher than that needed by the body of animals and humans; they cause damage and are also carcinogenic (Nyambura *et al.*, 2020). Heavy metals have an adverse impact on human health and the exposure to such metals increases because of modern industrialization and anthropogenetic activities. Air and water contamination via toxic metals is an ecological issue and all over the globe millions of individuals are influenced (Luo *et al.*, 2020). Heavy metals have a lethal impact on different organs of the body. Following are the toxic effects of heavy metals: cancer, birth defects, dysfunction of the immune system, skin lesions, disorders of the nervous system, vascular damage, dysfunction of the kidneys and gastrointestinal (GI) problems. There are cumulative effects on exposure to two or more than two metals (Gazwi *et al.*, 2020). When exposure to heavy metals occurs at a higher dosage, mainly, lead and mercury, it might cause severe problems like kidney failure, bloody diarrhoea and abdominal colic pain (Tsai *et al.*, 2017). The fact that numerous metals are carcinogenic for humans is one more significant consequence of chronic exposure.

DOI: 10.1201/9781003326281-4

Though the particular mechanism is uncertain, unusual alterations within the genome and gene expression are proposed as a fundamental process. Repair and synthesis of DNA is disrupted by carcinogenic metals like chromium, cadmium and arsenic (Koedrith *et al.*, 2013). Carcinogenicity and toxicity of heavy metals depend on the dose. Neuropsychiatric syndromes and DNA damage are caused in humans and animals when they are exposed to a high dose of metals (Gorini *et al.*, 2014). An understanding of the toxic mechanisms of diverse heavy metals enhances our knowledge of their detrimental effects on organs of the body and lead to best management of human and animal poisonings. The present chapter discusses heavy metals, the sources of such heavy metals, the mechanisms of toxicity related to heavy metals and the impact of heavy metals on the health of humans.

3.2 SOURCES OF HEAVY METALS IN WATER

Heavy metals enter into the surroundings by means of anthropogenic activities and by natural ways. Numerous heavy metal sources result from the use of insect or disease control agents applied to crops, sewage discharge, urban runoff, industrial effluents, soil erosion, mining, natural weathering of the crust of earth and several others (Morais *et al.*, 2012). Heavy metal sources which generally reach the environment (water) are summarized in Table 3.1.

TABLE 3.1

Sources of heavy metals in water

S/no.	Heavy metals	Sources	Effects on Human Health
1	Arsenic (As)	Atmospheric deposition, mining, pesticides, mining, smelting, metal hardening, paints, medicinal, pharmaceutical, wastewater, smelting of gold, lead, industrial waste, combustion of fossil fuel, industrial waste, textile, industrial dusts	Highly affects the dermal region (cancer), brain & cardiac problems. Bronchitis, dermatitis, poisoning
2	Lead (Pb)	Batteries, coal combustion, paint industry, industrial dust and fumes, solid waste, solid waste combustion and incineration, paints and pigments, explosive, ceramics and dishware, some types of PVC, pesticides, manufacturing of lead-acid batteries, paints and pigments	Serious effect on mental health (Alzheimer's disease), mental retardation in children, development delay, fatal infant encephalopathy, chronic damage to nervous system, liver, kidney damage
3	Mercury (Hg)	Coal combustion, fish, mining, paint industry, paper industry, volcanic eruption, mining, smelting and metallurgy, production of chemicals, industrial dust and fumes, industrial wastewater, incineration of municipal wastes, fertilizers, pesticides, electrical switches, mercury arc lamps, production of mercury products (batteries, thermometers, mercury amalgam), explosives	Sclerosis, blindness, Minamata disease, deafness, gastric problems, renal disorders, tremors, gingivitis, protoplasm poisoning, damage to the nervous system, spontaneous abortion
4	Cadmium (Cd)	Plastic, fertilizers, pesticides, phosphate fertilizer, electronics, pigments and paints, industrial and incineration dust and fumes, wastewaters, pesticides, battery, PVC products, colour pigments	Osteo-related problems, prostate cancer, lung diseases, renal issues, lung cancer, bone defects, kidney damage, bone marrow
5	Chromium (Cr)	Steel fabrication, electroplating, textile, mining and metallurgy, metal plating, rubber, industrial dust and fumes, tanning, leather industry, chemical industry, fertilizers	Lung disorders (bronchitis, cancer), renal and reproductive system, damage to the nervous system, irritability

Table 3.1 (Continued)

S/no.	Heavy metals	Sources	Effects on Human Health
6	Aluminium (Al)	Rock and soil leaching, minerals, solid wastes associated with industrial processes	Nervous system, Alzheimer's disease, Gehrig's disease
7	Iron (Fe)	High intake of iron, supplements & oral consumption	Vomiting, diarrhoea, abdominal pain, dehydration & lethargy

Adopted and modified from (Fahimirad, S. and Hatami, M., 2017. Heavy metal-mediated changes in growth and phyto-chemicals of edible and medicinal plants. In *Medicinal Plants and Environmental Challenges* (pp. 189–214). Springer, Cham; Jyothi, N.R., 2020. Heavy metal sources and their effects on human health. *Heavy Metals-Their Environmental Impacts and Mitigation;* Liu, Y. and Ma, R., 2020. Human health risk assessment of heavy metals in groundwater in the luan river catchment within the north China Plain. *Geofluids,* 2020; Balali-Mood, M., Naseri, K., Tahergorabi, Z., Khazdair, M.R. and Sadeghi, M., 2021. Toxic mechanisms of five heavy metals: mercury, lead, chromium, cadmium, and arsenic. *Frontiers in Pharmacology*, 12).

3.3 HEAVY METALS AND THEIR TOXICITY MECHANISMS

3.3.1 ARSENIC

It is considered a major risk factor for public health. Exposure sources are occupational or through polluted water and food. Arsenic has a long history of usage, either as a therapeutic product or as a metalloid substance. It has a notorious reputation as the king of poisons (Gupta *et al.*, 2017). Within the environment, it is present as a contaminant in water and food. Following are the forms of arsenic: arsine (AsH_3), organic, inorganic (As^{5+} and As^{3+}) and metalloid (As^0) (Shah *et al.*, 2010; Sattar *et al.*, 2016; Kuivenhoven and Mason, 2019). The primary absorption of As occurs in the small intestine. Inhalation and skin contact are other exposure routes. Following dispersal to numerous organs and tissues within the body such as neural tissue, muscles, liver, kidneys, heart and lungs, arsenic metabolism occurs to form dimethylarsinic acid (DMA) and monomethylarsonic acid (MMA) (Del Razo *et al.*, 1997; Ratnaike, 2003). Chronic and acute toxicity of As is associated with dysfunctions of several crucial enzymes. As compared to other heavy metals, the enzymes which contain sulphydryl groups are inhibited by As causing dysfunction. Furthermore, As binds to the lipoic acid moiety of enzymes and pyruvate dehydrogenase is inhibited by it. When pyruvate dehydrogenase is inactivated oxidative phosphorylation is inhibited and Krebs cycle is blocked. Consequently, production of ATP declines, and it results in cell damage (Shen *et al.*, 2013). Additionally, when the endothelium of the capillary is damaged by As vascular permeability is increased and it leads to circulatory collapse and vasodilation (Jolliffe *et al.*, 1991).

3.3.1.1 Mechanism of Toxicity

As far as biotransformation of As is concerned, the methylation of detrimental inorganic arsenic substances occurs through fungi, algae, bacteria and human beings to produce DMA and mono-methylarsonic acid (MMA). In such process of biotransformation, these inorganic arsenic species (iAs) are enzymatically transformed into methylated arsenicals that are the end-metabolites and a biomarker of chronic exposure of arsenic.

$$iAs\ (V) \rightarrow iAs\ (III) \rightarrow MMA\ (V) \rightarrow MMA\ (III) \rightarrow DMA\ (V)$$

Bio-methylation is a detoxification mechanism and methylated inorganic arsenic like DMA (V) and MMA (V) are the end products, which are released via urine and are a bio-indication of chronic exposure to arsenic. However, MMA (III) stays within the cell as a transitional product and is not excreted. As compared to other arsenicals, "monomethylarsonic acid (MMA III)", an intermediate product, is extremely toxic and is possibly responsible for carcinogenesis (i.e. arsenic-induced) (Singh *et al.*, 2007).

3.3.2 LEAD

Pb is an extremely lethal metal; extensive usage of lead in numerous regions of the globe causes health issues and results in ecological contamination. It is a metal which is silvery and bright, but marginally bluish in a dry atmosphere. On contact with air, it turns tarnish, thus making a composite compound mixture, dependending on the given circumstances (Sharma & Dubey, 2005). Domestic water, drinking water, smoking, food and industrial processes are the sources of exposure to Pb. Each year, much than 100,000–200,000 tonnes of lead is liberated from exhausts of vehicle within the US. Plants take some of it, some is fixed in the soil, and some is released into water bodies; hence, human exposure to lead is due to drinking water or food (Goyer, 1990). Pb is a very lethal heavy metal which interrupts with numerous physiological processes of plants and unlike another metals, like manganese, copper and zinc, biological functions are not performed by it. Plants with higher concentrations of lead increase reactive oxygen species (ROS), which damage the lipid membrane, which leads to an impairment of photosynthetic processes and chlorophyll and entire plant growth is therefore suppressed (Najeeb *et al.*, 2017). Some studies showed that Pb is able to inhibit tea plant growth by lowering biomass and degrade the quality of tea by altering its constituents (Yongsheng *et al.*, 2011). Even at lower levels, exposure to Pb causes enormous instability by uptake of ions via plants, leading to substantial metabolic alterations in photosynthetic capability and eventually inhibited plant growth.

3.3.2.1 Mechanism of Toxicity

Pb is toxic within living cells because it causes oxidative stress by an ionic mechanism. Numerous investigators have revealed that oxidative stress within cells is produced because of the imbalance between the generation of free radicals and the production of antioxidants for detoxifying reactive intermediates or for repairing the resultant damage. Antioxidants, for example, glutathione, existing within the cell defend it from free radicals like H_2O_2. Due to lead, antioxidant levels reduce, and the level of ROS increases. Glutathione occurs both in the oxidized glutathione disulphide (GSSG) state and in the reduced (GSH) state; the reduced glutathione form gives its reducing equivalents ($H^+ + e^-$) from its thiol groups of cysteine to ROS so as to provide stability. When glutathione peroxidase enzyme is present, glutathione (reduced form) readily attaches with another glutathione molecule after donating electrons and GSSG is formed. Under normal circumstances the oxidized form (GSSG) accounts for 10% and the reduced form (GSH) of glutathione accounts for 90% of the entire content of glutathione. However, under oxidative stress, the GSSG concentration increases the GSH level. Lipid peroxidation is one more biomarker of oxidative stress; meanwhile, the free radical gathers an electron from molecules of lipid existing within the cellular membrane, which causes peroxidation of the lipid (Flora *et al.*, 2008; Wadhwa *et al.*, 2012). When ROS is present in higher concentrations it might lead to structure-related damages to lipids, membranes, nucleic acid, proteins and cells and stress increases at the cellular level (Mathew *et al.*, 2011). The ionic mechanism of toxicity of lead happens primarily because of the capability of lead ions to substitute monovalent cations such as Na^+ and bivalent cations such as Fe^{2+}, Mg^{2+} and Ca^{2+}, which finally interrupts cellular metabolism. Numerous biological processes like the release of neurotransmitters, enzyme regulation, ionic transportation, apoptosis, maturation, protein folding, intra- and inter-cellular signalling, and cell adhesion are significantly changed by the ionic mechanism of toxicity of lead. Even at picomolar concentrations calcium is substituted by lead and protein kinase C is affected, which regulates memory storage and the excitation of neurons (Flora *et al.*, 2008).

3.3.3 MERCURY

Hg is present in the soil, water and air as organic mercury (usually ethyl or methyl mercury), inorganic mercury (Hg^{2+}, Hg^+) and metallic or elemental mercury (Hg^0) (Li *et al.*, 2017). At room temperature elemental mercury exists in the liquid state and can be evaporated readily to create vapour. As compared to the liquid form mercury vapour is more harmful. Breakage of containers causes Hg^0 spills and when larger concentrations of vapour of Hg are inhaled it becomes lethal. As compared to organic compounds, compounds of organic mercury like ethyl mercury (Et-Hg) or methyl mercury (Me-Hg) are much more lethal (Kungolos et al., 1999). Compounds of mercury have numerous uses in mining; for instance, it is used in industrial processes and gold extraction. In lamp making factories, mercury is utilized in the making of fluorescent light bulbs. Et-Hg and Me-Hg are utilized as fungicides for protecting plants from infections. Skin brightening creams contain an active ingredient in the form of mercury chloride, which are applied for removing spots and marks because of extreme melanin accumulation. Tyrosinase activity is inhibited by $HgCl_2$ and it is an enzyme which plays a role in the formation of melanin, by substituting the copper cofactor (Chen et al., 2020). Additionally, an organic compound which contains mercury termed as "thimerosal" is utilized as a preservative in vaccine vials which are multidose. From the lungs Hg^0 (vapour) is absorbed readily and dispersed throughout the body. Hg^0 can cross the placenta and the blood brain barrier (BBB); consequently, as compared to inorganic Hg its neurotoxicity is more. Within the body Hg^0 oxidation occurs and divalent Hg (Hg^{2+}) is produced. Placenta and BBB are not crossed by inorganic Hg. From the GI tract organic Hg is absorbed easily and dispersed all over the body. CH_3-Hg gets attached to thiol-comprising particles like cysteine (CH_3-Hg-Cys); so it can cross the BBB (Bridges and Zalups, 2017).

3.3.3.1 Mechanism of Toxicity

In all forms cellular functions are disturbed by inorganic mercury by changing the quaternary and tertiary protein's structure and it binds with selenohydryl and sulphydryl groups. Thus, sub-cellular structure and organ functions are impaired by Hg. Brain is the chief organ targeted by mercury vapour, but muscle function, endocrine function, immune function, renal function and peripheral nerve function are also disturbed (Berlin, 2003). Bronchiolitis and erosive bronchitis are caused on acute exposure to mercury vapour, which leads to respiratory problems, and together with this CNS (central nervous system) symptoms like erethism or tremor are also caused (Garnier et al., 1981). Neurological dysfunction is produced when chronic exposure to substantial dosages of mercury vapour occurs. Non-specific symptoms such as GI disturbance, weight loss, anorexia, fatigue and weakness are produced on exposure to it at a lower level. When exposure occurs at high levels it is linked with fine muscle fasciculations and mercurial tremor. Erethism might likewise be noticed; fatigue, depression, insomnia, loss of memory, emotional excitability, extreme behaviour and personality alterations and hallucinations and delirium are also noticed in extreme cases (Berglund et al., 1988). Profuse salivation and gingivitis also occur (Baldi et al., 1953; Berlin and Zalups, 2007).

Still some wards use Hg_2Cl_2 (Calomel) as a laxative. Though its absorption is poor, some of it gets converted to elemental mercury, which is absorbed and causes toxicity.

Kidneys and the GI tract are targeted by mercuric salts (typically $HgCl_2$), causing acute poisoning. Widespread precipitation of "enterocyte proteins" happens, with bloody diarrhoea, vomiting and abdominal pain with probable necrosis of mucosa of gut. Death is caused from hypovolemic or from septic shock or from peritonitis. Usually, renal tubular necrosis develops in patients with anuria (Barnes et al., 1980). Chronic poisoning is infrequent with Hg salts, and it frequently involves occupational exposure to mercury vapour. Toxicity of kidney comprises either autoimmune glomerulonephritis or renal tubular necrosis (Barnes et al., 1980). Immune dysfunctions comprise hypersensitivity reactions to mercury exposure, including dermatitis and asthma, numerous kinds of autoimmune conditions (de Vos et al., 2007) and NK cells (natural killer cells) suppression (Ilbäck et al., 1991) and disturbance of numerous other subpopulations of lymphocytes. With other kinds of Hg the dysfunction of the brain is less obvious. Dysfunction of the thyroid seems related to inhibition of 5′ deiodonases, with raised reverse T3 and reduced free T3 (Ellingsen et al., 2000). Spermatogenesis is inhibited when it is accumulated in the testicles (Rao and Sharma, 2001). Within the thigh muscle, capillary damage and atrophy occurs.

Throughout the body organic Hg undergoes a reaction with sulphydryl groups; consequently, it interferes with the function of subcellular or cellular structures. Hg interferes with DNA transcription and protein synthesis including synthesis of protein within the developing brain, with the disappearance of ribosomes and endoplasmic reticulum destruction (Berlin and Zalups, 2007). Evidence suggests that it disrupts several subcellular elements within the CNS and other organs, and it also has adverse effects on the synthesis of heme within the mitochondria (Fowler and Woods, 1977). Also, in numerous locations the integrity of the cell membrane is disrupted by it and free radicals are generated (Olivieri et al., 2000; Crespo-López et al., 2009). It also leads to disruption of neurotransmitters, and neural excitoxins are stimulated by it, which results in damage to the peripheral nervous system and numerous brain parts (Berlin and Zalups, 2007). The activity of NK cells is reduced by methyl mercury (Ilbäck et al., 1991; Park et al., 2000; Santarelli et al., 2006), together with an imbalance in ratios of Th2:Th1, which induces autoimmunity (Park et al., 2000; Tanigawa et al., 2004; de Vos et al., 2007). Mercury is correspondingly linked with disturbance in DNA repair (Berlin and Zalups, 2007; Crespo-López et al., 2009). Mercury shows affinity for sulphydryl groups of the "mitochondrial oxidative phosphorylation complex" (Gruenwedel and Cruikshank, 1979), which is associated with the destruction of mitochondrial membranes, leading to chronic fatigue.

3.3.4 CADMIUM

It is the seventh most lethal heavy metal. When humans adsorb such metals it accumulates in the body. In World War I Cd was used for the first time as a tin substitute and as a pigment in the paint industry. In the current situation, it is likewise being utilized in rechargeable batteries, for the production of alloys; it is also present in tobacco smoke. Approximately three-fourths of it is utilized as an electrode component in alkaline batteries; the residual portion is utilized as plastic stabilizers, platings, pigments and coatings. Exposure of humans to such metals occurs mainly through ingestion and inhalation and they could suffer from chronic and acute intoxication. Cd disseminated into the environment stays within sediments and soils for numerous years. These metals are progressively taken by plants due to which accumulation occurs in them and along the food web they are concentrated and finally they reach the human body. The Agency for Toxic Substances and Disease Registry reported that each year more than 500,000 of the workforce in the US gets exposed to lethal cadmium (Bernard, 2008; Mutlu et al., 2012). More than 11,000 hectares of area in China is polluted through Cd and each year more than 680 tonnes of Cd industrial waste is released into the environment. Compared to other countries, Cd exposure in the environment is more in China and Japan (Han et al., 2009). In vegetables and fruits, it is mainly found because of its higher soil-to-plant transfer rate (Satarug et al., 2010). It is a non-essential and extremely lethal heavy metal which is known for its adverse affects on the cellular enzymatic system. In plants it induces nutrient deficiency and also causes oxidative stress (Irfan et al., 2013).

3.3.4.1 Mechanism of Toxicity

Apoptosis, differentiation and proliferation of the cell is affected by Cd. Such actions are related with apoptotic induction, ROS production, and impaired DNA repair pathways (Rani et al., 2014). In the mitochondria Cd gets bound and at lower concentrations both oxidative phosphorylation and cellular respiration are inhibited (Patrick, 2003). It results in DNA-protein crosslinks within cell lines, breaks in DNA strands, exchange of sister chromatids and strand chromosomal aberrations. Potentially, chromosomal deletions and mutations are caused by cadmium (Joseph, 2009). Cd toxicity accounts for increased ROS generation like hydroxyl radicals, hydrogen peroxide and superoxide ions, the binding of sulphydryl groups and depletion in reduced glutathione (GSH). Antioxidant enzymes activity like copper/zinc dismutase, manganese superoxide dismutase and catalase are inhibited by Cd (Filipič, 2012). 33% of cysteine and zinc are concentrated in the metallothionein protein. It serves as a free radical scavenger of superoxide and hydroxyl radicals (Liu et al., 2009).

Usually, cells which contain metallothioneins show resistance to Cd toxicity. On the other hand, those cells which are unable to make metallothioneins are vulnerable to Cd intoxication (Han *et al.*, 2015). The cellular levels of Ca^{2+}, nitrogen-activated protein kinases (MRPKs) and caspase activities are modulated by Cd. Because of this, apoptosis is caused indirectly (Brama *et al.*, 2012). Production of ROS is also induced by Cd, which leads to oxidative stress. Such a strategy could demonstrate Cd's performance in terms of apoptotic cell death, carcinogenicity, and organ toxicity. Hepato-colic effects, kidney/liver lesions and Itai-Itai illness in humans is caused by cadmium (Bhattacharya *et al.*, 2016). It also causes hypertension, osteoporosis, prostatic adenocarcinoma, carcinoma, lesions of the kidney including mitochondrial and nuclear damage, enlargement and histological alterations; the antioxidant power of kidneys is reduced, which consequently disturbs the balance of minerals in the body. Moreover, skeletal diseases and sexual gland dysfunctions are also caused by it. Brain psychomotor function slows down because of the presence of Cd. Cd is toxic to cells because cadmium can dislocate vitamin E and C from their metabolically active regions. Calcium absorption is reduced in the intestine, while bone calcium dissolution is elevated, resulting in a condition that affects normal metabolic processes. Neuro-developmental toxicity is caused by Cd and it is also a disrupter of the endocrine system. Cd toxicity in fishes causes dysfunction and suppression of the immune system.

3.3.5 CHROMIUM

On earth it is the seventh most abundant element. Within the environment Cr exists in numerous oxidation states, i.e. it ranges from Cr^{2+} to Cr^{6+} (Rodríguez *et al.*, 2007). Cr exists in the following common forms: hexavalent (Cr^{6+}) and trivalent (Cr^{3+}). Both forms are toxic for plants, humans and animals (Mohanty & Kumar Patra, 2013). It is anthropogenically liberated into the environment through fertilizers and sewage (Ghani and Ghani, 2011). In the reduced state Cr(III) is non-soluble in water and immobile while the oxidized form Cr(VI) is extremely water soluble and hence mobile (Wolińska *et al.*, 2013). For human beings Cr(VI) is toxic (Gürkan *et al.*, 2017). Cr(III) exists in soil organic matter and aquatic habitats in the sulphate, hydroxide and oxide forms (Cervantes *et al.*, 2001). Cr is widely employed in industries like paper and pulp production, the production of chemicals, wood preservation, pigments and paints production, electroplating, metallurgy and tanning. Such industries cause chromium pollution and this has an adverse effect on ecological and biological species (Ghani and Ghani, 2011). A broad variety of agricultural and industrial practices raise the toxic levels in the atmosphere, raising concerns regarding chromium pollution (Zayed & Terry, 2003). Numerous compounds and contaminating heavy metals are discharged into water streams (Nath *et al.*, 2008). Because of the surplus oxygen in the environment, Cr(III) gets oxidized to Cr(VI), which is much more lethal and extremely water soluble (Cervantes *et al.*, 2001). The release of industrial wastes and contamination of the groundwater has significantly raised the chromium levels in soil (Bielicka *et al.*, 2005). Cr is continuously released into the environment because of modern agricultural practices. Due to the release of chromium the soil becomes polluted, and the soil-vegetable system and also the quality and yield of vegetables are affected (Duan *et al.*, 2010). When chromium is present in excess beyond the permissible range, it damages plants. Cr phytotoxicity could be the reason for depressed biomass, the inhibition of germination in seed, chlorosis of leaf and reduction in the growth of root. Biological processes are affected significantly by Cr toxicity in numerous plants like Citrullus, cauliflower, barley, wheat, maize and other vegetables. In plants, necrosis and chlorosis is caused by chromium toxicity (Ghani and Ghani, 2011). Cr toxicity also affects enzymes such as cytochrome oxidase, peroxidase and catalase (Nath *et al.*, 2008). In order to enter into the cell, chromium(III) uses a simpler diffusion process that doesn't rely on any particular membrane carrier. Cr(IV) can effortlessly pass through a cell membrane contrary to Cr(III) (Chandra & Kulshreshtha, 2004).

3.3.5.1 Mechanism of Chromium Toxicity

Usually, Cr(III) or trivalent Cr is harmless in the environment because of its weaker membrane permeability. On the other hand, Cr(VI) or hexavalent chromium penetrates the membrane much more actively and then by means of phagocytosis these chromates are taken. Cr(VI) is a stronger oxidizing agent and when its reduction occurs, it results in ephemeral species of tetravalent and pentavalent chromium, which are different from Cr(III). Glutathione carries out pentavalent form stabilization and thus intracellular Cr(VI) reduction is thought to be the detoxification process when reduction happens away from the targeted site. If intracellular Cr(VI) reduction happens near the targeted region, it might stimulate Cr. ROS (like the hydroxyl radical, hydrogen peroxide and super-oxide ion) are produced when reactions occur among biological reductants (such as ascorbate and thiols) and Cr(VI). This leads to oxidative stress within the cell and proteins and DNA are damaged (Stohs & Bagchi, 1995). Numerous studies reported that as compared to Cr(III), Cr(VI) is much more dangerous because Cr(VI) can enter the cell much more readily as compared to Cr(III). The International Agency for Research on Cancer has classified Cr(VI) as a group 1 human carcinogen because of its mutagenic properties (Zhitkovich, 2005).

3.3.6 Aluminium

On the earth's crust it is the third most common element. It is found in one oxidation state (+3) in the environment. Major consumption routes of Al in human beings are via dermal contact, ingestion and inhalation and aluminium containing drugs, beverages, food and drinking water. Aluminium naturally occurs in food. In humans, aluminium and compounds of aluminium get poorly absorbed, though the absorption rate has not been investigated. Arthritic pain, diarrhoea, vomiting, skin rashes, skin ulcers, mouth ulcers and nausea are the symptoms which indicate the occurrence of high aluminium concentrations in humans. In 1997 the WHO stated that the exposure to aluminium in human beings is responsible for causing Alzheimer's disease. Irritant dermatitis and contact dermatitis were noticed in individuals who were subjected to aluminium at their work-place. Aluminium has adverse effects on the nervous system. Loss of coordination and memory loss are some outcomes.

3.3.6.1 Mechanism of Toxicity

Toxic impacts of aluminium mostly arise from its pro-oxidant action in cellular lipids and pro-tein oxidation, free radical attack and oxidative stress (Exley, 2013). Secondary structures are formed from the protein polypeptides when ions of aluminium interact with these structures via the protein backbone, side chains and oxygen-comprising amino acids. This finally leads to dena-turation (Mujika *et al.*, 2018) or structural or conformational changes (Exley *et al.*, 2006) as within β-amyloid. Exposure to aluminium precipitates and aggregates β-amyloid, which is linked with Alzheimer's disease (Exley *et al.*, 2006), and such process might be accountable for dysneurogen-esis, neuronal death and the deposition of neurotic plaque. Exposure to aluminium is accountable for aggregation and fibrillation of human "islet amyloid polypeptide hormone (amylin)", which then leads to the formation of β-pleated sheet structures (Mirhashemi and Aarabi, 2011; Mirhashemi and Shahabaddin, 2011; Xu *et al.*, 2016), which might impair pancreatic β-cells (Sakamoto *et al.*, 2006). Ligands which are present on extracellular and intracellular surfaces might possibly associ-ate with aluminium, thereby inducing stimulatory or inhibitory effects (Exley and Birchall, 1992). When aluminium interacts with enzymes related to metabolism and other enzymes, it activates or inactivates these enzymes (Sushma *et al.*, 2007; Ohsaka and Nomura, 2016). Aluminium attaches with nucleotide phosphate groups like ATP and energy metabolism is therefore affected (Kawahara *et al.*, 2007). When hepatocytes are exposed to aluminium, it hinders production of ATP, encour-ages oxidation of proteins and lipids, damages the functioning of the Krebs cycle or the tricarboxylic acid cycle and also inhibits glycolysis (Mailloux *et al.*, 2006; Han *et al.*, 2013). Such metabolic issues might be accountable for body weight loss and reduced production performances (such as production of egg) in animals. Apart from this, exposure to aluminium disturbs homeostasis of iron and leads to iron overload (Contini *et al.*, 2007). Injury and oxidative stress are also caused by iron. Raised levels of iron within cells could cause oxidative stress to cells and are related to the development of neurodegenerative syndromes (Adzersen *et al.*, 2003; Deugnier, 2003). Aluminum exposure results in iron overload, which leads to apoptosis via ROS, DNA damage, and enhanced lipid peroxidation (Kell, 2009). Apoptosis of osteoblasts, lymphocytes and erythrocytes (eryptosis) is correspondingly activated by Al ions (Xu *et al.*, 2018; Yang *et al.*, 2018; Yu *et al.*, 2019). Within culture, aluminium causes apoptosis in osteoblasts by inhibiting expression of apoptotic Bcl-2 proteins, and it raises Bim, Bak and Bax expression. These are pro-apoptotic proteins (Xu *et al.*, 2018). Aluminium might lower synthesis of ferritin and transferrin receptor expression is increased, thus unsettling the usual transferrin receptor synthesis (Yamanaka *et al.*, 1999). Exposure to aluminium affects glutathione (GSH), glutathione peroxidase (GPx), catalase (CAT) and superoxide dismutase (SOD) activities (Campbell *et al.*, 1999). Irregular rises in concentrations of thiobarbituric acid reactive substances (TBARS) and malondialdehyde (MDA) occurred together with reduced anti-oxidant concentrations like CAT, SOD, GPx and GSH in tissues of rats on exposure to aluminium (Newairy *et al.*, 2009; Exley, 2013; Abd-Elhady *et al.*, 2013; Zhang *et al.*, 2016; Yu *et al.*, 2019). Changes in the function of gene and mutagenesis might result from the lethal action of aluminium with fluctuations in the expressions of transcription (Exley, 2013). When aluminium binds with DNA it influences the expression of neuronal genes (Lukiw *et al.*, 1998), disposing cells towards or triggering genotoxicity. and hence exposure to aluminium might lead to decrease in cell differentiation and proliferation (Sun *et al.*, 2015; Cao *et al.*, 2016; Li *et al.*, 2016; Yang *et al.*, 2016; Huang *et al.*, 2017). Aluminium toxicity impairs neurogenesis. Differentiation and proliferation of osteoblasts were repressed through aluminium (Sun *et al.*, 2015; Cao *et al.*, 2016; Huang *et al.*, 2017). Aluminium inhibits the BMP-2 signalling pathway, which inhibits the differentiation of osteoblasts (Yang *et al.*, 2016).

3.3.7 Iron

On the earth's crust it is the second most plentiful metal. Entire living beings require this element for survival and growth (Valko *et al.*, 2005). Algal species contain this as their crucial component; also, enzymes like catalase and cytochromes and proteins which carry out oxygen transportation like myoglobin and haemoglobin also contain iron (Vuori, 1995). The iron source in surface water is anthropogenetic and is associated with mining practices. Generally, the deep ocean contains 33.5×10^{-9} mg/L or 0.6 nM of dissolved iron. The groundwater level of iron is higher, i.e. 20 mg/L, while the freshwater level is lower, i.e. 5µg/L. Contamination of iron indirectly or directly affects the species like diversity of fish, periphyton and benthic invertebrates (Vuori, 1995). Iron precipitate can cause substantial injury through congestion and by hindering fish's respiration (Phippen *et al.*, 2008).

3.3.7.1 Mechanism of Iron Toxicity

When absorbed iron fails to attach with proteins an extensive variety of detrimental free radicals are made, which subsequently influence the iron concentrations in biological fluids and mammalian cells. Such circulating iron in the boundless state destroys biological fluids and the GI tract. An enormously high concentration of iron enters into the body and gets saturated. Brain, liver and heart cells are penetrated by such free irons. Free iron disrupts oxidative phosphorylation because of which ferrous iron is transformed into ferric iron, which produces hydrogen ions, raising metabolic acidity. Free iron could likewise lead to peroxidation of lipids, which causes extreme harm to microsomes, mitochondria and other cell organelles (Albretsen, 2006). Iron toxicity in cells accounts for tissue damage through cellular reducing and oxidizing processes and damages intracellular organelles like lysosomes and the mitochondria. An extensive variety of free radicals which are probably a consequence of cellular damage are formed through extra iron intake. Free radicals are produced by iron which attack DNA and damages the cell, causing malignant transformation and mutation that account for numerous illnesses (Grazuleviciene *et al.*, 2009).

3.4 HEALTH RISKS ASSOCIATED WITH HEAVY METALS

Following are the health risks associated with heavy metals.

3.4.1 ARSENIC EFFECTS

Arsenic has the following effects on human health.

3.4.1.1 Carcinogenicity

DNA methylation, modifications of histone, alterations in the expression of p53 proteins, DNA damage, epigenetic changes and decreased expression of p21 are caused by arsenic (Martinez *et al.*, 2011; Park *et al.*, 2015). Cancer risks are increased by arsenic poisoning as it attaches with DNA-binding proteins and the process of DNA repair is slowed down (García-Esquinas *et al.*, 2013).

3.4.1.2 Skin Toxicity

Chronic exposure to arsenic causes numerous skin illnesses, including hyperpigmentation and hyperkeratosis and various kinds of skin cancer. Continued exposure to arsenic causes most predominant skin alteration like hyperpigmentation. A disease named as Bowen's disease, which is a kind of initial skin cancer, is also caused by arsenic. Arsenic hyperkeratosis is frequently prevalent: it affects palms and soles; however, the back of the hands, arms, fingers, toes and legs could likewise be affected by it (Huang *et al.*, 2019).

3.4.1.3 Reproductive and Developmental Toxicity

Arsenic is recognized as a reproductive toxin in humans and in experimental animals it accounts for anomalies, predominantly abnormalities of the neural tube. Male reproduction is impaired by inorganic arsenic by lowering the weight of the testes, the amount of sperm within the epididymis and accessory sex organs. Apart from disturbing the production of sperm, gonadotropin and testosterone levels are likewise lowered by exposure to inorganic arsenic and the process of steroidogenesis is also disturbed. In females, arsenic is linked with higher prevalence of endometrial cancer (Salnikow and Zhitkovich, 2008). Angiogenesis of the endometrium is very important for development of the embryo. During pregnancy this process is impaired by exposure to arsenic causing spontaneous abortions, sterility, prematurity and subfertility, which are the symptoms of endometriosis (Milton *et al.*, 2017).

3.4.1.4 Genotoxicity

Many studies have shown significant inter-individual variability in the susceptibility to arsenic poisoning, and the genetic variables are acknowledged as the basic cause of these variabilities. Arsenic genotoxicity results in modification of DNA, which comprises exchange of sister chromatids, deletion, production of micronuclei, mutation and abnormalities of the chromosome (Roy *et al.*, 2018). Several studies were carried out for determining the process of arsenic's genotoxicity and its influence, which includes the induction of oxidative stress and interrupted DNA repair (Pierce *et al.*, 2012). Arsenic has no direct effect on DNA and is hence considered a weaker mutagen because, despite its lower mutagenicity, it influences the mutagenicity of subsequent carcinogens (Yin *et al.*, 2019).

3.4.1.5 Neurotoxicity

Consumption of arsenic causes cognitive impairment of the CNS. Numerous diseases are also caused by it like neurodegenerative illnesses and neurodevelopmental alterations. Arsenic poisoning likewise alters synaptic transmission and the balance of neurotransmitters (Garza-Lombó *et al.*, 2019). Moreover, arsenic causes neurotoxicity by boosting numerous apoptotic processes. Initially, apoptosis is induced in neural cells by arsenic and its methylated metabolites by means of MAPK signalling pathways which comprise p38, JNK or ERK2. Moreover, intracellular uptake of calcium is initiated by arsenic, which mediates apoptosis. In contrast, cellular apoptosis can be facilitated via autophagy stimulation through AMPK stimulation in addition to "mammalian target of rapamycin (mTOR)" inhibition (Garza-Lombó *et al.*, 2019).

3.4.2 LEAD

It has the following effects on human health.

3.4.2.1 Nephrotoxicity

Lead causes lethal effects in entire organ systems, though kidneys are greatly affected. Dysfunction of the proximal tubules is caused by acute lead nephropathy. Following are the characterization of chronic lead nephropathy: glomerulonephritis, renal failure, hyperplasia, tubules atrophy and interstitial fibrosis (Lentini *et al.*, 2017).

3.4.2.2 Carcinogenicity

It is a carcinogenic agent which damages DNA repair mechanisms, chromosomal sequence and structure and cellular tumour regulating genes via release of ROS. It also disrupts transcription by removing zinc from various regulatory proteins (Silbergeld *et al.*, 2000).

3.4.2.3 Hepatotoxicity

Lead also has toxic effects on liver cells. Exposure to lead causes oxidative stress, which results in liver damage. When organic solvents are combined with lead they likewise cause liver damage given that they have properties similar to lead (Farmand *et al.*, 2005; Malaguarnera *et al.*, 2012).

3.4.2.4 Cardiovascular Toxicity

Chronic or acute exposure to lead causes numerous irregularities in humans. Cardiac disease, atherosclerosis, thrombosis, hypertension and arteriosclerosis are caused when there is chronic exposure to lead and this occurs by a change in the renin-angiotensin system, which raises vasoconstrictor prostaglandins and lowers vasodilator prostaglandins, thereby disturbing vascular smooth muscle Ca^{2+} signalling. Arterial pressure is also raised on extensive exposure to lead over long periods (Hertz-Picciotto and Croft, 1993).

3.4.3 MERCURY

Mercury has the following effects on human health.

3.4.3.1 Nephrotoxicity

Tubular necrosis is caused when acute exposure of mercury occurs in kidneys and there are numerous symptoms, like hypotension, chills, vomiting, tremors, profuse salivation, abdominal pain, changed mental status and acute dyspnea. Apart from this, chronic mercury exposure damages the epithelium and necrosis is caused by it within the pars recta of the proximal tubules (Lentini *et al.*, 2017).

3.4.3.2 Carcinogenicity

The peroxidative action of mercury produces a substantial amount of ROS, which could result in pro-tumourigenic signalling and the growth of cancerous cells. DNA, lipids and cellular proteins are damaged by ROS, causing carcinogenesis and cellular damage (Reczek and Chandel, 2017; Zefferino et al., 2017).

3.4.3.3 Cardiovascular Toxicity

In human beings, hepatotoxicity, nephrotoxicity and neurotoxicity are caused by mercury. Mercury has also been found to have cardiovascular toxicity. Atherosclerosis, acute coronary failure and atherosclerotic lesions are caused by mercury (Yoshizawa et al., 2002). Mercury inactivates paraoxonase, which is the extracellular anti-oxidative enzyme and is associated with dysfunction of HDL (Gonzalvo et al., 1997; Salonen et al., 1999). Inactivation of this enzyme is directly associated with atherosclerosis development and coronary heart disease, acute myocardial infarction, carotid artery stenosis and cardiovascular disease risks (Kulka, 2016).

3.4.3.4 Skin Toxicity

Mercury-containing compounds and mercury are responsible for causing numerous skin diseases including acrodynia (pink illness). This is a common dermatologic illness and in this disease the skin get pink when it is exposed to heavy metals, predominantly mercury (Horowitz et al., 2002). Individuals who have undergone tattooing with the red tints of mercury sulphide and cadmium sulphide might suffer from inflammation limited to precise regions characteristically six months after undergoing tattooing. Mercury-containing compounds could cause acute contact dermatitis and it has the following symptoms: irritation, vesiculation, scaling and moderate swelling (Boyd et al., 2000).

3.4.4 CADMIUM

Cadmium has the following effects on human health.

3.4.4.1 Neurotoxicity

Cadmium causes neurotoxicity. Numerous neurodegenerative issues, including multiple sclerosis, Alzheimer's disease, Parkinson's disease and amyotrophic lateral sclerosis, are caused by cadmium (Branca et al., 2018). Many preclinical studies showed that cadmium strictly disturbs the functioning of the CNS (Marchetti, 2014) and the PNS (Miura et al., 2013), with numerous clinical symptoms, like mental retardation, learning incapacities, neurological disturbances, olfactory dysfunctions and peripheral neuropathy, together with motor function impairment and behaviour alterations in both children and adults. Moreover, numerous activities of the cell like cell proliferation and differentiation are influenced by it and cadmium poisoning causes cell death. Cadmium neurotoxicity results from death of neural cells by apoptosis, through abundant apoptosis-induction factors, including by inhibition of neuron gene expression and neurogenesis impairment.

3.4.4.2 Nephrotoxicity

Cadmium induces nephrotoxicity which causes symptoms like aminoaciduria, phosphaturia, Fanconi-like syndrome and glucosuria (Hazen-Martin et al., 1993; Reyes et al., 2013). When kidneys are directly exposed to cadmium, it affects proximal tubular epithelium and the outcomes are substantial levels of cadmium in urine, glucosuria, 32-microglobulinuria and aminoaciduria (Goyer, 1989). Extreme exposure could account for hypercalciuria, renal failure and renal tubular acidosis (Jacquillet et al., 2007).

3.4.4.3 Hepatotoxicity

In humans two tissues are targeted by cadmium: liver and the renal cortex (Bernard, 2004). Through acute cadmium exposure, it get accumulated in the liver and causes numerous hepatic dysfunctions. The redox balance of cells is altered by cadmium, which results in hepatocellular impairment and oxidative stress (Zalups, 2000). Both chronic and acute exposure to cadmium in the kidneys induce hepatotoxicity, causing liver failure.

3.4.4.4 Cardiovascular Toxicity

Cadmium is a carcinogenic and poisonous metal. It is responsible for inducing cardiac illnesses, bone disease and kidney disease. Lower to moderate exposure to cadmium causes heart failure and stroke (Peters *et al.*, 2010), myocardial infarction (Everett and Frithsen, 2008), chronic kidney disease (Hellström *et al.*, 2001), peripheral arterial disease (Navas-Acien *et al.*, 2004), carotid atherosclerosis (Messner *et al.*, 2009), diabetes (Schwartz *et al.*, 2003) and hypertension. Prospective studies indicated that cadmium was related to increased risk of cardiovascular death in the US population (Tellez-Plaza *et al.*, 2013, Tellez-Plaza *et al.*, 2012). Neurotoxicity is also caused by mercury.

3.4.5 CHROMIUM

It has the following effects on human health.

3.4.5.1 Hepatotoxicity

It is revealed by several investigations that Cr(VI) can cause liver damage and it can account for histopathological variations like hepatocytes steatosis, necrosis and parenchymatous degeneration. Lowering of antioxidant enzymatic action, DNA and RNA damage, DNA suppression, lipid peroxidation, raised levels of ROS and mitochondrial dysfunction, like impaired bioenergetics of mitochondria, apoptosis and arrest of cell growth, are all linked with hepatotoxicity of Cr(VI) (Hasanein and Emamjomeh, 2019).

3.4.5.2 Immunological Toxicity

Chromium causes numerous adverse effects on the human immune system. Faleiro et al. stated that when CoCrMo disc samples were utilized they blocked proliferation of lymphocytes. Humoral immune responses and the phagocytic activity of alveolar macrophages were lowered by higher doses of hexavalent chromium (Glaser *et al.*, 1985). Furthermore, two kinds of hypersensitivity reactions are induced by chromium: type IV (delayed type) and type I (anaphylactic type) (Bruynzeel *et al.*, 1988).

3.4.6 COPPER

It has the following effects on human health.

3.4.6.1 Neurotoxicity

Along with cadmium, arsenic and manganese, numerous heavy metals show neurotoxic effects. Additionally, zinc and copper, like iron, serve as obstacles for neuro-development once extreme quantity enters the brain (Prohaska, 2000). Wilson's disease is caused by extreme retention of copper. As in schizophrenia, it is the reason for neurobehavioural irregularities (Cai *et al.*, 2005). Experimental investigation by Ken-ichiro Tanaka and Masahiro Kawahara revealed that zinc-induced neurotoxicity is increased by copper (Tanaka and Kawahara, 2017).

3.4.6.2 Hepatotoxicity

Because of Wilson's disease, copper gets accumulated. Oxidative stress is caused by raised copper levels; hence, deposition of hepatic copper is not just pathogenic but also pathognomonic. In cholestatic liver illnesses the levels of hepatic copper are raised (Gross *et al.*, 1985; Yu *et al.*, 2019).

3.5 CONCLUDING REMARKS

In this chapter we discussed the effects of numerous heavy metals, like Fe, Al, Cr, Cd, Hg, Pb and As, on living beings particularly humans and the environment. Heavy metals enter the body in many ways, including through food, air or drinking water or infrequently through dermal contact. When heavy metals are absorbed they are retained and get accumulated in the human body. There are toxic effects when toxic metals bio-accumulate in numerous organs and tissues. Chronic and acute exposure to heavy metals cause toxicity. Numerous cellular processes are disturbed by heavy metals, including apoptosis, damage-repairing processes, differentiation, proliferation and growth. Epigenetic changes are promoted by toxic metals, which could affect gene expression.

Heavy metals are released into the environment either from anthropogenetic sources or naturally. Human exposure to heavy metals is either ecological, given that the metal occurs in nature, or via external sources. Different ingestion routes are there. It is essential to comprehend the way in which heavy metals enter into the body and the associated fatality rate and toxicity severity. Extreme levels can cause substantial harm to different body organs and induce osteoporosis, GI obstruction, carcinogenicity, respiratory disorders and neurological defects, etc. For countering toxicity numerous natural products and nanotechnological approaches were established. Numerous preventive approaches should be taken into consideration. Initially, to decrease heavy metal toxicity, the sufferers should be removed from the exposure source. Individuals must not employ chemical products wherever lethal heavy metals are used expansively. Also, self-alertness might help in this regard. If all these things are taken into consideration the health risks associated with heavy metals would reduce. Consequently, further investigation is needed for a better understanding of molecular processes and public health outcomes of human exposure to numerous toxic metals.

REFERENCES

Abd-Elhady, R.M., Elsheikh, A.M. and Khalifa, A.E., 2013. Anti-amnestic properties of Ginkgo biloba extract on impaired memory function induced by aluminum in rats. *International Journal of Developmental Neuroscience*, *31*(7), pp. 598–607.

Adzersen, K.H., Becker, N., Steindorf, K. and Frentzel-Beyme, R., 2003. Cancer mortality in a cohort of male German iron foundry workers. *American Journal of Industrial Medicine*, *43*(3), pp. 295–305.

Albretsen, J., 2006. The toxicity of iron, an essential element. *Veterinary Medicine-Bonner Springs Then Edwardsville-*, *101*(2), p. 82.

Alidadi, H., Tavakoly Sany, S.B., Zarif Garaati Oftadeh, B., Mohamad, T., Shamszade, H. and Fakhari, M., 2019. Health risk assessments of arsenic and toxic heavy metal exposure in drinking water in northeast Iran. *Environmental Health and Preventive Medicine*, *24*(1), pp. 1–17.

Baldi, G., Vigliani, E.C. and Zurlo, N., 1953. Mercury poisoning in hat industry. *La Medicina del Lavoro*, *44*(4), pp. 161–198.

Barnes, J.L., McDowell, E.M., McNeil, J.S., Flamenbaum, W. and Trump, B.F., 1980. Studies on the pathophysiology of acute renal failure. *Virchows Archiv B*, *32*(1), pp. 201–232.

Berglund, A., Pohl, L., Olsson, S. and Bergman, M., 1988. Determination of the rate of release of intra-oral mercury vapor from amalgam. *Journal of Dental Research*, *67*(9), pp. 1235–1242.

Berlin, M., 2003. Dental materials and health. *Statens Offentliga Utredningar*, Stockholm.

Berlin, M. and Zalups, R.K., 2007. Mercury. In *Handbook on the Toxicology of Metals*, G.F. Nordberg, B.A. Fowler, M. Nordberg and L.T. Friberg (Eds.) 943pp. New York: Elsevier.

Bernard, A., 2004. Renal dysfunction induced by cadmium: biomarkers of critical effects. *Biometals*, *17*(5), pp. 519–523.

Bernard, A., 2008. Cadmium & its adverse effects on human health. *Indian Journal of Medical Research*, *128*(4), p. 557.

Bhaskar, C.V., Kumar, K. and Nagendrappa, G., 2010. Assessment of heavy metals in water samples of certain locations situated around Tumkur, Karnataka, India. *E-Journal of Chemistry*, *7*(2), pp. 349–352.

Bhattacharya, P.T., Misra, S.R. and Hussain, M., 2016. Nutritional aspects of essential trace elements in oral health and disease: an extensive review. *Scientifica*, *2016*.

Bielicka, A., Bojanowska, I. and Wisniewski, A., 2005. Two faces of chromium-pollutant and bioelement. *Polish Journal of Environmental Studies*, *14*(1).

Boyd, A.S., Seger, D., Vannucci, S., Langley, M., Abraham, J.L. and King Jr, L.E., 2000. Mercury exposure and cutaneous disease. *Journal of the American Academy of Dermatology*, *43*(1), pp. 81–90.

Brama, M., Politi, L., Santini, P., Migliaccio, S. and Scandurra, R., 2012. Cadmium-induced apoptosis and necrosis in human osteoblasts: role of caspases and mitogen-activated protein kinases pathways. *Journal of Endocrinological Investigation*, *35*(2), pp. 198–208.

Branca, J.J.V., Morucci, G. and Pacini, A., 2018. Cadmium-induced neurotoxicity: still much ado. *Neural Regeneration Research*, *13*(11), p. 1879.

Bridges, C.C. and Zalups, R.K., 2017. Mechanisms involved in the transport of mercuric ions in target tissues. *Archives of Toxicology*, *91*(1), pp. 63–81.

Bruynzeel, D.P., Hennipman, G. and Van Ketel, W.G., 1988. Irritant contact dermatitis and chrome-passivated metal. *Contact Dermatitis*, *19*(3), pp. 175–179.

Cai, L., Li, X.K., Song, Y. and Cherian, M.G., 2005. Essentiality and toxicology of zinc and copper and its chelation therapy. *Current Medicinal Chemistry*, *12*(23), pp. 2753–2763.

Campbell, A., Prasad, K.N. and Bondy, S.C., 1999. Aluminum-induced oxidative events in cell lines: glioma are more responsive than neuroblastoma. *Free Radical Biology and Medicine*, *26*(9–10), pp. 1166–1171.

Cao, Z., Fu, Y., Sun, X., Zhang, Q., Xu, F. and Li, Y., 2016. Aluminum trichloride inhibits osteoblastic differentiation through inactivation of Wnt/β-catenin signaling pathway in rat osteoblasts. *Environmental Toxicology and Pharmacology*, *42*, pp. 198–204.

Cervantes, C., Campos-García, J., Devars, S., Gutiérrez-Corona, F., Loza-Tavera, H., Torres-Guzmán, J.C. and Moreno-Sánchez, R., 2001. Interactions of chromium with microorganisms and plants. *FEMS Microbiology Reviews*, *25*(3), pp. 335–347.

Chandra, P. and Kulshreshtha, K., 2004. Chromium accumulation and toxicity in aquatic vascular plants. *The Botanical Review*, *70*(3), pp. 313–327.

Chen, J., Ye, Y., Ran, M., Li, Q., Ruan, Z. and Jin, N., 2020. Inhibition of tyrosinase by mercury chloride: spectroscopic and docking studies. *Frontiers in Pharmacology*, *11*, p. 81.

Contini, M.D.C., Ferri, A., Bernal, C.A. and Carnovale, C.E., 2007. Study of iron homeostasis following partial hepatectomy in rats with chronic aluminum intoxication. *Biological Trace Element Research*, *115*(1), pp. 31–45.

Crespo-López, M.E., Macêdo, G.L., Pereira, S.I., Arrifano, G.P., Picanço-Diniz, D.L., do Nascimento, J.L.M. and Herculano, A.M., 2009. Mercury and human genotoxicity: critical considerations and possible molecular mechanisms. *Pharmacological Research*, *60*(4), pp. 212–220.

De Vos, G., Abotaga, S., Liao, Z., Jerschow, E. and Rosenstreich, D., 2007. Selective effect of mercury on Th2-type cytokine production in humans. *Immunopharmacology and Immunotoxicology*, *29*(3–4), pp. 537–548.

Del Razo, L.M., Garcia-Vargas, G.G., Albores, A., Vargas, H., Gonsebatt, M.E., Montero, R., Ostrosky-Wegman, P., Kelsh, M. and Cebrian, M.E., 1997. Altered profile of urinary arsenic metabolites in adults with chronic arsenicism: a pilot study. *Archives of Toxicology*, *71*(4), pp. 211–217.

Deugnier, Y., 2003. Iron and liver cancer. *Alcohol*, *30*(2), pp. 145–150.

Duan, N., Wang, X.L., Liu, X.D., Lin, C. and Hou, J., 2010. Effect of anaerobic fermentation residues on a chromium-contaminated soil-vegetable system. *Procedia Environmental Sciences*, *2*, pp. 1585–1597.

Egbueri, J.C. and Unigwe, C.O., 2020. Understanding the extent of heavy metal pollution in drinking water supplies from Umunya, Nigeria: an indexical and statistical assessment. *Analytical Letters*, *53*(13), pp. 2122–2144.

Ellingsen, D.G., Efskind, J., Haug, E., Thomassen, Y., Martinsen, I. and Gaarder, P.I., 2000. Effects of low mercury vapour exposure on the thyroid function in chloralkali workers. *Journal of Applied Toxicology: An International Journal*, *20*(6), pp. 483–489.

Everett, C.J. and Frithsen, I.L., 2008. Association of urinary cadmium and myocardial infarction. *Environmental Research*, *106*(2), pp. 284–286.

Exley, C., 2013. Human exposure to aluminium. *Environmental Science: Processes & Impacts*, *15*(10), pp. 1807–1816.

Exley, C., Begum, A., Woolley, M.P. and Bloor, R.N., 2006. Aluminum in tobacco and cannabis and smoking-related disease. *The American Journal of Medicine*, *119*(3), pp. 276-e9.

Exley, C. and Birchall, J.D., 1992. The cellular toxicity of aluminium. *Journal of Theoretical Biology*, *159*(1), pp. 83–98.

Farmand, F., Ehdaie, A., Roberts, C.K. and Sindhu, R.K., 2005. Lead-induced dysregulation of superoxide dismutases, catalase, glutathione peroxidase, and guanylate cyclase. *Environmental Research*, *98*(1), pp. 33–39.

Filipič, M., 2012. Mechanisms of cadmium induced genomic instability. *Mutation Research/Fundamental and Molecular Mechanisms of Mutagenesis, 733*(1–2), pp. 69–77.

Flora, S.J.S., Mittal, M. and Mehta, A., 2008. Heavy metal induced oxidative stress & its possible reversal by chelation therapy. *Indian Journal of Medical Research, 128*(4), p. 501.

Fowler, B.A. and Woods, J.S., 1977. Ultrastructural and biochemical changes in renal mitochondria during chronic oral methyl mercury exposure: the relationship to renal function. *Experimental and Molecular Pathology, 27*(3), pp. 403–412.

García-Esquinas, E., Pollán, M., Umans, J.G., Francesconi, K.A., Goessler, W., Guallar, E., Howard, B., Farley, J., Best, L.G. and Navas–Acien, A., 2013. Arsenic exposure and cancer mortality in a US-based prospective cohort: the strong heart study. *Cancer Epidemiology and Prevention Biomarkers, 22*(11), pp. 1944–1953.

Garnier, R., Fuster, J.M., Conso, F., Dautzenberg, B., Sors, C. and Fournier, E., 1981. Acute mercury vapour poisoning (author's transl). *Toxicological European Research. Recherche Europeenne en Toxicologie, 3*(2), pp. 77–86.

Garza-Lombó, C., Pappa, A., Panayiotidis, M.I., Gonsebatt, M.E. and Franco, R., 2019. Arsenic-induced neurotoxicity: a mechanistic appraisal. *JBIC Journal of Biological Inorganic Chemistry, 24*(8), pp. 1305–1316.

Gazwi, H.S., Yassien, E.E. and Hassan, H.M., 2020. Mitigation of lead neurotoxicity by the ethanolic extract of Laurus leaf in rats. *Ecotoxicology and Environmental Safety, 192*, p. 110297.

Ghani, A. and Ghani, A., 2011. Effect of chromium toxicity on growth, chlorophyll and some mineral nutrients of Brassica juncea L. *Egyptian Academic Journal of Biological Sciences, H. Botany, 2*(1), pp. 9–15.

Glaser, U., Hochrainer, D., Klöppel, H. and Kuhnen, H., 1985. Low level chromium(VI) inhalation effects on alveolar macrophages and immune functions in Wistar rats. *Archives of Toxicology, 57*(4), pp. 250–256.

Gonzalvo, M.C., Gil, F., Hernández, A.F., Villanueva, E. and Pla, A., 1997. Inhibition of paraoxonase activity in human liver microsomes by exposure to EDTA, metals and mercurials. *Chemico-Biological Interactions, 105*(3), pp. 169–179.

Gorini, F., Muratori, F. and Morales, M.A., 2014. The role of heavy metal pollution in neurobehavioral disorders: a focus on autism. *Review Journal of Autism and Developmental Disorders, 1*(4), pp. 354–372.

Goyer, R., 1989. Mechanisms of cadmium and lead nephropathy. *Toxicology Letters, 46*, pp. 153–162.

Goyer, R.A., 1990. Lead toxicity: from overt to subclinical to subtle health effects. *Environmental Health Perspectives, 86*, pp. 177–181.

Grazuleviciene, R., Nadisauskiene, R., Buinauskiene, J. and Grazulevicius, T., 2009. Effects of elevated levels of manganese and iron in drinking water on birth outcomes. *Polish Journal of Environmental Studies, 18*(5).

Gross Jr, J.B., Ludwig, J., Wiesner, R.H., McCall, J.T. and LaRusso, N.F., 1985. Abnormalities in tests of copper metabolism in primary sclerosing cholangitis. *Gastroenterology, 89*(2), pp. 272–278.

Gruenwedel, D.W. and Cruikshank, M.K., 1979. Effect of methylmercury(II) on the synthesis of deoxyribonucleic acid, ribonucleic acid and protein in HeLa S3 cells. *Biochemical Pharmacology, 28*(5), pp. 651–655.

Gupta, D.K., Tiwari, S., Razafindrabe, B.H.N. and Chatterjee, S., 2017. Arsenic contamination from historical aspects to the present. In *Arsenic Contamination in the Environment* (pp. 1–12). Cham: Springer.

Gürkan, R., Ulusoy, H.İ. and Akçay, M., 2017. Simultaneous determination of dissolved inorganic chromium species in wastewater/natural waters by surfactant sensitized catalytic kinetic spectrophotometry. *Arabian Journal of Chemistry, 10*, pp. S450–S460.

Han, S., Lemire, J., Appanna, V.P., Auger, C., Castonguay, Z. and Appanna, V.D., 2013. How aluminum, an intracellular ROS generator promotes hepatic and neurological diseases: the metabolic tale. *Cell Biology and Toxicology, 29*(2), pp. 75–84.

Han, J.X., Shang, Q. and Du, Y., 2009. Effect of environmental cadmium pollution on human health. *Health, 1*(03), p. 159.

Han, Y.L., Sheng, Z., Liu, G.D., Long, L.L., Wang, Y.F., Yang, W.X. and Zhu, J.Q., 2015. Cloning, characterization and cadmium inducibility of metallothionein in the testes of the mudskipper Boleophthalmus pectinirostris. *Ecotoxicology and Environmental Safety, 119*, pp. 1–8.

Hasanein, P. and Emamjomeh, A., 2019. Beneficial effects of natural compounds on heavy metal–induced hepatotoxicity. In *Dietary Interventions in Liver Disease* (pp. 345–355). Academic Press.

Hazen-Martin, D.J., Todd, J.H., Sens, M.A., Khan, W., Bylander, J.E., Smyth, B.J. and Sens, D.A., 1993. Electrical and freeze-fracture analysis of the effects of ionic cadmium on cell membranes of human proximal tubule cells. *Environmental Health Perspectives, 101*(6), pp. 510–516.

Hellström, L., Elinder, C.G., Dahlberg, B., Lundberg, M., Järup, L., Persson, B. and Axelson, O., 2001. Cadmium exposure and end-stage renal disease. *American Journal of Kidney Diseases, 38*(5), pp. 1001–1008.

Hertz-Picciotto, I. and Croft, J., 1993. Review of the relation between blood lead and blood pressure. *Epidemiologic Reviews*, *15*(2), pp. 352–373.

Horowitz, Y., Greenberg, D., Ling, G. and Lifshitz, M., 2002. Acrodynia: a case report of two siblings. *Archives of Disease in Childhood*, *86*(6), pp. 453–453.

Huang, H.W., Lee, C.H. and Yu, H.S., 2019. Arsenic-induced carcinogenesis and immune dysregulation. *International Journal of Environmental Research and Public Health*, *16*(15), p. 2746.

Huang, W., Wang, P., Shen, T., Hu, C., Han, Y., Song, M., Bian, Y., Li, Y. and Zhu, Y., 2017. Aluminum trichloride inhibited osteoblastic proliferation and downregulated the Wnt/β-catenin pathway. *Biological Trace Element Research*, *177*(2), pp. 323–330.

Hussain, S., Habib-Ur-Rehman, M., Khanam, T., Sheer, A., Kebin, Z. and Jianjun, Y., 2019. Health risk assessment of different heavy metals dissolved in drinking water. *International Journal of Environmental Research and Public Health*, *16*(10), p. 1737.

Ilbäck, N.G., Sundberg, J. and Oskarsson, A., 1991. Methyl mercury exposure via placenta and milk impairs natural killer (NK) cell function in newborn rats. *Toxicology Letters*, *58*(2), pp. 149–158.

Irfan, M., Hayat, S., Ahmad, A. and Alyemeni, M.N., 2013. Soil cadmium enrichment: allocation and plant physiological manifestations. *Saudi Journal of Biological Sciences*, *20*(1), pp. 1–10.

Jacquillet, G., Barbier, O., Rubera, I., Tauc, M., Borderie, A., Namorado, M.C., Martin, D., Sierra, G., Reyes, J.L., Poujeol, P. and Cougnon, M., 2007. Cadmium causes delayed effects on renal function in the offspring of cadmium-contaminated pregnant female rats. *American Journal of Physiology-Renal Physiology*, *293*(5), pp. F1450–F1460.

Jolliffe, D.M., Budd, A.J. and Gwilt, D.J., 1991. Massive acute arsenic poisoning. *Anaesthesia*, *46*(4), pp. 288–290.

Joseph, P., 2009. Mechanisms of cadmium carcinogenesis. *Toxicology and Applied Pharmacology*, *238*(3), pp. 272–279.

Kawahara, M., Konoha, K., Nagata, T. and Sadakane, Y., 2007. Aluminum and human health: its intake, bioavailability and neurotoxicity. *Biomedical Research on Trace Elements*, *18*(3), pp. 211–220.

Kell, D.B., 2009. Iron behaving badly: inappropriate iron chelation as a major contributor to the aetiology of vascular and other progressive inflammatory and degenerative diseases. *BMC Medical Genomics*, *2*(1), pp. 1–79.

Koedrith, P., Kim, H., Weon, J.I. and Seo, Y.R., 2013. Toxicogenomic approaches for understanding molecular mechanisms of heavy metal mutagenicity and carcinogenicity. *International Journal of Hygiene and Environmental Health*, *216*(5), pp. 587–598.

Krewski, D., Yokel, R.A., Nieboer, E., Borchelt, D., Cohen, J., Harry, J., Kacew, S., Lindsay, J., Mahfouz, A.M. and Rondeau, V., 2007. Human health risk assessment for aluminium, aluminium oxide, and aluminium hydroxide. *Journal of Toxicology and Environmental Health, Part B*, *10*(S1), pp. 1–269.

Kuivenhoven, M. and Mason, K., 2019. Arsenic (arsine) toxicity. StatPearls [Internet]. StatPearls Publishing.

Kulka, M., 2016. A review of paraoxonase 1 properties and diagnostic applications. *Polish Journal of Veterinary Sciences*.

Kungolos, A., Aoyama, I. and Muramoto, S., 1999. Toxicity of organic and inorganic mercury to Saccharomyces cerevisiae. *Ecotoxicology and Environmental Safety*, *43*(2), pp. 149–155.

Lentini, P., Zanoli, L., Granata, A., Signorelli, S.S., Castellino, P. and Dell'Aquila, R., 2017. Kidney and heavy metals-the role of environmental exposure. *Molecular Medicine Reports*, *15*(5), pp. 3413–3419.

Li, P., Luo, W., Zhang, H., Zheng, X., Liu, C. and Ouyang, H., 2016. Effects of aluminum exposure on the bone stimulatory growth factors in rats. *Biological Trace Element Research*, *172*(1), pp. 166–171.

Li, R., Wu, H., Ding, J., Fu, W., Gan, L. and Li, Y., 2017. Mercury pollution in vegetables, grains and soils from areas surrounding coal-fired power plants. *Scientific Reports*, *7*(1), pp. 1–9.

Liu, J., Qu, W. and Kadiiska, M.B., 2009. Role of oxidative stress in cadmium toxicity and carcinogenesis. *Toxicology and Applied Pharmacology*, *238*(3), pp. 209–214.

Lukiw, W.J., LeBlanc, H.J., Carver, L.A., McLachlan, D.R. and Bazan, N.G., 1998. Run-on gene transcription in human neocortical nuclei. *Journal of Molecular Neuroscience*, *11*(1), pp. 67–78.

Luo, L., Wang, B., Jiang, J., Fitzgerald, M., Huang, Q., Yu, Z., Li, H., Zhang, J., Wei, J., Yang, C. and Zhang, H., 2020. Heavy metal contaminations in herbal medicines: determination, comprehensive risk assessments, and solutions. *Frontiers in Pharmacology*, *11*, 595335.

Mailloux, R.J., Hamel, R. and Appanna, V.D., 2006. Aluminum toxicity elicits a dysfunctional TCA cycle and succinate accumulation in hepatocytes. *Journal of Biochemical and Molecular Toxicology*, *20*(4), pp. 198–208.

Malaguarnera, G., Cataudella, E., Giordano, M., Nunnari, G., Chisari, G. and Malaguarnera, M., 2012. Toxic hepatitis in occupational exposure to solvents. *World Journal of Gastroenterology: WJG*, *18*(22), p. 2756.

Marchetti, C., 2014. Interaction of metal ions with neurotransmitter receptors and potential role in neurodiseases. *Biometals*, 27(6), pp. 1097–1113.

Martin, S. and Griswold, W., 2009. Human health effects of heavy metals. *Environmental Science and Technology Briefs for Citizens*, 15, pp. 1–6.

Martinez, V.D., Vucic, E.A., Becker-Santos, D.D., Gil, L. and Lam, W.L., 2011. Arsenic exposure and the induction of human cancers. *Journal of Toxicology*, 2011.

Mathew, B.B., Tiwari, A. and Jatawa, S.K., 2011. Free radicals and antioxidants: a review. *Journal of Pharmacy Research*, 4(12), pp. 4340–4343.

Messner, B., Knoflach, M., Seubert, A., Ritsch, A., Pfaller, K., Henderson, B., Shen, Y.H., Zeller, I., Willeit, J., Laufer, G. and Wick, G., 2009. Cadmium is a novel and independent risk factor for early atherosclerosis mechanisms and in vivo relevance. *Arteriosclerosis, Thrombosis, and Vascular Biology*, 29(9), pp. 1392–1398.

Milton, A.H., Hussain, S., Akter, S., Rahman, M., Mouly, T.A. and Mitchell, K., 2017. A review of the effects of chronic arsenic exposure on adverse pregnancy outcomes. *International Journal of Environmental Research and Public Health*, 14(6), p. 556.

Mirhashemi, S.M. and Aarabi, M.H., 2011. To study various concentrations of magnesium and aluminium on amylin hormone conformation. *Pakistan Journal of Biological Sciences*, 14(11), p. 653.

Mirhashemi, S.M. and Shahabaddin, M.E., 2011. Evaluation of aluminium, manganese, copper and selenium effects on human islets amyloid polypeptide hormone aggregation. *Pakistan Journal of Biological Sciences*, 14(4), p. 288.

Miura, S., Takahashi, K., Imagawa, T., Uchida, K., Saito, S., Tominaga, M. and Ohta, T., 2013. Involvement of TRPA1 activation in acute pain induced by cadmium in mice. *Molecular Pain*, 9, pp. 1744–8069.

Morais, S., Costa, F. G. and Pereira, M. D. L., 2012. Heavy metals and human health. *Environmental Health–Emerging Issues and Practice*, 10(1), pp. 227–245.

Mujika, J.I., Dalla Torre, G., Formoso, E., Grande-Aztatzi, R., Grabowski, S.J., Exley, C. and Lopez, X., 2018. Aluminum's preferential binding site in proteins: sidechain of amino acids versus backbone interactions. *Journal of Inorganic Biochemistry*, 181, pp. 111–116.

Mutlu, A., Lee, B.K., Park, G.H., Yu, B.G. and Lee, C.H., 2012. Long-term concentrations of airborne cadmium in metropolitan cities in Korea and potential health risks. *Atmospheric Environment*, 47, pp. 164–173.

Najeeb, U., Ahmad, W., Zia, M.H., Zaffar, M. and Zhou, W., 2017. Enhancing the lead phytostabilization in wetland plant Juncus effusus L. through somaclonal manipulation and EDTA enrichment. *Arabian Journal of Chemistry*, 10, pp. S3310–S3317.

Nath, K., Singh, D., Shyam, S. and Sharma, Y.K., 2008. Effect of chromium and tannery effluent toxicity on metabolism and growth in cowpea (Vigna sinensis L. Saviex Hassk) seedling. *Research in Environment and Life Sciences*, 1(3), pp. 91–94.

Navas-Acien, A., Selvin, E., Sharrett, A.R., Calderon-Aranda, E., Silbergeld, E. and Guallar, E., 2004. Lead, cadmium, smoking, and increased risk of peripheral arterial disease. *Circulation*, 109(25), pp. 3196–3201.

Newairy, A.S.A., Salama, A.F., Hussien, H.M. and Yousef, M.I., 2009. Propolis alleviates aluminium-induced lipid peroxidation and biochemical parameters in male rats. *Food and Chemical Toxicology*, 47(6), pp. 1093–1098.

Nyambura, C., Hashim, N.O., Chege, M.W., Tokonami, S. and Omonya, F.W., 2020. Cancer and non-cancer health risks from carcinogenic heavy metal exposures in underground water from Kilimambogo, Kenya. *Groundwater for Sustainable Development*, 10, p. 100315.

Ohsaka, Y. and Nomura, Y., 2016. Rat white adipocytes activate p85/p110 PI3K and induce PM GLUT4 in response to adrenoceptor agonists or aluminum fluoride. *Acta Physiologica Hungarica*, 103(1), pp. 35–48.

Olivieri, G., Brack, C., Müller-Spahn, F., Stähelin, H.B., Herrmann, M., Renard, P., Brockhaus, M. and Hock, C., 2000. Mercury induces cell cytotoxicity and oxidative stress and increases β-amyloid secretion and tau phosphorylation in SHSY5Y neuroblastoma cells. *Journal of Neurochemistry*, 74(1), pp. 231–236.

Park, S.H., Araki, S., Nakata, A., Kim, Y.H., Park, J.A., Tanigawa, T., Yokoyama, K. and Sato, H., 2000. Effects of occupational metallic mercury vapour exposure on suppressor-inducer (CD4+ CD45RA+) T lymphocytes and CD57+ CD16+ natural killer cells. *International Archives of Occupational and Environmental Health*, 73(8), pp. 537–542.

Park, Y.H., Kim, D., Dai, J. and Zhang, Z., 2015. Human bronchial epithelial BEAS-2B cells, an appropriate in vitro model to study heavy metals induced carcinogenesis. *Toxicology and Applied Pharmacology*, 287(3), pp. 240–245.

Patrick, L., 2003. Toxic metals and antioxidants: part II. The role of antioxidants in arsenic and cadmium toxicity. *Alternative Medicine Review*, 8(2).

Peters, J.L., Perlstein, T.S., Perry, M.J., McNeely, E. and Weuve, J., 2010. Cadmium exposure in association with history of stroke and heart failure. *Environmental Research*, *110*(2), pp. 199–206.

Phippen, B., Horvath, C., Nordin, R. and Nagpal, N., 2008. Ambient water quality guidelines for iron: overview. Ministry of Environment Province of British Columbia. *British Columbia*, pp. 2–10.

Pierce, B.L., Kibriya, M.G., Tong, L., Jasmine, F., Argos, M., Roy, S., Paul-Brutus, R., Rahaman, R., Rakibuz-Zaman, M., Parvez, F. and Ahmed, A., 2012. Genome-wide association study identifies chromosome 10q24. 32 variants associated with arsenic metabolism and toxicity phenotypes in Bangladesh. *PLoS Genetics*, *8*(2), p. e1002522.

Prohaska, J.R., 2000. Long-term functional consequences of malnutrition during brain development: copper. *Nutrition*, *7*(16), pp. 502–504.

Rani, A., Kumar, A., Lal, A. and Pant, M., 2014. Cellular mechanisms of cadmium-induced toxicity: a review. *International Journal of Environmental Health Research*, *24*(4), pp. 378–399.

Rao, M.V. and Sharma, P.S.N., 2001. Protective effect of vitamin E against mercuric chloride reproductive toxicity in male mice. *Reproductive Toxicology*, *15*(6), pp. 705–712.

Ratnaike, R.N., 2003. Acute and chronic arsenic toxicity. *Postgraduate Medical Journal*, *79*(933), pp. 391–396.

Reczek, C.R. and Chandel, N.S., 2017. The two faces of reactive oxygen species in cancer. *Annual Review of Cancer Biology*, *1*, pp. 79–98.

Reyes, J.L., Molina-Jijón, E., Rodríguez-Muñoz, R., Bautista-García, P., Debray-García, Y. and Namorado, M.D.C., 2013. Tight junction proteins and oxidative stress in heavy metals-induced nephrotoxicity. *BioMed Research International*, *2013*.

Rodríguez, M.C., Barsanti, L., Passarelli, V., Evangelista, V., Conforti, V. and Gualtieri, P., 2007. Effects of chromium on photosynthetic and photoreceptive apparatus of the alga Chlamydomonas reinhardtii. *Environmental Research*, *105*(2), pp. 234–239.

Roy, J.S., Chatterjee, D., Das, N. and Giri, A.K., 2018. Substantial evidences indicate that inorganic arsenic is a genotoxic carcinogen: a review. *Toxicological Research*, *34*(4), pp. 311–324.

Sakamoto, T., Saito, H., Ishii, K., Takahashi, H., Tanabe, S. and Ogasawara, Y., 2006. Aluminum inhibits proteolytic degradation of amyloid β peptide by cathepsin D: a potential link between aluminum accumulation and neuritic plaque deposition. *FEBS Letters*, *580*(28–29), pp. 6543–6549.

Salnikow, K. and Zhitkovich, A., 2008. Genetic and epigenetic mechanisms in metal carcinogenesis and cocarcinogenesis: nickel, arsenic, and chromium. *Chemical Research in Toxicology*, *21*(1), pp. 28–44.

Salonen, J.T., Flather, M.D., Berger, A., Malin, R., Tuomainen, T.P., Nyyssönen, K., Lakka, T.A. and Lehtimäki, T., 1999. Polymorphism in high density lipoprotein paraoxonase gene and risk of acute myocardial infarction in men: prospective nested case-control study commentary: causality—the Achilles' heel of observational studies commentary: how highdensity lipoprotein protects against heart disease. *BMJ*, *319*(7208), pp. 487–489.

Santarelli, L., Bracci, M. and Mocchegiani, E., 2006. In vitro and in vivo effects of mercuric chloride on thymic endocrine activity, NK and NKT cell cytotoxicity, cytokine profiles (IL-2, IFN-γ, IL-6): role of the nitric oxide-L-arginine pathway. *International Immunopharmacology*, *6*(3), pp. 376–389.

Satarug, S., Garrett, S.H., Sens, M.A. and Sens, D.A., 2010. Cadmium, environmental exposure, and health outcomes. *Environmental Health Perspectives*, *118*(2), pp. 182–190.

Sattar, A., Xie, S., Hafeez, M.A., Wang, X., Hussain, H.I., Iqbal, Z., Pan, Y., Iqbal, M., Shabbir, M.A. and Yuan, Z., 2016. Metabolism and toxicity of arsenicals in mammals. *Environmental Toxicology and Pharmacology*, *48*, pp. 214–224.

Schwartz, G.G., Il'yasova, D. and Ivanova, A., 2003. Urinary cadmium, impaired fasting glucose, and diabetes in the NHANES III. *Diabetes Care*, *26*(2), pp. 468–470.

Shah, A.Q., Kazi, T.G., Baig, J.A., Arain, M.B., Afridi, H.I., Kandhro, G.A., Wadhwa, S.K. and Kolachi, N.F., 2010. Determination of inorganic arsenic species (As3+ and As5+) in muscle tissues of fish species by electrothermal atomic absorption spectrometry (ETAAS). *Food Chemistry*, *119*(2), pp. 840–844.

Shah, M.T., Suleman, M., Abdul Baqi, S., Sattar, A. and Khan, N., 2020. 1. Determination of heavy metals in drinking water and their adverse effects on human health. A review. *Pure and Applied Biology (PAB)*, *9*(1), pp. 96–104.

Sharma, P. and Dubey, R.S., 2005. Lead toxicity in plants. *Brazilian Journal of Plant Physiology*, *17*(1), pp. 35–52.

Shen, S., Li, X.F., Cullen, W.R., Weinfeld, M. and Le, X.C., 2013. Arsenic binding to proteins. *Chemical Reviews*, *113*(10), pp. 7769–7792.

Silbergeld, E.K., Waalkes, M. and Rice, J.M., 2000. Lead as a carcinogen: experimental evidence and mechanisms of action. *American Journal of Industrial Medicine*, *38*(3), pp. 316–323.

Singh, N., Kumar, D. and Sahu, A.P., 2007. Arsenic in the environment: effects on human health and possible prevention. *Journal of Environmental Biology*, 28(2), p. 359.

Stohs, S.J. and Bagchi, D., 1995. Oxidative mechanisms in the toxicity of metal ions. *Free Radical Biology and Medicine*, 18(2), pp. 321–336.

Sun, X., Cao, Z., Zhang, Q., Liu, S., Xu, F., Che, J., Zhu, Y., Li, Y., Pan, C. and Liang, W., 2015. Aluminum trichloride impairs bone and downregulates Wnt/β-catenin signaling pathway in young growing rats. *Food and Chemical Toxicology*, 86, pp. 154–162.

Sushma, N.J., Sivaiah, U., Suraj, N.J. and Rao, K.J., 2007. Aluminium acetate: role in oxidative metabolism of albino mice. *International Journal of Zoological Research*, 3(1), pp. 48–52.

Tanaka, K.I. and Kawahara, M., 2017. Copper enhances zinc-induced neurotoxicity and the endoplasmic reticulum stress response in a neuronal model of vascular dementia. *Frontiers in Neuroscience*, 11, p. 58.

Tanigawa, T., Takehashf, H. and Nakata, A., 2004. Naïve (CD4+ CD45RA+) T cell subpopulation is susceptible to various types of hazardous substances in the workplace. *International Journal of Immunopathology and Pharmacology*, 17(2_suppl), pp. 109–114.

Tellez-Plaza, M., Guallar, E., Howard, B.V., Umans, J.G., Francesconi, K.A., Goessler, W., Silbergeld, E.K., Devereux, R.B. and Navas-Acien, A., 2013. Cadmium exposure and incident cardiovascular disease. *Epidemiology*, 24(3), p. 421.

Tellez-Plaza, M., Navas-Acien, A., Menke, A., Crainiceanu, C.M., Pastor-Barriuso, R. and Guallar, E., 2012. Cadmium exposure and all-cause and cardiovascular mortality in the US general population. *Environmental Health Perspectives*, 120(7), pp. 1017–1022.

Tsai, M.T., Huang, S.Y. and Cheng, S.Y., 2017. Lead poisoning can be easily misdiagnosed as acute porphyria and nonspecific abdominal pain. *Case Reports in Emergency Medicine*, 2017.

Valko, M.M.H.C.M., Morris, H. and Cronin, M.T.D., 2005. Metals, toxicity and oxidative stress. *Current Medicinal Chemistry*, 12(10), pp. 1161–1208.

Vaziri, N.D., 2008. Mechanisms of lead-induced hypertension and cardiovascular disease. *American Journal of Physiology-Heart and Circulatory Physiology*, 295(2), pp. H454–H465.

Vuori, K.M., 1995, January. Direct and indirect effects of iron on river ecosystems. In *Annales Zoologici Fennici* (pp. 317–329). Finnish Zoological and Botanical Publishing Board.

Wadhwa, N., Mathew, B.B., Jatawa, S. and Tiwari, A., 2012. Lipid peroxidation: mechanism, models and significance. *International Journal of Current Science*, 3, pp. 29–38.

Wolińska, A., Stepniewska, Z. and Wlosek, R., 2013. The influence of old leather tannery district on chromium contamination of soils, water and plants.

Xu, F., Ren, L., Song, M., Shao, B., Han, Y., Cao, Z. and Li, Y., 2018. Fas-and mitochondria-mediated signaling pathway involved in osteoblast apoptosis induced by AlCl₃. *Biological Trace Element Research*, 184(1), pp. 173–185.

Xu, Z.X., Zhang, Q., Ma, G.L., Chen, C.H., He, Y.M., Xu, L.H., Zhang, Y., Zhou, G.R., Li, Z.H., Yang, H.J. and Zhou, P., 2016. Influence of aluminium and EGCG on fibrillation and aggregation of human islet amyloid polypeptide. *Journal of Diabetes Research*, 2016.

Yamanaka, K., Minato, N. and Iwai, K., 1999. Stabilization of iron regulatory protein 2, IRP2, by aluminum. *FEBS Letters*, 462(1–2), pp. 216–220.

Yang, X., Huo, H., Xiu, C., Song, M., Han, Y., Li, Y. and Zhu, Y., 2016. Inhibition of osteoblast differentiation by aluminum trichloride exposure is associated with inhibition of BMP-2/Smad pathway component expression. *Food and Chemical Toxicology*, 97, pp. 120–126.

Yang, X., Yu, K., Wang, H., Zhang, H., Bai, C., Song, M., Han, Y., Shao, B., Li, Y. and Li, X., 2018. Bone impairment caused by AlCl₃ is associated with activation of the JNK apoptotic pathway mediated by oxidative stress. *Food and Chemical Toxicology*, 116, pp. 307–314.

Yin, Y., Meng, F., Sui, C., Jiang, Y. and Zhang, L., 2019. Arsenic enhances cell death and DNA damage induced by ultraviolet B exposure in mouse epidermal cells through the production of reactive oxygen species. *Clinical and Experimental Dermatology*, 44(5), pp. 512–519.

Yongsheng, W., Qihui, L. and Qian, T., 2011. Effect of Pb on growth, accumulation and quality component of tea plant. *Procedia Engineering*, 18, pp. 214–219.

Yoshizawa, K., Rimm, E.B., Morris, J.S., Spate, V.L., Hsieh, C.C., Spiegelman, D., Stampfer, M.J. and Willett, W.C., 2002. Mercury and the risk of coronary heart disease in men. *New England Journal of Medicine*, 347(22), pp. 1755–1760.

Yu, H., Zhang, J., Ji, Q., Yu, K., Wang, P., Song, M., Cao, Z., Zhang, X. and Li, Y., 2019. Melatonin alleviates aluminium chloride-induced immunotoxicity by inhibiting oxidative stress and apoptosis associated with the activation of Nrf2 signaling pathway. *Ecotoxicology and Environmental Safety*, 173, pp. 131–141.

Zalups, R.K., 2000. Evidence for basolateral uptake of cadmium in the kidneys of rats. *Toxicology and Applied Pharmacology*, *164*(1), pp. 15–23.

Zayed, A.M. and Terry, N., 2003. Chromium in the environment: factors affecting biological remediation. *Plant and Soil*, *249*(1), pp. 139–156.

Zefferino, R., Piccoli, C., Ricciardi, N., Scrima, R. and Capitanio, N., 2017. Possible mechanisms of mercury toxicity and cancer promotion: Involvement of gap junction intercellular communications and inflammatory cytokines. *Oxidative Medicine and Cellular Longevity*, *2017*.

Zhang, Q., Cao, Z., Sun, X., Zuang, C., Huang, W. and Li, Y., 2016. Aluminum trichloride induces hypertension and disturbs the function of erythrocyte membrane in male rats. *Biological Trace Element Research*, *171*(1), pp. 116–123.

Zhitkovich, A., 2005. Importance of chromium – DNA adducts in mutagenicity and toxicity of chromium(VI). *Chemical Research in Toxicology*, *1*(18), pp. 3–11.

4 Nanotoxicology of Metal Oxides and Their Impacts on the Environment

Izza Taufiq, Fatima Tu Zahra and Ayesha Siddiqa

4.1 INTRODUCTION

Nanotoxicology is the rapidly expanding research area which involves the study of the safety concerns of nanoparticles (NPs), along with their wide implementations in various disciplines. The domain of nanotoxicology mainly determines the toxicity of nanomaterials along with their lethal effects on the ecosystem. NPs are small naturally occurring particles between 1 and 100 nm in size. These are the aggregates of many atomic particles of different elements with diameters in the nanomeric scale. NPs have distinct properties due to their quantum size effect and large surface area to volume ratio (Haddad and Seabra, 2012). NPs are mainly made up of inert elements. NPs have adverse effects on biological systems. Cells exposed to NPs undergo oxidative stress and DNA damage, leading to apoptosis (Gao et al., 2012). Moreover, exposure to toxic NPs causes alterations in the proliferation of cells (Corr et al., 2013). Nanotoxicology also includes the detailed study of the harmful effects of NPs on humans and the environment. The dynamic nature of nanomaterials in biological systems leads to chronic complexities. The development rate of new NPs possessing unique properties is increasing exponentially and has potential health risks for the environment (Rubilar et al., 2013).

The conventional methods for the synthesis of metal oxide NPs depends on physical and chemical techniques, which use expensive and dangerous chemicals having elevated energy input and harmful effects on the environment (Durán and Seabra, 2012). Formation of capped nanostructure proteins leads to biogenic synthesis of metallic nanomaterials. These capping agents stabilize the nanosystem leading to improved biocompatibility, thus preventing their aggregation (Seabra et al., 2013). Different NPs have been widely studied in the field of nanotoxicology. Some important metal oxide NPs include cobalt oxide, bismuth trioxide, copper oxide, iron oxide, silica, titanium dioxide, zinc oxide, tin oxide NPs. In the next section, the biogenic synthesis of the above-mentioned metal oxide NPs will be briefly described along with their various applications in the environment.

4.2 CLASSIFICATION OF NPS

NPs are divided into two major groups according to their nature and origin. They are either naturally produced or artificially synthesized by a physical or a chemical process. These are generated through mechanisms taking place in nature or by engineered activities (Griffin et al., 2017).

4.2.1 NATURALLY OCCURRING NPS

These NPs are found in volcanic eruptions, deep oceans, forest fires, abyssal, grit, silt, sand and crushed rocks. Naturally occurring NPs are known to be found in many living biota. They have reproducible structures and are biocompatible. They can be intracellular or extracellular. The most important ones in this class include viruses and liposomes (Barhoum et al., 2022).

DOI: 10.1201/9781003326281-5

4.2.2 Mechanism of Action of Protein NPs

Proteins NPs are widely used in several biological processes. Because of their high stability they are efficiently used for delivering drugs into biological systems. Apart from stability, protein NPs also perform many functions of biodegradation and bioaccumulation. Protein NPs are more favored due to their non-antigenic characteristics, lesser immunogenicity and toxicity; these NPs are more likely to be incorporated into different kinds of biopolymers.

Protein NPs are used for the delivery of non-soluble drugs into the cell through the process of endocytosis. These particles show more protection from various processes like toxicity, degradation by enzymes, phagocytic reactions and endocytosis and hence increase the life and stability of the drug being introduced, as shown in Figure 4.1 (Hong et al., 2020). These protein NPs are prepared through a complex physical and chemical method following a self-assembly method.

4.2.3 Engineered NPs

Such NPs are specifically manufactured by human beings. Engineered NPs have a particular morphology and geometry. Because of their definite shape and structure, they are easily made through chemical process. They are found in composition with other materials like a layer of gold or a central element such as gold covered with a shell made up of any other element like cobalt or silica (Matsoukas et al., 2015).

4.3 METAL OXIDE NPS

4.3.1 Cobalt Oxide (Co₃O₄) Nanocrystals

Cobalt oxide nanocrystals have various applications due to their desirable properties, i.e. optical magnetic and electrochemical capabilities. They can also be utilized as supercapacitors in devices used for storing energy. Thermal and solvothermal decomposition routes are the classical methods for the biogenic formation of Co_3O_4 nanocrystals. *Brevibacterium casei*, a marine bacterium, has a role in microbial synthesis of these NPs (Savi et al., 2021). Biogenic cobalt oxide NPs have proteins coated on their surface to conserve their identity. Moreover, these proteins also help in reducing agglomeration of isolated NPs (Iravani and Varma, 2020).

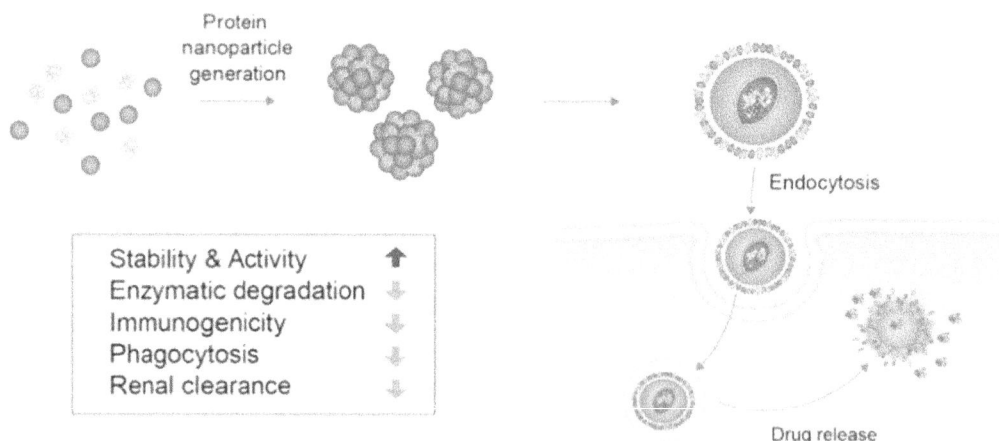

FIGURE 4.1 The endocytosis process used for the delivery of various protein NPs and insoluble drugs *in vivo* (Hong S, Choi DW, Kim HN, Park CG, Lee W, Park HH. Protein-Based Nanoparticles as Drug Delivery Systems. *Pharmaceutics*. 2020).

4.3.2 Bismuth Trioxide (Bi_2O_3) NPs

Bismuth trioxide (Bi_2O_3) NPs have gained importance because of their usage as a semiconductor due to their visible light sensitivity and increased photocatalytic activity. This metal oxide plays a role in water treatment. They are used as an optoelectronic material. They are obtained by the addition of organic or toxic solvents at a high temperature. In addition to conventional chemical methods, *Fusarium oxysporum* is used as an alternative method for the synthesis of these nanocrystals (Zulkifli et al., 2018). The formation of bismuth trioxide NPs by biogenic synthesis is more ecofriendly as compared to conventional chemical methods (Shahbazi et al., 2020).

4.3.3 Copper Oxide (CuO, Cu_2O) NPs

Copper oxide NPs have wide applications in electronic devices and optics and also act as an important antimicrobial agent. At room temperature, Cu_2O NPs of the size 10–20 nm were made by using Baker's yeast *Saccharomyces cerevisiae* (Adams et al., 2017). Copper oxide NPs are used as antibacterial and antifungal agents against *Escherichia coli* and *Aspergillus niger,* respectively. These were obtained by the reduction of copper sulfate using the reductase enzyme. Several fungal strains like *Penicillium aurantiogriseum*, *Penicillium citrinum* and *Penicillium waksmanii* which are obtained from soil are utilized in the biosynthesis of copper oxides (Adhikari et al., 2012).

4.3.4 Iron Oxide (Fe_2O_3, Fe_3O_4) Magnetic NPs

Iron oxide NPs have important applications in biomedical techniques including hyperthermia nuclear magnetic resonance imaging and precise drug delivery. Biogenic techniques are mainly applied for obtaining iron oxide NPs rather than classical chemical methods. *Shewanella strain* HN-41 is a bacterium used for the reduction of iron to iron oxide, where pyruvate acts as an electron donor (El-Temsah and Joner, 2012). The toxicity of biogenic and commercial iron oxide NPs was compared by observing hemagglutination and changes in morphology. The capping agents on the surface of iron oxide improve stability and prevent their aggregation.

4.3.5 Silica (SiO_2) NPs

Silica NPs have a vital role in the biomedical and engineering industry such as nanocarriers applied in drug delivery systems. The mycelium of *Fusarium oxysporum* secretes extracellular protein which in turn leads to the formation of silica NPs (Nel et al., 2006). Silica nanocomposites have a high level of cytotoxicity as they can damage skin cells upon direct exposure.

4.3.6 Titanium Dioxide (TiO_2) NPs

Titanium dioxide NPs have a wide range of applications in biomedicine, the environment and modern technology. *Lactobacillus* sp. has been reported to be the main source of production of titanium dioxide NPs at room temperature (Castiglione et al., 2011). Yeast cells or Lactobacillus undergo interaction with $TiO(OH)_2$ and result in the production of these NPs using carbon and nitrogen sources.

4.3.7 Zinc Oxide (ZnO)NPs

A probiotic microbe *Lactobacillus sporogenes* is utilized for the biogenic synthesis of ZnO NPs. These NPs can be used in the decontamination of highly toxic and corrosive hydrogen sulfide gas (Dumont et al., 2015).

4.3.8   Tin Oxide (SnO$_2$) NPs

The successful synthesis of tin oxide NPs was performed by a de novo biogenic method using extract from a flower *Saraca indica*. These NPs possess antioxidant and antibacterial properties against *E. coli* (Chae et al., 2014). Moreover, tin oxide NPs are utilized in various biomedical applications.

4.3.9   Gold (Au) NPs

Research has shown that small-sized gold NPs possess antimicrobial activity. These NPs have the potential to generate reactive oxygen species (ROS), which cause some bacteria to undergo oxidative stress (Mohammad and Hashmi, 2019). Due to their colloidal nature and radiant colors, gold NPs have been used by artists. These NPs possess important electronic and optical properties which can be varied by changing their physical characteristics like shape, size and surface composition (Huang et al., 2003).

4.3.10   Silver Oxide (Ag$_2$O) NPs

The most essential metallic oxide nanomaterials including silver oxide NPs have been widely utilized in nanoscience and biomedical applications (Zhang et al., 2016). The various physical, chemical and biological methods are utilized for the biogenic synthesis of silver NPs. Studies have determined that Ag NPs possess extensive multifunctional properties, for example, as anti-inflammatory, antiviral, anticancer and antimicrobial agents. Biological silver NPs express high solubility, high stability and high yield (Gurunathan et al., 2015b). Moreover, they have a well-defined size and morphology and are relatively less toxic and can therefore be used in green chemistry.

4.4   EFFECTS OF METAL OXIDE NPS

Several mechanisms are involved in the toxicity caused by metal oxide NPs. NPs interact with many components of the cell including DNA, various proteins and cell organelles like mitochondria and lysosomes. They can generate ROS, which affects the normal functioning of cells. This can lead to the production of DNA adducts, mitochondrial dysfunction, cellular damage, disruptions in the levels of ATP. These NPs get accumulated in the Golgi apparatus, which results in variations in proteins and have lethal effects on the cells.

4.4.1   Cobalt Oxide (Co$_3$O$_4$) Nanocrystals

Co$_3$O$_4$ nanocrystals release excessive cobalt ions that in turn activates NADPH oxidase and results in the generation of ROS (Savi et al., 2021). This induces oxidative stress on the lymphocytes. One experiment reported that cobalt oxide NPs are lethal to primary cells of the human immune system.

4.4.2   Bismuth Trioxide (Bi$_2$O$_3$) Nanocrystals

The toxicity of bismuth trioxide NPs has been studied for years, which shows negative impacts on human health and the ecosystem. The harmful effects of Bi$_2$O$_3$ NPs were observed in liver, kidney, lung and intestine cell cultures. Many cytotoxic and genotoxic effects were observed in these cultures, including apoptosis and necrosis of cells (Zulkifli et al., 2018).

4.4.3   Copper Oxide (CuO, Cu$_2$O) NPs

Copper oxide NPs are known to be highly genotoxic as they can cause DNA methylation and DNA fragmentation, along with damage to chromosomes and lipid peroxidation (Adams et al., 2017). They can also lead to the formation of micronucleus in cells, which can be determined by a mitochondrial activity assay. Moreover, they also have biocidal effects on mammalian cells.

4.4.4 Iron Oxide (Fe₂O₃, Fe₃O₄) Magnetic NPs

Fe_2O_3, Fe_3O_4 NPs have toxic effects on cells by disturbing the blood coagulation system and by reducing cell viability. These NPs can also cause inflammation, cell lysis and generation of ROS, leading to lipid peroxidation and DNA damage (El-Temsah and Joner, 2012). Magnetic iron oxide NPs accumulate in different organs and cause damage to liver, spleen, brain and lungs after inhalation.

4.4.5 Silica (SiO₂) NPs

The interaction of silica NPs with immunocompetent cells can induce immunotoxicity. This immunotoxicity includes cytotoxicity, genotoxicity and cellular dysfunctions (Nel et al., 2006). The major toxic mechanisms of these NPs are oxidative stress, autophagy and proinflammatory responses. Silica NPs are toxic to the cells of the immune system and cause direct damage to the cells through apoptosis and necrosis.

4.4.6 Titanium Dioxide (TiO₂) NPs

Titanium dioxide NPs have adverse effects on human health. Exposure to these NPs can occur through inhalation, ingestion and penetration of the skin. The main toxicity mechanism of TiO_2 NPs involves the generation of ROS in response to oxidative stress, genotoxicity, inflammation, cellular damage and carcinogenesis (Castiglione et al., 2011).

4.4.7 Zinc Oxide (ZnO) NPs

Exposure to ZnO NPs can lead to adverse effects on the environment and health. Extensive application of these NPs in different materials has increased risk for consumers, manufacturers and the atmosphere. Interaction with zinc oxide can cause oxidative DNA damage due to the generation of ROS, which induces genotoxicity and cytotoxicity (Dumont et al., 2015). Zinc oxide NPs are more harmful as compared to other metal oxides due to their ability of shedding ions.

4.4.8 Tin Oxide (SnO₂) NPs

Tin oxide NPs possess various toxic mechanisms that pose harmful effects on the human body especially on the respiratory system. The primary route for exposure to these NPs is inhalation through the nasal tract. However, these NPs might travel to nearby organs through the gastrointestinal tract (Chae et al., 2014). These NPs can result in inducing pathological lesions on the spleen, brain, kidney and liver. It has also been reported that they cause lung tumors in rats.

4.4.9 Gold (Au) NPs

Large amounts of gold NPs cause a significant decrease in red blood cells and the weight of the body. High deposition of gold NPs if administered orally in the human body can lead to lower levels of RBCs, spleen index and increased toxicity (Zhang et al., 2010). Multiple studies have shown the potential toxic effects of gold NPs, which include apoptotic cell death, growth and development inhibition in embryos and inflammatory immune responses (Gerber et al., 2013).

4.4.10 Silver Oxide (Ag₂O) NPs

In vivo studies have shown that exposure to Ag NPs can result in three main mechanisms of toxicity, i.e. DNA damage, induction of cytokines and oxidative stress. Some researches have also summarized that Ag NPs can lead to harmful effects in major organs. High doses of these NPs resulted in edema, epidermal hyperplasia and focal inflammation (Samberg et al., 2010).

4.5   APPLICATIONS OF METAL OXIDE NPS

As far as their applications are concerned, NPs are widely used in the fields of biotechnology and biomedicine. They have unique properties which can be utilized in biosensing, bioimaging, semiconductors, drug and gene delivery. Following are the effects and applications of various metal oxide NPs.

4.5.1   Cobalt Oxide (Co_3O_4) Nanocrystals

Surface functionalization and modification is favorable for their use in biomedicine and other industries (Savi et al., 2021). They can be utilized in superconductors, gas sensors, nanowires and biocatalysts (Iravani and Varma, 2020). The advantage of cobalt oxide NPs in environmental remediation includes degradation of dyes, dye waste and antibiotics. Other applications include photocatalytic degradation of hazardous waste in wastewater treatment using cobalt oxide nanocrystals (Anele et al., 2022).

4.5.2   Bismuth Trioxide ($B_{i2}O_3$) Nanocrystals

Due to their physicochemical properties and functions, bismuth trioxide NPs have extensive applications in the cosmetic, pharmaceutical, industrial and medical fields (Zhang et al., 2010). They can be used for cancer imaging, photoconduction and in solid oxide fuel cells (Shahbazi et al., 2020). The applications of bismuth trioxide NPs include industrial wastewater treatment by photocatalytic activity, degradation of dye molecules and eradication of actual organic pollutants (Trinh et al., 2019).

4.5.3   Copper Oxide (CuO, Cu_2O) NPs

Due to the desirable properties of copper oxide NPs, they have a major role in the research domain of biomedical sciences. It has been observed that the parameters and physical chemical properties help in biosensing at the commercial level. Copper oxide NPs have also been efficiently used for targeting both *in vivo* and *in vitro* environments leading to biogenic synthesis (Perreault et al., 2012). Copper and its compounds are widely used as fungicides, algicides, antibacterial and antifouling agents (Perelshtein et al., 2009).

4.5.4   Iron Oxide (Fe_2O_3, Fe_3O_4) Magnetic NPs

Fe_2O_3, Fe_3O_4 NPs are widely used in the biomedical, industrial and diagnostic fields. They also act as a carrier in drug and gene delivery and in molecular imaging by utilizing ultrasmall super paramagnetic particles of iron oxide (Attarilar et al., 2020). Iron oxide NPs have remarkable applications in several aspects of wastewater treatment such as removal of organic pollutants and heavy metals by adsorption, ionic exchange and degradation of harmful contaminants through oxidation (Gallo-Cordova et al., 2020).

4.5.5   Silica (SiO_2) NPs

Silica NPs have a great influence in the fields of pharmacy, biology and agricultural nanoproducts. Their physicochemical properties have vast uses in dental fillers, drug delivery, implants and dietary supplements (Dobrovolskaia and McNeil, 2007). Silicon dioxide is often used in industry due to its rare mechanical properties. Amorphous silica can be utilized in molecular imaging and gene carriers. Silica NPs have a vast range of environmental applications such as photocatalysis, environmental remediation and CO_2 capture. They are also known to convert CO to CO_2 and are involved in the removal of volatile organic compounds (Mafra et al., 2018).

4.5.6 Titanium Dioxide (TiO$_2$) NPs

TiO$_2$ NPs have vast applications in pharmaceutical products, food products and cosmetics such as sunscreens and toothpastes. These NPs also have immense industrial importance due to their usage as pesticides, for remediation of dyes, antibiotics and photocatalysts (Waghmode et al., 2019). Titanium dioxide is environmentally friendly because it has high chemical stability and non-toxicity. It can be used as an environment sanitizing agent due to its antifogging and self-cleaning properties (Waghmode et al., 2019).

4.5.7 Zinc Oxide (ZnO) NPs

Zinc oxide NPs are known to have various therapeutic effects including anticancer, antibacterial, antioxidant and immunomodulatory benefits (Attarilar et al., 2020). Furthermore, these NPs are also used in chemotherapeutic drugs and therefore have widespread applications in biomedicine. Zinc oxide NPs are mainly used for the treatment of wastewater, plastic and polluted air by using sunlight and artificial UV light. Zinc-based nanomaterials are the best known nanophotocatalysts due to their synthesis and characterization (Chakrabarti et al., 2020).

4.5.8 Tin Oxide (SnO$_2$) NPs

Tin oxide NPs have widespread applications in nanotechnology for drug delivery systems, electronics, cosmetics and antibacterial substances. In the field of nano-medicine, tin oxide NPs can be utilized for diagnostic purposes due to their physiochemical properties (Chae et al., 2014). Tin oxide NPs have an important role in pollutant detection, which can be harmful for the environment (Sayago et al., 2019).

4.5.9 Gold (Au) NPs

Gold NPs have a wide range of applications, which include electronics, conductors, photodynamic therapy, biosensors, probes, diagnostic markers and catalyst for chemical reactions (Thompson, 2007). Gold NPs have been shown to be effective for eradicating localized tumors. The high surface area to volume ratio enables these NPs to be used in surface coating. Different food items can be checked for their suitable consumption by using a colorimetric sensor made of Au NPs (Stuchinskaya et al., 2011). Gold NPs have broad spectrum applications in the fields of water purification, photothermal therapy, antimicrobial and the food industry. Gold NPs are known to be biocompatible due to their green synthesis (Küünal et al., 2018).

4.5.10 Silver Oxide (Ag$_2$O) NPs

Several reviews and studies emphasize the various applications of Ag NPs in the fields of biomedicine, healthcare industries, food storage and the environment. The antimicrobial activity of these NPs was reported against *E. coli*, in the bacterial cell walls (Sondi and Salopek-Sondi, 2004). Various applications include biosensing, bioimaging, water treatment, textiles and as a drug carrier. Silver NPs are found in food packaging materials, food storage containers, water purificants and detergents. They are used as surfactants and oxidizing and bleaching agents (Klasen, 2000).

4.6 SOURCES OF METAL OXIDE NPS

Metal oxide NPs are formed through various biotic and abiotic methods. Major sources of metal oxide NPs include biogenic synthesis through biomineralization and other electrochemical methods. There are various natural and anthropogenic sources of metal oxide NPs which are as follows:

TABLE 4.1

Effects and applications of various metal oxide NPs

Sr.#	Metal oxide NPs	Sources	Toxicological effects	Uses and applications	References
1.	Co_3O_4 NPs	Aqueous extracts of leaves, medicinal plants, biotic compounds	ROS production, oxidative stress on lymphocytes, lethal to immune cells	Microelectronics, biocatalysts, superconductors, gas sensors, electronic ceramics, nanowires	Savi et al. (2021)
2.	$B_{i2}O_3$ NPs	Smelting of ores, weathering of rocks, leaf extracts	Cytotoxicity, genotoxicity, apoptosis and necrosis of cells, cellular damage	Cosmetics, pharmaceuticals, biomedicine, electrochemical applications such as solid oxide fuel cells (SOFC), cancer imaging, photoconduction	Zhang et al. (2010)
3.	CuO, Cu_2O NPs	Wood preservatives, mineral supplements, weathering of rocks	DNA methylation, DNA fragmentation, lipid peroxidation, chromosomal aberrations, micronucleus formation	Biosensors, electrochemical sensors, plastic additives, coatings, textiles, catalysts, lubricants	Perreault et al. (2012)
4.	Fe_2O_3, Fe_3O_4 NPs	Sedimentary rocks, iron mineral ores, atmospheric dust, aeolian deposits	Reduced cell viability, cell lysis, inflammation, generation of ROS, lipid peroxidation, DNA damage	Magnetic imaging, nanowires, coatings, resonance imaging, environmental remediation, glass and ceramic industry	Attarilar et al. (2020)
5.	SiO_2 NPs	Sugarcane bagasse ash, clay mineral, quartz, opals and granite	Immunotoxicity, cytotoxicity, genotoxicity, cellular dysfunctions, oxidative stress, autophagy, proinflammatory responses	Drug and gene delivery, agro-nanoproducts, adsorbents, sensors, bioremediation, dental fillers, electric and thermal insulators	Dobrovolskaia and McNeil (2007)
6.	TiO_2 NPs	Sand, mineral ores, atmospheric dust, soil, volcanic eruption, sea water, fire smoke	Oxidative stress leading to ROS production, pulmonary inflammation, carcinogenesis, tumor formation, genotoxicity, cellular damage	Solar cells, food wraps, medicines, pharmaceuticals, photocatalysts, agriculture, antibacterial coatings, cosmetics	Waghmode et al. (2019)
7.	ZnO NPs	Leaf extracts, green synthesis, mineral ores, volcanic eruption	Oxidative DNA damage, apoptosis, inflammatory responses, mitochondrial membrane alterations, hematological alterations, genotoxicity	Medical and healthcare products such as anticancer, antibacterial and chemotherapeutic drugs, sunscreens, packaging, UV-protective textiles	Attarilar et al. (2020)
8.	SnO_2 NPs	Placer deposits, combustion, chemical reduction, mineral ores, fuel exhaust	Respiratory tract complications, pathological lesions on organs, tumor formation, DNA adduct formation	Drug delivery systems, cosmetics, electronics, antibacterial substances, solar cells, nano-medicine	Chae et al. (2014)

| 9. | Au NPs | Quartz, gold bearing mineral, gold mining, sedimentary rocks | Decrease in RBCs, decrease in body weight, lower spleen index, increased cytotoxicity, apoptosis, inflammatory immune responses | Electronic conductors, biosensors, diagnostic markers, probes, catalysts | Zhang et al. (2010) and Thompson (2007) |
| 10. | Ag_2O NPs | Industrial effluents, leaching of metal tailings, sewage, ore processing, atmospheric deposition | DNA damage, cytokine induction, oxidative stress, organ damage, edema, epidermal hyperplasia, focal inflammation | Biomedicine, healthcare products, food storage, antibacterial agents, bioimaging, drug carrier, water treatment | Sondi and Salopek-Sondi (2004) and Samberg et al. (2010) |

4.6.1 Natural Sources

For cobalt oxide, medicinal plants, aqueous extracts of leaves and biotic compounds are used for their synthesis (Shafey, 2020). Furthermore, smelting of mineral ores like bismite, bismuthinite are the major sources for the extraction of bismuth trioxide. For copper oxide NPs, wood preservatives, inorganic compounds, vitamin and mineral supplements and weathering of rocks are the main sources for metal oxide NPs production. Iron oxide NPs are extracted from sedimentary rocks, iron mineral ores and aeolian deposits found in atmospheric dust (Jung et al., 2007). Quarts and granites are considered to be an important source for silica oxide NPs. Similarly, the biotic sources of silica include mineral ores and sand mines. Titanium oxide NPs are obtained from sand, mineral ores, atmospheric dust, soil, volcanic eruptions, sea water and fire smoke.

4.6.2 Anthropogenic Sources

Anthropogenic NPs are made through human activities which are divided into two categories, i.e. the first category includes NPs without predetermined size and shows undefined chemistry. For example, diesel exhaust, coal fly ash, welding fumes and combustion particulates. The second category of anthropogenic NPs is also called as engineered NPs which have a specific size range of 1–100 nm. They are made up of pure materials with controlled surfaces. For example, carbon nanotubes, quantum dots, dendrimers, silver and gold nanocrystals (Sadik, 2013) (Figure 4.2).

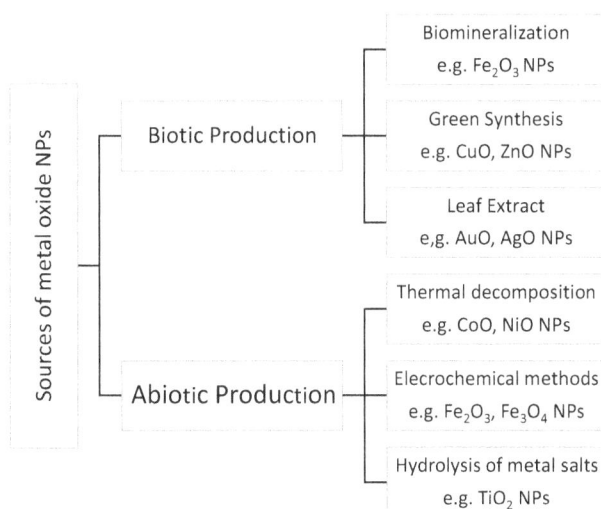

FIGURE 4.2 Sources of metal oxide NPs.

4.7 IMPACTS OF METAL OXIDE NPS ON THE ENVIRONMENT

Metal oxide NPs have various harmful impacts on the geological conditions of our environment. They can be categorized as ecotoxic due to their effects on aquatic and terrestrial organisms. Long-term exposure to these NPs can alter the morphology and properties of natural flora and fauna.

4.7.1 AQUATIC TOXICITY

Despite the wide range of applications of these nanomaterials, they can pose a serious threat to aquatic organisms (Moore, 2006; Ma et al., 2012b). Research studies have assessed the various risks due to the toxic effects of metallic oxide nanomaterials on the aquatic environment.

Several review papers have evaluated the lethal effects of nano ZnO on aquatic species (Cattaneo et al., 2009; Ma et al., 2012b). Recent studies have determined that lethal doses of zinc oxide upto the level 1.5 mg/L and 3.2 mg/L were most toxic to *Daphnia magna* (Heinlaan et al., 2008; Zhu et al., 2009). Another research has reported that 1 mg/L nano ZnO can affect the growth and reproduction of a marine amphipod, *Corophium volutator* (Fabrega et al., 2012).

Nano TiO_2 can exhibit toxicity toward various organisms if emitted in the aquatic environment (Kaegi et al., 2008; Windler et al., 2012). Many toxicological studies were carried out on aquatic organisms under optimum laboratory conditions. These studies suggested that levels of nano TiO_2 greater than 100 mg/L are toxic to *Daphnia magna* (Warheit et al., 2007). Moreover, several studies have reported the photo-induced toxicity of nano TiO_2 due to its high photoreactivity.

Exposure to these NPs can increase the risk of cellular and DNA damage to aquatic organisms and can be lethal. For example, one experiment tested the phototoxicity of 1 g/L nano TiO_2 under UV exposure of wavelength 365 nm on marine plankton (Matsuo et al., 2001). This caused fatal damage to the plankton species after exposure for 60–100 minutes. However, nano titanium dioxide has low toxicity in the absence of UV nano TiO_2 (Ma et al., 2012a).

4.7.2 TERRESTRIAL TOXICITY

Metal oxide NPs not only have harmful effects on aquatic environment but they also have significant impacts on terrestrial organisms. Studies have observed the toxic effects of nano TiO_2 on the organisms present on land. One case reported showed the harmful impacts of these NPs on soil invertebrates like earthworms and nematodes. The significant effects included DNA and mitochondrial damage and oxidative stress in earthworms after being exposed to toxic levels greater 1.0 g/kg in soil (Hu et al., 2010). Another experiment was carried out in which after a four-week exposure to nano TiO_2, it was observed that the exposure inhibited the reproduction of *Eisenia foetida* (Canas et al., 2011).

Other than earthworms nano TiO_2 ceased the reproduction and development in the nematodes *Caenorhabditis elegans* (Wang et al., 2009). Recent studies have determined that smaller-sized nano titanium dioxide was more toxic to terrestrial organisms than larger-sized nano titanium dioxide particles.

ZnO NPs have negative effects on the environment and health. Nano ZnO can get deposited in plants parts, i.e. roots, stems and leaves. One such case reported has shown the accumulation of ZnO NPs in velvet mesquite (*Prosopis juliflora-velutina*), leading to elevated levels of catalase and peroxidase (Hernandez-Viezcas et al., 2011). Nano zinc oxide has toxic effects on *Allium sepa,* which cause the inhibition of root elongation after being exposed to 5, 10, and 20 g/L lethal doses of nano ZnO (Ghodake et al., 2011).

Other metal oxide NPs like nano CuO and nano NiO affected the germination of seeds in cucumber, radish and lettuce. Moreover, nano Co_3O_4 influenced the root elongation of radish (Wu et al.,

2012). Lastly, studies suggest that the effects of these metal oxide NPs depend upon their physico-chemical properties and the type of terrestrial species. Most common effects on the plant species are slow seed germination, decreased biomass and elevated oxidative stress.

4.8 RELATIVE TOXICITY OF METAL OXIDE NPS

Various researches have shown the comparison between the toxic mechanisms of various metallic oxide NPs (Cho et al., 2013). According to the research, it has been reported that CuO is the highest toxic NP while TiO_2 has the lowest toxicity among all these metal oxide NPs (Baek et al., 2011). Many *in vivo* and *in vitro* cell line assays were performed to measure the relative cytotoxicity of these NPs. The ranking of toxicity was based on experiments performed on *E. coli*, *Bacillus subtilis* and *S. aureus* under both light radiations and dark conditions (Dasari et al., 2013). Following results were visualized:

$$ZnO > CuO > Co_3O_4 > Fe_2O_3 > TiO_2$$

The toxicity of these NPs was investigated and ranked in terms of ROS production, lipid peroxidation, oxidative stress, release of metal ions and amount of reduced glutathione. Furthermore, the lethal effects of these metal oxide NPs were also observed on *Lactuca* seeds, which had harmful impacts on the germination of seed and bioluminescence activity due to gene mutation (Ko et al., 2014). Copper and zinc oxide were determined as the most toxic NPs in both *in vivo* and *in vitro* assays.

4.9 BIOREMEDIATION STRATEGIES

There are three main strategies for bioremediation which are as follows:

- **Bioventing**: It is one of the most common methods for bioremediation. In this process, small-diameter wells are drilled into the soil that allows air ventilation where ground gases are produced by microbial action.
- Bioventing is used in the aerobic biodegradation of contaminants by indigenous microorganisms (Leu and Hou, 2020).
- **Biosparging**: It involves injection of air with high pressure into the soil. This increases concentration of oxygen in the groundwater, which is highly effective for tilling contaminated soil.
- Biosparging system is used with soil vapor extraction for *in situ* bioremediation (Sharma, 2019).
- **Bioaugmentation**: It is used to add exogenous species of microbes to the soil to make it biocompatible. The microbes are used to biodegrade specific soil and groundwater contaminants. It works in conjunction with biosparging and bioventing applications.
- A consortium of bacteria uses 2,4-dichlorophenol at a laboratory scale utilizing synthetic wastewater as a medium for bioaugmentation (Nzila et al., 2016).

4.9.1 NANOBIOREMEDIATION

NPs play an important role in remediating the environment by removing various contaminants from the biosphere. They are environmentally friendly as they reduce biodegradable waste (Abbasi et al., 2012). Thus, these nanomaterials significantly clean up the air, water, soil around us. This process of remediation is done by detection of the spill leading to remediation and finally pollution prevention of the environment. These NPs cause breakdown of various organic and inorganic wastes present in the environment (Anuradha et al., 2015). It is observed that nano-sized iron oxide is used as an absorbent for different kinds of contaminants. The high reactivity of iron oxide NPs makes it a brilliant remediator for cleaning up the environment. However, the small size, the crystalline structure, solubility affects the reactivity (Braunschweig et al., 2013).

4.9.2 Wastewater Treatment

Various metal oxide NPs are known for wastewater treatment. These NPs having specific physical, chemical and attributes are used for adsorption, degradation through catalysts, decomposition and finally the filtration of water from various industrial effluents. Iron oxide nanocrystals along with activated charcoal are efficiently used for the removal of toxic chromium from water (Abuhatab and Al-Zerei, 2020). Similarly, various metal oxide NPs in conjugation with other elements enhance their remediating ability. A magnetic polymer of cobalt in conjugation with nylon is used to remove lead from water. Because of the stability and recyclable nature, metal oxide NPs work effectively in improving the hydrosphere (Ahsan et al., 2020).

4.10 METAL OXIDE NPS AND MICROBES

The use of microorganisms in the field of nanotechnology results in several advantages for the environment. Copper oxide NPs synthesized from *Escherichia* sp. are used for degrading various kinds of dye (Secundo, 2013). Silica NPs assisted with Actinomycetes is used for removing industrial effluents. These nanocrystals show high photocatalytic degradation. Iron oxide NPs from *Aspergillus tubingensis* are used to remove several kinds of heavy metals present in contaminated water (Mohanraj et al., 2020).

4.11 CONCLUSION

The above-mentioned metal oxide NPs have different biogenic synthesis pathways and thus possess various toxic mechanisms. Both traditional and modern physical and chemical techniques have been utilized in the investigation of their toxicity toward humans and the environment (Golinska et al., 2014). This chapter summarizes and compares some vital aspects of metal oxide NPs including effects, applications and their impacts on the environment. Some applications include bioremediation, drug and gene delivery systems, biomedicine, pharmaceuticals, cosmetics, electrical appliances and biocatalysts. These NPs have also been used in the treatment and diagnosis of several diseases (Landsiedel et al., 2010). Despite the advantages, these NPs pose different harmful threats to humans including cellular damage, DNA damage, formation of ROS, fatal diseases, inflammation and tumor formation (Dasari et al., 2013).

Moreover, many physicochemical properties have also been related with these metal oxide NPs and these properties have their advantages and disadvantages for the environment. Although the biological synthetic methods for these NPs are environmentally friendly, researchers have reported that metal oxide NPs can be highly toxic to humans and the environment. In a nutshell, it is mandatory to first investigate the potential toxicity of these NPs before utilizing them in any diagnostic method so that they pose no harmful threat to the human body. The studies suggest that nanotoxicity of all these metal oxides is most commonly related to the release of toxic ions from these metallic oxides (Goix et al., 2014). Finally, many experimental methods have compared the toxic levels of NPs and different studies have reported their various applications in the fields of modern technology and biomedicine (Ko and Kong, 2014).

4.12 FUTURE PERSPECTIVE

Metal oxide NPs can have several long-term applications in the research of future nanotoxicology. Cobalt oxide nanocrystals cam be used in the degradation of dyes, dye waste and antibiotics. Bismuth trioxide NPs can be utilized in the eradication of actual organic pollutants. Similarly, copper oxide NPs can play an important role as fungicides, algicides and pesticides. Iron oxide NPs can have remarkable applications in several aspects of wastewater treatment. Moreover, nanocrystals of silica can have vast applications in environmental remediation.

Titanium dioxide can act as environmentally friendly NPs because of their high chemical stability and non-toxicity. Zinc oxide NPs can be best used as chemotherapeutic drugs and biomedicine. Tin oxide NPs can be used as pollutant detectors. Gold nanocrystals can have broad spectrum roles in the fields of water purification and photothermal therapy. Silver oxide NPs can be best utilized as surfactants and bleaching agents. We hope that these future prospects of various NPs will be shown by practical evidence.

REFERENCES

Adams, L. K., Lyon, D. Y., Alvarez, P. J. J. 2006. Comparative eco-toxicity of nanoscale TiO_2 SiO_2 and ZnO water suspensions. *Water Res.*, 40, 3527–3532.

Ahamed, M., Akhtar, M. J., Khan, et al. 2019. Oxidative stress mediated cytotoxicity and apoptosis response of bismuth oxide (Bi_2O_3) nanoparticles in human breast cancer (MCF-7) cells. *Chemosphere*, 216, 823–831.

Ali, M. E., Mustafa, S., Hashim, U., Man, Y. C., & Foo, K. L. 2012. Nanobioprobe for the determination of pork adulteration in burger formulations. *Journal of Nanomaterials*, 2012, 8–8.

Arbab, A. S., Yocum, G. T., Rad, A. M., et al. 2005. Labeling of cells with ferumoxides–protamine sulfate complexes does not inhibit function or differentiation capacity of hematopoietic or mesenchymal stem cells. *NMR Biomed.*, 18, 553–559.

Aschberger, K., Micheletti, C., Sokull-Kluettgen, B., et al. 2011. Analysis of currently available data for characterising the risk of engineered nanomaterials to the environment and human health – lessons learned from four case studies. *Environ. Int.*, 37, 1143–1156.

Attarilar, S., Yang, J., Ebrahimi, M., et al. 2020. The toxicity phenomenon and the related occurrence in metal and metal oxide nanoparticles: a brief review from the biomedical perspective. *Front. Bioeng. Biotechnol.*, 8, 822.

Brown, S. D., Nativo, P., Smith, J., et al. 2010. Gold nanoparticles for the improved anticancer drug delivery of the active component of oxaliplatin. *J. Am. Chem. Soc.*, 132(13), 4678–4684.

Brunner, T. J., Wick, P., Manser, P., et al. 2006. In vitro cytotoxity of oxide nanoparticles: comparison to asbestos, silica, and the effect of particle solubility. *Environ. Sci. Technol.*, 40, 4374–4381.

Cañas-Carrell, J. E., Li, S., Parra, A. M., & Shrestha, B. 2014. Metal oxide nanomaterials: health and environmental effects. *In Health and environmental safety of nanomaterials*, (pp. 200–221). Woodhead Publishing.

Carlson, C., Hussain, S. M., Schrand, A.M., et al. 2008. Unique cellular interaction of silver nanoparticles: size-dependent generation of reactive oxygen species. *J. Phys. Chem. B*, 112, 13608–13619.

Chae, H., Song, D., Lee, Y. G., Son, T., Cho, W., Pyun, Y. B., ... & Kang, Y. S. 2014. Chemical effects of tin oxide nanoparticles in polymer electrolytes-based dye-sensitized solar cells. *The Journal of Physical Chemistry C*, 118(30), 16510–16517.

Chattopadhyay, S., Dash, S. K., Tripathy, S., Das, B., Mandal, D., Pramanik, P., & Roy, S. 2015. Toxicity of cobalt oxide nanoparticles to normal cells; an in vitro and in vivo study. *Chemico-biological interactions*, 226, 58–71.

Chernousova, S., Epple, M. 2013. Silver as antibacterial agent: Ion, nanoparticle, and metal. *Angew. Chem. Int. Ed.*, 52, 1636–1653.

Dang, Y., Zhang, Y., Fan, L., Chen, H., & Roco, M. C. 2010. Trends in worldwide nanotechnology patent applications: 1991 to 2008. *Journal of nanoparticle research*, 12, 687–706.

De, M., Ghosh, P. S., Rotello, V. M. 2008. Applications of nanoparticles in biology. *Adv. Mater.*, 20(22), 4225–4241.

Duan, X. P., Li, Y. P. 2013. Physicochemical characteristics of nanoparticles affect circulation, biodistribution, cellular internalization, and trafficking. *Small*, 9, 1521–1532.

Gurunathan, S., Han, J. W., Kim, E. S., et al. 2015a. Reduction of graphene oxide by resveratrol: a novel and simple biological method for the synthesis of an effective anticancer nanotherapeutic molecule. *Int. J. Nanomed.*, 10, 2951–2969.

Gurunathan, S., Kalishwaralal, K., Vaidyanathan, R., et al. 2009. Biosynthesis, purification and characterization of silver nanoparticles using Escherichia coli. *Colloids Surf. B Biointerfaces*, 74, 328–335.

Gurunathan, S., Park, J. H., Han, J. W., et al. 2015b. Comparative assessment of the apoptotic potential of silver nanoparticles synthesized by *Bacillus tequilensis* and *Calocybe indica* in MDA-MB-231 human breast cancer cells: targeting p53 for anticancer therapy. *Int. J. Nanomed.*, 10, 4203–4222.

Huang, D., Liao, F., Molesa, S., et al. 2003. Plastic-compatible low resistance printable gold nanoparticle conductors for flexible electronics. *J. Electrochem. Soc.*, 150(7), G412.

Jo, D. H., Kim, J. H., Lee, T. G., et al. 2015. Size, surface charge, and shape determine therapeutic effects of nanoparticles on brain and retinal diseases. *Nanomedicine*, 11, 1603–1611.

Li, L., Hu, J., Yang, W., et al. 2001. Band gap variation of size- and shape-controlled colloidal CdSe quantum rods. *Nano Lett.*, 1, 349–351.

Li, W. R., Xie, X. B., Shi, Q. S., et al. 2010. Antibacterial activity and mechanism of silver nanoparticles on Escherichia coli. *Appl. Microbiol. Biotechnol.*, 8, 1115–1122.

Li, C. Y., Zhang, Y. J., Wang, M., et al. 2014. In vivo real-time visualization of tissue blood flow and angiogenesis using Ag_2S quantum dots in the NIR-II window. *Biomaterials*, 35, 393–400.

Lin, P. C., Lin, S., Wang, P. C., et al. 2014. Techniques for physicochemical characterization of nanomaterials. *Biotechnol. Adv.*, 32, 711–726.

Mukherjee, P., Ahmad, A., Mandal, D., et al. 2001. Fungus-mediated synthesis of silver nanoparticles and their immobilization in the mycelial matrix: a novel biological approach to nanoparticle synthesis. *Nano Lett.*, 1, 515–519.

Murdock, R. C., Braydich-Stolle, L., Schrand, A. M., et al. 2008. Characterization of nanomaterial dispersion in solution prior to in vitro exposure using dynamic light scattering technique. *Toxicol. Sci.*, 101, 239–253.

Peng, G., Tisch, U., Adams, O., Hakim, M., Shehada, N., Broza, Y. Y., ... & Haick, H. 2009. Diagnosing lung cancer in exhaled breath using gold nanoparticles. *Nature nanotechnology*, 4(10), 669–673.

Perrault, S. D., Chan, W. C. W. 2010. In vivo assembly of nanoparticle components to improve targeted cancer imaging. *Proc. Nat. Acad. Sci.*, 107(25), 11194–11199.

Pleus, R. 2012. Nanotechnologies-guidance on physicochemical characterization of engineered nanoscale materials for toxicologic assessment. ISO, Geneva, Switzerland.

Sapsford, K. E., Tyner, K. M., Dair, B. J., et al. 2011. Analyzing nanomaterial bioconjugates: a review of current and emerging purification and characterization techniques. *Anal. Chem.*, 83, 4453–4488.

Savi, M., Bocchi, L., Cacciani, F., Vilella, R., Buschini, A., Perotti, A., ... & Zaniboni, M. 2021. Cobalt oxide nanoparticles induce oxidative stress and alter electromechanical function in rat ventricular myocytes. *Particle and Fibre Toxicology*, 18(1), 1–17.

Seabra, A. B., Durán, N. (2015). Nanotoxicology of metal oxide nanoparticles. *Metals*, 5(2), 934–975.

Sharma, V. K., Yngard, R. A., Lin, Y. 2009. Silver nanoparticles: green synthesis and their antimicrobial activities. *Adv. Colloid Interface*, 145, 83–96.

Shi, H., Magaye, R., Castranova, V. et al. 2013. Titanium dioxide nanoparticles: a review of current toxicological data. *Part. Fibre Toxicol.*, 10, 15.

Sondi, I., Salopek-Sondi, B. 2004. Silver nanoparticles as antimicrobial agent: a case study on E. coli as a model for Gram-negative bacteria. *J. Colloid Interface Sci.*, 275, 177–182.

Staquicini, F. I., Ozawa, M. G., Moya, C. A., et al. 2011. Systemic combinatorial peptide selection yields a non-canonical iron-mimicry mechanism for targeting tumors in a mouse model of human glioblastoma. *J. Clin. Investig.*, 121, 161–173.

Stensberg, M. C., Wei, Q., McLamore, E. S., et al. 2011. Toxicological studies on silver nanoparticles: challenges and opportunities in assessment, monitoring and imaging. *Nanomedicine*, 6(5), 879–898.

Stuchinskaya, T., Moreno, M., Cook, M. J., et al. 2011. Targeted photodynamic therapy of breast cancer cells using antibody phthalocyanine gold nanoparticle conjugates. *Photochem. Photobiol. Sci.*, 10(5), 822.

Thompson, D. T. 2007. Using gold nanoparticles for catalysis. *Nano Today*, 2(4), 40–43.

Yang, H., Liu, C., Yang, D., et al. 2009. Comparative study of cytotoxicity, oxidative stress and genotoxicity induced by four typical nanomaterials: the role of particle size, shape and composition. *J. Appl. Toxicol.*, 29(1), 69–78.

Zulkifli, Z. A., Razak, K. A., Rahman, W. N. W. A., & Abidin, S. Z. (2018, August). Synthesis and characterisation of bismuth oxide nanoparticles using hydrothermal method: the effect of reactant concentrations and application in radiotherapy. *In Journal of Physics: Conference Series* (Vol. 1082, p. 012103). IOP Publishing.

5 Theory of Membranes
Types of Membranes, Working Principle, and Comparison

Aamir Khan and Sher Jamal Khan

5.1 INTRODUCTION

Achieving environmental sustainability via treatment of wastewater is one of the key considerations in the United Nation's Sustainable Development Goals (SDGs). Water shortage, water scarcity, and excessive water pollution caused by industrial processes have become significant global or universal concerns in the recent past. Climate change, escalating worldwide population, and swift urbanization have also fast-tracked the increasing need/demand for fresh water (Adam et al., 2022).

Among the various pollutants of concern, the major ones are heavy metals as they cannot be simply degraded like other pollutants using biological processes. Heavy metals are principally described as those elements that have atomic weights from 63.5 to 200.6 and a corresponding density greater than $5 \, g/m^3$ (Abdullah et al., 2019). Some heavy metals like copper (Cu), iron (Fe), zinc (Zn), manganese (Mn), and cobalt (Co) play crucial roles in biochemical processes of the human body. However, unwarranted exposure to such metals can lead to lethal impacts. Further, heavy metals like cadmium (Cd), arsenic (As), mercury (Hg), lead (Pb), and chromium (Cr) are quite toxic, at even very trace levels like in ppb, because these metals are non-degradable and are capable of bio-accumulating in the human body (Landaburu-Aguirre et al., 2006). Under normal conditions, the human body is able to tolerate metals at trace/minute amounts without being subjected to serious health problems. Yet, long-term exposure to such heavy metals is able to trigger the accumulation of toxins in the body, which may lead to malfunctioning of body systems and ultimately fatality. Hence, maximum contamination levels (MCLs) have been put in place by the World Health Organization (WHO) to guarantee that zero or just the bare minimum limit of toxic heavy metals is permitted in a water source (Jabłońska and Siedlecka, 2015).

Heavy metal contaminants found in wastewater are usually removed through various technologies. Such techniques include ion exchange, chemical precipitation, adsorption, electrochemical treatment, coagulation-flocculation, and membrane filtration (Zainuddin et al., 2022). A comparison of the pros and cons of such technologies is given in Table 5.1 (Gautam et al., 2012).

Recently, the removal of heavy metals utilizing membrane-based processes has become more popular because of the compact design of reactors and lower sludge handling issues. Various types of membranes are generally used for water and wastewater treatment including low pressure-driven membranes (microfiltration (MF), ultrafiltration (UF), distillation), high pressure-driven membranes (nanofiltration and reverse osmosis (RO)), osmotic pressure-driven membranes like forward osmosis and other membranes like those used in electrodialysis, etc. The primary aspects to consider for selection and usage of membranes are pore size, pore size distribution, degree of hydrophilicity, surface charge, flow rate, and the existence of functional groups that support the separation processes (Shaheen et al., 2022).

DOI: 10.1201/9781003326281-6

TABLE 5.1

Comparison of treatment technologies for heavy metal removal (Gautam et al., 2012)

Technology/treatment	Advantages	Disadvantages
Chemical precipitation	Simplicity of the method Not metal discriminating Low-priced capital costs	Large quantity of metal sludge High sludge disposal costs High maintenance expenses
Coagulation and flocculation	Inactivation capability of bacteria Fine sludge dewatering and settling attributes	Chemicals consumption Increased sludge volume production
Ion exchange	Metal selective Restricted pH tolerance High regeneration capability	High capital costs High maintenance costs
Flotation	Metal selective Modest retention times Removal of small particulates	High capital costs High operation and maintenance costs
Membrane filtration	Less solid waste generation Less chemicals consumption Modest space requirements Possible to be metal discriminating	High capital costs High operation and maintenance costs Membrane fouling/clogging Limited flow rates
Electrochemical treatment	Moderately metal selective No chemical required; can be engineered to tolerate suspended solids	High capital costs Production of hydrogen gas (in some processes) Filtration process for flocs
Adsorption	Wide variety of target pollutants Possibly selective depending on the adsorbent High capacity Fast kinetics	Performance dependent on adsorbent type Physical or chemical activation done to enhance its sorption capacity

5.2 BASIC TERMINOLOGIES

There are some basic terminologies or parameters, the understanding of which is very important to totally understand the detailed application of membranes for heavy metal removal or wastewater treatment. These terminologies are discussed below.

5.2.1 TRANS-MEMBRANE PRESSURE (TMP)

TMP is described as the pressure gradient of the membrane, or the mean feed pressure excluding the permeate pressure. The feed pressure is usually measured at the initial/entry point of a membrane module and permeate pressure is measured at the membrane outlet.

$$\text{TMP} = \left(\frac{P_f + P_r}{2} \right) - P_p$$

Where P_f is the feed pressure in KPa, P_r is the retentate pressure in KPa, and P_p is the permeate pressure in KPa.

5.2.2 FLUX

The amount/quantity of permeate produced from a unit area of membrane surface per unit time is known as flux through the membrane. It is represented by the letter "J" and is represented in units of liters per squared meter per hour, and is abbreviated as LMH. Flux across a membrane depends on membrane permeability, TMP, and the degree of fouling.

$$J = Q/A$$

Where J is the membrane flux, Q is the permeate flow rate, and A is the effective surface area.

Flux is dependent on pressure and temperature as temperature affects the water viscosity. Temperature changes are hence accommodated by calculating flux at a standard temperature, known as standard flux, J_s.

$$J_s = J_m \left(\frac{\mu_m}{\mu_s} \right)$$

Where J_s is the standard flux (L/m^2 h), J_m is the flux at measured temperature (L/m^2 h), μ_m is the dynamic viscosity of water at measured temperature (kg/m s), and μ_s is the dynamic viscosity of water at standard temperature of 20°C (kg/m s) (Crittenden, 2005).

Flux is usually standardized for pressure by determining specific flux, which means the flux at a standard temperature over the trans-membrane pressure. The specific flux is also known as membrane permeability once clean water is being passed/filtered through a new, virgin membrane (Crittenden, 2005).

$$J_{sp} = J_s /TMP$$

Where J_{sp} is the specific flux at standard temperature (L/m^2 h bar).

5.2.3 RECOVERY

It is very common in membrane processes to refer to the efficiency of the treatment process by referring to its percent recovery. The percent recovery of a membrane system is the proportion of the feed water that passes through the membrane pores and becomes permeate.

$$Recovery = \frac{F_p}{F_f} \times 100$$

Where F_p is the permeate flow rate (L/h) and F_f is the feed flow rate (L/h).

Recovery of a membrane system is basically a comparison between the amount of water that flows into the treatment system and the amount of water that flows out as permeate and is delivered to the distribution system.

5.2.4 REJECTION COEFFICIENT

Rejection coefficient or the solute rejection coefficient is the percentage of solids retained by the membrane compared to solids in the incoming feed stream.

$$R = \frac{C_f - C_p}{C_f}$$

where R is the rejection coefficient, C_f is the concentration of solids in the feed (mg/L), and C_p is the concentration of solids in the permeate (mg/L).

From the above equation, it is evident that if the concentration in the permeate is zero, R will be equal to 1, showing complete rejection of solids and hence a perfectly selective membrane.

As concentration in permeates increases, it means the membrane has become permeable and therefore the selectivity has reduced. It can hence be deduced that lower values of R show poor membrane performance with no treatment at all if R becomes zero.

5.3 MEMBRANE MODULES

Membrane module is basically the hardware in which the membrane is placed/arranged/fitted to separate the incoming feed stream into retentate and permeate streams. These modules are briefly discussed below.

5.3.1 Flat Sheet/Plate and Frame

Flat sheet modules consist of a selective flat sheet of membrane on the top or above and a flat plate at the bottom or lower side, in the middle of which a web-like material is positioned to provide room for permeate removal. This material is generally known as spacer. On the other face of the flat plate, a second sheet of membrane and a similar web-like substance is placed to form a sandwich-shaped module. Such module is basically known as a plate-and-frame module (Figure 5.1). Flat sheets with cross-flow pattern modules are the best pick where blocking/fouling is the foremost concern (Cui et al., 2010).

5.3.2 Spiral Wound

This module contains an envelope of membranes and spacers wound over a perforated collection tube. This module is housed within a tubular pressure vessel. The feed stream passes axially down the module over the membrane envelope. A part of the feed permeates/penetrates into the membrane envelope, from where it moves spirally toward the center and escapes through the central collection tube (Figure 5.2). These modules have relatively huge packing densities (surface area to volume ratio) and hence less capital cost (Scott, 1995).

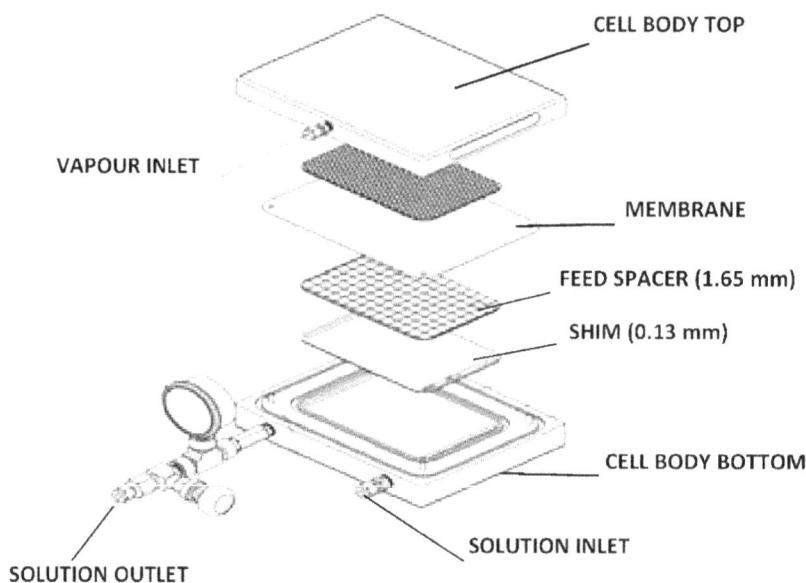

FIGURE 5.1 Flat sheet membrane module (Berdasco et al., 2017).

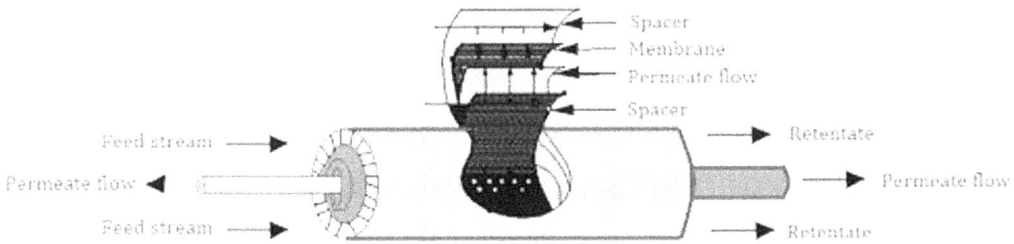

FIGURE 5.2 Spiral wound module (Pal, 2015).

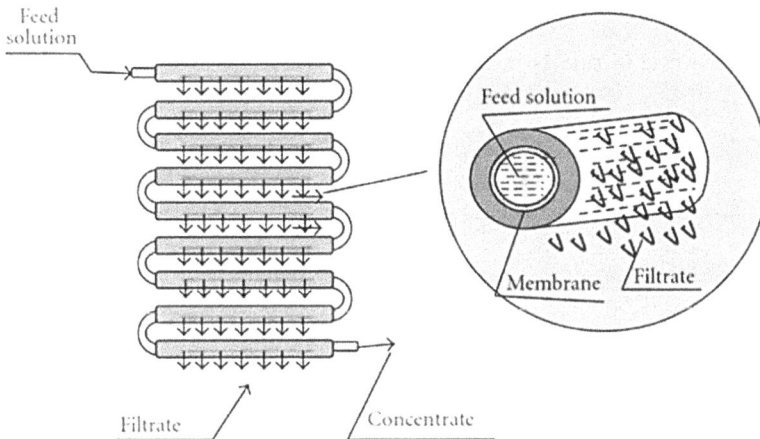

FIGURE 5.3 Tubular module (Alhathal Alanezi et al., 2012).

5.3.3 TUBULAR

A conventional tubular membrane system consists of a large number of tubes assembled in a shell-and-tube arrangement, in which the feed liquid is injected through all of the tubes. The permeate after passing through each of the tubes gets collected in a single collection unit. The tubes comprise a porous fiberglass or paper support with the membrane developed on the interior of the tubes (Figure 5.3). Because of their large core diameters, these modules are useful for dealing with the feeds containing pretty large particles (Berdasco et al., 2017).

5.3.4 CAPILLARY

The capillary membrane type module comprises a huge number of capillaries gathered/assembled all together in a single pressure vessel. Feed water flows inside of the capillaries and travels across the pores in the capillaries, departing from the top of the vessel as permeate. As the feed travels down the capillary section, it exits the other end of the capillary tube as retentate or concentrate (Balster, 2015).

5.3.5 HOLLOW FIBER

In a hollow fiber type membrane module, the feed is provided either within or outside of the hollow fiber and the resulting permeate goes through the fiber wall to the other edge of the fiber. These fibers have the composition of an asymmetric membrane, in which the active or skin layer faces the feed solution (Figure 5.4). A bunch of hollow fibers are encased in a pressure vessel, and the open edges of U-shaped fibers are connected to the head plate using epoxy (Ismail et al., 2019).

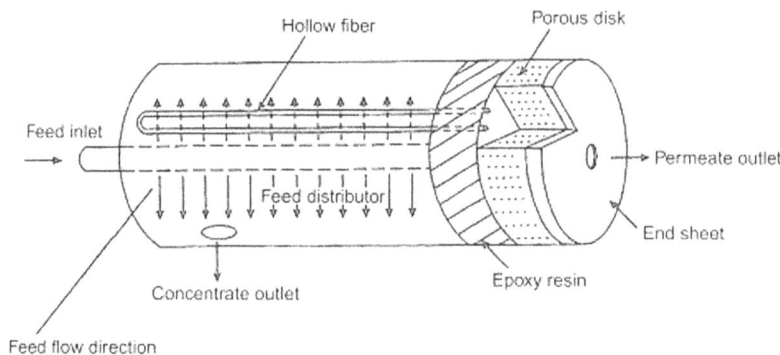

FIGURE 5.4   Hollow fiber module (Ismail et al., 2019).

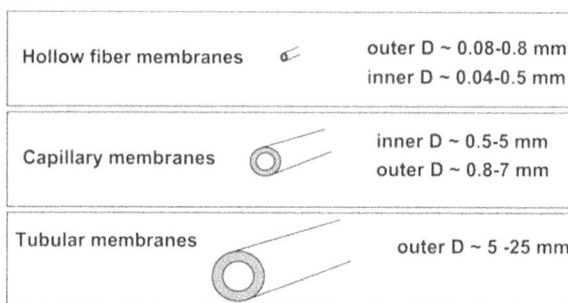

FIGURE 5.5   Comparison of tube type membranes (Balster, 2015).

TABLE 5.2
Comparison of membrane modules (Sridhar et al., 2007)

Type of module	Plate and frame	Spiral wound	Tubular	Capillary	Hollow fiber
Diameter	–	50–200 mm	5–15 mm	0.5–5 mm	<0.5 mm
Packing density (m²/m³)	30–500	200–800	30–200	600–1,200	500–9,000
Ease of cleaning	Good	Fair	Excellent	Fair	Poor
Relative cost	High	Low	High	Low	Low
Resistance to fouling	Good	Moderate	Very Good	Moderate	Poor

The tube or duct type membranes are almost the same, i.e., tubular, capillary, and hollow fiber modules. The difference lies in the diameter of the tubes and hence variable packing density, as depicted in Figure 5.5 (Balster, 2015).

Table 5.2 shows a comparison of the various membrane modules employed for water and wastewater management.

5.4 ION EXCHANGE MEMBRANES

Ion exchange membranes (IEMs) have a huge potential in various applications as well as play an imminent role in addressing the energy along with environment-related problems. Over the previous decade, the utilization of IEMs has stoked considerable research thought in terms of preparation, materials, and applications, owing to their educational and industrial values (Strathmann et al., 2013).

IEMs normally comprise hydrophobic substrates, non-movable/immobilized ion-functionalized groups, and mobile counter-ions. These ion-functionalized materials should be permeable to oppositely charged ions known as counter-ions but impermeable toward ions carrying same charge known as co-ions, hence creating a perm selectivity. IEMs can hence be classified into two groups, depending on their ion-functionalized groups and ion selectivity. Anion exchange membranes (AEMs) contain positively charged functional groups like $-NR_3^+$, $-NR_2H^+$ and permit the passage of anions while preventing the passage of cations. On the other hand, cation exchange membranes (CEMs) contain negatively charged functional groups like $-SO_3^-$, $-COO^-$, $-PO_3^{2-}$ and hence allow the passage of positive ions or cations while rejecting the passage of anions (Swanckaert et al., 2022).

5.4.1 WORKING PRINCIPLE

In IEMs, the transport of components usually occurs due to the driving forces of concentration and electric potential gradients. However, the two types of ions, anions and cations, go in opposing directions under an electric potential gradient. IEMs are hence considered in terms of the quantity of charge transported rather than the extent of material transported (Ahmad et al., 2018).

CEMs normally include sulfonic acid groups, sulfonamides, phosphoric acid, and azole derivatives. A wide variety of polymer materials like poly ether ketone (PEK), poly ether sulfone (PES), polybenzimidazole (PBI), poly phenylene, polyimide (PI), polyvinylidene fluoride (PVDF), and polyphosphazene have been reported as the pillars of CEMs (Ran et al., 2017).

AEMs are commonly made from positively charged polyelectrolytes and are intended to allow/conduct anions while being impervious to neutral molecules or cations (Ran et al., 2017).

IEMs are widely being used for water and wastewater management in both industrial and municipal systems. The technique provides various advantages over other treatment technologies: it is environmentally friendly, it can deliver a high flow rate of treated water, and it also has low operational and maintenance cost. Alongside these advantages, there are some disadvantages associated with these membranes, such as iron fouling, calcium sulfate fouling, adsorption of organics on the membrane matrix, bacterial or microbial contamination, and chlorine contamination (Ran et al., 2017).

5.4.2 EFFICIENCY AND TARGET POLLUTANTS

Due to a wide variety of ion functional groups and the selective transfer of ionic species and blocking of neutral species, IEMs are commonly applied in separation and transfer/transport applications or technologies. These are the crucial components in some conventional processes, like electrodialysis, diffusion dialysis, and bipolar membrane electrodialysis. Lately, IEMs have been expanded to novel applications linked with energy transformation and production, involving fuel cells, reverse electrodialysis, and redox flow batteries (Ran et al., 2017).

5.5 PRESSURE-DRIVEN MEMBRANES

Pressure-driven membrane processes are divided into two major categories: low-pressure-driven and high-pressure-driven processes. The high-pressure-driven membrane processes are RO and nanofiltration (NF), whereas low-pressure membranes are UF and MF. These processes are emerging as crucial components of water treatment systems all over the world and are being used specifically for post treatment (Pervez et al., 2022). The huge amount of capital required and huge operating costs of these membrane systems is because of the high pressures required to remove dissolved impurities like small organic molecules and monovalent ions. NF and RO are quite effective at eradicating dissolved ions and organic molecules. However, very high pressures of around 100–1,000 psi are required to operate these membranes. On the other hand, MF and UF membranes require much lower operating pressure, i.e., 5–60 psi. However, they are not efficient in eliminating dissolved ions and organic molecules (Diallo, 2014).

5.5.1 WORKING PRINCIPLE

A generic membrane process splits a feed stream into retentate and permeate using a membrane. Pressure-driven membrane techniques use the pressure difference between the feed stream and the permeate side as the major driving force for transporting the solvent through the membrane, which acts as a barrier to solid particles. Particles and dissolved impurities are retained or removed based on properties like size, shape, and charge. The removal efficiency is thus expressed by the rejection coefficient R, which ranges from 0 (for complete permeation) to 1 (for complete rejection). In full-scale installations, the ratio between the permeate and the feed stream flow (recovery) normally ranges from 50% to 90%, but, in general, it is around 80% (Van Der Bruggen et al., 2003).

Table 5.3 shows the advantages and disadvantages of pressure-driven membrane processes.

5.5.2 EFFICIENCY AND TARGET POLLUTANTS

Pressure-driven membranes are quite successful for the removal of pollutants from wastewater streams. Different types of pollutants removed by these membranes are shown in Figure 5.6.

RO membranes are very valuable for the removal of heavy metals from water as well as wastewater. Regis et al. (2022) have reported arsenic removal of around 95% using RO and NF membranes.

Table 5.4 shows some applications of pressure-driven membrane processes in the treatment of wastewater and their removal efficiencies.

5.6 GRAVITY-DRIVEN MEMBRANES

5.6.1 WORKING PRINCIPLE

Usually, membranes are avoided because of the biofouling formed over the membrane surface, which makes the use of membranes for industrial applications quite challenging. The biofilm that forms on the membrane exterior was, however, discovered to be beneficial in creating a stable flux for a lengthier filtration period (Barambu et al., 2021). As a consequence of the accumulated

---

TABLE 5.3

Advantages and disadvantages of pressure-driven membranes (Ahmad et al., 2018)

Membrane process	Advantages	Disadvantages
Microfiltration (MF)	• Compact modules • Low capital costs • No degradation due to heating • Less pretreatment	• High operational costs • Fouling of low-molecular-weight organics
Ultrafiltration (UF)	• Effective in treating oily microemulsions • Low energy consumption and high efficiency • Less pretreatment	• Lower flux • Increased fouling rates
Nanofiltration (NF)	• Better removal efficiencies due to smaller pore size • Can remove divalent ions including hardness • Compact S/W modules	• Higher energy consumption • Needs good pretreatment • More fouling rate than MF and UF
Reverse osmosis (RO)	• Can remove all solids and ions • Best for heavy metal removal • Reuse of product water for different purposes	• Very high pressure requirements • Increased fouling rates • Needs very effective pretreatment

Microfiltration   Ultrafiltration   Nanofiltration Reverse Osmosis
10 – 0.1 µm   0.1 – 0.01 µm   0.01 – 0.001 µm   < 0.001 µm

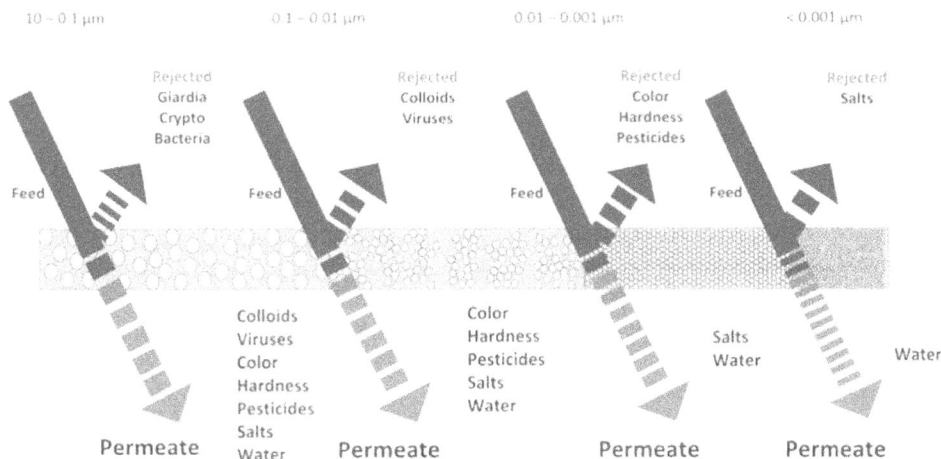

Rejected
Giardia
Crypto
Bacteria

Rejected
Colloids
Viruses

Rejected
Color
Hardness
Pesticides

Rejected
Salts

Feed   Feed   Feed   Feed

Colloids   Color
Viruses   Hardness
Color   Pesticides   Salts
Hardness   Salts   Water   Water
Pesticides   Water
Salts
Permeate Water   Permeate   Permeate   Permeate

FIGURE 5.6   Pressure-driven membranes (Czarny et al., 2017).

TABLE 5.4
Pressure-driven membrane processes for wastewater treatment (Obotey Ezugbe and Rathilal, 2020)

Pressure-driven membrane process	Wastewater treated	Reported results	References
UF	Vegetable oil factory	COD (91%), Total suspended solids (TSS) (100%), TOC (87%), PO_4^{3-} (85%), Cl^- (40%)	Mohammadi and Esmaeelifar (2004)
MF coupled with RO	Urban wastewater	Pesticides and pharmaceuticals removed to meet discharge limits	Rodriguez-Mozaz et al. (2015)
MF	Municipal wastewater (disinfection and phosphorus removal)	Contaminants removed to below detection limit	Dittrich et al. (1996)
MF	Synthetic emulsified oily wastewater	95% removal of organics	Wang et al. (2009)
NF coupled with RO	Dumpsite's leachate	95% recovery of water	Rautenbach et al. (2000)
UF	Poultry and slaughterhouse wastewater	Chemical oxygen demand (COD) and biological oxygen demand (BOD) removal of >94%, fats removal of 99%, TSS (98%)	Yordanov (2010)
NF	Textile	COD (57%), salinity (30%), color (100%)	Ellouze et al. (2012)
UF coupled with RO	Metal finishing industry	90%–99% removal of all contaminants	Petrinic et al. (2015)
UF coupled with RO	Oily wastewater	Oil and grease (100%), COD and TOC (98%), TDS (95%), turbidity (100%)	Salahi et al. (2011)
UF coupled with NF/RO	Phenolic wastewater from paper mill	Phenol (94.9%), COD (95.5%)	Sun et al. (2015)

biofouling on the membrane exterior during gravity-driven membrane (GDM) filtration, a well-permeable biofilm is formed. It has been demonstrated that it can stimulate a stable flux, allowing for farseeing operation without the need for membrane fouling supervision. GDM filtration is a dead-end gravity-driven purification/filtration system that uses a simple setup to perform at a very low TMP. It can be used in both submerged and side-stream configurations just like conventional membrane bioreactors (MBRs).

The flux stabilization method is the key to long-term operation of GDM filtration not including fouling. The shape and nature of the biofilm layer generated on the membrane shell as a result of the accumulation of rejected microorganisms, biomass, as well as particulate matter in the feed wastewater by the membrane are primarily responsible for this stable flux. Not only does GDM provide a consistent flux, but it has also been demonstrated to increase permeate quality during filtration. This property along with performance of the biofilm layer are significant in GDM filtration systems for two reasons:

- The biofilm is bioactive, allowing organic compounds to be biodegraded by a bioactive layer, which increases removal.
- The biofilm layer likewise serves as a derived/secondary membrane that helps to increase the rejection and removal efficiency by trapping additional organic along with inorganic substances with its porous structure.

5.6.2 EFFICIENCY AND TARGET POLLUTANTS

GDM systems are energy-efficient processes because of their gravity-driven nature. When comparing GDM to other techniques, the reduced energy usage or complete lack of external energy has been frequently noted as a benefit. Certainly, in most residential water purification systems, the scheme is handled manually, requiring no external energy. In large-scale gravity-driven drinking water production systems, however, a pumping system becomes necessary to carry the influent water up to the system's intake (Barambu et al., 2021). The existence of biofilms in the GDM process can help reduce a variety of pollutants, including polysaccharides, humic acids, proteins, algal toxins, assimilable organic carbon, biopolymers, and microcystins (Pronk et al., 2019).

5.7 OSMOTIC DRIVEN/FORWARD OSMOSIS (FO) MEMBRANES

5.7.1 WORKING PRINCIPLE

FO is an osmotic-pressure-driven membrane process that utilizes the principle of osmosis to induce water transport. In the FO system, feed flows on one side/end of the membrane, while a draw solution with greater total dissolved solids (TDS) or salinity flows on the opposite side of the membrane. Osmotic pressure is created by the difference in TDS between the two sides, which causes water to flow from the feed mixture through the membrane and into the draw solution while keeping the pollutants in the feed stream. The water penetrates through a semi-permeable barrier/membrane, but solute molecules are rejected. The draw solution dilutes as the water passes over the membrane, while the feed solution concentrates, resulting in a concentrated effluent (Figure 5.7). The entire process can be carried out without the use of any additional hydraulic pressure. A simple salt/water solution or a material specifically designed for some application like fertigation can be used as the draw solution (Nagy, 2019). A successive separation unit is required to re-concentrate the draw solute for reuse and to produce clean water for probable reuse/reclaim in different applications (Jafarinejad, 2021).

5.7.2 EFFICIENCY AND TARGET POLLUTANTS

The FO system has received a lot of attention in recent years for water treatment, particularly for heavy metal ion removal, because of its low polluting propensity, low energy cost, and environmental

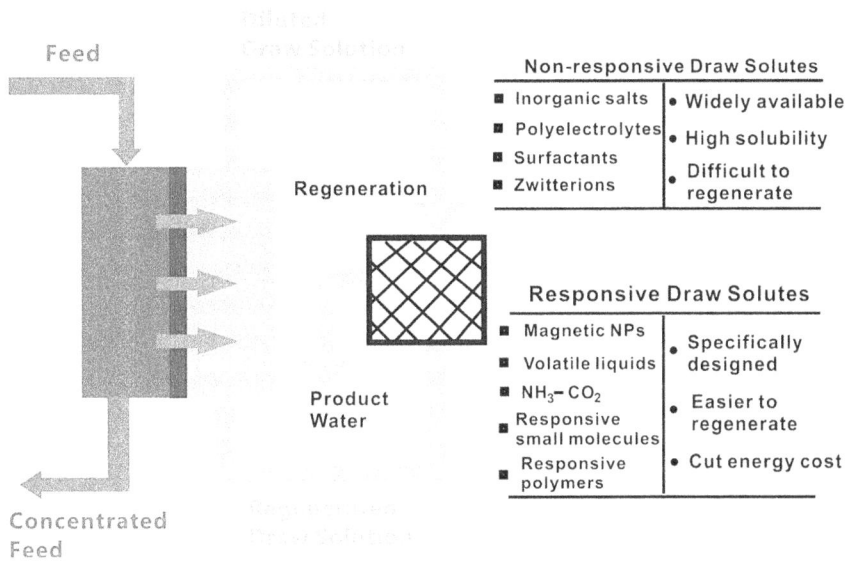

FIGURE 5.7 FO process (Cai and Hu, 2016).

friendliness. Recent studies have proven that FO membranes are very efficient for the removal of heavy metals. FO membranes modified with zeolite have high heavy metals rejection, i.e., 96%, specifically for copper, nickel, and lead (Qiu and He, 2019).

FO has proven to be a very effective treatment option for municipal wastewater treatment. In such wastewater and metropolitan runoff treatment, chemical oxygen demand (COD) rejection close to 99% is achieved. The retention of trace metals like (Cu, Pb, Mn, Cr, Zn, Ni) has also been reported to be 98%–100% (Linares et al., 2013).

5.8 METAL ORGANIC FRAMEWORKS

5.8.1 WORKING PRINCIPLE

Metal organic frameworks (MOFs) are a modern type of organic-inorganic composite type material, gaining popularity due to their advantages, which include substantial specific surface area, a regular and customizable pore structure, and excellent thermal and chemical durability (Deng et al., 2021). These are crystalline porous organic-inorganic hybrid materials composed of a regular array of positively charged metal ions bordered by organic "linker" molecules. The metal ions form nodes that link or connect the linkers' arms to form a recurring, cage-like structure. MOFs thus have an extremely large internal surface area due to their hollow structure (Jia et al., 2022). The substrates normally used to make MOF membranes are usually inorganic materials and polymers as stated in Table 5.5. Thus, MOF membranes are categorized into two major categories: pure MOF membranes (comprising MOF layers and a porous inorganic substrate) and mixed matrix membranes (MMMs), which are created by mixing MOF particles together with an organic matrix (Safaei et al., 2019). In these types of membranes, MOF particles are applied as the scattered phase, and the organic matrix is employed as the continuous phase to guarantee the continuity of the membranes (Deng et al., 2021). Table 5.5 shows common substrates used for manufacturing MOF membranes.

The membrane has a quite small pore size and hence separation is centered on the sieving mechanism of the membrane pores, given that the pore size of the membrane is smaller than the pollutant size, the particle can be trapped (Safaei et al., 2019).

TABLE 5.5

Substrates used for preparation of MOF membranes (Deng et al., 2021)

Substrate type	Substrate material type
Inorganic substrate	ZnO, SiO$_2$, Al$_2$O$_3$, TiO$_2$, graphene oxide
Polymer substrate	PVDF, PAN, cellulose acetate, PVA, PDMS, nylon
Metal-mesh substrate	Copper mesh, mesh of stainless steel, nickel mesh

5.8.2 Efficiency and Target Pollutants

MOF membranes are widely used in gas separation processes, pervaporation and the removal of heavy metals in wastewater and organic wastewaters like dyeing wastewater, oily wastewater (Jia et al., 2022). At present, these membranes are largely used in wastewater remediation (He et al., 2017). The pore size of these membranes can be as tiny as 0.2 nm or as huge as 10 nm, which allows MOF membranes to be utilized both in nanofiltration processes and for desalination purposes such as RO membranes. Even for saline water with a concentration of around 10% by weight, the desalination rate of these membranes is around 99.8% (Cavka et al., 2008).

5.9 THERMALLY DRIVEN/MEMBRANE DISTILLATION

5.9.1 Working Principle

Membrane distillation (MD) is a potential technique for treating saline water and wastewater with high rejection coefficients that traditional technologies cannot produce. In MD, only vapor molecules pass through a microporous hydrophobic membrane, which is a thermally induced or driven separation process. The vapor pressure difference caused by the temperature difference across the membrane surface is the driving force in the MD process (Alkhudhiri and Hilal, 2018).

MD is a separation method in which two aqueous solutions at different temperatures are separated by a microporous hydrophobic membrane. The hydrophobicity of these membranes limits liquid mass transfer, resulting in the formation of a liquid-gas interface. The temperature difference/gradient on the membrane causes a difference in vapor pressure, causing volatile components in the source mix to evaporate via pores and be transported to the compartment with low vapor pressure via diffusion and/or convection in the compartment with high vapor pressure, where they are condensed in the cold liquid/vapor phase. Water vapor is transferred over the membrane for feed waters that solely include non-volatile components, such as salts, and demineralized water can be produced on the distillation side and an additional concentrated saline flow on the supply or feed side (Martínez et al., 2003). The particular module design determines how the vapor pressure differential is formed across the membrane. The most typical arrangement, direct contact membrane distillation (DCMD), has a condensation liquid (generally pure water) in direct connection with the membrane on the permeate side. Alternatively, the evaporated solvent can be collected on a condensation surface separated from the membrane by an air gap (AGMD) or a vacuum (VMD), or discharged using a cold, inert sweep gas (SGMD). Condensation of vapor molecules occurs outside the membrane module in the latter two circumstances. The sort of driving force, along with the membrane's water-repellent characteristic, theoretically allows for complete rejection of non-volatile components such as ions, colloids, and macro molecules (Mahmoudi and Akbarzadeh, 2018). Figure 5.8 shows the basic module configurations for different types of MD.

5.9.2 Efficiency and Target Pollutants

MD offers a number of appealing properties, including lower operating temperatures than traditional processes and the fact that the solution (i.e., feed water) does not have to be heated to boiling

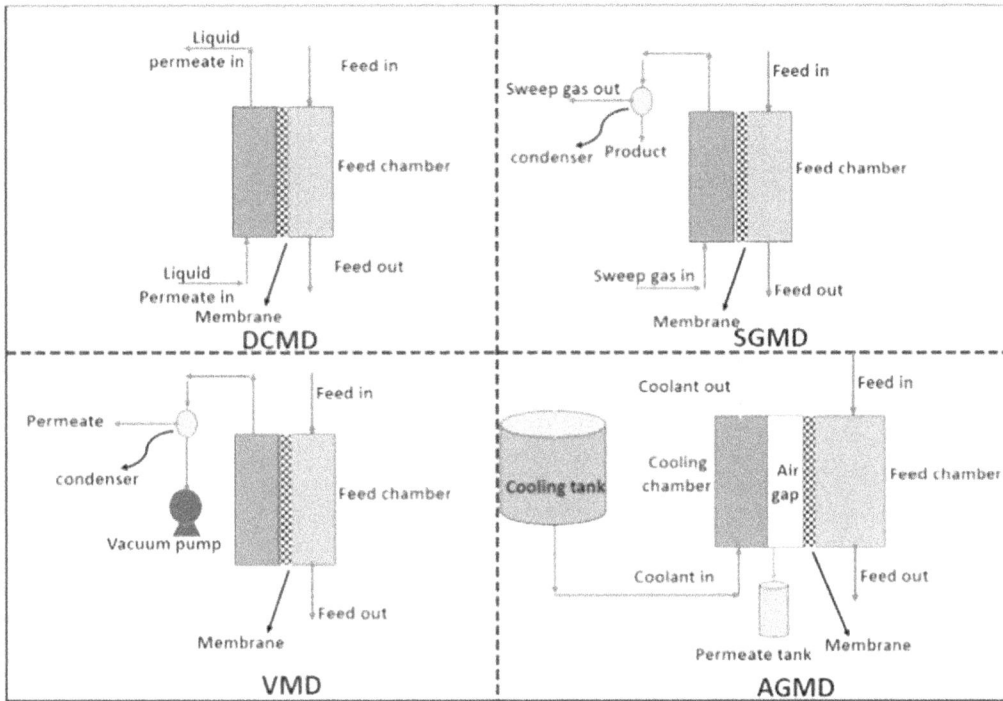

FIGURE 5.8 Membrane distillation configurations (Aghaeinejad-Meybodi and Ghasemzadeh, 2017).

point. Furthermore, in MD, the hydrostatic pressure is lesser than in pressure-driven membrane systems like RO. As a result, MD can be a cost-effective approach that also requires fewer membrane properties. In this regard, less costly materials such as plastic, for example, can be employed to ease corrosion issues (Alkhudhiri and Hilal, 2018). Furthermore, the membrane pore size essential for MD is higher than that necessary for other membrane methods like RO. As a result, the MD process is less prone to blocking/fouling. The MD system can be combined with other separation processes, such as UF or a RO unit, to create an integrated separation system. MD also has the capacity to use alternative energy sources such as solar power (Criscuoli and Drioli, 1999).

5.10 ELECTRICALLY DRIVEN MEMBRANES/ELECTRODIALYSIS

5.10.1 WORKING PRINCIPLE

Electrodialysis (ED) is a membrane process for desalinating saltwater or brackish water. ED is based on the use of an electric current to move dissolved salts across a stack of cationic and anionic membranes, resulting in a diluted stream. It may also be used to measure charged species concentrations in aqueous solutions (Pourcelly, 2000). The operating principle of an ED system is shown in Figure 5.9.

A cation and an anion exchange membrane are alternatively positioned between two electrodes in a traditional ED stack. Under the influence of an electrical potential difference, ions in a concentrate stream are enriched, whereas ions in a dilute stream are depleted (Ghyselbrecht et al., 2013).

5.10.2 EFFICIENCY AND TARGET POLLUTANTS

The desalination of saline and brackish water and seawater has traditionally been the primary use of conventional ED. For the generation of potable water from brackish water sources, ED and RO are currently competing processes (Strathmann, 2010). When TDS feed concentrations are less

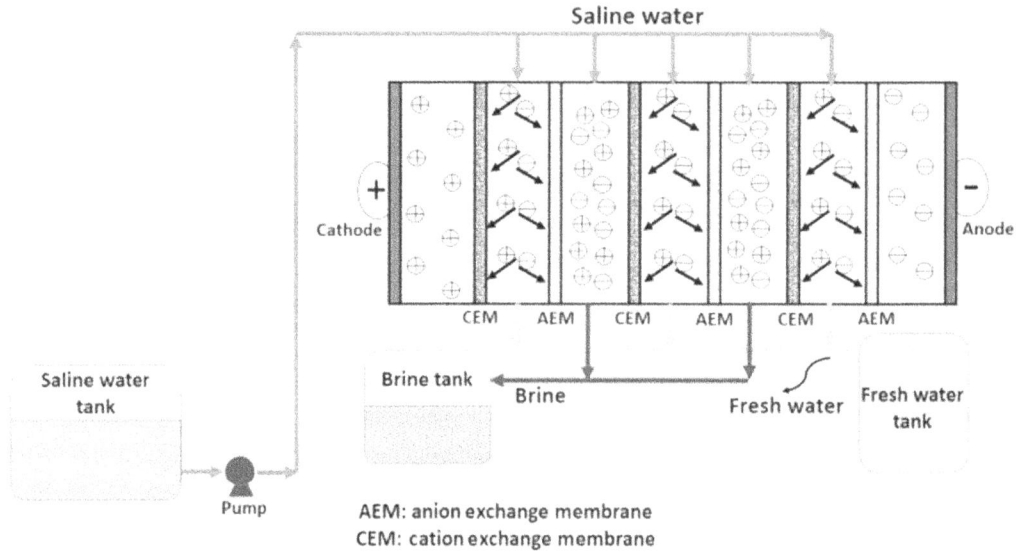

FIGURE 5.9 Operation principle of electrodialysis (Aghaeinejad-Meybodi and Ghasemzadeh, 2017).

than 3,000 ppm or when substantial feed recoveries are required, electrodialysis is the more cost-effective option. However, all particles larger than 10 μm, as well as hardness, colloidal matter, large organic anions, iron, and manganese oxides, must be removed during the pretreatment process (Moran, 2018).

5.11 MEMBRANE BIOREACTORS

5.11.1 Working Principle

MBRs are a widely used wastewater treatment method. They are a mix of a biological process like suspended growth process and physical liquid-solid separation like MF and UF. The breakdown of biomass takes place inside the bioreactor tank, while the separation of treated wastewater from microorganisms is completed in a membrane module, which therefore excludes the requirement for a secondary clarification unit (Iorhemen et al., 2016). In comparison to traditional systems, a MBRs can treat substantially greater sludge concentrations while using significantly less reactor volume. The membrane can be positioned either outside the biological basin (side stream) or inside the basin (submerged). Continuous cross-flow circulation along the membranes is required for side-stream systems. This is accomplished using both tubular and flat sheet membranes. However, in submerged systems, effluent is extracted under pressure. Hollow fiber or flat sheet membranes are commonly used in this configuration (Al-Asheh et al., 2021). The two basic configurations of MBRs are depicted in Figure 5.10.

5.11.2 Efficiency and Target Pollutants

MBRs have long been seen as a potential solution in a variety of different fields, including biofuel generation, food processing, medicines, and a variety of other small-scale applications. Fouling at membrane surfaces, on the other hand, remains the most significant barrier to the full commercialization of this technology. MBRs being a combination of biological and filtration systems can remove organics as well as other pollutants that fall within the pore size dispersal/distribution of the membrane used (Li et al., 2022).

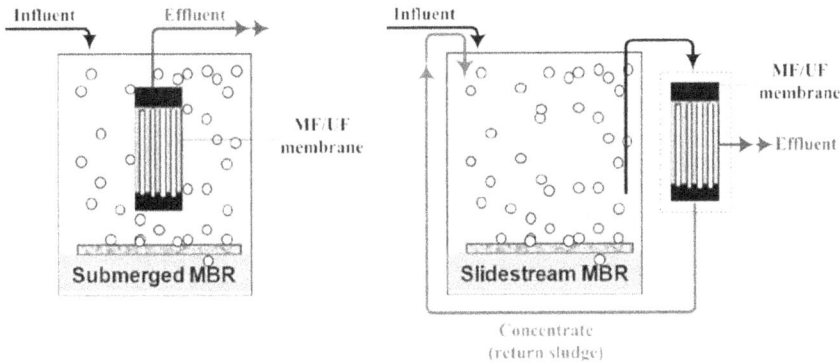

FIGURE 5.10   MBR configurations (Hai et al., 2016).

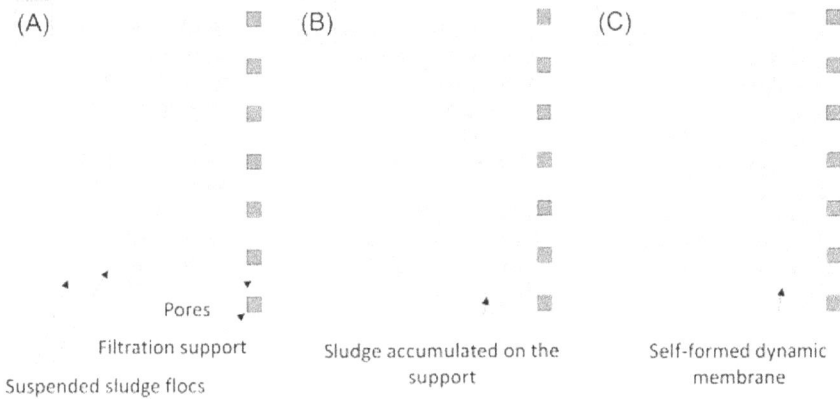

FIGURE 5.11   Dynamic membrane's self-formation: (A) startup of the process; (B) first phase of cake layer formation; (C) dynamic stability of the cake layer (Pollice and Vergine, 2020).

5.12   SELF-FORMING DYNAMIC MEMBRANE BIOREACTOR

5.12.1   Working Principle

The invention of the self-forming dynamic membrane bioreactor (SFD-MBR), in which "temporary" filtering layers are generated on a supportive surface composed of a relatively coarse surface, is a fresh attempt toward a more sustainable MBR process. The dynamic membrane in these systems is made up of activated sludge dangling in mixed liquor that tends to deposit on the supportive medium, resulting in effluent quality equivalent to that of traditional UF-based MBRs (Fan and Huang, 2002).

Filtration in the SFD-MBR begins with a mesh or similar filtering support made of inert material with pore sizes ranging from 20 to 100 microns. The filtering module can be immersed in the bioreactor or inserted side stream into specialized cartridges, depending on the arrangement. In either situation, filtration results in the deposition of activated sludge on the supporting material's surface, resulting in the creation of a biological layer (the dynamic membrane) with increased filtering capacity relative to the support (Figure 5.11). With the aim of restricting the thickness/width of the dynamic membrane and keeping a stable filtering boundary, the hydrodynamic circumstances (due to aeration, mixing, and scouring) should be correctly tuned to promote biomass separation/detachment from the outer cake layer (Pollice and Vergine, 2020).

5.12.2 Efficiency and Target Pollutants

Benefits of SFD-MBRs are primarily reduced capital costs, higher fluxes, easy maintenance, and fewer energy constraints with respect to conventional MBRs. Table 5.6 summarizes some findings of using SFD-MBRs.

5.13 GRAPHENE-BASED MEMBRANES

5.13.1 Working Principle

A distinct layer of carbon atoms forms a honeycomb web/network in graphene. The hexagonal configuration of graphene is supported by planar sp^2 hybridization of carbon atoms, each of which forms three strong bonds with three surrounding atoms. A minimal overlap among the valence bands (VBs) and conduction bands (CBs) gives graphene a practically microscopic, small band gap. This permits electrons to flow via the hexagonal structure of graphene very freely and swiftly, around 100–200 times quicker than electrons can move via silicon. Graphene can also absorb photons through the visible spectrum and beyond, transforming energy into electrical current, thanks to its practically microscopic narrow band gap (Peplow, 2013).

Due to the 2D structure and potential for large-scale manufacture, graphene equivalents, such as graphene oxide (GO), have lately received interest in the field of membranes. There are pristine zones/regions (non-oxidized) comprising carbon atoms in GO membranes that form nearly frictionless 2D nanochannels, allowing water along with other small molecules to pass between the layers (Nair et al., 2012). The presence of many oxygen-rich functional groups in GO is considered to enhance the interlayer distance, which affects the water flow and the membranes' capacity to retain pollutants. As a result, the oxidized areas are ideal for molecular transport. Furthermore, oxygenated groups are able to establish hydrogen bonds in conjunction with gases or water and electrostatic reactions/interactions with ionic components due to their hydrophilic nature, improving the selective penetration capabilities of GO (Liu et al., 2015).

TABLE 5.6
Selected studies on SFD-MBRs (Pollice and Vergine, 2020)

Configuration; mesh; pore dia (μm); flux (LMH); control/cleaning	Type of wastewater; loading rate (gCOD/L/day); SRT (days); MLSS (g/L)	Effluent turbidity (NTU); effluent TSS (mg/L)	References
Aerobic			
Flat sheet; nylon; 30, 50–150; air flush	Municipal; 0.2–1.2; N/A; 4–7	8; less than 12	Fuchs et al. (2005)
Tubular; acrylic, polyester, and nylon; n/a; 16; water jet, brush, and air flush	Municipal; 1.7–1.8; 26; 5–8	1.1–1.6; 2.1–2.9	Zahid and El-Shafai (2011)
Flat sheet type; nylon; 50; 40; water jet	Agro and industrial (canning); 1.2; 25; 3–4	0.5–25 (2.7 median); 1–50	Vergine et al. (2020)
Anaerobic			
Flat sheet type; polypropylene; 10; 2.6; biogas purging and regular backwash (every 3 min)	Synthetic ww (urea, milk, starch, others); 2; 20 and 40; 5.0–6.5	Less than 20; less than 10	Ersahin et al. (2014)
Tubular; silk; 100; 6–25; n/a	Synthetic molasses and sewage sludge; 5–17 (VFA production); n/a; 5–60 gVFA/L	n/a; less than 3 (after cleaning 3–20)	Liu et al. (2019)

5.13.2 EFFICIENCY AND TARGET POLLUTANTS

Figure 5.12 shows different types of pollutants that can be removed using graphene-based membranes.

Catalytic membranes used for water treatment are typically preferable to powdered catalysts (which require separation) and non-catalytic membranes (in which the retentate must be treated). Graphene-based catalytic membranes, despite their early stages of research, are worthy materials that commonly comprise metals/metal oxides, such as Fe_3^+, TiO_2, and Ag^+, for use in a range of advanced oxidation technologies (AOTs), including electrocatalysis, photocatalysis, and persulfate (PS)/peroxymonosulfate (PMS) activation catalysis. Blending reaction (i.e., AOTs) and separation (i.e., membrane) procedures are advantageous not only because the target pollutants are oxidized, but also because membrane fouling is reduced and membrane reusability is extended (Pedrosa et al., 2021).

5.14 THIN-FILM COMPOSITE (TFC) MEMBRANES

14.1 WORKING PRINCIPLE

A TFC membrane is a semi-permeable membrane intended to be used in water purification and desalination systems. TFC membranes are molecular sieves made up of two or more than two layered materials that are formed into a film. The three-layer structure provides strong rejection of unwanted elements (such as salts), a high filtering rate, and superior mechanical strength. The high rejection rate is due to the polyamide top layer, which was chosen for its water permeability and relative impermeability to different dissolved contaminants such as salt ions and other tiny, unfilterable molecules (Alsayed and Ashraf, 2021).

Various membranes, such as TFC membranes and thin-film nanocomposite (TFN) membranes, are manufactured for desalination and other comparable purposes (TFNs). TFNs are an interfacial polymerization-based modification of existing TFCs (IP). The alteration involves the incorporation of thin nanoparticles (NPs) hooked onto a polyamide (PA)-thick layer on top of the TFC membrane

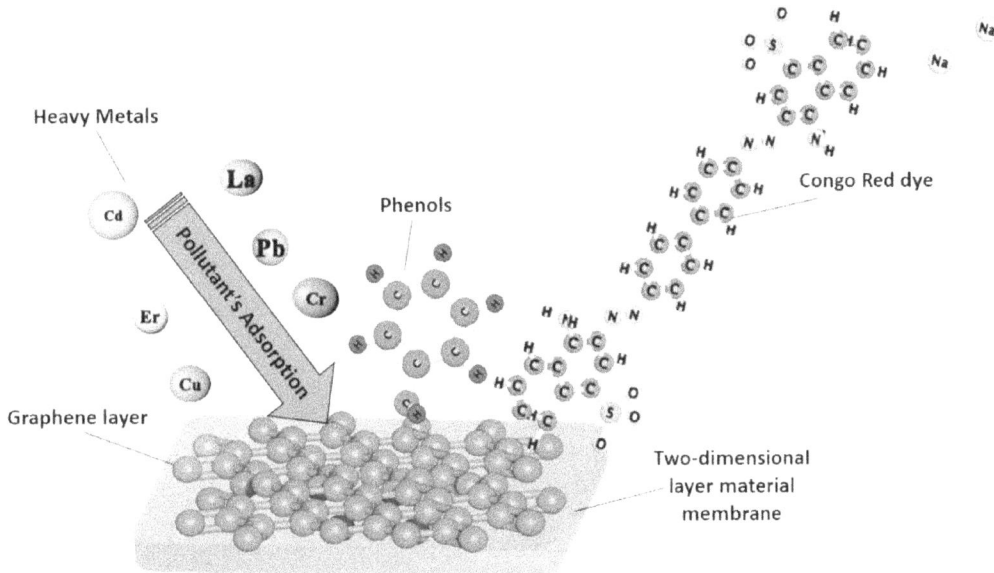

FIGURE 5.12   Adsorption by a graphene-based membrane (Khraisheh et al., 2021).

so as to improve its performance (Jeong et al., 2007). TFC membranes are the forerunners of TFN membranes. Asymmetrical membranes with a thick upper layer and a porous support built of various materials are known as composite membranes. Initially, commercially used RO membranes were made from two types of membranes: first PA and second cellulose acetate (CA) (Li et al., 2013). PA TFCs are made up of two layers: a porous substrate layer (typically comprising polysulfone) and a thin polyamide layer produced on top of it (Figure 5.13). The top layer contributes to the membrane's permeability, while the porous sub-layer offers mechanical strength and support (Lind et al., 2010). The benefit of having two layers comprising different compounds is that each layer may be produced or altered separately to improve the membrane's overall performance. TFC membranes outperform CA membranes in terms of salt rejection, water flow, and biological assault resistance, as well as being able to work over a larger pH range (1–11) and temperature range (0°C–45°C) (Nambi Krishnan et al., 2022).

14.2 Efficiency and Target Pollutants

By including NPs into the polymer matrix, membranes may be made to absorb heavy metals from water. Daraei et al. (2012) developed a technique to eradicate copper from aqueous solutions by integrating PANI/Fe_3O_4 NPs within the PES matrix. Similarly, Gohari et al. (2013) employed Fe-Mn binary oxide (FMBO) to remove arsenic from wastewaters. Xu et al. (2022) studied heavy metal removal through a TFC-nanofiltration membrane and observed 97% rejection for $MgCl_2$, 98% for Zn^{2+}, Ni^{2+}, 96%–98% for Cu^{2+}, >94% for Ca^{2+}, and >85% for Pb^{2+} with a very high efficiency for water reclamation.

5.15 MXENE-BASED MEMBRANES

5.15.1 Working Principle

MXenes (titanium carbide ($Ti_3C_2T_x$), a novel family of 2D materials) were first created in 2011 by Drexel University researchers and have become very popular nanomaterials since then due to their

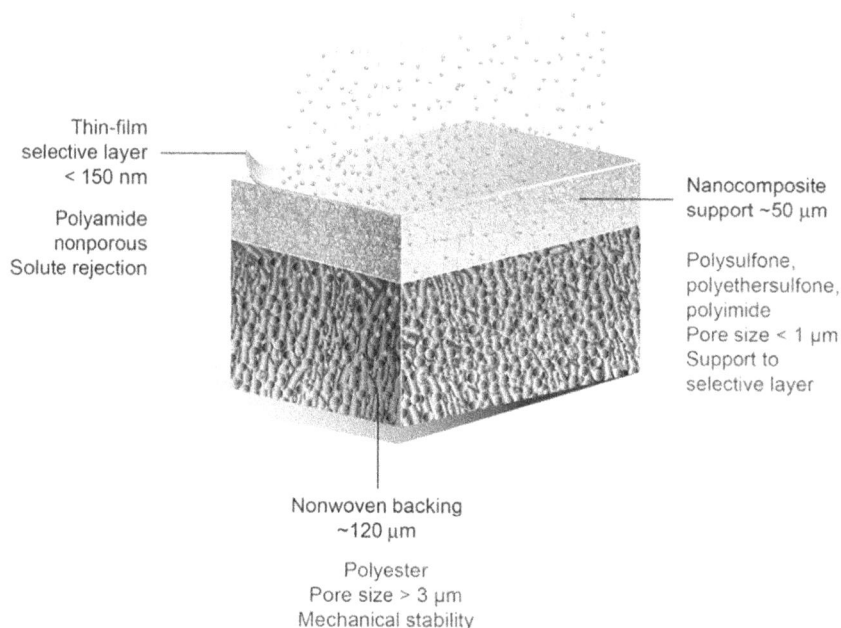

FIGURE 5.13   Schematic of a TFC membrane (Idarraga-Mora et al., 2018).

unique properties, which include excellent flexibility, large surface area, outstanding stability, excellent oxidation resistance, high electrical/thermal conductivity, and hydrophilicity (Al-Hamadani et al., 2020). MXene carries a negative charge and is hydrophilic, making it perfect for ion sieving and water filtration in aquatic environments. Moreover, the surface functions allow MXene nanosheets to be chemically fabricated to control interlayer channels and improve permeability by separating distinct molecules of varying sizes. According to the size-exclusion principle, laminated membranes may sieve bigger solutes while permitting smaller solutes to pass through the solutes if charged; however, the Donnan exclusion theory can be used to separate them. Ions with negative charges are rejected by MXene because of electrostatic contact, but ions with positive charges are enabled. As a result, manipulating/modifying the surface chemistry of MXene may be used to design transport performance and separation performance (Huang et al., 2021). Figure 5.14 shows the basic types of MXene-based membranes.

5.15.2 EFFICIENCY AND TARGET POLLUTANTS

The enormous surface area, plentiful surface terminations, electron vibrancy, and hydrophilic nature of MXenes results in making them exceptionally attractive as adsorbents for eradicating heavy metals from solutions (Othman et al., 2022). Because of their huge specific surface area, plenty of functional -O and -OH groups, and flexible surface chemistry, adsorption performance of MXenes may effectively compete with or surpass that of other nanomaterial adsorbents. The presence of numerous functional groups on the MXene exterior surface not only provides sites for heavy metal surface complexation along with ion exchange, but also acts as a reducing agent for some metals (Zou et al., 2016). Wang along with other co-workers studied the removal efficiency of heavy metal ions and found rejection of Pb^{2+}, Cu^{2+}, and Cd^{2+} up to 99.5% using hydroxylated MXene membranes (Wang et al., 2021).

16 CHALLENGES AND LIMITATIONS

Over the last decade, membrane filtering techniques have progressively acquired prominence in environmental engineering separations. The most common problems involved with using membranes

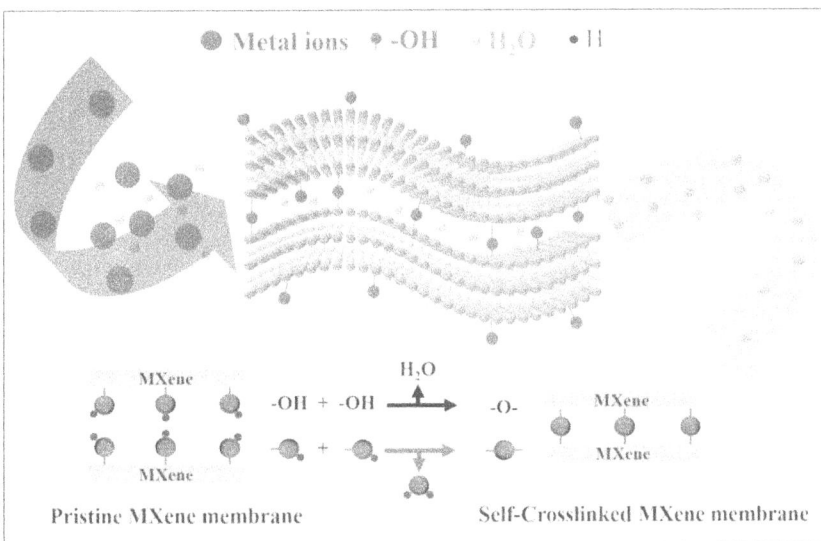

FIGURE 5.14 MXene membrane for metal removal (Lu et al., 2019).

are pretreatment requirements, fouling leading to concentration polarization, energy requirements, capital costs, specifically cost of membrane modules, maintenance costs like membrane cleaning involving back washing and chemical cleaning (Akkoyunlu et al., 2021).

Before membrane fouling can be removed or controlled, it is critical to identify membrane foulants and fouling processes. Clear fundamental knowledge of membrane fouling and how to reduce it are critical, and it may be achieved through precise process design and appropriate operating conditions. The important preventative step in minimizing membrane fouling and maintaining high membrane productivity is to maintain optimum operating conditions, an exact pretreatment procedure, appropriate cleaning solutions or processes, and appropriate membrane selection for particular membrane applications (Akkoyunlu et al., 2021). Fouling is, however, the most significant stumbling block to the prevalent application of membrane separation techniques, because it can result in high operating and maintenance expenditures, lessened productivity and permeate quality, and recurrent membrane renewal (Zularisam et al., 2010).

Another commercialization barrier is the cost of MBR processes, which varies depending on the kind of membrane and application. Membrane bioreactos have the potential to reduce manufacturing expenses by reducing the number of unit processes for products that need high-level purity, like lactic acid. Yet, the price of membrane modules, aeration strategies, extra equipment, along with energy needs must all be considered. The costs may be broken down into three categories: capital expenditures (CAPEX), operating expenditures (OPEX), and material and utility costs (Helmi and Gallucci, 2020).

Concentration polarization is a special case of fouling that has a detrimental impact on membrane-based systems' overall efficiency. Rejected solutes tend to collect on the membrane surface, where their concentration steadily rises. On the bulk side, a concentration build-up occurs. It's possible that a diffusive flow back to the bulk solution will occur resulting in reduced flux (Koseoglu et al., 2018).

17 CONCLUSIONS

Natural energy resources are insufficient to meet everyone's needs. Because of the growing population, depleting natural resources, restricted natural resources, and environmental concerns, it is necessary to concentrate on replicating, recycling, and reusing processes. With the growing economy, the discharge and reuse standards are becoming more stringent. Such standards can only be met using membrane technology for treating domestic and industrial wastewaters. The conventional membrane processes have higher energy requirements and are more susceptible to fouling; hence, novel membrane materials are now available to be utilized for dealing with these problems. Membrane-based procedures have a number of advantages that make them an appealing alternative for water purification and desalination. Despite the complexity of product water, RO, NF, FO, MD, and their combined systems may be completely utilized as novel purification options for product water. Nonetheless, further progress is needed to overcome the existing limits of these techniques. The overall membrane separation performance for wastewater treatment is determined by membrane characteristics. As a result, membrane surface modification has been intensively researched in order to improve membrane rejection, permeability, and antifouling properties. Furthermore, much of the research only looked at one component of progress. Mass transport resistance has been degraded because of membrane modification by surface coating to increase antifouling qualities, for example. It is critical to understand and overcome the limits in current membrane development in order to speed up commercialization. The laboratory procedures should be streamlined so that the whole process costs less and is highly flexible for industrial applications.

For desalination, stand-alone membrane filtration systems like as RO, FO, and MD are viable options. However, to deal with more complex process waters, more technological innovation is required. The capability to treat wastewater is improved by combining two or more membrane processes with conventional treatment techniques. The hybridization procedure enhances the overall performance while lowering energy consumption and increasing productivity.

Novel membranes that include NPs are proving to be promising techniques for overcoming the current problems faced by full-scale membrane applications. Examples of such techniques include MXene-based membranes, graphene-based membranes, thin-film composites embedded with NPs, SFD-MBRs, etc. The study should not merely focus on inventing new membrane materials along with procedures in the future, but also demonstrate the efficacy of present technologies at commercial and industrial levels.

Despite the present progress in this subject, additional in-depth research and investigations are still required to improve performance and boost technological industrialization.

ACRONYMS/ABBREVIATIONS

SDGs	Sustainable development goals
IEM	Ion exchange membrane
CEM	Cation exchange membrane
AEM	Anion exchange membrane
MBR	Membrane bioreactor
MF	Microfiltration
RO	Reverse osmosis
UF	Ultrafiltration
NF	Nanofiltration
FO	Forward osmosis
ED	Electrodialysis
MD	Membrane distillation
AGMD	Air gap membrane distillation
DCMD	Direct contact membrane distillation
TMP	Trans-membrane pressure
AOT	Advanced oxidation technologies
GO	Graphene oxide
NPs	Nanoparticles
SFD-MBR	Self-forming dynamic membrane bioreactor
MOF	Metal organic framework

REFERENCES

Abdullah, N., Yusof, N., Lau, W. J., Jaafar, J. & Ismail, A. F. 2019. Recent trends of heavy metal removal from water/wastewater by membrane technologies. *Journal of Industrial and Engineering Chemistry*, 76, 17–38.

Adam, M. R., Othman, M. H. D., Kurniawan, T. A., Puteh, M. H., Ismail, A. F., Khongnakorn, W., Rahman, M. A. & Jaafar, J. 2022. Advances in adsorptive membrane technology for water treatment and resource recovery applications: A critical review. *Journal of Environmental Chemical Engineering*, 10, 107633.

Aghaeinejad-Meybodi, A. & Ghasemzadeh, K. 2017. Chapter 8 - Silica membrane application for desalination process. In: Basile, A. & Ghasemzadeh, K. (eds.) *Current Trends and Future Developments on (Bio-) Membranes* (pp. 181–216). Elsevier.

Ahmad, N. A., Goh, P., Abdul Karim, Z. & Ismail, A. 2018. Thin film composite membrane for oily waste water treatment: Recent advances and challenges. *Membranes*, 8, 86.

Akkoyunlu, B., Daly, S. & Casey, E. 2021. Membrane bioreactors for the production of value-added products: Recent developments, challenges and perspectives. *Bioresource Technology*, 341, 125793.

Al-Asheh, S., Bagheri, M. & Aidan, A. 2021. Membrane bioreactor for wastewater treatment: A review. *Case Studies in Chemical and Environmental Engineering*, 4, 100109.

Al-Hamadani, Y. A. J., Jun, B.-M., Yoon, M., Taheri-Qazvini, N., Snyder, S. A., Jang, M., Heo, J. & Yoon, Y. 2020. Applications of MXene-based membranes in water purification: A review. *Chemosphere*, 254, 126821.

Alhathal Alanezi, A., Sharif, A., Sanduk, M. & Khan, A. 2012. Experimental investigation of heat and mass transfer in tubular membrane distillation module for desalination. *International Scholarly Research Network Chemical Engineering*, 2012, 8.

Alkhudhiri, A. & Hilal, N. 2018. 3 - Membrane distillation—Principles, applications, configurations, design, and implementation. In: Gude, V. G. (ed.) *Emerging Technologies for Sustainable Desalination Handbook* (pp. 55–106). Butterworth-Heinemann.

Alsayed, A. F. M. & Ashraf, M. A. 2021. 2 - Modified nanofiltration membrane treatment of saline water. In: Samui, P., Bonakdari, H. & Deo, R. (eds.) *Water Engineering Modeling and Mathematic Tools* (pp. 25–44). Elsevier.

Balster, J. 2015. Capillary membrane module. In: Drioli, E. & Giorno, L. (eds.) *Encyclopedia of Membranes* (pp. 1–2). Springer Berlin Heidelberg.

Barambu, N. U., Marbelia, L., Bilad, M. R. & Arahman, N. 2021. 9 - Gravity-driven membrane filtration for decentralized water and wastewater treatment. In: Samui, P., Bonakdari, H. & Deo, R. (eds.) *Water Engineering Modeling and Mathematic Tools* (pp. 177–185). Elsevier.

Berdasco, M., Coronas, A. & Vallès, M. 2017. Theoretical and experimental study of the ammonia/water absorption process using a flat sheet membrane module. *Applied Thermal Engineering*, 124, 477–485.

Cai, Y. & Hu, X. M. 2016. A critical review on draw solutes development for forward osmosis. *Desalination*, 391, 16–29.

Cavka, J. H., Jakobsen, S., Olsbye, U., Guillou, N., Lamberti, C., Bordiga, S. & Lillerud, K. P. 2008. A new zirconium inorganic building brick forming metal organic frameworks with exceptional stability. *Journal of the American Chemical Society*, 130, 13850–13851.

Criscuoli, A. & Drioli, E. J. D. 1999. Energetic and exergetic analysis of an integrated membrane desalination system. *Desalination*, 124, 243–249.

Crittenden, J. C., Rhodes Trussell, R., Hand, D. W., Howe, K. J. & Tchobanoglous, G. 2005. *Water Treatment: Principles and Design / MWH*, revised by John C. Crittenden et al. (pp. 1–1868). John Wiley & Sons.

Cui, Z. F., Jiang, Y. & Field, R. W. 2010. Chapter 1 - Fundamentals of pressure-driven membrane separation processes. In: Cui, Z. F. & Muralidhara, H. S. (eds.) *Membrane Technology* (pp. 1–18). Butterworth-Heinemann.

Czarny, J., Präbst, A., Spinnler, M., Biek, K. & Sattelmayer, T. 2017. Development and simulation of decentralised water and energy supply concepts – Case study of rainwater harvesting at the Angkor Centre for conservation of biodiversity in Cambodia. *Journal of Sustainable Development of Energy, Water and Environment Systems*, 5, 626–644.

Daraei, P., Madaeni, S. S., Ghaemi, N., Salehi, E., Khadivi, M. A., Moradian, R. & Astinchap, B. J. J. O. M. S. 2012. Novel polyethersulfone nanocomposite membrane prepared by PANI/Fe_3O_4 nanoparticles with enhanced performance for Cu (II) removal from water. *Journal of Membrane Science*, 415, 250–259.

Deng, Y., Wu, Y., Chen, G., Zheng, X., Dai, M. & Peng, C. 2021. Metal-organic framework membranes: Recent development in the synthesis strategies and their application in oil-water separation. *Chemical Engineering Journal*, 405, 127004.

Diallo, M. S. 2014. Chapter 15 - Water treatment by dendrimer-enhanced filtration: Principles and applications. In: Street, A., Sustich, R., Duncan, J. & Savage, N. (eds.) *Nanotechnology Applications for Clean Water (Second Edition)* (pp. 143–155). William Andrew Publishing.

Dittrich, J., Gnirss, R., Peter-Fröhlich, A. & Sarfert, F. 1996. Microfiltration of municipal wastewater for disinfection and advanced phosphorus removal. *Water Science and Technology*, 34, 125–131.

Ellouze, E., Tahri, N. & Amar, R. B. 2012. Enhancement of textile wastewater treatment process using nanofiltration. *Desalination*, 286, 16–23.

Ersahin, M. E., Ozgun, H., Tao, Y. & Van Lier, J. B. J. W. R. 2014. Applicability of dynamic membrane technology in anaerobic membrane bioreactors. 48, 420–429.

Fan, B. & Huang, X. 2002. Characteristics of a self-forming dynamic membrane coupled with a bioreactor for municipal wastewater treatment. *Environmental Science & Technology*, 36, 5245–5251.

Fuchs, W., Resch, C., Kernstock, M., Mayer, M., Schoeberl, P. & Braun, R. J. W. R. 2005. Influence of operational conditions on the performance of a mesh filter activated sludge process. *Water Research*, 39, 803–810.

Gautam, R., Chattopadhyaya, M. & Sharma, S. 2012. Biosorption of heavy metals: Recent trends and challenges. *Wastewater Reuse and Management* 2012, 305–322.

Ghyselbrecht, K., Huygebaert, M., Van Der Bruggen, B., Ballet, R., Meesschaert, B. & Pinoy, L. J. D. 2013. Desalination of an industrial saline water with conventional and bipolar membrane electrodialysis. *Desalination*, 318, 9–18.

Gohari, R. J., Lau, W., Matsuura, T., Ismail, A. J. S. & Technology, P. 2013. Fabrication and characterization of novel PES/Fe–Mn binary oxide UF mixed matrix membrane for adsorptive removal of As (III) from contaminated water solution. *Separation and Purification Technology*, 118, 64–72.

Hai, F., Alturki, A., Nguyen, L., Price, W. & Nghiem, L. 2016. Removal of trace organic contaminants by integrated membrane processes for water reuse applications. 2016, 533.

He, Y., Tang, Y. P., Ma, D. & Chung, T.-S. J. J. O. M. S. 2017. UiO-66 incorporated thin-film nano-composite membranes for efficient selenium and arsenic removal. *Journal of Membrane Science*, 541, 262–270.

Helmi, A. & Gallucci, F. J. P. 2020. Latest developments in membrane (bio) reactors. *Processes*, 8, 1239.

Huang, L., Ding, L. & Wang, H. 2021. MXene-based membranes for separation applications. *Small Science*, 1, 2100013.

Idarraga-Mora, J., Childress, A., Friedel, P., Ladner, D., Rao, A. & Husson, S. 2018. Role of nanocomposite support stiffness on TFC membrane water permeance. *Membranes*, 8, 111.

Iorhemen, O. T., Hamza, R. A. & Tay, J. H. 2016. Membrane bioreactor (MBR) technology for wastewater treatment and reclamation: Membrane fouling. *Membranes*, 6, 33.

Ismail, A. F., Khulbe, K. C. & Matsuura, T. 2019. Chapter 5 - RO membrane module. In: Ismail, A. F., Khulbe, K. C. & Matsuura, T. (eds.) *Reverse Osmosis* (pp. 117–141). Elsevier.

Jabłońska, B. & Siedlecka, E. 2015. Removing heavy metals from wastewaters with use of shales accompanying the coal beds. *Journal of Environmental Management*, 155, 58–66.

Jafarinejad, S. 2021. Forward osmosis membrane technology for nutrient removal/recovery from wastewater: Recent advances, proposed designs, and future directions. *Chemosphere*, 263, 128116.

Jeong, B.-H., Hoek, E. M., Yan, Y., Subramani, A., Huang, X., Hurwitz, G., Ghosh, A. K. & Jawor, A. J. J. O. M. S. 2007. Interfacial polymerization of thin film nanocomposites: A new concept for reverse osmosis membranes. *Journal of Membrane Science*, 294, 1–7.

Jia, S., Ji, D., Wang, L., Qin, X. & Ramakrishna, S. 2022. Metal–organic framework membranes: Advances, fabrication, and applications. *Small Structures*, 3, 2100222.

Khraisheh, M., Elhenawy, S., Almomani, F., Al-Ghouti, M., Hassan, M. K. & Hameed, B. H. 2021. Recent progress on nanomaterial-based membranes for water treatment. *Membranes*, 11, 995.

Koseoglu, H., Guler, E., Harman, B. I. & Gonulsuz, E. 2018. Chapter 2 - Water flux and reverse salt flux. In: Sarp, S. & Hilal, N. (eds.) *Membrane-Based Salinity Gradient Processes for Water Treatment and Power Generation* (pp. 57–86). Elsevier.

Landaburu-Aguirre, J., García, V., Pongrácz, E. & Keiski, R. 2006. Applicability of membrane technologies for the removal of heavy metals. *Desalination*, 200, 272–273.

Li, D., Wang, H. J. F. N. M. & Treatment, M. F. W. 2013. Thin film nanocomposite membranes for water desalination. *Journal of Industrial and Engineering Chemistry*, 116, 163–194.

Li, R., Kadrispahic, H., Koustrup Jørgensen, M., Brøndum Berg, S., Thornberg, D., Mielczarek, A. T. & Bester, K. 2022. Removal of micropollutants in a ceramic membrane bioreactor for the post-treatment of municipal wastewater. *Chemical Engineering Journal*, 427, 131458.

Linares, R. V., Li, Z., Abu-Ghdaib, M., Wei, C.-H., Amy, G. & Vrouwenvelder, J. S. J. J. O. M. S. 2013. Water harvesting from municipal wastewater via osmotic gradient: An evaluation of process performance. *Journal of Membrane Science*, 447, 50–56.

Lind, M. L., Eumine Suk, D., Nguyen, T.-V., Hoek, E. M. J. E. S. & Technology. 2010. Tailoring the structure of thin film nanocomposite membranes to achieve seawater RO membrane performance. *Environmental Science & Technology*, 44, 8230–8235.

Liu, G., Jin, W. & Xu, N. J. C. S. R. 2015. Graphene-based membranes. *Chemical Society Reviews*, 44, 5016–5030.

Liu, H., Wang, L., Zhang, X., Fu, B., Liu, H., Li, Y. & Lu, X. J. J. O. H. M. 2019. A viable approach for commercial VFAs production from sludge: Liquid fermentation in anaerobic dynamic membrane reactor. *Journal of Hazardous Materials*, 365, 912–920.

Lu, Z., Wei, Y., Deng, J., Ding, L., Li, Z.-K. & Wang, H. 2019. Self-crosslinked MXene ($Ti_3C_2T_x$) membranes with good antiswelling property for monovalent metal ion exclusion. *ACS Nano*, 13, 10535–10544.

Mahmoudi, F. & Akbarzadeh, A. 2018. 5 - Sustainable desalination by permeate gap membrane distillation technology. In: Gude, V. G. (ed.) *Emerging Technologies for Sustainable Desalination Handbook* (pp. 157–204). Butterworth-Heinemann.

Martínez, L., Florido-Díaz, F., Hernández, A., Prádanos, P. J. S. & Technology, P. 2003. Estimation of vapor transfer coefficient of hydrophobic porous membranes for applications in membrane distillation. *Separation and Purification Technology*, 33, 45–55.

Mohammadi, T. & Esmaeelifar, A. 2004. Wastewater treatment using ultrafiltration at a vegetable oil factory. *Desalination*, 166, 329–337.

Moran, S. 2018. Chapter 7 - Clean water unit operation design: Physical processes. In: Moran, S. (ed.) *An Applied Guide to Water and Effluent Treatment Plant Design* (pp. 69–100). Butterworth-Heinemann.

Nagy, E. 2019. Chapter 17 - Forward osmosis. In: Nagy, E. (ed.) *Basic Equations of Mass Transport through a Membrane Layer (Second Edition)* (pp. 1–536). Elsevier.

Nair, R., Wu, H., Jayaram, P., Grigorieva, I. & Geim, A. J. S. 2012. Unimpeded permeation of water through helium-leak–tight graphene-based membranes. *Science*, 335, 442–444.

Nambi Krishnan, J., Venkatachalam, K. R., Ghosh, O., Jhaveri, K., Palakodeti, A. & Nair, N. 2022. Review of thin film nanocomposite membranes and their applications in desalination. *Frontiers in Chemistry*, 10, 781372.

Obotey Ezugbe, E. & Rathilal, S. 2020. Membrane technologies in wastewater treatment: A review. *Membranes*, 10, 89.

Othman, Z., Mackey, H. R. & Mahmoud, K. A. 2022. A critical overview of MXenes adsorption behavior toward heavy metals. *Chemosphere*, 295, 133849.

Pal, P. 2015. Chapter 4 - Arsenic removal by membrane filtration. In: Pal, P. (ed.) *Groundwater Arsenic Remediation* (pp. 105–177). Butterworth-Heinemann.

Pedrosa, M., Figueiredo, J. L. & Silva, A. M. T. 2021. Graphene-based catalytic membranes for water treatment – A review. *Journal of Environmental Chemical Engineering*, 9, 104930.

Peplow, M. J. N. 2013. Graphene: The quest for supercarbon. *Nature*, 503, 327–329.

Pervez, M. N., Mahboubi, A., Uwineza, C., Zarra, T., Belgiorno, V., Naddeo, V. & Taherzadeh, M. J. 2022. Factors influencing pressure-driven membrane-assisted volatile fatty acids recovery and purification- A review. *Science of the Total Environment*, 817, 152993.

Petrinic, I., Korenak, J., Povodnik, D. & Hélix-Nielsen, C. 2015. A feasibility study of ultrafiltration/reverse osmosis (UF/RO)-based wastewater treatment and reuse in the metal finishing industry. *Journal of Cleaner Production*, 101, 292–300.

Pollice, A. & Vergine, P. 2020. 10 - Self-forming dynamic membrane bioreactors (SFD MBR) for wastewater treatment: Principles and applications. In: Mannina, G., Pandey, A., Larroche, C., Ng, H. Y. & Ngo, H. H. (eds.) *Current Developments in Biotechnology and Bioengineering* (pp. 1–396). Elsevier.

Pourcelly, G. 2000. Electrodialysis: Ion exchange. In: Wilson, I. D. (ed.) *Encyclopedia of Separation Science* (pp. 2235–2244). Academic Press.

Pronk, W., Ding, A., Morgenroth, E., Derlon, N., Desmond, P., Burkhardt, M., Wu, B. & Fane, A. G. 2019. Gravity-driven membrane filtration for water and wastewater treatment: A review. *Water Research*, 149, 553–565.

Qiu, M. & He, C. 2019. Efficient removal of heavy metal ions by forward osmosis membrane with a poly-dopamine modified zeolitic imidazolate framework incorporated selective layer. *Journal of Hazardous Materials*, 367, 339–347.

Ran, J., Wu, L., He, Y., Yang, Z., Wang, Y., Jiang, C., Ge, L., Bakangura, E. & Xu, T. 2017. Ion exchange membranes: New developments and applications. *Journal of Membrane Science*, 522, 267–291.

Rautenbach, R., Linn, T. & Eilers, L. 2000. Treatment of severely contaminated waste water by a combination of RO, high-pressure RO and NF — Potential and limits of the process. *Journal of Membrane Science*, 174, 231–241.

Regis, A. O., Vanneste, J., Acker, S., Martínez, G., Ticona, J., García, V., Alejo, F. D., Zea, J., Krahenbuhl, R., Vanzin, G. & Sharp, J. O. 2022. Pressure-driven membrane processes for boron and arsenic removal: pH and synergistic effects. *Desalination*, 522, 115441.

Rodriguez-Mozaz, S., Ricart, M., Köck-Schulmeyer, M., Guasch, H., Bonnineau, C., Proia, L., De Alda, M. L., Sabater, S. & Barceló, D. 2015. Pharmaceuticals and pesticides in reclaimed water: Efficiency assessment of a microfiltration–reverse osmosis (MF–RO) pilot plant. *Journal of Hazardous Materials*, 282, 165–173.

Safaei, M., Foroughi, M. M., Ebrahimpoor, N., Jahani, S., Omidi, A. & Khatami, M. J. T. T. I. A. C. 2019. A review on metal-organic frameworks: Synthesis and applications. *TrAC Trends in Analytical Chemistry*, 118, 401–425.

Salahi, A., Badrnezhad, R., Abbasi, M., Mohammadi, T. & Rekabdar, F. 2011. Oily wastewater treatment using a hybrid UF/RO system. *Desalination and Water Treatment*, 28, 75–82.

Scott, K. 1995. Introduction to Membrane Separations. In: Scott, K. (ed.) *Handbook of Industrial Membranes* (pp. 1–895). Elsevier Science.

Shaheen, A., Albadi, S., Zhuman, B., Taher, H., Banat, F. & Almarzooqi, F. 2022. Photothermal air gap membrane distillation for the removal of heavy metal ions from wastewater. *Chemical Engineering Journal*, 431, 133909.

Sridhar, S., Smitha, B. & Aminabhavi, T. 2007. Separation of carbon dioxide from natural gas mixtures through polymeric membranes - A review. *Separation and Purification Reviews*, 36, 113–174.

Strathmann, H. J. D. 2010. Electrodialysis, a mature technology with a multitude of new applications. *Desalination*, 264, 268–288.

Strathmann, H., Grabowski, A. & Eigenberger, G. 2013. Ion-exchange membranes in the chemical process industry. *Industrial & Engineering Chemistry Research*, 52, 10364–10379.

Sun, X., Wang, C., Li, Y., Wang, W. & Wei, J. 2015. Treatment of phenolic wastewater by combined UF and NF/RO processes. *Desalination*, 355, 68–74.

Swanckaert, B., Geltmeyer, J., Rabaey, K., De Buysser, K., Bonin, L. & De Clerck, K. 2022. A review on ion-exchange nanofiber membranes: Properties, structure and application in electrochemical (waste)water treatment. *Separation and Purification Technology*, 287, 120529.

Van Der Bruggen, B., Vandecasteele, C., Van Gestel, T., Doyen, W. & Leysen, R. 2003. A review of pressure-driven membrane processes in wastewater treatment and drinking water production. *Environmental Progress*, 22, 46–56.

Vergine, P., Salerno, C., Berardi, G., Pappagallo, G. & Pollice, A. J. N. B. 2020. The self-forming dynamic membrane bioreactor (SFD MBR) as a suitable technology for agro-industrial wastewater treatment. *New Biotechnology*, 56, 87–95.

Wang, Y., Chen, X., Zhang, J., Yin, J. & Wang, H. 2009. Investigation of microfiltration for treatment of emulsified oily wastewater from the processing of petroleum products. *Desalination*, 249, 1223–1227.

Wang, S., Wang, F., Jin, Y., Meng, X., Meng, B., Yang, N., Sunarso, J. & Liu, S. 2021. Removal of heavy metal cations and co-existing anions in simulated wastewater by two separated hydroxylated MXene membranes under an external voltage. *Journal of Membrane Science*, 638, 119697.

Xu, S.-J., Shen, Q., Luo, L.-H., Tong, Y.-H., Wu, Y.-Z., Xu, Z.-L. & Zhang, H.-Z. 2022. Surfactants attached thin film composite (TFC) nanofiltration (NF) membrane via intermolecular interaction for heavy metals removal. *Journal of Membrane Science*, 642, 119930.

Yordanov, D. J. B. J. O. A. S. 2010. Preliminary study of the efficiency of ultrafiltration treatment of poultry slaughterhouse wastewater. *Bulgarian Journal of Agricultural Science*, 16, 700–704.

Zahid, W. M. & El-Shafai, S. A. J. B. T. 2011. Use of cloth-media filter for membrane bioreactor treating municipal wastewater. *Bioresource Technology*, 102, 2193–2198.

Zainuddin, N. A., Doumin, J. & Puasa, S. W. 2022. Removal of heavy metal from wastewater by using acetylene gas production sludge. *Materials Today: Proceedings*, 63, S391–S399.

Zou, G., Guo, J., Peng, Q., Zhou, A., Zhang, Q. & Liu, B. J. J. O. M. C. A. 2016. Synthesis of urchin-like rutile titania carbon nanocomposites by iron-facilitated phase transformation of MXene for environmental remediation. *Journal of Materials Chemistry A*, 4, 489–499.

Zularisam, A., Ismail, A. & Sakinah, M. J. J. O. A. S. 2010. Application and challenges of membrane in surface water treatment. *Journal of Applied Sciences*, 10, 380–390.

Section 2

Membranes for heavy metals removal

6 Biopolymeric Membranes for Heavy Metal Removal from Water and Wastewater

Amna Siddique, Urwa Mahmood, Ahsan Nazir,
Hafiz Affan Abid, Tanveer Hussain, and Sharjeel Abid

6.1 INTRODUCTION

Heavy metals are usually defined as inorganic impurities with high molecular weight and density, which pose various environmental threats (Briffa, Sinagra and Blundell, 2020). These heavy metals have become one of the primary causes of environmental pollution in the modern era due to extensive industrialization. The most common heavy metals include nickel (Ni), lead (Pb), mercury (Hg), chromium (Cr), zinc (Zn), arsenic (As), cadmium (Cd), and copper (Cu). The interaction of even a tiny quantity of these heavy metals could cause severe damage to plants, animals, and humans.

6.1.1 ORIGIN OF HEAVY METALS IN DRINKING WATER

Being present naturally on the earth's crust, heavy metals can easily contaminate surface water. Sources of water contamination from heavy metals include soil erosion of metal ions, atmospheric deposition, metal corrosion and leaching of heavy metals, and metal evaporation from water resources to groundwater and soil. Other natural processes like soil erosion of metal ions, atmospheric deposition, corrosion of metals, volcanic activities, leaching of heavy metals, and the evaporation of metals from water resources to groundwater also increase the accumulation of heavy metal ions in the surroundings. However, due to industrialization, the leaching of heavy metals in the surrounding water resources is becoming a threat to the global population. A recent study on Nigerian water showed that various heavy metals, including zinc, lead, chromium, iron, magnesium, and others, were present in samples, which the authors related to industrialization (Izah, Chakrabarty and Srivastav, 2016). Heavy metals are posing a severe threat to globalization. In developing countries, exposure to heavy metals through contaminated drinking water is increasing (Chowdhury *et al.*, 2016). On the other hand, developed nations also face the same concern as heavy metal contamination in drinking water has been reported by various researchers (Dong, Zhang and Quan, 2020). The standard drinking water parameters provided by the World Health Organization (WHO) are given in Table 6.1 (Akale *et al.*, 2017).

Some researchers have reported another way of human interaction with heavy metals. The groundwater used for rice cultivation absorbs heavy metals from water, and then when humans eat this rice, it becomes a pathway for human exposure to metals (Zhang *et al.*, 2019).

6.1.2 ORIGIN OF HEAVY METALS IN WASTEWATER

Although heavy metals are found throughout the earth's crust, other sources of heavy metals include extreme environmental pollution from activities such as manufacturing processes and mining, petroleum combustion, coal burning in power plants, metal processing in refineries, nuclear

DOI: 10.1201/9781003326281-8

TABLE 6.1
Standard of drinking water (WHO)

Sr. no.	Parameter	Maximum permissible limit
1	Al	0.2 mg/L
2	Ca	75 mg/L
3	Fe	0.3 mg/L
4	Mg	50 mg/L
5	NO_3	50 mg/L
6	pH	6.5–8
7	TDS	1000 mg/L
8	*E. coli*	0

power stations, paper processing, textiles, industrial manufacturing and use of microelectronics, and agricultural and household use of metals and metal-comprising compounds. Industrial wastewater is rich in heavy metals due to different processes that lead to the release of these heavy metals. Therefore, industries are the primary source of heavy metals which go into wastewater (Hubeny *et al.*, 2021). Inadequate wastewater treatment is causing a considerable heavy metal load in the environment. Furthermore, landfills and waste dumpings also add to the heavy metal accumulation in the surroundings.

6.1.3 WHY TO REMOVE HEAVY METALS?

Heavy metal ions are known to pollute the atmosphere, soil, and water system and are toxic even in shallow concentrations. These metal ions cannot be broken down easily. These heavy metals' exposure routes to humans include consumption of contaminated food, drinking contaminated water, residency in industrial areas, and direct contact with skin (Masindi and Muedi, 2018). However, not all heavy metals are unnecessary. A few of them are essential for everyday activities; for example, Zn is one of the essential heavy metals required by the human body. Zinc deficiency could cause serious health issues and a poor immune system (Roohani *et al.*, 2013). Contrarily, lead (Pb) metal, if accumulated in the body, causes irrecoverable damage, including reproductive system failure, nervous system failure, renal system failure, premature birth, and other complications (Wani, Ara and Usmani, 2015). Thus, removing these heavy metals from drinking water and waste streams is vital.

6.2 METHODS FOR HEAVY METAL REMOVAL

Various methods reported in the literature efficiently remove the heavy metal impurities from drinking water and wastewater. The most widely utilized methods include adsorption, chemical precipitation, reverse osmosis (RO), electrochemical treatment, solvent extraction, and membrane filtration (Wołowiec *et al.*, 2019). Each of these methods has its limitations and advantages. For example, adsorption was reported to have high removal efficiency, low operation cost, easy handling, and simple treatment by restoring the adsorbed heavy metal ion. However, toxic sludge is generated that requires disposal, which is a major problem. Chemical precipitation method is known as a mature method. Hydroxide precipitation is broadly used in industries due to its simple application and is economically sound. However, its process efficiency decreases at pH greater than 11, which is one of its limitations.

The RO process is reported to have high rejection efficiency; however, fouling and damage of membranes are the major limitations of the RO process. A redox reaction (oxidation-reduction) in electrochemical treatment leads to water refining by separating metal ions. This treatment has the

Heavy metal removal using Bio-polymer

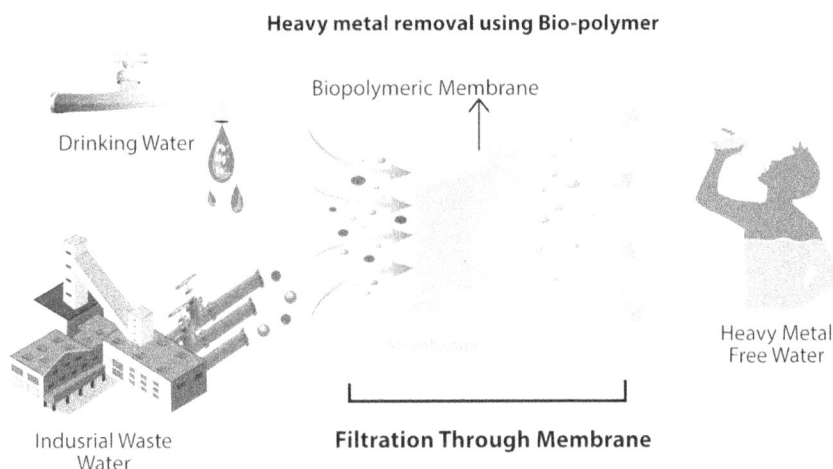

FIGURE 6.1   An illustration of heavy metal removal using membrane technology.

drawbacks of high energy utilization and the challenge of large-scale applications at lower energy consumption (Qasem, Mohammed and Lawal, 2021). However, membrane technology has proven cost-effective, efficient, and easy to use.

6.2.1   MEMBRANE TECHNOLOGY FOR HEAVY METAL REMOVAL

There are various ways to define membranes. The polymer membrane is defined as a thick film of a polymeric network where the permeate diffuses through the membranes with the help of pressure, concentration, or other potential gradients (Jose, Kappen and Alagar, 2018). There are various membrane technologies, each with their pros and cons. However, due to its cost-effectiveness, membrane technology is preferred for clean drinking water and wastewater, as shown in Figure 6.1.

6.2.1.1   RO Membranes

RO is a water distillation process in which a semipermeable membrane (pore size 0.5–1.5 nm) is used to remove unwanted and large size particles and allow the passage of only smaller molecules. This process can isolate 95%–99% of charged particles and inorganic salts. The RO process applies pressure (20–70 bar) more significant than the osmotic pressure of the feed solution to reverse the typical osmosis process. The molecular size of the solutes removed usually is in the range of 0.00025–0.003 μm. This is a compact process with high rejection efficiency. The RO process has been used to remove heavy metal ions, including Cu^{2+}, Cr^{6+}, and Ni^{2+}, from wastewater with 98.7562% removal efficiency. However, the RO system has some drawbacks, such as polluted and degraded membranes. Recently, industrial wastewaters from Costerfield mining operations have been purified using an RO separation system, with an average removal efficiency of 10%, 48%, 66%, 82%, and 95% for Fe^{3+}, Zn^{2+}, As^{3+}, Ni^{2+}, and Sb^{3+}, respectively (Qasem, Mohammed and Lawal, 2021).

6.2.1.2   Nanofiltration Membranes

As the name suggests, nanofiltration membranes are membranes having nanopores in their structure. Due to the nanopores, heavy metals cannot pass through the tiny (nano) spaces, and the efficacy of nanofiltration membranes is higher than that of microfiltration membranes. Researchers have used polyester-curcumin nanofiltration membranes to remove heavy metals from water and reported a high removal capacity of 99.88% for Fe^{2+}, 98.72% for Cu^{2+}, 99.61% for Pb^{2+}, 99.31% for Mn^{2+}, 99.11% for Zn^{2+}, and 99.51% for Ni^{2+} (Li *et al.*, 2017).

6.3 BIOPOLYMERS FOR MEMBRANE

Among various polymers used for membrane technologies, biopolymers offer unmatched advantages, including an environmentally friendly nature and an inherent ability to attract heavy metals. As the name suggests, biopolymers are polymers directly or indirectly produced by nature in plants, animals, and other living organisms. Biodegradable materials are entirely of bio-origin or are chemically synthesized using bio-origin molecules as substrates/raw materials (Hernández, Williams and Cochran, 2014; Smith, Moxon and Morris, 2016). Due to their availability in abundant quantities, nontoxicity, and biocompatibility, they are the best choice for various industrial applications (Bozell and Petersen, 2010; Mohan *et al.*, 2016).

6.4 CLASSIFICATION OF BIOPOLYMERS

There are various methods to classify biopolymers due to their abundance and properties. However, biopolymers are usually categorized into four distinct types, i.e., (i) natural biopolymers, which are derived purely from biomass such as chitin, starch, and cellulose; (ii) synthetic biopolymers, which are synthesized using microbial fermentation; (iii) synthetic biopolymers, which are chemically synthesized from microbial fermentation such as polyhydroxyalkanoates and polyesters; and (iv) synthetic biopolymers, which are chemically produced from non-renewable resources like petrochemicals such as polyethylene and nylon (Shankar and Rhim, 2018).

6.5 PROPERTIES AND APPLICATIONS

Biopolymers exhibit fascinating properties, including biodegradability, nontoxicity, biocompatibility, renewability, and abundance in nature. Furthermore, various biopolymers naturally attract or adsorb heavy metals, making them suitable for water purification systems. Therefore, due to their various properties, biopolymers have found enormous applications in different fields of life science, including food, packaging, medical, hygiene, textiles, cosmetics, pharmaceutical, water filtration, and others (Sahana and Rekha, 2018; Hameed *et al.*, 2020; Raza, Khalil and Abid, 2020; Gough *et al.*, 2021; Hussain, Ramakrishna and Abid, 2021; Raza, Anwar and Abid, 2021; Waqas *et al.*, 2022). Water filtration for removing heavy metals is becoming necessary for drinking water and wastewater due to the toxicity of these heavy metals.

6.6 THE INHERENT AFFINITY OF BIOPOLYMERS FOR HEAVY METALS

Biopolymers are complex molecules with different functional groups that can interact with heavy metals, facilitate adsorption, and even reduce them (Choque-Quispe *et al.*, 2022). For example, chitosan is a polysaccharide with a hydroxyl and an amide functional group, which makes it a perfect candidate for removing heavy metals from contaminated waters. Similarly, alginate polysaccharide biopolymers have hydroxyl and carboxyl functional groups, providing the surface with an extra negative charge and enhancing the power to interact with heavy metals (Omer, 2021).

6.7 BIOPOLYMERS TO REMOVE CONTAMINANT METAL IONS

Biopolymers have a natural affinity for heavy metals. Biosorption is the best alternative used for heavy metal removal and agriculture waste treatment. Biopolymer-based biosorbents are the focus of most researchers due to their natural abundance and cost-effective attributes. Furthermore, specific physicochemical treatments are carried out to improve the metal absorption capacity of biopolymers. Biosorption pollution removal process from water gained importance in the early 1990s. Various biopolymers are used for the said purpose, including the following.

6.7.1 CELLULOSE

Cellulose is an abundantly available natural material (Choi *et al.*, 2020). Between 10^{11} and 10^{12} tons/year is formed by photo synthesis (Rol *et al.*, 2019). The cellulose polymer has a large no. of hydroxyl groups, which act as active sites to bind metal ions (O'Connell, Birkinshaw and O'Dwyer, 2008). It contains microfibrils of nano-sized diameter. This feature makes it a potential starting material for manufacturing membranes. Cellulose-based polymeric membranes are used in various industries, including water purification. Cellulose molecular structure comprises many hydroxyl groups, enabling its chemical modifications for various end applications (Grigoray *et al.*, 2014; Lv *et al.*, 2018; Xu *et al.*, 2019).

6.7.2 CHITIN

Chitin is a biopolymer obtained from the shells of crabs and shrimps. Chitosan is obtained from chitin, which contains about 20% residual sugar. For example, N-acetyl glucosamine and chitin contain 20% of the glucosamine chitosan residuals, meaning they are both not pure. The physical structure of chitosan plays a vital role in defining its biological characteristics. Chitosan must be dissolved to activate its antibacterial characteristics and other biological actions. The chemical nature of chitosan is different from that of chitin and cellulose due to amino groups. The functional groups in the chemical structure of biopolymers allow them to attract oppositely charged metal ions due to free electrons in amino linkages and hydroxyl groups (Sánchez., Butter and Rivas, 2020).

6.7.3 CHITOSAN

Chitosan is a biopolymer having a natural adsorption tendency for heavy metals (Hudson and Smith, 1998). The use of chitosan for the cleanup of heavy metals from water is highly desirable due to its availability, low cost, ecofriendliness, and modification for multiple applications. The availability of specific binding functional groups increases the heavy metal ion removal capacity from water. Specifically, the presence of amine groups in the chitosan structure makes heavy metal binding possible.

Chitosan blended with N-succinic and N-propylphosphonyl (NPP) to form polysulfone ultrafiltration membranes (Kumar *et al.*, 2013) has shown the property of improved water solubility, heavy metals adsorption capacity, and efficient binding of heavy metals (Kumar *et al.*, 2014). The blending of chitosan with CMC (carboxymethyl cellulose) to form carboxymethyl-chitosan showed greater efficiency for Cd^{2+} and Cr^{2+}. Effectively incorporating carboxylic groups enhanced membrane adsorption and complexation behavior toward these heavy metals (Hena, 2010; Borsagli *et al.*, 2015).

The effectiveness of chitosan membranes with regard to heavy metal removal improves significantly when chitosan is coated on a functionalized polymeric support (Kołodyńska, 2011). Chitosan can be coated on polymeric materials like polyvinylidene fluoride (Elizalde, 2018), polyethersulfone (Lusiana *et al.*, 2020) (PES), and polysulfone (Zailani *et al.*, 2021). This strategy introduces hydrophilic groups to hydrophobic polymers and induces a positive surface charge onto the polymeric membranes. Chitosan-coated clay particles support the removal of tungsten (W) from water. The coating of chitosan on clay changes the surface charge from negative to positive and point of zero charge (PZC) from 2.8 to 5.8 (Srinivasan, 2011). This coated clay shows better efficiency in removing heavy metals than natural clay because of positively charged sites (Gecol *et al.*, 2006).

6.7.4 SODIUM ALGINATE

Sodium alginate comprises many functional groups, including active hydroxyl and carboxyl groups, which react with heavy metals (Dong *et al.*, 2019). Sodium alginate modified with a sulfonic functional group has been successfully used for effluent treatment (Fatin-Rouge *et al.*, 2006; Petrovič and Simonič, 2016). Sodium alginate nanocomposites can capture heavy metal ions like Cd^{2+}, Pub^{2+}, Cu^{2+}, Cr^{2+}, and Co^{2+} (Wang *et al.*, 2018c).

The standard method of preparing sodium alginate membranes is a chemical modification (Thakur *et al.*, 2018). Sodium alginate-based membrane manufacturing and adsorption mechanism are comprehensively prepared and evaluated for membrane adsorption and were studied by Aburabie *et al.* (2019). Researchers prepared sodium alginate composite membranes using the reaction-induced phase separation method, and the prepared and modified membranes showed efficient results.

Recently, researchers have shown dual filtration efficiency of alginate membranes for heavy metal ions and organic pollutants (Esmaeili and Aghababai Beni, 2015). Besides sodium alginate, calcium alginate-based coated materials have also been used for adsorption of heavy metals (Wang *et al.*, 2016). The ionic membranes produced with sodium alginate and PVA have great potential to capture Cr ions from wastewater (Chen *et al.*, 2010).

6.7.5 POLYLACTIC ACID

Polylactic acid (PLA) is derived from naturally occurring organic acid (lactic acid), and it is included in the class of synthetic linear aliphatic thermoplastic polyesters (Castro-Aguirre, 2016). Because of the biocompatibility of PLA, it has numerous applications in tissue engineering and drug delivery systems (Chuan, 2020; ShujunYuab *et al.*, 2021). Recently it has gained attention for capturing heavy metal ions through ultrafiltration membranes due to its hydrophilic and nontoxic behavior. The hydroxyapatite nanoparticles incorporated PLA membranes were fabricated by E. Shokri et al. to efficiently remove Pb^{2+} and As^{2+} from drinking water. The Pb^{2+} ion uptake capacity from water was calculated (Shokri *et al.*, 2021). It is established that the material used as membrane adsorbents exhibited a removal efficiency of 100% as compared to previously employed techniques, and an efficiency of 93% for As^{2+}. The high adsorption capacity was attributed to the strong electrostatic interactions between the membrane's metal ion and phosphate functional groups (Chatterjee and De, 2015).

6.7.6 POLYCAPROLACTONE

Polycaprolactone (PCL) is a promising material to fabricate biocompatible membranes for heavy metals. It is mainly used owing to its excellent properties such as low glass transition temperature and a melting point around 60°C (Croisier *et al.*, 2012).

PCL and cellulose nanofiber membranes have shown suitable mechanical, adsorption, and filtration properties to remove contaminants from water (Carpenter, de Lannoy and Wiesner, 2015). A composite membrane has been prepared using electrospinning agave cellulose and blending it with PCL, which showed better final properties (Yalcinkaya, 2019). The agave biomaterial is translated into a sustainable cellulose and polycaprolactone composites membrane that can filter heavy metals like Fe^{3+} and Cr^{2+} with a good filtration system. The electrospun membrane removed 99% and 75% of Cr^{2+} and Fe^{3+}, respectively (Hinestroza *et al.*, 2020).

The cellulose acetate (CA) and PCL nanostructured membranes have been developed to adsorb Pb^{2+} from water. The low/cost-benefit ratio of CA has made it suitable for use both alone and in combination with other materials like PCL. The presence of different functional groups such as -COOH, -SO_3H, and -NH_2 groups on the backbone of CA provides sites for metal sequestering and adsorption from water in the form of a membrane. The developed membrane has adsorption abilities for electropositive ions in conjunction with Langmuir and Freundlich isotherm models (Aquino *et al.*, 2018).

6.7.7 GELATIN

Gelatin is a sustainable water-soluble biopolymer with excellent properties related to nontoxicity, biodegradability, and economical production. Different types of active groups, especially amino acids, in their structure can sorbate metal ions and serve as a template to prepare functional materials for heavy metal removal from contaminated water (Zhou *et al.*, 2017). The oxidation of rich amino groups in gelatin macromolecules forms nitrite functional groups. The use of gelatin as a reducing and stabilizing agent has led to the preparation of gold-gelatin composites (Zhang *et al.*, 2009) as well as a reductant and stabilizing agent for the stable synthesis of manganese dioxide (MnO_2) (Wang and Li, 2002). Adding functional groups tends to increase the sorbent adsorption capacity (Wanga *et al.*, 2018 a,b,c).

Porous gelatin membranes are prepared by Pickering emulsion templating, which are stabilized with h-BNNS for ultrafiltration of Fe^{2+} ions to produce promising porous membranes to remove heavy metals. The results have shown that the pore size distribution of the h-BNNS/gelatin membranes for Fe^{2+} capturing decreases with longer crosslinking times, so the crosslinking time must be optimized for targeted filtration of iron particles from water (Mateur Nafti Molka *et al.*, 2020).

The gelatin-TPA, gelatin-FA, and gelatin-GTA crosslinked membranes have been developed to extract heavy metals like Cu^{2+}, Zn^{2+}, and Fe^{2+} by ion diffusion from aqueous acid solutions. The prepared membranes established better transmission and capturing properties from strong acidic contaminated water and low concentrations of Cu^{2+} ions (Kamal, Pochat-Bohatier and Sanchez-Marcano, 2017).

6.8 COMPOSITE ELECTROSPUN MEMBRANES

A single material cannot be multifunctional. Thus, a composite combines or amplifies the characteristics of various materials or limits the deficiencies of individual components (Zare-Gachi *et al.*, 2020). Composite membranes are fabricated using two or more ingredients, each with different characteristics. Combining these, a new material with remarkable attributes can be achieved (Sagitha *et al.*, 2019). Polymeric composite membranes won the battle against other membranes in terms of unique morphologies, various structures, higher filtration efficiencies, and physiochemical properties (Hai *et al.*, 2021). These membranes are fabricated through solution casting, etching, phase inversion, stretching, interfacial polymerization, and electrospinning.

Conventional techniques have drawbacks: less reproducibility, complicated process, low selectivity, low porosity, less surface roughness, directly reducing the surface area, and low permeability (Goh and Ismail, 2020). Electrospinning is a remarkably versatile and flexible technique for fabricating nanomembranes (Davoodi *et al.*, 2021; Patel and Hota, 2022). Electrospinning is used to fabricate various polymeric nanofibers, having diameters ranging from 2 nanometers to a few micrometers, through a high voltage source (Ray *et al.*, 2019). Moreover, a few crucial parameters, as shown in Figure 6.2, significantly affect the fabrication and morphology of nanofibers (Bagbi, Pandey and Solanki, 2018).

Different electrospinning configurations are used to fabricate composite membranes (Haider, Haider and Kang, 2015; Choi *et al.*, 2020), including:

• Single-needle electrospinning
• Multi-needle electrospinning
• Coaxial electrospinning
• Needleless electrospinning
• Melt electrospinning

To improve the biopolymers' performance, additional functionalities can also be introduced to electrospun membranes (Ray *et al.*, 2019). The desired attributes for heavy metal removal can be achieved by blending different polymers, incorporating different additives or fillers into the polymer's dope, coating, crosslinking, in situ growth of nanostructures, introducing active species, etc. (Jose Varghese *et al.*, 2019). The potential of electrospun nanofibrous membranes for functionalization makes them the best candidate to deal with the arising heavy metal contamination in water

- Molecular weight
- Viscosity
- Polymer concentration
- Conductivity
- Surface tension
- Solvent properties

- Voltage
- Flow rate
- Electrode to collector voltage

- Temperature
- Humidity

Solution Parameters

Process Parameters

Ambient Conditions

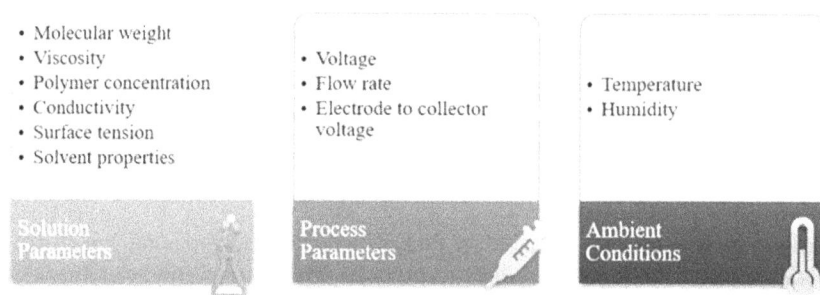

FIGURE 6.2   Crucial electrospinning parameters.

(Snowdon and Liang, 2020). The functionalized electrospun membranes can selectively adsorb heavy metal ions, depending on the adsorbate and adsorbent interactions. Additionally, adsorbents based on these membranes offer ease of recycling without causing any secondary pollution, as in the case of nanoparticles (Wang *et al.*, 2014).

Xiao et al. developed composite nanofibers composed of PVA and polyacrylic acid (PAA) to remove Cu^{2+} and reported 91% efficiency in an aqueous solution as the PAA contains carboxylic groups to capture heavy metal ions. At the same time, PVA provides hydrophilicity (Xiao *et al.*, 2010). The authors reported their work of crosslinking PVA nanofibers using glutaraldehyde (GLA) vapors and a solution method to remove Cu^{2+} and Pb^{2+} ions. Results showed the adsorption uptake of solution crosslinked was higher, i.e., 28 and 161 mg/g, respectively (Ullah *et al.*, 2020).

He et al. grafted maleic anhydride on PVA followed by the condensation reaction to graft octa-amino-POSS on PVA molecular chains. Electrospun nanofibrous composites of PVA/octa-amino-POSS and PVA-g-POSS were prepared and it was reported that the adsorption of Pb^{2+} and Cu^{2+} was higher by PVA-g-POSS as compared to octa-amino-based nanofibers. It was concluded that the adsorption was physical and chemical, but chemisorption dominates over physio adsorption (He *et al.*, 2021). PVA and PEI composite nanofibers were prepared for the removal of Cr^{6+} in batch experimentations. The adsorption was pH-dependent, and 150 mg/g uptake capacity was calculated via the Langmuir model. The amino group in the nanofibers serves as the capturing species for Cr^{6+} and attenuates the reduction of Cr^{6+} (Zhang *et al.*, 2020). PVA and sodium alginate (SA) composite nanofibers with the ratio of 40/60 were fabricated using electrospinning and used as adsorbents for Cd^{2+} removal. The adsorption capacity obtained was 93 mg/g with the Langmuir isotherm and it best fits the Pseudo II-order kinetic model (Ebrahimi *et al.*, 2019).

Habiba et al. explored the adsorption of electrospun PVA and chitosan membrane under the effect of deacetylation. It was found that the membrane containing chitosan with a higher deacetylation degree showed higher affinity toward Cr^{6+} owing to the higher number of amino groups. However, the lower deacetylation degree chitosan showed a higher affinity toward Fe^{3+} ions. Higher hydrolysis time results in a higher deacetylation degree and lowers the molecular weight of chitosan. It disrupts the polymer chains, which causes hindrance to the adsorption process (Habiba *et al.*, 2017). The authors studied the crosslinking of PVA electrospun nanofibers using different agents, i.e., citric acid, maleic acid, and PAA. The nanofibrous membranes of PAA and PVA have better performance for the adsorption of Cu^{2+} and Zn^{2+} (Truong *et al.*, 2017). Table 6.2 highlights some recent reports of heavy metal removal using membranes.

Chitpong and Husson used the electrospinning technique to prepare macroporous CA membranes to efficiently recover cadmium ions from wastewater (Chitpong and Husson, 2017). The prepared membranes were modified for desired properties. The membrane showed a high capacity for Cd^{2+} removal due to functional groups on the membrane.

Choi et al. developed thiol-functionalized cellulose nanofibrous membranes to efficiently remove heavy metal ions from water by adsorption (Choi *et al.*, 2020). Langmuir isotherms were used to investigate the adsorption capacities of the metal ions. Maximum adsorption of Pb^{2+}, Cd^{2+}, and Cu^{2+} ions were 22, 45.9, and 49 mg/g, respectively. According to Ritchie et al., incorporating nanoparticles can increase the adsorption capacity of the membranes. Based on this, Ibrahim et al. synthesized cellulose membranes by surface coating to eliminate heavy metals from wastewater (Ibrahim et al., 2019). Radzyminska-Lenarcik & Ulewicz also used cellulose triacetate-based polymeric membranes for the removal of Zn^{2+} and Mn^{2+} (Radzyminska-Lenarcik and Ulewicz, 2019). The membrane was prepared by pouring a specified quantity of the solution containing cellulose triacetate, 1-alkyl imidazole, and plasticizer into the membrane. These studies highlight the importance of surface functionalization on sustainable, nontoxic, and biocompatible cellulosic constituents to broaden their applicability for water purification applications.

6.9 MECHANISM OF ELECTROSPUN NANOFIBERS FOR HEAVY METAL REMEDIATION

The porous structure of nanofibers enables them to be a leading microfiltration medium. Because of the synergistic effect of microfiltration and strong interactions toward selective heavy metal ions the electrospun membrane is considered to be optimal (Choi et al., 2020). The nanofibers serve their purpose by following the adsorption mechanism, which is further accompanied by several processes, such as ion exchange, Van der walls forces, electrostatic interactions between the toxic heavy metal ions and the nanofibrous membrane (Zhu et al., 2021). The adsorption process is the transfer progression, meaning heavy metal ions from wastewater/aqueous solution get deposited in no electrospun membrane by chemical or physical adsorption (Agrawal et al., 2021).

As the first step, heavy metal ions from wastewater/aqueous solution start transferring toward the surface of the nanofibrous membrane. Then, metal ions diffuse from the external surface within the nanofibers, which is considered the "rate-limiting step." At last, the chemical or physical interactions occur between the nanofibers and heavy metal ions (Vo et al., 2020). Further, the composite or hybrid biopolymeric membranes based explicitly on electrospun nanofibers for removing heavy metal ions are discussed in the upcoming section.

Electrospun nanofibrous pure chitosan-based membranes were fabricated on a PET spun-bonded nonwoven substrate to treat the water's Cr^{6+}, Pb^{2+}, Cd^{2+}, and Cu^{2+} contaminations. The membrane was crosslinked using GLA and wound on a spiral module. The results showed that the developed membrane has higher selectivity toward Cr^{6+} than toward Pb^{2+}, Cd^{2+}, and Cu^{2+} ions in a dead-end filtration assembly (Li et al., 2017). Authors fabricated PVA and chitosan composite membranes via electrospinning for the selective adsorption of Cd^{2+} and Pb^{2+}. The composite chitosan-PVA membrane showed good kinetic performance toward both metal ions with a maximum adsorption uptake of 148 and 266 mg/g for Cd^{2+} and Pb^{2+}. The adsorption of heavy metals was followed by the Langmuir isotherm model and Pseudo II-order kinetics (Karim et al., 2019). Langmuir models are best suited for adsorption (Phan et al., 2019). Ma et al. developed a coaxial electrospun membrane with a core-shell structure composed of PCL-chitosan and CA in the core. The developed core-shell structure membrane showed higher adsorption than chitosan powder. Chemisorption was observed as a rate-controlling step, with the adsorption capacity for Cr^{6+} being 126 mg/g (Ma et al., 2019).

Christou et al. studied uranium adsorption using a chitosan and polyvinylpyrrolidone (PVP) nanofibrous membrane. Owing to the presence of carbonyl groups and the fibrous structure of the membrane, the affinity toward uranium increases, increasing the adsorption capacity, i.e., 167 g/kg for Cr^{6+} (Christou et al., 2019). Jiang et al. prepared chitosan nanofibers loaded with polyacrylic acid sodium (PAAS) and subjected them to an annealing treatment at a high temperature to increase solvents' mechanical properties and resistance. The sorption capacity for Cr^{6+} increases compared to pure chitosan powder, i.e., 78 mg/g, due to the availability of chelating ligands (Jiang et al., 2018).

Li et al. developed polycaprolactam and chitosan filter paper based on electrospun nanofibers. The adsorption behavior with three different means was studied, and their respective capacities are

80, 81, and 114 mg/g in shaking adsorption, static adsorption, and vacuum filtration, respectively. The authors investigated that when the adsorbent solution was forced to pass through the whole exposed area in vacuum filtration, it led to more interaction between the chelation sites and Cr^{6+} (Li et al., 2016). Zia et al. prepared poly(L-lactic acid) nanofibers with a porous surface to target Cu^{2+} ions. The electrospun nanofibers were then subjected to acetone to coat chitosan on a nanofibrous membrane through a direct inversion technique. Chitosan causes equilibrium to be achieved in minimum time due to its hydroxide and aminic groups. The highest adsorption uptake was 111 mg/g at 25°C and pH 7 within 10 minof contact time. The Langmuir isotherm and Pseudo II-order kinetics explain the adsorption process best (Zia et al., 2020).

6.10 FUNCTIONALIZED COMPOSITE MEMBRANES

Over the past few decades, much work has been performed to improve nanofibrous membranes' performance properties (such as metal ion selectivity, adsorption dose, and equilibrium time) (Cheng et al., 2019). For this, different biopolymers blended with synthetic ones are utilized, possessing different functionalities such as hydroxyl, amino, carboxyl, sulfonic, thiol, etc. (Huang et al., 2013; Wang et al., 2014). Moreover, the surface of polymers can also be immobilized with functional groups to enhance adsorption ability (Pereao et al., 2017).

Wang et al. reported a cellulose-based nanofiltration membrane to purify drinking water (Wang et al., 2020). A unique layered structure comprising three layers was fabricated, which significantly improved the firmness of the membrane (increased from 5 to 36 h) and the rejection rate of $MgSO_4$ (from 66.3% to 75.6%), which indicated that the developed, cellulosic membrane was helpful for the said purpose.

Phan et al., prepared a composite membrane of chitosan and CA using a co-solvent, i.e., acetic acid and trifluoroacetic, and treated the membrane with Na_2CO_3. Na_2CO_3 treatment converted the CA into cellulose via deacetylation and induced the chitosan neutralization process. The chitosan-cellulose membrane possessed better adsorption behavior with the capacity of 57, 112, and 39 mg/g for Pb^{2+}, Cu^{2+}, and As^{5+}, respectively. Yang et al., used functionalized membranes to remove various heavy metals from water. The group grafted poly-ethylenimine (PEI) and poly(glycidyl methacrylate) (PGMA) on the surface of the electrospun chitosan membrane. It was observed that equilibrium was achieved in about 60 min with sorption capacities of 68, 69, and 138 mg/g for Co^{2+}, Cu^{2+}, and Cr^{6+}. The membrane possessed good stability and reproducibility (Haykin, 2005). Choi et al., functionalized CA nanofibers with the thiol group via deacetylation of CA nanofibers, followed by the esterification of thiol. The Langmuir isotherm explained that the adsorption capacity depends on the adsorbate dose, and the maximum capacities were calculated as 45, 22, and 49 mg/g for Cd^{2+}, Pb^{2+}, and Cu^{2+} (Choi et al., 2020).

6.11 NANO-ENHANCED MEMBRANES

Nanotechnology is an emerging field for treating heavy metals using different chemically and physically modified nanostructures. Nanostructures offer extraordinary properties, specificity, and efficiency, in marvelous morphologies and chemical structures. These morphologies and chemical functionalities allow quenching the heavy metals from the contaminated water.

Sharma et al. fabricated a chitosan and polyvinyl alcohol (PVA) composite nanofibrous membrane loaded with cerium to adsorb As^{3+} in water. It was reported that the developed composite membrane had effectively removed As^{3+} with an adsorption capacity of 18 mg/g as calculated by the Langmuir isotherm (Sharma et al., 2014). Shooto et al. electrospun pristine PVA and incorporated antimony, lanthanum, and strontium MOFs within PVA benzene tetra-carboxylate nanofibers. The fabricated novel nanofibers followed the Pseudo II-order kinetics and Langmuir model. The adsorption reaction was spontaneous and exothermic, with 50, 92, and 58 mol adsorption capacity per gram (Shooto et al., 2016).

TABLE 6.2
Summary of biopolymeric membranes used for heavy metal removal

Membrane	Type of metal ions	Removal efficiency/ rejection/adsorption capacity	Water type	References
Chitosan/polyvinyl alcohol/polyether sulfone (PES) nanofibrous membrane	Cr^{6+} and Pb^{2+} ions	Cr^{6+} 509.7 mg/g and Pb^{2+} 525.8 mg/g	Aqueous solution	Koushkbaghi et al. (2018)
Poly(vinyl alcohol)/ tetraethyl orthosilicate (PVA/ TEOS) hybrid membrane	Pb^{2+}	61.62 mg/g	Aqueous solution	Irani, Keshtkar and Mousavian (2012)
Regenerated cellulose (RC) membrane	Cu^{2+} and Zn^{2+}	70% rejection for copper and 93% for zinc	Aqueous solution	Ennigrou, Ben Sik Ali and Dhahbi (2014)
Nanometric ethylene diamine-poly-ethylenimine graphene oxide framework membranes	Mg^{2+}, Pb^{2+}, Ni^{2+}, Cd^{2+} and Zn^{2+}	High rejection rate	Wastewater	Zhang, Zhang and Chung (2015)
P84 polyimide nanofiltration hollow fiber membranes	Cd^{2+}	Rejection 94%	Wastewater	Gao et al. (2016)
Diethylenetriamine (DETA)-anchored RC ultrafiltration membranes	Pb^{2+} and Cu^{2+} ions	Max. removal of lead and copper ions was 87% and 83%, respectively		Madaeni, Heidary and Salehi (2013)
Quaterinized chitosan-incorporated regenerated cellulose membrane	Arsenate and chromate	A 73.0% removal efficiency of As(V) and a 94.0% of Cr(VI)	Aqueous solution	Rivas (2015)
Amine-functionalized MCM-41 ultrafiltration membrane	Cr^{6+} and Cu^{2+}	Cr adsorption capacity 2.8 mg/g, and Cu adsorption capacity is 3.7 mg/g	Aqueous solution	Bao et al. (2015)
CA/poly ethylene glycol (PEG) membrane	Fe^{2+}	–	Wastewater	Căprărescu et al. (2020)
Chitosan-silver ions incorporated-CA/ PEG membrane	Fe^{2+}	–	Wastewater	Căprărescu et al. (2020)
Acrylo nitrile-dimethyl sulfoxide copolymers/ marigold flower extract natural-based functional membrane	Ni^{2+}	67%	Wastewater	Caprarescu et al. (2019)

(Continued)

TABLE 6.2 (*Continued*)

Summary of biopolymeric membranes used for heavy metal removal

Membrane	Type of metal ions	Removal efficiency/ rejection/adsorption capacity	Water type	References
RC membrane	Pb^{2+}	About 90% of rejection	Wastewater	Madaeni and Heidary (2011)
Polyvinylpyrrolidone modified poly ether sulfone membranes	Cu^{2+} and Fe^{2+}	Maximum rejection 99%	Wastewater	Jasiewicz and Pietrzak (2013)
Poly(acrylic acid) functionalized Cellulose nanofiber membranes	Cd^{2+}	–	Wastewater	Chitpong and Husson (2017)
CA membranes	For removal of monovalent and divalent ions in water	Rejection rate for monovalent 40% and 60%, rejection rate for divalent 70% and 90%	Wastewater	Choi, Fukushi and Yamamoto (2007)
Poly(methacrylic acid) modified CA nonwoven membrane	Cu^{2+}, Hg^{2+} and Cd^{2+}	The adsorption capacity of Hg^{2+} was 4.8 mg/g		Tian *et al.* (2011)
Cellulose-based polyamide nanofiltration membrane	$MgSO_4$	The rejection rate of about 9.3%	Drinking water	Wang *et al.* (2020)

Researchers used electrospun PVA nanofibers incorporated with nano-zeolite (NaX) as adsorbents for Cd^{2+} and Ni^{2+} removal. Results indicated that the composite nanofibers have more affinity toward Cd^{2+} than toward Ni^{2+} ions, and monolayer adsorption was measured as 838 and 342 mg/g. The as-prepared nanofibrous composite membrane has potential recyclability for up to five cycles with promising efficiency (Rad *et al.*, 2014). Wu et al. prepared mesoporous PVA-incorporated SiO_2 nanofibrous membrane functionalized with thiol for Cu^{2+} ions. The reported surface area of the composite membrane was >290 m²/g and 489 mg/g (at 303 K), which is higher than for the pure PVA membrane (Wu *et al.*, 2010). Roque-Ruiz et al. developed a nanofibrous membrane composed of polycaprolactone (PCL) and PVA embedded with metakaolin (MK) and kaolin (Kao) clays. The Kao_5PCL_{10} membrane exhibited the highest efficiency for Cd^{2+}, Pb^{2+}, and Cr^{3+}, while the MK_5PVA_{10} membrane showed the highest efficiency for Cu^{2+} (Roque-Ruiz *et al.*, 2016).

Chitosan possesses abundant amino groups making it a suitable candidate for heavy metal adsorption. Moreover, the performance of chitosan can also be enhanced by combining it with different polymers and nanomaterials (Bui, Park and Lee, 2017). Wang et al. reported permutit-incorporated chitosan and PVA electrospun membrane. The operating conditions include 15 kV voltage, 3 µL/min feed rate with a 12-cm tip to collector distance, and 150 rpm speed. The membrane containing 1% permit performs better in Cr^{6+} with maximum adsorption uptake of 208 mg/g at acid pH 3. The adsorption followed Langmuir and Pseudo II-order kinetic models (Wang *et al.*, 2018).

Razzaz et al. used the chitosan nanomembrane via two modes: incorporating and coating TiO_2 nanoparticles. The Redlich described the adsorption process for both membranes – Peterson and Pseudo I-order kinetic models. Coated and incorporated membranes' adsorption capacities for Pb^{2+} and Cu^{2+} were 475, 579, and 526, 710 mg/g, respectively, while maintaining the performance characteristics up to five cycles. Further, the selectivity of chitosan-TiO_2 membranes toward Cu^{2+} is higher than for Pb^{2+} (Razzaz *et al.*, 2016). ZabihiSahebi et al. synthesized a composite of TiO_2, SWCNT, and Fe_3O_4 and loaded it in a chitosan and CA blend. This solution was then subjected to electrospinning at 20 kV voltage, 10 cm distance from the tip to the collector, and 0.25 ml/h flow rate. The developed adsorbent membrane was suggested to be applicable for the adsorption process in lower concentrations of Cr^{6+} and As^{5+} ions, while the photocatalytic reduction process is preferable for high concentrations. Furthermore, the active sites for adsorption were increased in the reported membrane, which becomes feasible for hydrogen bonding, Van der Walls, and π-π interactions (ZabihiSahebi *et al.*, 2019).

Koushkbaghi et al. fabricated a mixed matrix membrane composed of PVA and chitosan nanofibers as a dual-layer structure filled with Fe_3O_4 nanoparticles. The adsorption uptake was achieved at 525 and 509 mg/g for Pb^{2+} and Cr^{6+}, respectively, and possessed excellent reusability in the adsorption process and membrane separation (Koushkbaghi *et al.*, 2018). Bozorgi et al. synthesized the PVA and chitosan membrane loaded with ZnO-APTES through electrospinning and solution casting for Ni^{2+} and Cd^{2+} adsorption. The nanofibrous membrane indicated higher adsorption than the casted membrane, and the adsorption process followed double-exponential kinetics. The adsorption capacity for Ni^{2+} and Cd^{2+} was measured as 0.851 and 1.239 mmol/g with a nanofibrous membrane; however, in the case of the casted membrane, 0.474 and 0.625 mmol/g, respectively, were measured (Bozorgi *et al.*, 2018).

Alharbi et al. prepared chitosan and polyacrylonitrile (PAN) with ZnO and TiO_2 nanoparticles through electrospinning in a dual structure. TiO_2-incorporated chitosan-PAN nanofibers showed better sorption capacity for Cd^{2+} and Pb^{2+}, i.e., 160 and 127 mg/g, due to the increased surface area associated with incorporating TiO_2 nanoparticles and a dual-layer structure. In the case of pristine chitosan/PAN nanofibers, the sorption capacities were 110 and 83 mg/g for Cd^{2+} and Pb^{2+}. ZnO-incorporated nanofibers have 133 and 126 mg/g adsorption capacities for Cd^{2+} and Pb^{2+} (Alharbi *et al.*, 2020).

6.12   CONCLUSIONS

From this chapter, it can be concluded that with increasing industrialization, the water resources are being heavily contaminated with undesired heavy metals, causing a threat to the human race. Among various techniques to remove heavy metals from water, membrane technologies are preferred due to their cost and high filtration efficiency. However, conventional membranes are produced using synthetic non-biodegradable polymers, which is also a concern for the environment. Using biopolymers to synthesize membranes could provide additional benefits of enhanced heavy metal adsorption, eco-friendly nature, and cost-effectiveness. After the service life of the biopolymer membranes, these membranes do not build up in the environment, being biodegradable. Thus, biopolymers are unbeatable candidates for fabricating heavy metal filtration membranes.

REFERENCES

Aburabie, J. *et al.* (2019) 'Alginate-based membranes: Paving the way for green organic solvent nanofiltration', *Journal of Membrane Science*, 596(2), p. 117615.

Agrawal, S. *et al.* (2021) 'Synthesis and water treatment applications of nanofibers by electrospinning', *Processes*, 9(10). doi: 10.3390/pr9101779.

Akale, A.T. *et al.* (2017) 'Groundwater quality in an upland agricultural watershed in the sub-humid Ethiopian highlands', *Journal of Water Resource and Protection*, 09(10), pp. 1199–1212. doi: 10.4236/jwarp.2017.910078.

Alharbi, H.F. *et al.* (2020) 'Electrospun bilayer PAN/chitosan nanofiber membranes incorporated with metal oxide nanoparticles for heavy metal ion adsorption', *Coatings*, 10(3), p. 285. doi: 10.3390/COATINGS10030285.

Aquino, R.R. *et al.* (2018) 'Adsorptive removal of lead (Pb^{2+}) ion from water using cellulose acetate/polycaprolactone reinforced nanostructured membrane', in *The 4th International Conference on Water Resource and Environment*, pp. 1–8. Kaohsiung City.

Bagbi, Y., Pandey, A. and Solanki, P.R. (2018) *Electrospun Nanofibrous Filtration Membranes for Heavy Metals and Dye Removal, Nanoscale Materials in Water Purification.* Elsevier Inc. doi: 10.1016/B978-0-12-813926-4.00015-X.

Bao, Y. *et al.* (2015) 'Application of amine-functionalized MCM-41 modified ultrafiltration membrane to remove chromium(VI) and copper(II)', *Chemical Engineering Journal*, 281(vi), pp. 460–467. doi: 10.1016/j.cej.2015.06.094.

Borsagli, F.G.M. *et al.* (2015) 'O-carboxymethyl functionalization of chitosan: Complexation and adsorption of Cd(II) and Cr(VI) as heavy metal pollutant ions', *Reactive & Functional Polymers*, 97, pp. 37–47.

Bozell, J.J. and Petersen, G.R. (2010) 'Technology development for the production of biobased products from biorefinery carbohydrates—The US Department of Energy's "top 10" revisited', *Green Chemistry*, 12(4), pp. 539–555. doi: 10.1039/b922014c.

Bozorgi, M. *et al.* (2018) 'Performance of synthesized cast and electrospun PVA/chitosan/ZnO-NH$_2$ nano-adsorbents in single and simultaneous adsorption of cadmium and nickel ions from wastewater', *Environmental Science and Pollution Research*, 25(18), pp. 17457–17472. doi: 10.1007/S11356-018-1936-Z/TABLES/6.

Briffa, J., Sinagra, E. and Blundell, R. (2020) 'Heavy metal pollution in the environment and their toxicological effects on humans', *Heliyon*, 6(9), p. e04691. doi: 10.1016/J.HELIYON.2020.E04691.

Bui, V.K.H., Park, D. and Lee, Y.C. (2017) 'Chitosan combined with ZnO, TiO$_2$ and Ag nanoparticles for antimicrobial wound healing applications: A mini review of the research trends', *Polymers*, 9(1), p. 21. doi: 10.3390/POLYM9010021.

Caprarescu, S. *et al.* (2019) 'Removal of nickel ions from synthetic wastewater using copolymers/natural extract blend membranes', *Romanian Journal of Physics*, 64(9–10), pp. 1–10.

Căprărescu, S. *et al.* (2020) 'Biopolymeric membrane enriched with chitosan and silver for metallic ions removal', *Polymers*, 12(8). doi: 10.3390/polym12081792.

Carpenter, A.W., de Lannoy, C.-F. and Wiesner, M.R. (2015) 'Cellulose nanomaterials in water treatment technologies', *Journal of Environmental Science and Technology*, 49, pp. 5277–5287.

Castro-Aguirre, E. (2016) 'Poly(lactic acid)—Mass production, processing, industrial applications, and end of life', *Advanced Drug Delivery Reviews*, 107, pp. 333–366.

Chatterjee, S. and De, S. (2015) 'Adsorptive removal of arsenic from groundwater using a novel high flux polyacrylonitrile (PAN)–laterite mixed matrix ultrafiltration membrane', *Environmental Science: Water Research & Technology*, 1, pp. 227–243.

Chen, J.H. *et al.* (2010) 'Cr(III) ionic imprinted polyvinyl alcohol/sodium alginate (PVA/SA) porous composite membranes for selective adsorption of Cr(III) ions', *Journal of Chemical Engineering*, 165, pp. 465–473.

Cheng, C. *et al.* (2019) 'Electrospun nanofibers for water treatment', *Electrospinning: Nanofabrication and Applications*, pp. 419–453. doi:10.1016/B978-0-323-51270-1.00014-5.

Chitpong, N. and Husson, S.M. (2017) 'Polyacid functionalized cellulose nanofiber membranes for removal of heavy metals from impaired waters', *Journal of Membrane Science*, 523, pp. 418–429. doi: 10.1016/j.memsci.2016.10.020.

Choi, H.Y. *et al.* (2020) 'Thiol-functionalized cellulose nanofiber membranes for the effective adsorption of heavy metal ions in water', *Carbohydrate Polymers*, 234, p. 115881. doi: 10.1016/j.carbpol.2020.115881.

Choi, J.H., Fukushi, K. and Yamamoto, K. (2007) 'A submerged nanofiltration membrane bioreactor for domestic wastewater treatment: The performance of cellulose acetate nanofiltration membranes for long-term operation', *Separation and Purification Technology*, 52(3), pp. 470–477. doi: 10.1016/j.seppur.2006.05.027.

Choque-Quispe, D. *et al.* (2022) 'Heavy metal removal by biopolymers-based formulations with native potato starch/nopal mucilage', *Revista Facultad de Ingenieria*, (103), pp. 44–50. doi: 10.17533/udea.redin.20201112.

Chowdhury, S. *et al.* (2016) 'Heavy metals in drinking water: Occurrences, implications, and future needs in developing countries', *Science of the Total Environment*, 569–570, pp. 476–488. doi: 10.1016/J.SCITOTENV.2016.06.166.

Christou, C. *et al.* (2019) 'Uranium adsorption by polyvinylpyrrolidone/chitosan blended nanofibers', *Carbohydrate Polymers*, 219, pp. 298–305. doi: 10.1016/J.CARBPOL.2019.05.041.

Chuan, D. (2020) 'Stereocomplex poly(lactic acid)-based composite nanofiber membranes with highly dispersed hydroxyapatite for potential bone tissue engineering', *Composites Science and Technology*, 192, p. 108107.

Croisier, F. *et al.* (2012) 'Mechanical testing of electrospun PCL fibers acta', *Acta Biometer*, 8, pp. 218–224.

Davoodi, P. *et al.* (2021) 'Advances and innovations in electrospinning technology', *Biomedical Applications of Electrospinning and Electrospraying*, pp. 45–81. doi: 10.1016/B978-0-12-822476-2.00004-2.

Dong, W., Zhang, Y. and Quan, X. (2020) 'Health risk assessment of heavy metals and pesticides: A case study in the main drinking water source in Dalian, China', *Chemosphere*, 242, p. 125113. doi: 10.1016/J. CHEMOSPHERE.2019.125113.

Dong, Y. *et al.* (2019) 'Mxene/alginate composites for lead and copper ion removal from aqueous solutions', *RSC Advances*, 9, pp. 29015–29022.

Ebrahimi, F. *et al.* (2019) 'Fabrication of nanofibers using sodium alginate and poly(vinyl alcohol) for the removal of Cd^{2+} ions from aqueous solutions: Adsorption mechanism, kinetics and thermodynamics', *Heliyon*, 5(11), p. e02941. doi: 10.1016/J.HELIYON.2019.E02941.

Elizalde, C.N.B. (2018) 'Fabrication of blend polyvinylidene fluoride/chitosan membranes for enhanced flux and fouling resistance', *Separation and Purification Technology*, 190, pp. 68–76.

Ennigrou, D.J., Ben Sik Ali, M. and Dhahbi, M. (2014) 'Copper and zinc removal from aqueous solutions by polyacrylic acid assisted-ultrafiltration', *Desalination*, 343, pp. 82–87. doi: 10.1016/j.desal.2013.11.006.

Esmaeili, A. and Aghababai Beni, A. (2015) 'Novel membrane reactor design for heavy-metal removal by algi-nate nanoparticles', *Journal of Industrial and Engineering Chemistry*, 26, pp. 122–128. doi: 10.1016/j. jiec.2014.11.023.

Fatin-Rouge, N. *et al.* (2006) 'Removal of some divalent cations from water by membrane-filtration assisted with alginate', *Water Research*, 40(6), pp. 1303–1309. doi: 10.1016/j.watres.2006.01.026.

Gao, J. *et al.* (2016) 'Green modification of outer selective P84 nanofiltration (NF) hollow fiber membranes for cadmium removal', *Journal of Membrane Science*, 499, pp. 361–369. doi: 10.1016/j.memsci.2015.10.051.

Gecol, H., *et al.* (2006) 'Biopolymer coated clay particles for the adsorption of tungsten from water', *Desalination*, 197(1–3), pp. 165–178.

Goh, P.S. and Ismail, A.F. (2020) 'Nanocomposite membrane fabrication', *Nanocomposite Membranes for Gas Separation*, pp. 125–162. doi: 10.1016/B978-0-12-819406-5.00004-6.

Gough, C.R. *et al.* (2021) 'Biopolymer-based filtration materials', *ACS Omega*, 6(18), pp. 11804–11812. doi: 10.1021/ACSOMEGA.1C00791/ASSET/IMAGES/ACSOMEGA.1C00791.SOCIAL.JPEG_V03.

Grigoray, O. *et al.* (2014) 'Photoresponsive cellulose fibers by surface modification with multifunctional cel-lulose derivatives', *Carbohydrate Polymers*, 111, pp. 280–287. doi: 10.1016/j.carbpol.2014.04.089.

Habiba, U. *et al.* (2017) 'Effect of deacetylation on property of electrospun chitosan/PVA nanofibrous mem-brane and removal of methyl orange, Fe(III) and Cr(VI) ions', *Carbohydrate Polymers*, 177, pp. 32–39. doi: 10.1016/J.CARBPOL.2017.08.115.

Hai, A. *et al.* (2021) 'Smart polymeric composite membranes for wastewater treatment', *Smart Polymer Nanocomposites*, pp. 313–350. doi: 10.1016/B978-0-12-819961-9.00010-4.

Haider, A., Haider, S. and Kang, I.-K. (2015) 'A comprehensive review summarizing the effect of electrospin-ning parameters and potential applications of nanofibers in biomedical and biotechnology', *Arabian Journal of Chemistry*, 11(8), pp. 1165–1188. doi: 10.1016/J.ARABJC.2015.11.015.

Hameed, M. *et al.* (2020) 'Moxifloxacin-loaded electrospun polymeric composite nanofibers-based wound dressing for enhanced antibacterial activity and healing efficacy', *International Journal of Polymeric Materials and Polymeric Biomaterials*, 70(17), pp. 1–9. doi: 10.1080/00914037.2020.1785464.

Haykin, S. (2005) 'Cognitive radio: Brain-empowered wireless communications', *IEEE Journal on Selected Areas in Communications*, 23(2), pp. 201–220.

He, Y. *et al.* (2021) 'Fabrication of PVA nanofibers grafted with octaamino-POSS and their application in heavy metal adsorption', *Journal of Polymers and the Environment*, 29(5), pp. 1566–1575. doi: 10.1007/S10924-020-01865-X/TABLES/1.

Hena, S. (2010) 'Removal of chromium hexavalent ion from aqueous solutions using biopolymer chitosan coated with poly 3-methyl thiophene polymer', *Journal of Hazardous Materials*, 181(1), pp. 474–479.

Hernández, N., Williams, R.C. and Cochran, E.W. (2014) 'The battle for the "green" polymer. Different approaches for biopolymer synthesis: Bioadvantaged vs. bioreplacement', *Organic and Biomolecular Chemistry*, 12(18), pp. 2834–2849.

Hinestroza, H.P. *et al.* (2020) 'Nanocellulose and polycaprolactone nanospun composite membranes and their potential for the removal of pollutants from water', *Molecules*, 25, pp. 1–13.

Huang, F. *et al.* (2013) 'Preparation of amidoxime polyacrylonitrile chelating nanofibers and their application for adsorption of metal ions', *Materials*, 6(3), pp. 969–980. doi: 10.3390/MA6030969.

Hubeny, J. *et al.* (2021) 'Industrialization as a source of heavy metals and antibiotics which can enhance the antibiotic resistance in wastewater, sewage sludge and river water', *PLOS ONE*, 16(6), p. e0252691. doi: 10.1371/JOURNAL.PONE.0252691.

Hudson, S.M. and Smith, C. (1998) 'Polysaccharides: Chitin and chitosan: Chemistry and technology of their use as structural materials', in *Biopolymers from Renewable Sources*, D. I. Kaplan (Ed.). Berlin: Springer, pp. 96–118.

Hussain, T., Ramakrishna, S. and Abid, S. (2021) 'Nanofibrous drug delivery systems for breast cancer: A review', *Nanotechnology*, 33(10), p. 102001. doi: 10.1088/1361-6528/AC385C.

Ibrahim, Y. *et al.* (2019) 'Synthesis of super hydrophilic cellulose-alpha zirconium phosphate ion exchange membrane via surface coating for the removal of heavy metals from wastewater', *Science of The Total Environment*, 690, pp. 167–180. doi: 10.1016/j.scitotenv.2019.07.009.

Irani, M., Keshtkar, A.R. and Mousavian, M.A. (2012) 'Preparation of poly(vinyl alcohol)/tetraethyl orthosilicate hybrid membranes modified with TMPTMS by sol-gel method for removal of lead from aqueous solutions', *Korean Journal of Chemical Engineering*, 29(10), pp. 1459–1465. doi: 10.1007/s11814-012-0022-3.

Izah, S.C., Chakrabarty, N. and Srivastav, A.L. (2016) 'A review on heavy metal concentration in potable water sources in Nigeria: Human health effects and mitigating measures', *Exposure and Health*, 8(2), pp. 285–304. doi: 10.1007/S12403-016-0195-9/TABLES/6.

Jasiewicz, K. and Pietrzak, R. (2013) 'Metals ions removal by polymer membranes of different porosity', *The Scientific World Journal*, 2013. doi: 10.1155/2013/957202.

Jiang, M. *et al.* (2018) 'Removal of heavy metal chromium using cross-linked chitosan composite nanofiber mats', *International Journal of Biological Macromolecules*, 120, pp. 213–221. doi: 10.1016/J.IJBIOMAC.2018.08.071.

Jose, A.J., Kappen, J. and Alagar, M. (2018) 'Polymeric membranes: Classification, preparation, structure physiochemical, and transport mechanisms', *Fundamental Biomaterials: Polymers*, pp. 21–35. doi: 10.1016/B978-0-08-102194-1.00002-5.

Jose Varghese, R. *et al.* (2019) 'Introduction to nanomaterials: Synthesis and applications', *Nanomaterials for Solar Cell Applications*, pp. 75–95. doi: 10.1016/B978-0-12-813337-8.00003-5.

Kamal, O., Pochat-Bohatier, C. and Sanchez-Marcano, J. (2017) 'Development and stability of gelatin cross-linked membranes for copper(II) ions removal from acid waters', *Separation and Purification Technology*, 183, pp. 153–161.

Karim, M.R. *et al.* (2019) 'Composite nanofibers membranes of poly(vinyl alcohol)/chitosan for selective lead(II) and cadmium(II) ions removal from wastewater', *Ecotoxicology and Environmental Safety*, 169, pp. 479–486. doi: 10.1016/J.ECOENV.2018.11.049.

Kołodyńska, D. (2011) 'Chitosan as an effective low-cost sorbent of heavy metal complexes with the polyaspartic acid', *Chemical Engineering Journal*, 173(2), pp. 520–529. doi: 10.1016/j.cej.2011.08.025.

Koushkbaghi, S. *et al.* (2018) 'Aminated-Fe$_3$O$_4$ nanoparticles filled chitosan/PVA/PES dual layers nanofibrous membrane for the removal of Cr(VI) and Pb(II) ions from aqueous solutions in adsorption and membrane processes', *Chemical Engineering Journal*, 337(vi), pp. 169–182. doi: 10.1016/j.cej.2017.12.075.

Kumar, R. *et al.* (2013) 'Synthesis and characterization of novel water soluble derivative of chitosan as an additive for polysulfone ultrafiltration membrane', *Journal of Membrane Science*, 440, pp. 140–147.

Kumar, R. *et al.* (2014) 'Preparation and evaluation of heavy metal rejection properties of polysulfone/chitosan, polysulfone/N-succinyl chitosan and polysulfone/N-propylphosphonyl chitosan blend ultrafiltration membranes', *Desalination*, 350, pp. 102–108.

Li, L. *et al.* (2017) 'Removal of Cr(VI) with a spiral wound chitosan nanofiber membrane module via dead-end filtration', *Journal of Membrane Science*, 544, pp. 333–341. doi: 10.1016/J.MEMSCI.2017.09.045.

Li, Z. *et al.* (2016) 'Preparation of chitosan/polycaprolactam nanofibrous filter paper and its greatly enhanced chromium(VI) adsorption', *Colloids and Surfaces A: Physicochemical and Engineering Aspects*, 494, pp. 65–73. doi: 10.1016/J.COLSURFA.2016.01.021.

Lusiana, R.A. *et al.* (2020) 'Permeability improvement of polyethersulfone-polietylene glycol (PEG-PES) flat sheet type membranes by tripolyphosphate-crosslinked chitosan (TPP-CS) coating', *International Journal of Biological Macromolecules*, 152, pp. 633–644.

Lv, J. *et al.* (2018) 'Improvement of antifouling performances for modified PVDF ultrafiltration membrane with hydrophilic cellulose nanocrystal', *Applied Surface Science*, 440, pp. 1091–1100. doi: 10.1016/j.apsusc.2018.01.256.

Ma, L. *et al.* (2019) 'Electrospun cellulose acetate–polycaprolactone/chitosan core–shell nanofibers for the removal of Cr(VI)', *Physica Status Solidi(a)*, 216(22), p. 1900379. doi: 10.1002/PSSA.201900379.

Madaeni, S.S. and Heidary, F. (2011) 'Improving separation capability of regenerated cellulose ultrafiltration membrane by surface modification', *Applied Surface Science*, 257(11), pp. 4870–4876. doi: 10.1016/j.apsusc.2010.12.128.

Madaeni, S.S., Heidary, F. and Salehi, E. (2013) 'Co-adsorption/filtration of heavy metal ions from water using regenerated cellulose UF membranes modified with DETA ligand', *Separation Science and Technology*, 48(9), pp. 1308–1314. doi: 10.1080/01496395.2012.735741.

Masindi, V. and Muedi, K.L. (2018) 'Environmental contamination by heavy metals', *Heavy Metals*. doi: 10.5772/INTECHOPEN.76082.

Mateur Nafti Molka *et al.* (2020) 'Porous gelatin membranes obtained from pickering emulsions stabilized with h-BNNS: Application for polyelectrolyte-enhanced ultrafiltration', *Membranes*, 10, pp. 144–161.

Mohan, S. *et al.* (2016) 'Biopolymers – Application in nanoscience and nanotechnology', *Recent Advances in Biopolymers*. doi: 10.5772/62225.

O'Connell, D.W., Birkinshaw, C. and O'Dwyer, T.F. (2008) 'Heavy metal adsorbents prepared from the modification of cellulose: A review', *Bioresource Technology*, 99(15), pp. 6709–6724. doi: 10.1016/j.biortech.2008.01.036.

Omer, S. (2021) 'Heavy metal removal by alginate based agriculture and industrial waste nanocomposites', *Properties and Applications of Alginates*. doi: 10.5772/INTECHOPEN.98832.

Patel, S. and Hota, G. (2022) 'Electrospun polymer composites and ceramics nanofibers: Synthesis and environmental remediation applications', *Design, Fabrication, and Characterization of Multifunctional Nanomaterials*, pp. 503–525. doi: 10.1016/B978-0-12-820558-7.00019-4.

Pereao, O.K. *et al.* (2017) 'Electrospinning: Polymer nanofibre adsorbent applications for metal ion removal', *Journal of Polymers and the Environment*, 25(4), pp. 1175–1189. doi: 10.1007/s10924-016-0896-y.

Petrovič, A. and Simonič, M. (2016) 'Removal of heavy metal ions from drinking water by alginate-immobilised *Chlorella sorokiniana*', *International Journal of Environmental Science and Technology*, 13(7), pp. 1761–1780. doi: 10.1007/s13762-016-1015-2.

Phan, D.N. *et al.* (2019) 'Fabrication of electrospun chitosan/cellulose nanofibers having adsorption property with enhanced mechanical property', *Cellulose*, 26(3), pp. 1781–1793. doi: 10.1007/S10570-018-2169-5/TABLES/2.

Qasem, N.A.A., Mohammed, R.H. and Lawal, D.U. (2021) 'Removal of heavy metal ions from wastewater: A comprehensive and critical review', *npj Clean Water*, 4(1). doi: 10.1038/s41545-021-00127-0.

Rad, L.R. *et al.* (2014) 'Removal of Ni^{2+} and Cd^{2+} ions from aqueous solutions using electrospun PVA/zeolite nanofibrous adsorbent', *Chemical Engineering Journal*, 256, pp. 119–127. doi: 10.1016/J.CEJ.2014.06.066.

Radzyminska-Lenarcik, E. and Ulewicz, M. (2019) 'The application of polymer inclusion membranes based on CTA with 1-alkylimidazole for the separation of zinc(II) and manganese(II) ions from aqueous solutions', *Polymers*, 11(2). doi: 10.3390/polym11020242.

Ray, S.S. *et al.* (2019) 'Electrospinning: A versatile fabrication technique for nanofibrous membranes for use in desalination', *Nanoscale Materials in Water Purification*, pp. 247–273. doi: 10.1016/B978-0-12-813926-4.00014-8.

Raza, Z.A., Anwar, F. and Abid, S. (2021) 'Sustainable antibacterial printing of cellulosic fabrics using an indigenous chitosan-based thickener with distinct natural dyes', *International Journal of Clothing Science and Technology*, 33(6), pp. 914–928. doi: 10.1108/IJCST-01-2020-0005/FULL/XML.

Raza, Z.A., Khalil, S. and Abid, S. (2020) 'Recent progress in development and chemical modification of poly(hydroxybutyrate)-based blends for potential medical applications', *International Journal of Biological Macromolecules*, 160, pp. 77–100. doi: 10.1016/J.IJBIOMAC.2020.05.114.

Razzaz, A. *et al.* (2016) 'Chitosan nanofibers functionalized by TiO_2 nanoparticles for the removal of heavy metal ions', *Journal of the Taiwan Institute of Chemical Engineers*, 58, pp. 333–343. doi: 10.1016/J.JTICE.2015.06.003.

Rivas, L. (2015) 'Quaternised chitosan in conjunction with ultrafiltration membranes to remove arsenate and chromate ions'. *Polymer Bulletin*, 72, pp. 1365–1377. doi: 10.1007/s00289-015-1341-4.

Rol, F. *et al.* (2019) 'Recent advances in surface-modified cellulose nanofibrils', *Progress in Polymer Science*, 88, pp. 241–264. doi: 10.1016/j.progpolymsci.2018.09.002.

Roohani, N. *et al.* (2013) 'Zinc and its importance for human health: An integrative review', *Journal of Research in Medical Sciences : The Official Journal of Isfahan University of Medical Sciences*, 18(2), p. 144.

Roque-Ruiz, J.H. *et al.* (2016) 'Preparation of PCL/clay and PVA/clay electrospun fibers for cadmium (Cd^{2+}), chromium (Cr^{3+}), copper (Cu^{2+}) and lead (Pb^{2+}) removal from water', *Water, Air, and Soil Pollution*, 227(8), pp. 1–17. doi: 10.1007/S11270-016-2990-0/FIGURES/10.

Sagitha, P. *et al.* (2019) 'Development of nanocomposite membranes by electrospun nanofibrous materials', *Nanocomposite Membranes for Water and Gas Separation*, pp. 199–218. doi: 10.1016/B978-0-12-816710-6.00008-0.

Sahana, T.G. and Rekha, P.D. (2018) 'Biopolymers: Applications in wound healing and skin tissue engineering', *Molecular Biology Reports*, 45(6), pp. 2857–2867. doi: 10.1007/S11033-018-4296-3/TABLES/2.

Sánchez, J., Butter, B. and Rivas, B.L. (2020) 'Biopolymers applied to remove metal ions through ultrafiltration. A review', *Journal of the Chilean Chemical Society*, 65(4), pp. 5004–5010. doi: 10.4067/S0717-97072020000405004.

Shankar, S. and Rhim, J.W. (2018) 'Bionanocomposite films for food packaging applications', *Innovative Food Processing Technologies: A Comprehensive Review*, pp. 234–243. doi: 10.1016/B978-0-12-815781-7.21875-1.

Sharma, R. *et al.* (2014) 'Electrospun chitosan–polyvinyl alcohol composite nanofibers loaded with cerium for efficient removal of arsenic from contaminated water', *Journal of Materials Chemistry A*, 2(39), pp. 16669–16677. doi: 10.1039/C4TA02363C.

Shokri, E. *et al.* (2021) 'Biopolymer-based adsorptive membrane for simultaneous removal of cationic and anionic heavy metals from water', *International Journal of Environmental Science and Technology*. doi: 10.1007/s13762-021-03592-9.

Shooto, N.D. *et al.* (2016) 'Novel PVA/MOF nanofibres: Fabrication, evaluation and adsorption of lead ions from aqueous solution', *Nanoscale Research Letters*, 11(1), pp. 1–13. doi: 10.1186/S11671-016-1631-2/FIGURES/8.

ShujunYuab *et al.* (2021) 'Recent advances in metal-organic framework membranes for water treatment: A review', *Science of The Total Environment*, 800, p. 149662.

Smith, A.M., Moxon, S. and Morris, G.A. (2016) 'Biopolymers as wound healing materials', *Wound Healing Biomaterials*, 2, pp. 261–287. doi: 10.1016/B978-1-78242-456-7.00013-1.

Snowdon, M.R. and Liang, R.L. (2020) 'Electrospun filtration membranes for environmental remediation', *Nanomaterials for Air Remediation*, pp. 309–341. doi: 10.1016/B978-0-12-818821-7.00016-6.

Srinivasan, R. (2011) 'Advances in application of natural clay and its composites in removal of biological, organic, and inorganic contaminants from drinking water', *Advances in Material Science and Engineering*, 17(3), pp. 385–403.

Thakur, S. *et al.* (2018) 'Recent progress in sodium alginate based sustainable hydrogels for environmental applications', *Journal of Cleaner Production*, 198, pp. 143–159.

Tian, Y. *et al.* (2011) 'Electrospun membrane of cellulose acetate for heavy metal ion adsorption in water treatment', *Carbohydrate Polymers*, 83(2), pp. 743–748. doi: 10.1016/j.carbpol.2010.08.054.

Truong, Y.B. *et al.* (2017) 'Functional cross-linked electrospun polyvinyl alcohol membranes and their potential applications', *Macromolecular Materials and Engineering*, 302(8), p. 1700024. doi: 10.1002/MAME.201700024.

Ullah, S. *et al.* (2020) 'Stabilized nanofibers of polyvinyl alcohol (PVA) crosslinked by unique method for efficient removal of heavy metal ions', *Journal of Water Process Engineering*, 33, p. 101111. doi: 10.1016/J.JWPE.2019.101111.

Vo, T.S. *et al.* (2020) 'Heavy metal removal applications using adsorptive membranes', *Nano Convergence*, 7(1), pp. 1–26. doi: 10.1186/S40580-020-00245-4.

Wang, D. *et al.* (2020) 'A cellulose-based nanofiltration membrane with a stable three-layer structure for the treatment of drinking water', *Cellulose*, 27(14), pp. 8237–8253. doi: 10.1007/s10570-020-03325-0.

Wang, P. *et al.* (2018a) 'Adsorption of hexavalent chromium by novel chitosan/poly(ethylene oxide)/permutit electrospun nanofibers', *New Journal of Chemistry*, 42(21), pp. 17740–17749. doi: 10.1039/C8NJ03899F.

Wang, X., Ge, J., Si, Y., Ding, B. (2014). Adsorbents Based on Electrospun Nanofibers. In Ding, B., Yu, J. (eds.), *Electrospun Nanofibers for Energy and Environmental Applications*. Nanostructure Science and Technology. Springer, Berlin, Heidelberg. https://doi.org/10.1007/978-3-642-54160-5_19

Wang, X. and Li, Y. (2002) 'Rational synthesis of alpha-MnO$_2$ single-crystal nanorods', *Chemical Communications*, pp. 764–765.

Wanga, X. *et al.* (2018b) 'Preparation of dumbbell manganese dioxide/gelatin composites and their application in the removal of lead and cadmium ions', *Journal of Hazardous Materials*, 350, pp. 46–54.

Wang, Z. *et al.* (2016) 'Macroporous calcium alginate aerogel as sorbent for Pb^{2+} removal from water media', *Journal of Environmental Chemical Engineering*, 4, pp. 3185–3192.

Wang, Z. *et al.* (2018c) 'Recent advances in nanoporous membranes for water purification', *Nanomaterials*, 8, p. 65.

Wani, A.L., Ara, A. and Usmani, J.A. (2015) 'Lead toxicity: A review', *Interdisciplinary Toxicology*, 8(2), p. 55. doi: 10.1515/INTOX-2015-0009.

Waqas, M. *et al.* (2022) 'Silver sulfadiazine loaded nanofibers for burn infections'. *International Journal of Polymeric Materials and Polymeric Biomaterials*, 72(7), pp. 517–523. doi: 10.1080/00914037.2022.2032701.

Wołowiec, M. *et al.* (2019) 'Removal of heavy metals and metalloids from water using drinking water treatment residuals as adsorbents: A review', *Minerals*, 9(8), p. 487. doi: 10.3390/MIN9080487.

Wu, S. *et al.* (2010) 'Effects of poly(vinyl alcohol) (PVA) content on preparation of novel thiol-functionalized mesoporous PVA/SiO$_2$ composite nanofiber membranes and their application for adsorption of heavy metal ions from aqueous solution', *Polymer*, 51(26), pp. 6203–6211. doi: 10.1016/J.POLYMER.2010.10.015.

Xiao, S. *et al.* (2010) 'Fabrication of water-stable electrospun polyacrylic acid-based nanofibrous mats for removal of copper(II) ions in aqueous solution', *Journal of Applied Polymer Science*, 116(4), pp. 2409–2417. doi: 10.1002/APP.31816.

Xu, X. *et al.* (2019) 'Modified cellulose membrane with good durability for effective oil-in-water emulsion treatment', *Journal of Cleaner Production*, 211, pp. 1463–1470. doi: 10.1016/j.jclepro.2018.11.284.

Yalcinkaya, F. (2019) 'Preparation of various nanofiber layers using wire electrospinning system', *Arabian Jornal of Chemistry*, 12, pp. 6162–6172.

ZabihiSahebi, A. *et al.* (2019) 'Synthesis of cellulose acetate/chitosan/SWCNT/Fe$_3$O$_4$/TiO$_2$ composite nanofibers for the removal of Cr(VI), As(V), methylene blue and Congo red from aqueous solutions', *International Journal of Biological Macromolecules*, 140, pp. 1296–1304. doi: 10.1016/J.IJBIOMAC.2019.08.214.

Zare-Gachi, M. *et al.* (2020) 'Improving anti-hemolytic, antibacterial and wound healing properties of alginate fibrous wound dressings by exchanging counter-cation for infected full-thickness skin wounds', *Materials Science and Engineering: C*, 107, p. 110321. doi: 10.1016/J.MSEC.2019.110321.

Zhang, J., Gu, M. and Zheng, T. (2009) 'Synthesis of gelatin-stabilized gold nanoparticles and assembly of carboxylic single-walled carbon nanotubes/au composites for cytosensing and drug uptake', *Journal of Analytical Chemistry*, 81, pp. 6641–6648.

Zhang, S. *et al.* (2020) 'Chromate removal by electrospun PVA/PEI nanofibers: Adsorption, reduction, and effects of co-existing ions', *Chemical Engineering Journal*, 387, p. 124179. doi: 10.1016/J.CEJ.2020.124179.

Zhang, T. *et al.* (2019) 'Heavy metals in human urine, foods and drinking water from an e-waste dismantling area: Identification of exposure sources and metal-induced health risk', *Ecotoxicology and Environmental Safety*, 169, pp. 707–713. doi: 10.1016/J.ECOENV.2018.10.039.

Zhang, Y., Zhang, S. and Chung, T.S. (2015) 'Nanometric graphene oxide framework membranes with enhanced heavy metal removal via nanofiltration', *Environmental Science and Technology*, 49(16), pp. 10235–10242. doi: 10.1021/acs.est.5b02086.

Zhou, Z. *et al.* (2017) 'Sorption performance and mechanisms of arsenic(V) removal by magnetic gelatinmodified biochar', *Journal of Chemical Engineering*, 314, pp. 223–231.

Zhu, F. *et al.* (2021) 'A critical review on the electrospun nanofibrous membranes for the adsorption of heavy metals in water treatment', *Journal of Hazardous Materials*, 401, p. 123608. doi: 10.1016/J.JHAZMAT.2020.123608.

Zia, Q. *et al.* (2020) 'Porous poly(L-lactic acid)/chitosan nanofibres for copper ion adsorption', *Carbohydrate Polymers*, 227, p. 115343. doi: 10.1016/J.CARBPOL.2019.115343.

Zailani, M.Z., *et al.* (2021) 'Immobilizing chitosan nanoparticles in polysulfone ultrafiltration hollow fibre membranes for improving uremic toxins removal', *Journal of Environmental Chemical Engineering*, 9(6), p. 106878.

7 Bioremediation of Heavy Metals Using the Symbiosis of Hydrophytes and Metal-Resistant Bacteria

Hera Naheed Khan and Muhammad Faisal

7.1 INTRODUCTION

Our planet Earth is in peril; environmental pollution has jeopardized our home and made it insecure for the generations to come. Anthropogenic activities have notched up the level of habitat and ecological degradation at an alarming rate (Martín et al. 2015). Technological advancements, urbanization, precarious farming practices and industrialization are some of the key factors that lead to pollution. Industrial effluents that are rich in hazardous chemical compounds and heavy metals are one of the most dangerous components that pose major health threats (Akpor et al. 2014). Heavy metals are extremely toxic to not just humans but all life forms. Metal pollutants pose toxicities in humans, plants, microorganisms, fungi, algae, cyanobacteria, etc. Although their importance as important modulators of biological functions can't be denied, yet increasing levels are pernicious to ecosystems and their inhabitants. Major metal pollutants are discharged in the environment through industrial activities; they enter water bodies contaminating them and leach to the nearby soils spoiling land too. Minute quantities of these metals can also prove to be hazardous for living organisms (Clemens and Ma 2016). Heavy metals like cadmium, chromium, copper, nickel, lead, selenium, silver, zinc and arsenic are the most frequently encountered pollutants. They tend to stay in the water and soil for long intervals and the only way to get rid of them is by detoxifying them. Their ionic forms are extremely dangerous. They not just amputate biological functions but are responsible for growth defects, mental retardation, infertility and other developmental abnormalities in humans, and cause major growth defects in plants too by negatively regulating their biological functions (Jan et al. 2015). All heavy metals are particularly essential for one biological function or another; they mainly play a role by acting as enzyme catalysts: however, elevated levels above a certain threshold value prove to be extremely dangerous for all life forms. Their elevated levels usually interfere in metabolic reactions of humans, plants and even microorganisms. Their accumulation in plants leads to their transfer through the food chain to other living beings and metal contamination goes on increasing. Therefore, it is the need of the hour to take immediate steps to prevent heavy metal induction in our ecosystems, because their buildup causes oxidative stress due to production and release of metal ions and free radicals with the potential to destroy and kill healthy cells. Fighting metal pollution has been an ongoing war since many decades. Overt industrialization and urbanization are key components of this environmental adulteration. Leaders in these are tanning industries, wood processing units, dying industries, paper and pulp production units, and fertilizer units. Utilizing modern agricultural reforms has also played a significant role in polluting our agricultural land and associated water bodies. A recent report by the Environment Protection Authority

DOI: 10.1201/9781003326281-9

(EPA) mentions that 40% of the heavily contaminated and polluted areas consist of heavy metals and organic pollutants (Olaniran et al. 2013).

Remediation of heavy metals is considered more complex than remediation of organic pollutants because the latter can be transformed to carbon dioxide and water (Sharma et al. 2018); however, for metal removal the only way is their detoxification and immobilization to limit their bioavailability. Also, their presence impedes degradation of other organic and inorganic pollutants. A variety of physical, chemical, biological and physiochemical and physiobiological methods are available for heavy metal detoxification and removal.

Despite the availability of such advance techniques, biological remediation methods are considered the most friendly of all. They are cheaper, safer and more efficient than conventional methods. This chapter focuses on the detailed analysis of bioremediation techniques employed for cleaning metal-polluted water bodies.

7.2   HEAVY METALS: MAJOR POLLUTANTS

Heavy metals are metallic elements with high density within the range of 3.5–7 g/cm^3. Heavy metals have biological significance, as they are important for biological functions but only in trace amounts. They are classified as the number one pollutant contaminating our soil and water bodies. They are indigenous to the Earth's crust but their elevated levels in the environment is due to their discharge from industrial plants, mining wastes, landfill leaches, municipal wastewater, urban runoff, electroplating, electronic and metal-finishing industries. Their extensive use has led to their accumulation in water and land, which has serious consequences. The water we drink, the food we eat all contain traces of heavy metals, which are extremely dangerous. Heavy metal pollution of land and water allows chemicals to enter both the terrestrial and the aquatic food chain. Their accumulation over time and non-degradable nature allows them to persist in the environment over decades and centuries. They pose some serious toxicity issues to life forms: humans, animals and all other life forms. Their toxicities are related to their inability to get metabolized within the body and hence their levels concentrate and gradually pile up. Adverse effects of metal contamination to humans have been extensively reported in literature; Table 7.1 outlines the major health threats posed by common heavy metals:

7.3   REMEDIATION OF HEAVY METALS

The cleaning of contaminated waters is necessary as water resources are limited; the increase in population growth, industrial development and urbanization has put a drag on the availability of clean water. An estimated 80% of the water worldwide returns to the environment without treatment (Weerasekara 2017) Life can't exist without water so it is the need of the hour to preserve and conserve water and try initiating strategies to clean water. A number of conventional methods (Figure 7.1) are available for treating polluted waters. Commonly used physicochemical remediation strategies include the following:

7.3.1   CHEMICAL PRECIPITATION

It is a widely used technique to treat metal-contaminated waters; like the name suggests, it refers to the conversion of soluble pollutants to insoluble forms and their subsequent removal from the liquid via filtration process (Nomanbhay and Palanisamy 2005). The process of precipitation is enhanced by the addition of chemical coagulants to help increase particle size via aggregation and the addition of aggregators is influenced by the acid base ratio of the water. Commonly used flocculants for metal removal are hydrogen peroxide or lime after adjusting the pH to 11 for the water being treated

TABLE 7.1

Toxicity of commonly encountered heavy metals to living organisms

Metals	Route of entry	Biological effect	Reference
Arsenic (As)	Ingestion, inhalation or skin contact	Inhibition of mitochondrial enzymes, DNA repair, chromosomal aberrations & carcinogenic. Arsenic keratosis	Substances and Registry (2000) and Jaishankar et al. (2014)
Cadmium (Cd)	Inhalation & ingestion	Pulmonary & gastrointestinal irritant, kidney damage and DNA damage, osteoporosis (skeletal damage)	Bernard (2008) and Jaishankar et al. (2014)
Chromium (Cr)		Skin irritant, ulceration with liver & kidney damage. Group 1 human carcinogen (research on cancer)	Jaishankar et al. (2014) and Pandey and Madhuri (2014)
Lead (Pb)	Inhalation, skin and ingestion	Acute or chronic lead poisoning, a carcinogen (EPA), birth defects, psychosis, mental defects, autism, allergies & hyperactivity	Martin and Griswold (2009) and Jaishankar et al. (2014)
Mercury (Hg)	Ingestion	Mercury poisoning (acrodynia), depression, memory loss, increased blood pressure	Martin and Griswold (2009) and Jaishankar et al. (2014)
Aluminum (Al)	Inhalation, ingestion & dermal contact	Accumulation leads to bone & brain damage, secondary hyperparathyroidism, aluminum-induced osteomalacia	Cannata Andia (1996) and Jaishankar et al. (2014)

(Akpor and Muchie 2010). Metals are usually precipitated in the form of hydroxides as represented in the following equation (Kurniawan et al. 2006):

$$\text{Metal}^{2+}(\text{dissolved metal}) + 2(\text{OH})^- \leftrightharpoons \text{Metal}(\text{OH})_2(\text{precipitated metal hydroxide})$$

Calcium hydroxide (lime) can be used to precipitate metals with up to 1,000 mg/ml concentrations. Benefits of using this treatment are its low cost, the use of inexpensive equipment and safety. However, this method has not been productive enough to remove metals within the stringent limit set by the US-EPA of less than 1 mg/ml (Nichols 1974); second, it also requires large quantities of chemicals for the metal reduction process (Jüttner et al. 2000). Other disadvantages include sludge production, its disposal, inefficient precipitation rate and the environmental impacts of sludge and precipitants involved in the process (Wingenfelder et al. 2005; Kurniawan et al. 2006).

7.3.2 COAGULATION-FLOCCULATION

Coagulation-flocculation is a method very similar in principle to precipitation employed by Charerntanyarak (1999) for the efficient removal of zinc, manganese and cadmium ions from

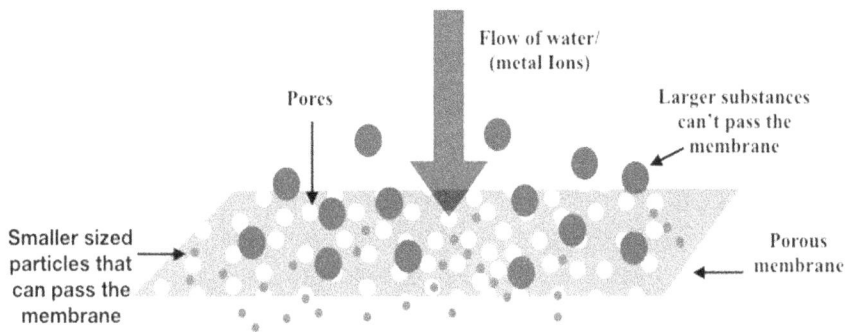

FIGURE 7.1 Filtration technique for wastewater treatment.

wastewater. This process, however, in principle impairs colloidal particles using coagulants that settle down via sedimentation (Wang et al. 2007). Particle size is increased by the formation of floccules of unstable particles; the optimum pH for carrying out this process is 11 and ferric salts are used as flocculants (Zhao et al. 2011). This process can help remove effluents with metal concentration of less than 100 mg/l and greater than 1,000 mg/l. Despite major advantages of sludge settling and stability, it is limited due to its functional expensiveness.

7.3.3 Ion Exchange Method

Another method for the successful removal of heavy metals from water is the ion exchange method; it involves the exchange of ions (charged particles) between a solid phase and a liquid phase, and the ion exchange occurs between the water molecules and charged ions attached onto an insoluble substance, usually resins (artificial) or naturally available inorganic zeolites. This process is quite similar in function to biosorption and has been predominantly used for recovery of valuable metals also. The ion exchange interaction occurs between the sulfonic acidic functional groups of the resins and metals as demonstrated (Technical Memorandum Standards for Effluents Discharged into Drainage and Sewerage Systems, Inland and Coastal Water 2005; Akpor and Muchie 2010):

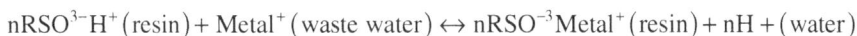

$$nRSO^{3-}H^+\left(resin\right)+Metal^+\left(waste\ water\right)\leftrightarrow nRSO^{-3}Metal^+\left(resin\right)+nH+\left(water\right)$$

7.3.4 Filtration

Membrane filtration is a universally available method for separation of unwanted materials from bulk solutions. A general layout of how filtration works is shown in Figure 7.2. Depending on the size of the particles that need to be separated filtration can be of the following types:

7.3.4.1 Ultrafiltration

The pore size of the membrane ranges from 5 to 20 nm and it can separate compounds based on molecular weight (1,000–100,000 Da). So the pore size allows the passage of molecules that are smaller than 20 nm and of low molecular weight (Sablani et al. 2001). Studies carried out by Juang and Shiau reported the ability of chitosan-based membrane filtration for removal of Cr(II) and Zn(II); the filtration membrane YM10 made from cellulose was used as the ultrafilter. The membrane was able to filter Cu(II) with 100% capacity and Zn(II) with 95% capacity within pH range of 8.5–9.5. Chitosan-embedded membranes showed a better rate of metal removal of about 6%–7%, which is more efficient than when membranes are used alone. The presence of amino groups on the chitosan helps in enhancing metal binding capacity (Juang and Shiau 2000).

Ultrafiltration has the capacity to remove metals with almost 90% efficiency with concentrations ranging from 10 to 112 mg/l within the pH ranges of 5–9.5 & 2–5 bar of pressure. But its major drawback lies in the fouling of filtration membranes, which directly affects its performance and metal removal capacity. Also, these membranes are biodegradable and hence are not suitable for long-time use; so it tends to become expensive (Kurniawan et al. 2006).

7.3.4.2 Nanofiltration

This method of heavy metal removal is slightly advanced and more competitive than the ultra-filtration (UF) technique. It is a method similar to the reverse osmosis (RO) technique. However, it is cheaper than RO due to its ability to operate at a lower pressure and higher flow rate. It is a filtration process that allows filtration using an organic porous membrane with pore size in the range of 0.1–10 nm (most notably 1–2 nm) (Abhanga et al. 2013). The filtration occurs due to the difference in pressure on either side of the membrane. Substances with molecular weight less than 200 dalton can pass the membrane while larger-sized molecules can't permeate. The membrane also selects and restricts the passage of multivalent ions while allowing the passage of monovalent ions (Pandya 2015).

A comparison between removal capacity of copper and cadmium ions was performed using RO and nanofiltration (NF) – the starting concentration of metals for both the techniques was 200 mg/l; RO was able to filter Cu(II) with 98% efficiency and Cd(II) with 99% efficiency while NF was able to remove Cu(II) & Cd(II) with 90% and 97% efficiency, respectively (Qdais and Moussa 2004). Although both the techniques are able to remove metals with almost similar efficiency, yet NF is preferred over RO due to lower operational costs. Figure 7.1 provides an overview of the filtration technique employed to treat wastewater.

7.3.5 Reverse Osmosis

This is a pressure-driven process where wastewater carrying the unwanted heavy metals passes through the membrane that has a pore size of around 0.1 nm. RO is considered most efficient with rejection efficiency of almost 97% for metal concentrations ranging between 21 and 200 mg/l. The working conditions greatly affect the working potential of the membrane; a pH range of range of 3–11 and a pressure of 4.5–15 bar are considered to be the most optimal with pressure being the deciding factor. Other benefits of RO are its non-biodegradable nature, the strength of the membrane, its stability and resistance toward chemical agents and its ability to resist high temperature. Other benefits of RO encompass a high water flux rate, increased salt rejection and protection against biological agents, mechanical stability & strength, resistance to chemicals and temperature (Kurniawan et al. 2006). However it's slightly expensive and sometimes more prone to fouling due to its small pore size; hence, it is not preferred over other methods.

7.4 BIOREMEDIATION

The most advanced and promising technology for the efficient expulsion of heavy metals from environmental soil and waters is bioremediation. Microorganisms are metabolically diverse and they have developed multiple strategies to help them adapt to unfavorable habitats. They do this by establishing mechanisms of detoxifying heavy metals and other undesirable compounds through a variety of processes like biosorption or biomineralization. They can also accumulate them through the process of bioaccumulation or convert them to non-toxic states via biotransformation; all these processes allow us to exploit microorganisms as land cleaners (Dixit et al. 2015).

Microorganisms (bacteria) uptake metals either actively or passively; i.e. they either accumulate it within their bodies or adsorb it to their surfaces (Dixit et al. 2015; Ojuederie and Babalola 2017). The cell surfaces of microorganisms have cell walls that are rich in polysaccharides, proteins and

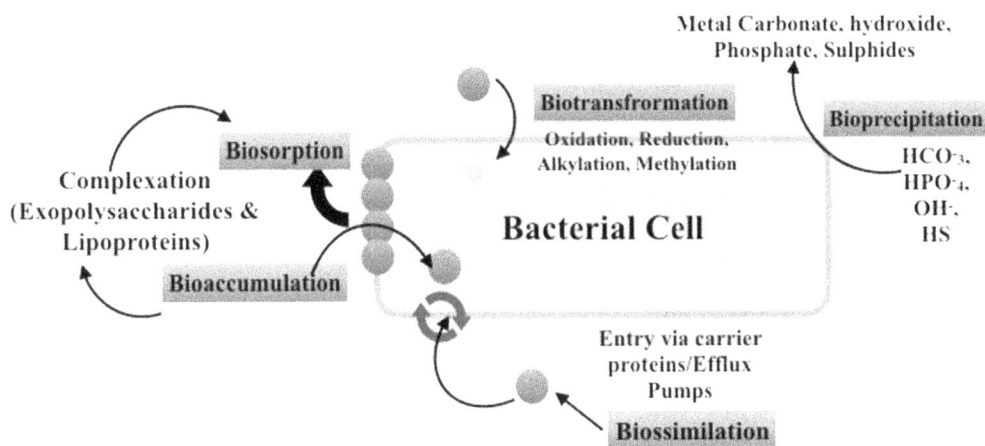

FIGURE 7.2 Different mechanisms used by microbes to remove/detoxify heavy metals.

lipids and these contain a variety of functional groups that can help in binding and holding metal ions to them (Ayangbenro and Babalola 2017a). Biosorption seems to be the most practical bioremediation method because for bioaccumulation of metals microorganisms require extra nutrients specially increase BOD (biological oxygen demand) and COD (chemical oxygen demand). They are able to perform metal removal by reducing or oxidizing them; they can also immobilize metals, transform them to less toxic states or volatilize them. Although heavy metals can be toxic to the cell itself, microbes tend to adapt themselves by generating protective outer coverings, biofilms that are hydrophobic and prevent uptake of toxic substances within the cells (Prabhakaran et al. 2016). The most commonly employed mechanism for metal removal from within the cells is through energy-driven efflux pumps that make prokaryotes resistant to variety of metals (Voica et al. 2016). Figure 7.2 outlines some of the commonly used mechanisms for heavy metal removal used by microbes while a list of these techniques is mentioned in Table 7.2. Surface adsorption/biosorption is discussed in detail below:

7.4.1 SURFACE ADSORPTION/BIOSORPTION

The most common and popular mechanism for metal removal from contaminated sites is via their adsorption outside cell walls/boundaries. This mechanism specially depends on the secretion of extracellular polymeric substances (EPS); they are known to complex metals through proton exchanges or precipitation techniques. The EPS biomass is rich in charged freely available functional groups like phosphate, carbonyl, hydroxyl, uronic acid, sulfate groups, etc., that readily attach with metal ions (Gupta and Diwan 2017; Ojuederie and Babalola 2017). These functional groups and other groups rich in peptidoglycan layers of gram positive bacteria like teichoic acid, glycoproteins, lipoproteins, di-aminopimelic acid and lipopolysaccharides in gram negative bacteria are all involved in biosorption (Ayangbenro and Babalola 2017b; Yue et al. 2015).

Bacteria are excellent bioadsorbants; similarly, fungi can also bioaccumulate heavy metals and can detoxify them via biotransformation so they are also good recipients of bioremediation. *Saccharomyces cerevisiae,* the commonly available baker's yeast, is famous for its biosorbing potential of Zn(II) and cadmium (Cd II) heavy metals. They store metals within their hyphae and mycelia. Alae are also considered primary biosorbants with sorption capacity of 15.3%–84.6% as reported by Mustapha and Normala (2015). Algal proteins and polysaccharides contribute toward the biosorption capacity in algae (Zeraatkar et al. 2016).

TABLE 7.2
Some commonly used bioremediation techniques employed by microorganisms

Process	Mechanism	Metals	Bacteria	References
Intracellular sequestration	Complexation with different compounds within the cell cytosol	Cd, Cu & Zn	*Pseudomonas putida*	Higham et al. (1986)
Bioaccumulation	Extracellular sequestration (insoluble form in periplasm)	Cu, Fe, S, Mn, Cr & U	*Pseudomonas syringae*, Geobacter spp., Desulfuromonas spp., *G. metallireducens*	Cha and Cooksey (1991), Gavrilescu (2004) and Bruschi and Goulhen (2007)
Biomethylation	Addition of methyl groups to metals (CH_3), which changes their solubility and volatility (aerobic & anaerobic)	Hg, As, Se, Sn, Pb & Te	Bacillus spp., Pseudomonas spp.	Gadd (2004) and Ramasamy and Banu (2007)
Transformation (redox)	Oxidation or reduction of metals can lead to their transformation to less toxic states	Fe, U, Cr, Se etc.	Iron & sulfur reducing bacteria	Bisht and Harsh (2014)

Saccharomyces cerevisiae biosorbs Zn(II) and Cd(II) heavy metals by using the ion exchange method (Chen and Wang 2007). *Cunninghamella elegans* has also been reported as a promising organism for the treatment of metals released from textile industrial wastewaters (Tigini et al. 2010). A detailed list of microbial biosorbants has been provided in Table 7.3.

7.5 PHYTOREMEDIATION

Other than bacteria plants have been exploited for their capacity of environment cleaning for decades; plants as hydrophytes can remove toxic compounds, heavy metals (Ali et al. 2013) and other dangerous substances from the water via multiple techniques. Hydrophytes are the plants that grow in water. They are of different types depending on their submerged nature. The use of plants for remediation of contaminated lands and water bodies has been emphasized over the years due to its cost-effectiveness, efficiency and environment-friendly nature (Dixit et al. 2015). The efficiency is specifically dependent on the metal uptake capacity of the recipient plant and its survival post remediation (Khan and Faisal 2018). Plants with the ability to accumulate around 1,000 mg/kg or 10,000 mg/kg are referred to as hyperaccumulators (Krämer 2010; Wu et al. 2010; Khan and Faisal 2018). A good candidate of phytoremediation should possess the following qualities: high development rate, increased threshold limit to withstand metals, resistance to pests and insects, a complex and thick root and shoot system (Pascal-Lorber and Laurent 2011), the ability to resist multiple metals rather than just one – also, it should not be an ornamental plant; rather, it should not be attractive to herbivores and other predators (Wolfenbarger and Phifer 2000). This property ensures that no metal accumulation occurs at higher trophic levels (Khan and Faisal 2018). Phytoremediation is also subdivided into the following types: phytoextraction, which refers to the uptake of pollutants and their transportation to the shoots (Bhargava et al. 2012); phytofiltration, also referred to as rhizofiltration, which is the use of plant roots to remove pollutants from contaminated land and water (Rawat et al. 2012); phytostabilization, which is a process that immobilizes metals and makes them unavailable in the environment preventing their migration and toxicity up the food chain (Cheraghi et al. 2011), which is primarily done by

TABLE 7.3
Microbial biosorbants and their biosorption capacity

Group	Organism	Metals biosorped	Biosorption capacity	References
Bacteria	*Bacillus cereus* immobilized on alginate surfaces	Hg	104.1 mg/g	Sinha et al. (2012)
	Aneurinibacillus aneurinilyticus	As(III)	500 mg/l	Dey et al. (2016)
	B. licheniformis	Cd	159 mg/g	Zouboulis
		Cr(VI)	62 mg/g	et al. (2004)
	Desulfovibrio desulfuricans (immobilize on zeolite)	Cu	98.2 mg/g	Ayangbenro and
		Ni	90.1 mg/g	Babalola (2017a)
		Cr(VI)	99.8 mg/g	
	Micrococcus luteus DE2008	Cu	408 mg/g	Puyen et al. (2012)
		Pb	1965 mg/g	
	Pseudomonas aeruginosa	Co	8.92 mg/g	Kang et al. (2005)
		Ni	8.26 mg/g	
		Cr(III)	6.42 mg/g	
	Pseudomonas veronii 2E	Cd	50%	Vullo et al. (2008)
		Zn	50%	
		Cu	40%	
	P. fluorescence	Cr(VI)	40.8 mg/l	Remade (1990)
	Azotobacter sp.	Pb	310 mg/g	Remade (1990)
	Aeromonas sp.			
	Xanthomonas compestris	Cu	7.3×10^8	Remade (1990)
	Enterobacter Cloaceae B1	Cd	64.17%	Banerjee et al. (2015)
		Pb	95.25%	
		Ni	36.77%	
Fungi	*Aspergillus parasitica*	Pb(II)	4.46×10^{-4} mol g^{-1}	Tunali et al. (2006)
	Cephalosporium aphidicola	Pb(II)	4.02×10^{-4} mol g^{-1}	Akar et al. (2007)
	Hymenoscyphus ericae, Neocosmospora vasinfecta and *Verticillum terrestre*	Hg(II)	50%–80%	Kelly et al. (2006)
	Asperigillus niger	Pb	34.4 mg/g	Zeng et al. (2015)
		Cu	28.7 mg/g	
	Penicillium chrysogenum	Ni	260 mg/g	Martins et al. (2015)
		Pb	204 mg/g	
	Saccharomyces cerevisiae	Pb	270 mg/g	Hrynkiewicz and
		Hg	64.2 mg/g	Baum (2014)
	Asparagopsis armata	Cd	32.3 mg/g	Romera et al. (2007)
		Ni	17.7 mg/g	
		Zn	21.6 mg/g	
		Cu	21.3 mg/g	
		Pb	63.7 mg/g	

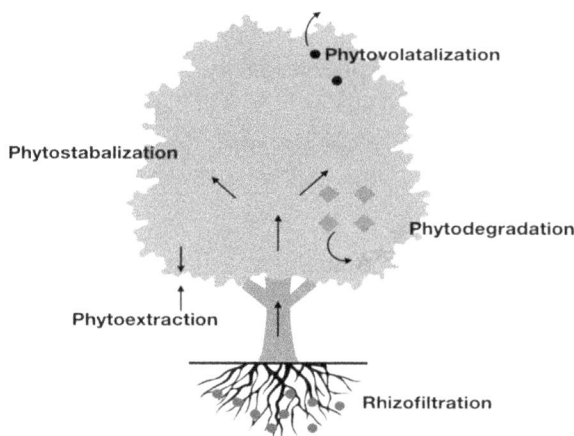

FIGURE 7.3   Different methods of phytoremediation.

complexation with surface receptors, or degradation by plant enzymes like dehalogenases, pre-cipitation or metal reduction. Phytovolatilization is another process of plant-mediated remedia-tion that leads to conversion of metals especially Hg and selenium (Se) to gaseous states (Karami and Shamsuddin 2010).

Phytoremediation has achieved a lot of importance over the years because of its low cost and safe nature; however, the criteria of hyperaccumulation differs depending on the metal type, i.e. it is calculated to be 100 mg/kg for Cd and 1,000 mg/kg for copper, cobalt, chromium and lead (Dixit et al. 2015) (Figure 7.3).

7.5.1 PHYTOREMEDIATION USING HYDROPHYTES & MACROPHYTES

Hydrophytes are those plant types that have a tendency to grow in water; they are considered to be an excellent pollutant removal machinery especially for the treatment of contaminated waters because of their capacity to uptake contaminants with increased efficiency. They are specifically favored over terrestrial plants because of their small size, higher biomass, rapid growth and higher metal uptake capacity (Khan and Faisal 2018). Water or aquatic plants that are not microscopic but can be observed with the naked eye are referred to as macrophytes; they are further classified as submerged, emergent or free-floating macrophytes (Rai 2009).

A list of macrophytes has been mentioned in Table 7.4:

7.6 METAL DETOXIFICATION IN PLANTS

Plants take up heavy metals by the mechanisms mentioned above; however, if plants don't pos-sess any mechanism of metal detoxification the plants would eventually die due to metal toxicity. Metals are classified into two categories: redox-active heavy metals that include Cr, Cu, Mn and Fe and non-redox heavy metals that comprise Cd, Ni, Hg, Zn and Al (Bücker-Neto et al. 2017). Redox metals generate reactive oxygen species (ROS) leading to damage to DNA and cell death via oxidation (Emamverdian et al. 2015) while non-redox metals cause toxicity by activating ROS-producing enzymes (Bielen et al. 2013; Emamverdian et al. 2015). Plants react against this toxicity induced by heavy metals through their physical barriers like trichomes, cuticles and cell wall and mycorrhizal association also serves as an important physical barrier (Emamverdian et al. 2015; Silva and Matos 2016). Second, plants inhibit ROS production by actively secreting

TABLE 7.4
Common heavy metals and their phytoaccumulators

Heavy metals	Macrophytes	References
Hg	*Azolla pinnata* (R.) Br, *Arcocornia fruticosa* (L.) A.J. Scott, *Halimione portulacoides* (L.) Aellen and *Spartina maritima* (Curtis) Fernald	Canario et al. (2007) and Newete and Byrne (2016)
As	*Lemna gibba* L. and *Lemna minor*	Newete and Byrne (2016)
Cd	*Azolla pinnata* (R.) Br. and *Bacopa monnieri* (L.) Pernnell	Newete and Byrne (2016)
Pb	*Leersia hexandra*, *Equisetum ramosisti*, *Juncus effuses* and *C. demersum*	Newete and Byrne (2016)
Cr	*Callitriche cophocarpa* (water-starwort), *Salvinia herzogii* and *Pistia stratiotes*, *Lemna* spp.	Maine et al. (2004) and Augustynowicz et al. (2010)

anti-oxidants like ascorbate, glutathione, alkaloids, tocopherols and some enzymes that degrade these ROS like catalase, glutathione reductase, etc. (Rastgoo et al. 2011; Singh et al. 2016). Another defense system utilized by plants is the secretion of the enzyme phytochelatin synthase that has the ability to bind to lethal concentrations of heavy metals (Solanki and Dhankhar 2011; Gupta et al. 2013; Jan and Parray 2016); another mechanism employed by plants is the secretion of metallothioneins (Du et al. 2012), which are small-sized metal binding proteins rich in cysteine (Guo et al. 2013; Ojuederie and Babalola 2017) and can bind to a broad range of heavy metals like copper, nickel, mercury, cadmium, thallium, chromium, zinc, lead and arsenic, etc. They sequester the metals, maintain homeostasis and also help in repair mechanisms (Guo et al. 2013); proline is an osmolyte, radical scavenger and stabilizer for macromolecules (Ehsanpour et al. 2012).

7.7   SYMBIOTIC ASSOCIATION BETWEEN HYDROPHYTES AND BACTERIA

A class of bacteria that reside within the plant roots and are in a symbiotic association with the plants are regarded as epiphytic bacteria, and as they reside within the roots they are also called as rhizospheric bacteria. These bacterial species are mostly involved in improving plant growth, yield and protection from environmental hazards; hence, they are broadly classified as plant growth promoting rhizospheric bacteria (PGPR). They can be intracellular rhizobacteria (bacteria that reside within the root nodules) or extracellularly existing rhizobacteria (that are not physically attached to the plants). They enhance plant growth by providing access to nutrients that are easier for plant roots to absorb; they also release growth factors especially phytohormones like auxins (indole 3-acetic acid (IAA)) and provide protection against pathogens. Basically, growth enhancement by PGPRs is via two approaches: direct and indirect. In the direct approach bacteria release compounds that help in enhancement of plant growth while in the indirect approach the PGPRs act as agents that prevent infestation of plants with pests and insects and help in detoxification of heavy metals (Glick 2012).

7.7.1   DIRECT APPROACH

In this approach the bacteria secrete the following compounds like siderophores for sequestration, phosphate solubilization agents and 1-aminocyclopropane-1-carboxylate (ACC) deaminase synthesis (Table 7.5); these compounds allow the plants to fight unfavorable conditions and hence help in growth enhancement (Donot et al. 2012; Ahemad and Kibret 2014).

TABLE 7.5
Mechanism employed by PGPRs in plant growth promotion

Compounds secreted	Effects/benefits	Bacteria	References
Siderophores	Metal chelator proteins that are produced under metal limiting conditions especially for Fe; under Fe deficiency the bacterial siderophores make iron available to the plants by removing it from soil, hence help in overcoming Fe deficiency	*Pseudomonas* spp., *Bacillus megaterium, Micobacterium, Bravibacterium* sp. *Kluyvera ascorbata*	Ahmed and Holmstrom (2014)
Phosphate solubilization	A macronutrient required by plants, bacteria that reside in the rhizosphere convert phosphorous into a soluble form which is readily uptaken by plants; hence these bacteria are called phosphate-solubilizing bacteria	*Flavobacterium, Bacillus, Burkholderia, Erwinia, Azotobacter, Microbacterium*, etc.	Choudhary et al. (2016) and Bhattacharyya and Jha (2012)
Auxin production	IAA is an important plant hormone; secreted by almost 80% of rhizobacteria they enhance plant growth by improving root growth and protection against abiotic stresses. IAA production enhances dense root growth with increased root hair production. This particularly increases the surface area for more enhanced nutrient uptake	*Achromobacter xylosooxidans, Brevibacillus brevis, Xanthomonas* spp., *Azomonas* spp., *Sphingomonas* spp., *Mycobacterium* spp.	Spaepen and Vanderleyden (2011), Tak et al. (2013) and Ahemad and Kibret (2014)
ACC deaminase production	Plants under stress conditions release excess ethylene, which leads to decrease in plant growth and death. ACC deaminase is involved in degradation of ethylene to help plants fight abiotic stresses like metal contamination. PGPR helps to stimulate synthesis of this enzyme		Milošević et al. (2012) and Glick (2014)

7.8 PLANT AND PGPR-MEDIATED HEAVY METAL DETOXIFICATION

As PGPRs are associated with plant roots, their role in plant growth and detoxification of heavy metals and the stress generated by them is undeniable. These microorganisms have a synergistic effect on metal remediation when used with plants. Bacteria that reside in the rhizosphere help in sequestration of heavy metals and therefore facilitate bioremediation; they particularly boost phy-toextraction abilities by changing metal dissolution, its translocation within the plant, a decrease in rhizospheric pH and production of chelators like siderophores (Tak et al. 2013). Plant siderophores' binding ability to iron is lower compared to microbial siderophores; hence, in metal-contaminated sites this efficiency is reduced manifold so the siderophores released by bacteria (PGPR) is used by plants to sequester iron (Glick 2003). A study conducted by Burd et al. (1998) showed that when the nickel-resistant bacterial strain *Kluyvera ascorbata* and *Brassica juncea,* common name Indian mustard, which is a nickel-hyperaccumulating plant, were co-cultivated, stunted plant growth due to elevated nickel levels in the soil was reversed. The bacterium did not affect the metal intake or consumption capacity nor did it affect its accumulation in the plant; it only reduced ethylene produc-tion in the plants treated with nickel in contrast to plants that were not co-cultivated with bacteria. In simple terms, bacteria only protects the plant against nickel-related stress that leads to ethylene production. Bacteria perform this by releasing ACC deaminase, which lowers the amount of ethyl-ene, thus protecting plants against nickel-induced stress (Burd et al. 1998).

In another study by the same group (Burd et al. 2000), the same bacterium *K. ascorbata* SUD165 when grown in heavy metal-contaminated sites led to the production of a mutant that over-produced siderophores; it was referred to as *K. ascorbata* SUD165/26. They used this over-producing mutant and inoculated it into seeds of tomato, canola and Indian mustard growing in heavy metal-contaminated sites. Metal stress usually creates iron deficiency in surrounding rhizospheric regions; as mentioned previously, plant siderophores are ineffective in sequestering iron in lower concentrations – hence, it leads to iron deficiency in plants, which leads to chlorosis and stress-mediated release of ethylene. When these plants were inoculated with the siderophore over-producing mutant, iron deficiency and ethylene-caused stress and metal-caused inhibition of plant development were overcome. So in this instance also bacteria not just helped in overcoming metal contamination but also supported hyper-accumulator plant growth by protecting it against metal stress (Glick 2003).

Another study highlights the negative effects conferred by arsenic to arsenic-resistant strains; however, in this study scientists produced bioengineered seeds that were transformed with bacte-rial ACC deaminase: transgenic seeds showed a 70% germination rate compared to non-transgenic seeds, whose germination rate was only 25% (Nie et al. 2002). Here also the inhibitory effects of ethylene were combated by the bacterial ACC deaminase. A comparison between transgenic cano-la's rate of arsenic accumulation showed it to be fourfold more than that of non-transgenic canola, indicating the role of bacteria in making phytoremediation more efficient (Glick 2003).

This and many other studies indicate the positive effects metal-resistant bacterial strains confer to hydrophytes when co-cultivated together; although a hyperaccumulating plant will grow in the pres-ence of heavy metals and accumulate and degrade it too, the efficiency of the process is reduced, i.e. plant growth is usually effected with a reduction in yield and development as mentioned above. This metal-related toxicity can be overcome when a metal-resistant strain grows in close vicinity to the plant and enhances this remediation process. The bacteria help in the release of phytohormones and antibiotics that contribute to enhanced plant growth under metal stress conditions. So bacteria pro-vide a supportive environment for phytoremediation, improving plants reaction to metal exposure.

7.9 FUTURE PROSPECTS

A better understanding of bacterial involvement can help to improve bioremediation techniques. The future of bioremediation is using engineered plants and bacteria, genetically modified organ-isms (GMOs) carrying genes that not only enhance growth but also confer the organism with the ability to remove unwanted metals is the new paradigm. Genes that code for antioxidant enzymes and oxidants like phytochelatins enhance the ability of plants to accumulate metals and tolerate high metal concentrations (Divya and Kumar 2011; Mani and Kumar 2014). A study by Mani and Kumar (2014) shows modified rice and tobacco plants carrying *merA* genes are able to resist up to ten times more mercury concentrations than their non-transgenic controls. Many transgenic plants species have been produced using genetic engineering and recombinant DNA technology; some commonly known transgenic plants include *Arabidopsis thaliana* (a plant able to tolerate extremely high salt concentrations), tobacco (*Nicotiana tabaccum*), *Brassica* species and variants, *botrytis* and *Lycopersicon esculentum,* which are commonly used for phytoremediation (Chen and Wilson 1997). Genetically modified plants carrying genes for ACC deaminase and possessing the ability to translocate mercury to the roots and shoots by expression of *merA* and *merB* genes (Arshad et al. 2007), which are also able to degrade other organic pollutants, have been reported. Plants with the ability to efficiently grow have been produced using bacterial strains (Singh et al. 2011).

However, a possible limitation of this dissemination of GMOs, i.e. transgenic plants and microbes, is the likelihood of horizontal gene transfer that can lead to the generation of resistant bacterial strains through easy transfer of resistant genes between transgenic bacteria/plants and indigenous microorganisms. This would, however, be a possible threat to the appearance of multidrug resistant strains, a possibility that is already known. To overcome this a suicidal gene should be introduced in transgenic organisms or they should be programmed to die once the bio-agent has performed its function (Jan et al. 2014).

7.10   CONCLUSION

Phytoremediation and bioremediation using plants and bacteria to remediate heavy metal pollutants from contaminated sites is an extremely cheap, environmentally safe and efficient process; however, if we synergistically apply the two, i.e. bacteria and plants, to decontaminate sites and detoxify heavy metals the rate of metal removal is enhanced manifold. This process can further be scaled up if genetically engineered plants and bacteria are used that carry the resistant genes and help to clean contaminated waters. However, more advancements in GMOs are required to overcome one of its major drawbacks, i.e. the dissemination of resistant genes in the environment, to make it a feasible bioremediation tool.

REFERENCES

Abhanga R, Wanib K, Patilc V, Pangarkara B, Parjanea S (2013) Nanofiltration for recovery of heavy metal ions from waste water-a review. *Liver* 1:0.60–65.60.
Ahemad M, Kibret M (2014) Mechanisms and applications of plant growth promoting rhizobacteria: current perspective. *J King Saud Univ Sci* 26 (1):1–20.
Ahmed E, Holmstrom SJM (2014) Siderophores in environmental research: roles and applications. *Microb Biotechnol* 7 (3):196–208. doi:10.1111/1751-7915.12117.
Akar T, Tunali S, Cabuk A (2007) Study on the characterization of lead (II) biosorption by fungus Aspergillus parasiticus. *Appl Biochem Biotechnol* 136 (3):389–405. doi:10.1007/s12010-007-9032-8.
Akpor OB, Muchie M (2010) Remediation of heavy metals in drinking water and wastewater treatment systems: processes and applications. *Int J Phys Sci* 5 (12):1807–1817.
Akpor OB, Ohiobor GO, Olaolu D (2014) Heavy metal pollutants in wastewater effluents: sources, effects and remediation. *Adv Biosci Bioeng* 2 (4):37–43.
Ali H, Khan E, Sajad MA (2013) Phytoremediation of heavy metals-concepts and applications. *Chemosphere* 91 (7):869–881. doi:10.1016/j.chemosphere.2013.01.075.
Arshad M, Saleem M, Hussain S (2007) Perspectives of bacterial ACC deaminase in phytoremediation. *Trends Biotechnol* 25 (8):356–362.
Augustynowicz J, Grosicki M, Hanus-Fajerska E, Lekka M, Waloszek A, Koloczek H (2010) Chromium(VI) bioremediation by aquatic macrophyte Callitriche cophocarpa Sendtn. *Chemosphere* 79 (11):1077–1083. doi:10.1016/j.chemosphere.2010.03.019.
Ayangbenro AS, Babalola OO (2017a) A new strategy for heavy metal polluted environments: a review of microbial biosorbents. *Int J Environ Res Public Health* 14 (1):94. doi:10.3390/ijerph14010094.
Ayangbenro AS, Babalola OO (2017b) A new strategy for heavy metal polluted environments: a review of microbial biosorbents. *Int J Env Res Public Health* 14 (1):94. doi:10.3390/ijerph14010094.
Banerjee G, Pandey S, Ray AK, Kumar R (2015) Bioremediation of heavy metals by a novel bacterial strain Enterobacter cloacae and its antioxidant enzyme activity, flocculant production, and protein expression in presence of lead, cadmium, and nickel. *Water Air Soil Pollut* 226 (4):91.

Bernard A (2008) Cadmium & its adverse effects on human health. *Indian J Med Res* 128 (4):557.

Bhargava A, Carmona FF, Bhargava M, Srivastava S (2012) Approaches for enhanced phytoextraction of heavy metals. *J Environ Manage* 105:103–120. doi:10.1016/j.jenvman.2012.04.002.

Bhattacharyya PN, Jha DK (2012) Plant growth-promoting rhizobacteria (PGPR): emergence in agriculture. *World J Microbiol Biotechnol* 28 (4):1327–1350. doi:10.1007/s11274-011-0979-9.

Bielen A, Remans T, Vangronsveld J, Cuypers A (2013) The influence of metal stress on the availability and redox state of ascorbate, and possible interference with its cellular functions. *Int J Mol Sci* 14 (3):6382–6413. doi:10.3390/ijms14036382.

Bisht J, Harsh N (2014) Utilizing Aspergillus niger for bioremediation of tannery effluent. *Octa J Env Res* 2 (1):77–81.

Bruschi M, Goulhen F (2007) New bioremediation technologies to remove heavy metals and radionuclides using Fe(III)-, sulfate- and sulfur-reducing bacteria. In: Singh SN, Tripathi RD (eds.), *Environmental Bioremediation Technologies*. Springer, pp. 35–55. https://doi.org/10.1007/978-3-540-34793-4_2

Bücker-Neto L, Paiva ALS, Machado RD, Arenhart RA, Margis-Pinheiro M (2017) Interactions between plant hormones and heavy metals responses. *Genet Mol Biol* 40 (1):373–386.

Burd GI, Dixon DG, Glick BR (1998) A plant growth-promoting bacterium that decreases nickel toxicity in seedlings. *Appl Environ Microbiol* 64 (10):3663–3668.

Burd GI, Dixon DG, Glick BR (2000) Plant growth-promoting bacteria that decrease heavy metal toxicity in plants. *Can J Microbiol* 46 (3):237–245.

Canario J, Caetano C, Vale C, Cesario R (2007) Evidence for elevated production of methylmercury in salt marshes. *Environ Sci Technol* 41 (21):7376–7382. doi:10.1021/es071078j.

Cannata Andia J (1996) Aluminium toxicity: its relationship with bone and iron metabolism. *Nephrol Dial Transplant* 11 (supp3):69–73.

Cha JS, Cooksey DA (1991) Copper resistance in Pseudomonas syringae mediated by periplasmic and outer membrane proteins. *Proc Natl Acad Sci U S A* 88 (20):8915–8919. doi:10.1073/pnas.88.20.8915.

Charerntanyarak L (1999) Heavy metals removal by chemical coagulation and precipitation. *Water Sci Technol Health Care* 39 (10–11):135–138.

Chen C, Wang JL (2007) Characteristics of Zn^{2+} biosorption by Saccharomyces cerevisiae. *Biomed Environ Sci* 20 (6):478–482.

Chen S, Wilson DB (1997) Genetic engineering of bacteria and their potential for Hg^{2+} bioremediation. *Biodegradation* 8 (2):97–103.

Cheraghi M, Lorestani B, Khorasani N, Yousefi N, Karami M (2011) Findings on the phytoextraction and phytostabilization of soils contaminated with heavy metals. *Biol Trace Elem Res* 144 (1–3):1133–1141. doi:10.1007/s12011-009-8359-0.

Choudhary DK, Varma A, Tuteja N (2016) *Plant-Microbe Interaction: An Approach to Sustainable Agriculture*. Springer

Clemens S, Ma JF (2016) Toxic heavy metal and metalloid accumulation in crop plants and foods. *Ann Rev Plant Biol* 67:489–512.

Dey U, Chatterjee S, Mondal NK (2016) Isolation and characterization of arsenic-resistant bacteria and possible application in bioremediation. *Biotechnol Rep* 10:1–7.

Divya B, Kumar MD (2011) Plant-microbe interaction with enhanced bioremediation. *Res J Biotechnol* 6 (1):72–79.

Dixit R, Wasiullah, Malaviya D, Pandiyan K, Singh UB, Sahu A, Shukla R, Singh BP, Rai JP, Sharma PK, Lade H, Paul D (2015) Bioremediation of heavy metals from soil and aquatic environment: an overview of principles and criteria of fundamental processes. *Sustainability* 7 (2):2189–2212. doi:10.3390/su7022189.

Donot F, Fontana A, Baccou JC, Schorr-Galindo S (2012) Microbial exopolysaccharides: main examples of synthesis, excretion, genetics and extraction. *Carbohydr Polym* 87 (2):951–962. doi:10.1016/j.carbpol.2011.08.083.

Du J, Yang JL, Li CH (2012) Advances in metallotionein studies in forest trees. *Plant Omics* 5 (1):46–51.

Ehsanpour A, Zarei S, Abbaspour J (2012) The role of over expression of P5CS gene on proline, catalase, ascorbate peroxidase activity and lipid peroxidation of transgenic tobacco (Nicotiana tabacum L.) plant under in vitro drought stress. *J Cell Mol Res* 4 (1):43–49.

Emamverdian A, Ding Y, Mokhberdoran F, Xie Y (2015) Heavy metal stress and some mechanisms of plant defense response. *Sci World J*. 2015:756120.

Gadd GM (2004) Microbial influence on metal mobility and application for bioremediation. *Geoderma* 122 (2–4):109–119. doi:10.1016/j.geoderma.2004.01.002.

Gavrilescu M (2004) Removal of heavy metals from the environment by biosorption. *Eng Life Sci* 4 (3):219–232. doi:10.1002/elsc.200420026.

Glick BR (2003) Phytoremediation: synergistic use of plants and bacteria to clean up the environment. *Biotechnol Adv* 21 (5):383–393. doi:10.1016/S0734-9750(03)00055-7.

Glick BR (2012) Plant growth-promoting bacteria: mechanisms and applications. *Scientifica* 2012:963401. doi:10.6064/2012/963401.

Glick BR (2014) Bacteria with ACC deaminase can promote plant growth and help to feed the world. *Microbiol Res* 169 (1):30–39. doi:10.1016/j.micres.2013.09.009.

Guo JL, Xu LP, Su YC, Wang HB, Gao SW, Xu JS, Que YX (2013) ScMT2-1-3, a metallothionein gene of sugarcane, plays an important role in the regulation of heavy metal tolerance/accumulation. *Biomed Res Int* 2013: Artn 904769. doi:10.1155/2013/904769.

Gupta P, Diwan B (2017) Bacterial Exopolysaccharide mediated heavy metal removal: a review on biosynthesis, mechanism and remediation strategies. *Biotechnol Rep* 13:58–71. doi:10.1016/j.btre.2016.12.006.

Gupta DK, Huang HG, Corpas FJ (2013) Lead tolerance in plants: strategies for phytoremediation. *Env Sci Pollut Res* 20 (4):2150–2161. doi:10.1007/s11356-013-1485-4.

Higham DP, Sadler PJ, Scawen MD (1986) Cadmium-binding proteins in Pseudomonas putida: pseudothioneins. *Environ Health Perspect* 65:5–11. doi:10.1289/ehp.86655.

Hrynkiewicz K, Baum C (2014) Application of microorganisms in bioremediation of environment from heavy metals. In: Malik A, Grohmann E, Akhtar R (eds.), *Environmental Deterioration and Human Health*. Springer, pp. 215–227. https://doi.org/10.1007/978-94-007-7890-0_9

Jaishankar M, Tseten T, Anbalagan N, Mathew BB, Beeregowda KN (2014) Toxicity, mechanism and health effects of some heavy metals. *Interdiscip Toxicol* 7 (2):60–72.

Jan AT, Azam M, Ali A, Haq QMR (2014) Prospects for exploiting bacteria for bioremediation of metal pollution. *Criti Rev Environ Sci Technol Health Care* 44 (5):519–560.

Jan AT, Azam M, Siddiqui K, Ali A, Choi I, Haq QM (2015) Heavy metals and human health: mechanistic insight into toxicity and counter defense system of antioxidants. *Int J Mol Sci* 16 (12):29592–29630.

Jan S, Parray JA (2016) *Approaches to Heavy Metal Tolerance in Plants*. Springer,

Juang RS, Shiau RC (2000) Metal removal from aqueous solutions using chitosan-enhanced membrane filtration. *J Membr Sci* 165 (2):159–167. doi:10.1016/S0376-7388(99)00235-5.

Jüttner K, Galla U, Schmieder H (2000) Electrochemical approaches to environmental problems in the process industry. *Electrochimica Acta* 45 (15–16):2575–2594.

Kang S, Lee J, Kim K (2005) Metal removal from wastewater by bacterial sorption: kinetics and competition studies. *Environ Technol* 26 (6):615–624.

Karami A, Shamsuddin ZH (2010) Phytoremediation of heavy metals with several efficiency enhancer methods. *Afr J Biotechnol* 9 (25):3689–3698.

Kelly DJA, Budd K, Lefebvre DD (2006) The biotransformation of mercury in pH-stat cultures of microfungi. *Can J Bot* 84 (2):254–260. doi:10.1139/B05-156.

Khan HN, Faisal M (2018) Phytoremediation of industrial wastewater by hydrophytes. In: Ansari A, Gill S, Gill R, Lanza G, Newman L (eds.), *Phytoremediation*. Springer, pp. 179–200. https://doi.org/10.1007/978-3-319-99651-6_8

Krämer U (2010) Metal hyperaccumulation in plants. *Annu Rev Plant Biol* 61:517–534.

Kurniawan TA, Chan GY, Lo W-H, Babel S (2006) Physico–chemical treatment techniques for wastewater laden with heavy metals. *Chem Eng J* 118 (1–2):83–98.

Maine MA, Sune NL, Lagger SC (2004) Chromium bioaccumulation: comparison of the capacity of two floating aquatic macrophytes. *Water Res* 38 (6):1494–1501. doi:10.1016/j.watres.2003.12.025.

Mani D, Kumar C (2014) Biotechnological advances in bioremediation of heavy metals contaminated ecosystems: an overview with special reference to phytoremediation. *Int J Environ Sci Technol* 11 (3):843–872. doi:10.1007/s13762-013-0299-8.

Martín JR, De Arana C, Ramos-Miras J, Gil C, Boluda R (2015) Impact of 70 years urban growth associated with heavy metal pollution. *Environ Pollut* 196:156–163.

Martin S, Griswold W (2009) Human health effects of heavy metals. *Environ Sci Technol Briefs Citizens* 15:1–6.

Martins LR, Lyra FH, Rugani MM, Takahashi JA (2015) Bioremediation of metallic ions by eight Penicillium species. *J Environ Eng* 142 (9):C4015007.

Milošević N, Marinković J, Tintor B (2012) Mitigating abiotic stress in crop plants by microorganisms. *Matica Srp Proc Nat Sci* (123):17–26.

Mustapha MU, Normala H (2015) Microorganisms and biosorption of heavy metals in the environment: a review paper. *J Microb Biochem Technol* 7:253–256.

Newete SW, Byrne MJ (2016) The capacity of aquatic macrophytes for phytoremediation and their disposal with specific reference to water hyacinth. *Environ Sci Pollut Res* 23 (11):10630–10643. doi:10.1007/s11356-016-6329-6.

Nichols C (1974) *Development Document for Effluent Limitations Guidelines and New Source Performance Standards for the Steam Electric Power Generating Point Source Category*. Environmental Protection Agency.

Nie L, Shah S, Rashid A, Burd GI, Dixon DG, Glick BR (2002) Phytoremediation of arsenate contaminated soil by transgenic canola and the plant growth-promoting bacterium Enterobacter cloacae CAL2. *Plant Physiol Biochem* 40 (4):355–361.

Nomanbhay, SM, Palanisamy K (2005) Removal of heavy metal from industrial wastewater using chitosan coated oil palm shell charcoal. *Electronic Journal of Biotechnology*, 8:43–54.

Ojuederie OB, Babalola OO (2017) Microbial and plant-assisted bioremediation of heavy metal polluted environments: a review. *Int J Env Res Public Health* 14 (12):1504. doi:10.3390/ijerph14121504.

Olaniran AO, Balgobind A, Pillay B (2013) Bioavailability of heavy metals in soil: impact on microbial biodegradation of organic compounds and possible improvement strategies. *International Journal of Molecular Sciences* 14 (5):10197–10228.

Pandey G, Madhuri S (2014) Heavy metals causing toxicity in animals and fishes. *Res J Anim Vet Fishery Sci* 2 (2):17–23.

Pandya JA (2015) Nanofiltration for recovery of heavy metal from waste water. *Liver* 1:0–30.

Pascal-Lorber S, Laurent F (2011) Phytoremediation techniques for pesticide contaminations. In *Alternative Farming Systems, Biotechnology, Drought Stress and Ecological Fertilisation*. Sustainable Agriculture Reviews, vol 6, pp. 77–105. Springer. https://doi.org/10.1007/978-94-007-0186-1_4

Prabhakaran P, Ashraf MA, Aqma WS (2016) Microbial stress response to heavy metals in the environment. *RSC Adv* 6 (111):109862–109877. doi:10.1039/c6ra10966g.

Puyen ZM, Villagrasa E, Maldonado J, Diestra E, Esteve I, Solé A (2012) Biosorption of lead and copper by heavy-metal tolerant Micrococcus luteus DE2008. *Bioresour Technol* 126:233–237.

Qdais HA, Moussa HJD (2004) Removal of heavy metals from wastewater by membrane processes: a comparative study. *Desalination* 164 (2):105–110.

Rai PK (2009) Heavy metal phytoremediation from aquatic ecosystems with special reference to macrophytes. *Crit Rev Environ Sci Technol* 39 (9):697–753. doi:10.1080/10643380801910058.

Ramasamy K, Banu SP (2007) Bioremediation of metals: microbial processes and techniques. In: Singh SN, Tripathi RD (eds.), *Environmental Bioremediation Technologies*. Springer, pp. 173–187. https://doi.org/10.1007/978-3-540-34793-4_7

Rastgoo L, Alemzadeh A, Afsharifar A (2011) Isolation of two novel isoforms encoding zinc-and copper-transporting P1B-ATPase from Gouan (Aeluropus littoralis). *Plant Omics J* 4 (7):377–383.

Rawat K, Fulekar M, Pathak B (2012) Rhizofiltration: a green technology for remediation of heavy metals. *Intl J Inno Biosci* 2 (4):193–199.

Remade J (1990) The cell wall and metal binding. In: *Biosorption of Heavy Metals*. CRC Press Boca Raton, pp. 83–92.

Romera E, Gonzalez F, Ballester A, Blazquez ML, Munoz JA (2007) Comparative study of biosorption of heavy metals using different types of algae. *Bioresour Technol* 98 (17):3344–3353. doi:10.1016/j.biortech.2006.09.026.

Sablani SS, Goosen MFA, Al-Belushi R, Wilf M (2001) Concentration polarization in ultrafiltration and reverse osmosis: a critical review. *Desalination* 141 (3):269–289. doi:10.1016/S0011-9164(01)85005-0.

Sharma S, Tiwari S, Hasan A, Saxena V, Pandey LM (2018) Recent advances in conventional and contemporary methods for remediation of heavy metal-contaminated soils. *3 Biotech* 8 (4):216.

Silva P, Matos M (2016) Assessment of the impact of Aluminum on germination, early growth and free proline content in Lactuca sativa L. *Ecotoxicol Environ Saf* 131:151–156. doi:10.1016/j.ecoenv.2016.05.014.

Singh JS, Abhilash P, Singh H, Singh RP, Singh D (2011) Genetically engineered bacteria: an emerging tool for environmental remediation and future research perspectives. *J Gene* 480 (1–2):1–9.

Singh A, Prasad SM, Singh S, Singh M (2016) Phytoremediation potential of weed plants' oxidative biomarker and antioxidant responses. *Chem Ecol* 32 (7):684–706. doi:10.1080/02757540.2016.1182994.

Sinha A, Pant KK, Khare SK (2012) Studies on mercury bioremediation by alginate immobilized mercury tolerant Bacillus cereus cells. *Int Biodeterior Biodegrad* 71:1–8.

Solanki R, Dhankhar R (2011) Biochemical changes and adaptive strategies of plants under heavy metal stress. *Biologia* 66 (2):195–204. doi:10.2478/s11756-011-0005-6.

Spaepen S, Vanderleyden J (2011) Auxin and plant-microbe interactions. *Cold Spring Harb Perspect Biol* 3 (4):a001438. doi:10.1101/cshperspect.a001438.

Substances AfT, Registry D (2000) *Toxicological Profile for Arsenic TP-92/09*. Centre for Disease Control Georgia.

Tak HI, Ahmad F, Babalola OO (2013) Advances in the application of plant growth-promoting rhizobacteria in phytoremediation of heavy metals. In: Whitacre D (eds.), *Reviews of Environmental Contamination and Toxicology Volume 223*. Springer, pp. 33–52. https://doi.org/10.1007/978-1-4614-5577-6_2

Technical Memorandum Standards for Effluents Discharged into Drainage and Sewerage Systems, Inland and Coastal Water (2005). Hong Kong.

Tigini V, Prigione V, Giansanti P, Mangiavillano A, Pannocchia A, Varese GC (2010) Fungal biosorption, an innovative treatment for the decolourisation and detoxification of textile effluents. *Water* 2 (3):550–565. doi:10.3390/w2030550.

Tunali S, Akar T, Ozcan AS, Kiran I, Ozcan A (2006) Equilibrium and kinetics of biosorption of lead(II) from aqueous solutions by Cephalosporium aphidicola. *Sep Purif Technol* 47 (3):105–112. doi:10.1016/j.seppur.2005.06.009.

Voica DM, Bartha L, Banciu HL, Oren A (2016) Heavy metal resistance in halophilic bacteria and archaea. *FEMS Microbiol Lett* 363 (14): Artn fnw146. doi:10.1093/femsle/fnw146.

Vullo DL, Ceretti HM, Daniel MA, Ramirez SAM, Zalts A (2008) Cadmium, zinc and copper biosorption mediated by Pseudomonas veronii 2E. *Bioresour Technol* 99 (13):5574–5581. doi:10.1016/j.biortech.2007.10.060.

Wang LK, Hung Y-T, Shammas NK (2007) *Advanced Physicochemical Treatment Technologies*. Springer.

Weerasekara P (2017) The United Nations world water development report 2017 wastewater. *Future Food: J Food Agric Soc* 5 (2):80–81.

Wingenfelder U, Hansen C, Furrer G, Schulin R (2005) Removal of heavy metals from mine waters by natural zeolites. *Environ Sci Technol Health Care* 39 (12):4606–4613.

Wolfenbarger LL, Phifer PR (2000) The ecological risks and benefits of genetically engineered plants. *Science* 290 (5499):2088–2093.

Wu G, Kang H, Zhang X, Shao H, Chu L, Ruan C (2010) A critical review on the bio-removal of hazardous heavy metals from contaminated soils: issues, progress, eco-environmental concerns and opportunities. *J Hazard Mater* 174 (1–3):1–8. doi:10.1016/j.jhazmat.2009.09.113.

Yue ZB, Li Q, Li CC, Chen TH, Wang J (2015) Component analysis and heavy metal adsorption ability of extracellular polymeric substances (EPS) from sulfate reducing bacteria. *Bioresour Technol* 194:399–402. doi:10.1016/j.biortech.2015.07.042.

Zeng XF, Wei SH, Sun LN, Jacques DA, Tang JX, Lian MH, Ji ZH, Wang J, Zhu JY, Xu ZX (2015) Bioleaching of heavy metals from contaminated sediments by the Aspergillus niger strain SY1. *J Soils Sed* 15 (4):1029–1038. doi:10.1007/s11368-015-1076-8.

Zeraatkar AK, Ahmadzadeh H, Talebi AF, Moheimani NR, McHenry MP (2016) Potential use of algae for heavy metal bioremediation, a critical review. *J Environ Manage* 181:817–831. doi:10.1016/j.jenvman.2016.06.059.

Zhao Y, Gao B, Shon H, Cao B, Kim J-H (2011) Coagulation characteristics of titanium (Ti) salt coagulant compared with aluminum (Al) and iron (Fe) salts. *J Hazard Mater* 185 (2–3):1536–1542.

Zouboulis A, Loukidou M, Matis K (2004) Biosorption of toxic metals from aqueous solutions by bacteria strains isolated from metal-polluted soils. *Process Biochem* 39 (8):909–916.

8 Comparison of Different Membrane Technologies for Boron Removal from Seawater

Noura Najid, Sanaa Kouzbour, Bouchaib Gourich,
Mohamed Chaker Necibi, and Azzedine Elmidaoui

8.1 INTRODUCTION

Current global population growth and rapid industrial development are leading to increased demand for fresh water, such that many regions of the world have already experienced or are experiencing water stress (Hunt *et al.*, 2021). Many common fresh waters, like lakes and rivers, have been overused and misused. These resources are not only reducing but they are also becoming saline. In addition, it has been reported that almost 2.7 billion people are exposed to dangerous diseases because of water pollution (Boretti and Rosa, 2019). Therefore, there is an inherent need for additional fresh water resources. Actually, desalinated seawater has been turned into an unlimited resource of drinking water to surmount water shortages and satisfy the increasing demand for fresh water (Bint El Hassan *et al.*, 2018). It was estimated that the current number of seawater desalination plants in the world is more than 15,000, yielding more than 56 million m^3/d of drinking water out of a worldwide drinking water production capacity of approximately 500 million m^3/d (Najid *et al.*, 2021b).

Generally, the reverse osmosis (RO) process, the principal technique adopted for seawater desalination, utilizes a semi-impermeable membrane that permits only water molecules to pass through it and retains solutes and microorganisms owing to the high pressure, in the range of 50–75 bars, which overcomes the osmotic pressure of the water, and it is approximately 26–29 bars according to the salinity of the seawater (Maddah *et al.*, 2018). The RO process has been growing in interest since the 1960s as a result of improvements in membrane materials and technologies, reduced energy consumption, and its potential to separate essentially all total dissolved solids (TDS) from seawater and to satisfy the regulations set by the World Health Organization (WHO) (Filippini *et al.*, 2019). In fact, different small contaminants or uncharged species, such as boron, are a real challenge facing the use of RO membranes (Cho *et al.*, 2015). In fact, boron in seawater is found naturally in concentrations of 3.5–9 mg/L and diffuses like water through RO membranes since it is uncharged (Kabay, Güler and Bryjak, 2010). This is a thorny issue, particularly in areas where seawater is the primary source of water. Boron is a poisonous element in many water systems. Its uptake and leaching into groundwater and surface water cause an increase in the bioconcentration index, which endangers humans, animals, and even crop quality (Moseman, 1994). Nevertheless, boron is a tremendous micronutrient for human, animal, and plant growth (Xu *et al.*, 2007). Hence, the permissible standard value for boron content in drinking water has been defined, respectively, as 2.4 mg/L and 1 mg/L by the WHO and the European Union (EU) (Tu, Nghiem and Chivas, 2010). Many treatment techniques for boron removal still face challenges to achieve this level for. Actually, a multi-pass process with pH increase is the most extensively applied process for seawater deboronation to achieve compliance with stringent regulatory requirements. Nevertheless, this technology requires higher capital and operating expenses and therefore lowers the cost-effectiveness. Thus, intensive efforts are needed to increase boron removal efficiency. In addition, diverse boron rejection membrane techniques have been suggested including forward osmosis (FO), nanofiltration

(NF), membrane distillation (MD), electrodialysis, electrode ionization (EDI), and adsorption membrane filtration (AMF) (Al Haddabi *et al.*, 2014; Darwish *et al.*, 2020).

This chapter aims to provide a comparative assessment of the above-mentioned processes that have been performed for boron rejection from seawater, both stand-alone and hybrid, with an evaluation of their corresponding benefits and shortcomings. Additionally, a cost discussion of the previously mentioned deboronation technologies has also been analyzed. The last part of this chapter is focused on the life cycle assessment (LCA) method by summarizing the recent literature on the environmental implications of a seawater RO desalination system with boron disposal to identify the main findings and knowledge gaps. As a matter of fact, energy consumption, water intake, and brine discharge are the major environmental aspects that will be addressed in detail together with proposed remediation options.

8.2   BORON DISTRIBUTION IN THE ENVIRONMENT

Boron (B) is an omnipresent element that can be found in nature in the form of two stable isotopes, ^{10}B and ^{11}B, the second representing 80.9% of natural boron (Alharati, 2018). Boron is broadly dispersed in the earth's lithosphere and it is present in granitic rocks and clay-rich marine sediments under natural crumbling action. According to the rock origin, boron content in the earth's crust ranges from about 1 to 500 mg/kg (Kochkodan, Darwish and Hilal, 2015). Boron does not exist naturally in elemental form. It is worth mentioning that the borate ores mined mainly contain the minerals kernite, tincal, and ulexite. Kernite is used to produce boric acid, sodium borate has been produced by tincal, and ulexite has been considered the main ingredient in the manufacture of a variety of glasses and ceramics (Elevli, Yaman and Laratte, 2022). The latter are exploited by approximately 2 M tons on a yearly basis in several countries including the United States, Italy, Spain, Tibet, Chili, and Turkey (Helvacı and Palmer, 2017).

Furthermore, other natural contributors of boron are volcanic emissions (almost 17.10^6 to 22.10^6 kg of boron annually) as boric acid $(B(OH)_3)$ and boron trifluoride (BF_3) (Conseil canadien des ministres de l'environnement, 2009). Hence, boron levels in the vicinity of volcanic zones are important. Tectonic activity gives rise to boron in the environment, which is in the form of borate and is potentially soluble; as such, it would make sense to look for boron in hydrological systems, particularly in the oceans that lie above the boundaries of tectonic plates (Argust, 1998; Parks and Edwards, 2005). It has been reported that boron is the tenth most plentiful element in oceanic salts with a mean content of 4.5 mg/L, but it can be more than 9 mg/L in several areas like the Mediterranean Sea (Güler *et al.*, 2015). Boron may also be transferred from the seas to the atmosphere by a process in which boron salts are released into the atmosphere at the interface between water and air, and by volatilization of boron from the seas as boric acid (Argust, 1998; Helvaci, 2017). It was noticed that more than 4 M tons of boron are emitted to the air annually from ocean sources (Schlesinger and Vengosh, 2016).

Soil systems also constitute a substantial environmental component in the material pathway of boron (Argust, 1998). Boron may penetrate soil through deposition from the atmosphere, alteration of rocks, decomposition of organic matter, leaching from landfills, and from the application of boron-containing fertilizers (Degryse, 2017). In fact, there are various factors that affect the residence time of boron in the soil. Among these are the demand for boron for plant growth, the quantity of rainwater that removes or adds boron from the soil profile, the activity of clays that can act as boron adsorption/desorption sites, and the pH of the soil (Argust, 1998). Boron can also be industrially emitted because of soap and detergents production, glass manufacturing, biomass, and disposal burnings (Najid *et al.*, 2021b). In 2019, the glass and ceramics industries located in the north-central United States and the eastern United States continued to be the largest domestic users of boron products, representing about 80% of all borate intake (Turkbay *et al.*, 2022). However, it was deemed that the industrial supplies rejected less boron into the environment than natural erosion (Butterwick, de Oude and Raymond, 1989). It is true that boron exhibits commercial importance in several sectors and can be considered an essential element for plant growth and a micronutrient for animals and humans. However, it is found to be a toxic element in many water supplies. Its accumulation threatens human and animal health as well as crop quality.

8.3 BENEFITS AND ADVERSE IMPACTS OF BORON ON PLANTS GROWTH AND ANIMAL AND HUMAN HEALTH

Boron is a very important element for plant growth, which is why it is commonly used as a fertilizer in the form of borax (Rahmawati, 2011). It contributes to the formation of cell walls, the conversion of sugar or energy into growth materials, pollination, and seed formation. Nonetheless, it has been identified as a deleterious and poisonous component at elevated levels (Maddah *et al.*, 2018). It was first detected in Eilat, Israel, in 1997, when farmers were complaining about the poisoning of their cultures from irrigation wastewater loaded with boron (>0.3 mg/L) (Najid *et al.*, 2021b). Similarly, for animals, boron has a potential positive effect on the metabolism of various enzymes. It improves bone density and embryonic development when added to the feed. Nevertheless, a high supply of boron can impair cell structure function and lead to death (Abdelnour *et al.*, 2018). Boron has also been shown to be an important nutrient for humans. The typical daily food intake of adults has been pegged at 1.5 mg (Health and Ecological Criteria Division, Office of Science and Technology, 2008).

In addition, there is a growing tendency for boron to be used in a number of areas including the agriculture, glass, detergent, and ceramics industries, due to the importance of its raw material and the variety of its composition. As a result, WHO has set a boron threshold of 2.4 mg/L in drinking water since 2011 (Hilal, Kim and Somerfield, 2011). Based on the toxicity of boron, a guideline of 1 mg/L has been advised by the EU, however (Dolati *et al.*, 2017). This guideline differs among nations based on natural characteristics, plant tolerability, and health of humans and ecology. Table 8.1 lists the global and regional standard cut-off values for boron in drinking water. Indeed, the chemical properties of boron as a chemical element in aqueous solutions are addressed in the subsequent section.

TABLE 8.1
Standards for boron in drinking water in several countries (Ezechi *et al.*, 2014; Isa *et al.*, 2014; Najid *et al.*, 2021b)

Countries	Standards levels (mg/L)	Setting year
WHO	2.4	2011
EU	1	1998
Canada	5	2003
USA	6	2009
Australia	4	2004
Morocco	0.3	2007
Israel	0.3	2004
Japan	1.5	2000
Egypt	0.5	2007
Kuwait	0.5	1999
Saudi Arabia	0.5	2000
Iraq	0.1	2001
Jordan	1	2005
Sudan	0.2	2006
Syrian Arab Republic	0.3	2006

8.4 BORON AND ITS CHEMICAL PROPERTIES

Boron is the only nonmetallic (metalloid) component in its column (group) of the periodic table, whose valence electron configuration $(1s)^2 (2s)^2 (2p)^1$ makes it possess a deficient electron in a vacant p-orbital (Tang *et al.*, 2017b). As such, its chemical properties differ from those of aluminum, gallium, indium, and thallium. Indeed, boric acid can behave as a weak acid. Nevertheless, with just three electrons in the valence shell, boron does not respect the octet rule and boric acid is hence not a proton donor. Also, the small size of its ionic radius makes it difficult to remove electrons from the shell (Wolska and Bryjak, 2013). Boron has no natural redox chemistry, so the fractionation between the two isotopes is almost entirely controlled by the distribution in solution between trigonal or tetrahedral species (Figure 8.1) (Kochkodan, Darwish and Hilal, 2015).

In aqueous media, dissolved boron is found in different forms in accordance with the boron content and the pH. At contents ≤216 mg/L, boron is present primarily as boron mononuclear boric acid and hydroxyborate ions ($B(OH)_3$ and $B(OH)_4^-$). At a higher level and pH, soluble polyborate ions are formed like $B_3O_3(OH)_4^-$, $B_4O_5(OH)_4^-$, $B_5O_6(OH)_4^-$, $B_3(OH)_{10}^-$, $B_2O(OH)_6^-$ (Najid *et al.*, 2021b). Figure 8.2 shows the polyborate ion distribution as a function of pH (Kochkodan, Darwish and Hilal, 2015). The boron content in seawater is around 4.5 mg/L; only $B(OH)_3$ and $B(OH)_4^-$ species are present (Figure 8.3). Hence, boric acid with acid dissociated constant of 5.8×10^{-10} at 20°C is hydrolyzed as follows (equation 8.1) (Rioyo *et al.*, 2018):

$$B(OH)_3 + H_2O \leftrightarrow B(OH)_4^- + H^+ \quad pK_a = 9.23 \tag{8.1}$$

FIGURE 8.1 The molecular structures of (a) boric acid and (b) hydroxyborate ions (Liu *et al.*, 2013).

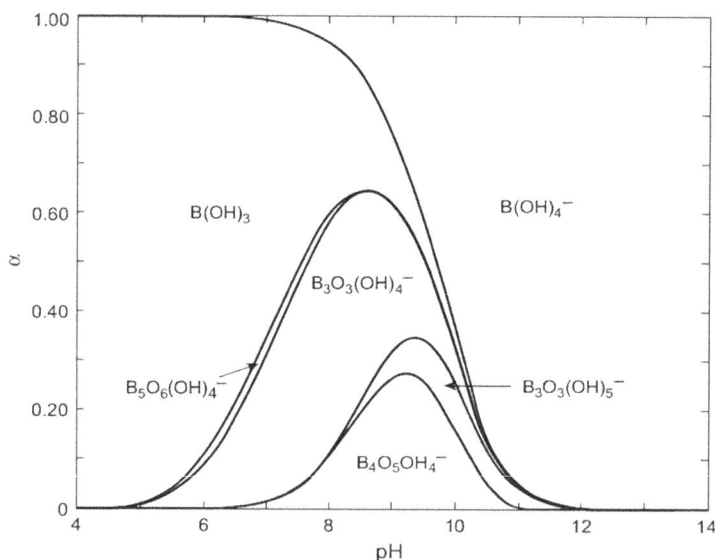

FIGURE 8.2 Distribution of polyborate ions according to pH in 0.4 M of boric acid (Kochkodan, Darwish and Hilal, 2015).

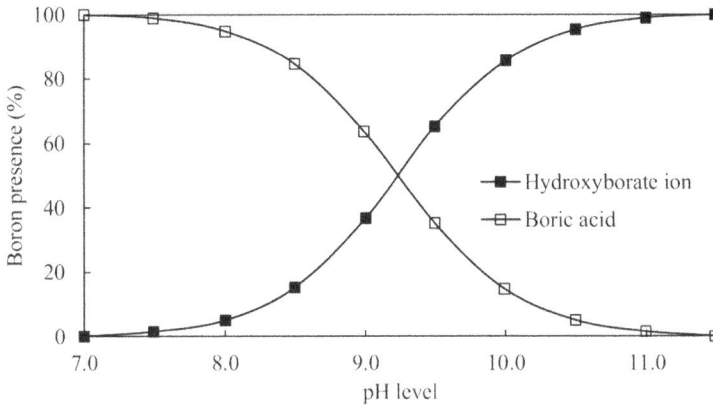

FIGURE 8.3 Boron species distribution based on pH (Redondo, Busch and De Witte, 2003).

At a low pH, boric acid is not properly hydrated because of its large crystalline radius. Therefore, it occurs in a non-dissociated form (Koseoglu *et al.*, 2008). Conversely, the completely hydrated form is prevalent at pH values above the dissociation constant pK_a, as shown in Figure 8.3 (Redondo, Busch and De Witte, 2003). It should be noted that a pK_a of 9.23, which corresponds to a dilute solution at atmospheric pressure and 20°C, can be altered by many factors, such as salinity, temperature, and pressure (Najid *et al.*, 2021a). Indeed, the pK_a of boric acid has been shown to increase with increasing temperature from 9.08 to 9.38 at 10°C–50°C, respectively, and decreases from 9.23 to 8.6 when salinity increases from 0 to 4×10^4 mg/L (Kochkodan, Darwish and Hilal, 2015). The authors also stated that when the pressure augments from 0 to 6×10^3 bar, pK_a changes from 9.23 to 11.23 (Tu, Nghiem and Chivas, 2010). In any case, this does not matter because such high pressures are not used for membrane separation, in particular for seawater RO.

As mentioned above, the presence of boron in seawater has received worldwide attention due to its toxicity to human health and aquatic systems. Improper elimination of this component has resulted in non-compliance with WHO guidelines for drinking water and irrigation water Consequently, different membrane technologies have demonstrated their potential to reject boron from seawater.

8.5 MEMBRANE TECHNOLOGIES FOR BORON REMOVAL FROM SEAWATER

8.5.1 Reverse Osmosis

RO is the most widespread and proven technology for desalination of seawater for drinking water needs. Nevertheless, it has some weaknesses regarding boron removal. At pH > 9, boron, which exists as a negatively charged borate ion $B(OH)_4^-$, gets easily discharged by RO membranes thanks to size exclusion and Donnan repulsion (Tang *et al.*, 2017b). However, the weakly hydrated form is difficult to separate under neutral and/or acidic conditions because of the absence of electrostatic repulsion between the membrane surface and the non-dissociated boric acid of neutral charge (Hilal, Kim and Somerfield, 2011). In spite of the ability of RO to eliminate ionic salts (up to 99%), RO is unable to attain a high release level of boron (Hyung and Kim, 2006). The state of the art of RO process indicated that the boron removal rate was only 40%–60% over a pH range of 5–9, respectively (Prats, Chillon-Arias and Rodriguez-Pastor, 2000).

The release of boron from seawater relies heavily on the RO membrane type (Tang *et al.*, 2017b). Commonly, the majority of RO processes are equipped with cellulosic membranes that are made up of a selective layer and a porous sublayer (Ibrahim, Isloor and Farnood, 2020). In fact, it has been found in various reports that the water flux diminishes due to the hydrophobic nature of RO cellulosic membranes (Ghaseminezhad, Barikani and Salehirad, 2019). Currently, multiple efforts have been made to enhance the filtration efficiency by increasing the hydrophilicity of the membrane surface (Ibrahim, Isloor and Farnood, 2020). For example, Asim *et al.* (2018) combined cross-linked polyvinyl alcohol membranes with polyethylene oxide and polypropylene (0.04 wt%) to block copolymer and nanocrystalline cellulose arabic (NCC/GuA) conjugates for boron elimination. The experimental outcomes showed satisfactory boron rejection as high as 92% when NCC (0.06 wt%) and GuA (0.04 wt%) were added. This can be explained by the hydrophilic character of the prepared membrane, causing an increase in the permeation flux rate (from 8.6 to 21.3 L/m²/h). This happened as the conjugates provide the active sites in charge of hydrogen bonding. Eventually, this results in a tight packing that prevents the diffusion of boron molecules.

It might be worth noting that despite the different trials that have been conducted for the preparation and development of the cellulose membrane, polyamide thin film composite (TFC) membranes dominate the entire RO market (Tang *et al.*, 2017b). Recently, a new selective tight membrane has been fabricated with great promise to reduce boron content in the RO permeate with up to 96% rejection rate (Freger *et al.*, 2015) (Table 5.1). For instance, Ruiz-García, León, and Ramos-Martín (2019a) have evaluated the boron disposal performance of two commercially available RO membranes from the company Toray (TM820S-400 and TM820L-440). The TM820L-440 membrane maintained the boron concentration in the permeate between 0.25 and 0.75 mg/L during the operating time, while the TM820S-400 membrane maintained the boron concentration in the permeate between 0.5 and 2 mg/L. These new manufactured membranes were designed taking into consideration their low affinity toward boron, high affinity toward water and membrane pore sealing for boric acid size exclusion (Taniguchi *et al.*, 2004). Overall, boron rejection is greatly linked to its permeability across the membrane. The solute permeability (cm/s) is given by the formula below (Freger *et al.*, 2015; Ruiz-García, León and Ramos-Martín, 2019b) (equation 8.2):

$$P_s = \frac{\mathrm{Diff}k}{\partial} \tag{8.2}$$

where Diff and k are, respectively, the diffusion (cm²/s) and partitioning coefficient of the solute in the membrane, and ∂ is the membrane thickness (cm).

TABLE 8.2
Examples of manufactured SWRO membranes for boron rejection

Manufactured membrane	Operational conditions	Boron rejections (%)	References
Toray TM800K	5.52 Mpa, 25°C, pH 8, 32 g/L of NaCl, recovery rate 8%, 5 mg/L of boron	96	Toray Membrane: Innovation by Chemistry (2021)
Toray TM800V	5.52 Mpa, 25°C, pH 8, 32 g/L of NaCl, recovery rate 8%, 5 mg/L of boron	92	
Toray TM800M	5.52 Mpa, 25°C, pH 8, 32 g/L of NaCl, recovery rate 8%, 5 mg/L of boron	95	
Hydranautics SWC4 B	5.52 Mpa, 25°C, pH 7, 32 g/L of NaCl, recovery rate 10%, 5 mg/L of boron	95	Hydranautics – A Nitto Group Company (2021)
Hydranautics SWC4+8040	5.52 Mpa, 25°C, pH 7, 32 g/L of NaCl, recovery rate 10%, 5 mg/L of boron	93	
Hydranautics SWC6 MAX	4.1 MPa, 25°C, pH 7, 32 g/L of NaCl, recovery rate 10%, 5 mg/L of boron	91	
Dow FILMTEC™ SW30–8040	5.52 MPa, 25°C, pH 8, 32 g/L of NaCl, recovery rate 8%, 5 mg/L of boron	92	DOW Filmtec Reverse Osmosis Membranes (2021)
Dow FILMTEC™ SW30XHR-400i	5.52 MPa, 25°C, pH 8, 32 g/L of NaCl, recovery rate 8%, 5 mg/L of boron	93	
Dow FILMTEC™ SW30HRLE-370/34i	5.52 MPa, 25°C, pH 8, 32 g/L of NaCl, recovery rate 8%, 5 mg/L of boron	91	

In addition, high membrane cross-linking has been reported to be a key determinant of membrane perm-selectivity (Ibrahim, Isloor and Farnood, 2020). For example, Baransi-Karkaby, Bass and Freger (2019) studied the improvement of discharge of multi-micro-pollutants such as boric acid in ESPA1-2521 RO membranes that were modified utilizing surfactant-reinforced surface polymerization. The membrane was modified by grafting with 2 mM poly(glycidyl methacrylate) and 0.045 mM Triton during 30 min. The findings showed an enhanced release of boric acid solutes and salts with a slight loss of flow (<25%), due to the coating layer that allowed the use of the high selectivity of the polyamide layer by sealing and caulking minor flaws and random blocking of the non-selective passing of boric acid solutes, instead of eliminating the solutes directly.

To meet the boron content standards for drinking water, a single-stage system is not satisfactory to reduce the boron level (Najid *et al.*, 2021a). In contrast, in two-pass systems that combine brackish water reverse osmosis (BWRO) and seawater reverse osmosis (SWRO) units, an adjustment in pH is highly recommended to convert the undissociated form of boron to the ionic form (Figure 8.4). The addition of sodium hydroxide can reject 80%–95% of the boron (Freger *et al.*, 2015).

Boron rejection mainly depends on different parameters such as pH, temperature, pressure, and feed salinity, as shown in Table 8.3. Indeed, the effect of pH and temperature on boron removal from RO membranes is clearly highlighted. It significantly impacts boron and borate transport and permeabilities (Hyung and Kim, 2006). Among the current transport models, the solution-diffusion model (Wijmans and Baker, 1995; Al-Obaidani *et al.*, 2008) is the most widely used because of its simplicity of application and its capacity to produce results that closely resemble real operational scenarios (Kucera, 2015). Equation (8.3) gives the solute transport through RO membranes based on the solution-diffusion model.

$$J_s = P_s \left(C_{ms} - C_{ps} \right) \tag{8.3}$$

where J_s is the solute flux (kg/m^2/s), C_{ms} and C_{ps} are, respectively, the solute content near the membrane surface and in the permeate (kg/m^3), and P_s is the solute permeability (m/s). Indeed, boron has different P_s coefficients depending on whether its prevailing form in the feed water is boric acid or metaborate ions (Hyung and Kim, 2006). C_{ms} is determined on the basis of the concentration

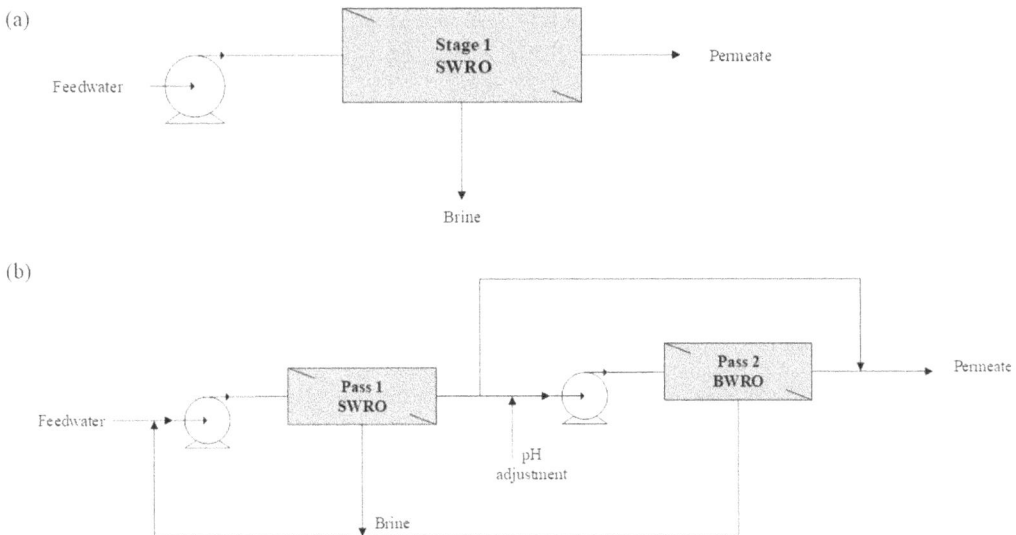

FIGURE 8.4 Schematic illustrations of (a) the single-pass design and (b) the two-pass design (based on Sassi and Mujtaba (2013) and Tang *et al.* (2017b)).

polarization phenomena that link the concentration of the feed brine (C_{fbs}) to C_{ms} via the polarization factor (PF). In the steady state, PF is calculated by the equation (8.4):

$$PF = \frac{C_{ms} - C_{ps}}{C_{ms} - C_{fbs}} = e^{\left(\frac{J_w}{k_s}\right)} \tag{8.4}$$

where J_w is the permeation flux within the RO membrane (L/m^2/s) and k_s is the mass transfer coefficient of solute (m/s).

8.5.2 FORWARD OSMOSIS

FO has emerged as an attractive alternative process for boron removal in recent years. It works by the osmotic pressure difference across a semi-permeable membrane, with a net flow of water from a dilute feed solution (FS) to a high concentration draw solution (DS) (Tang *et al.*, 2017b). In fact, FO membranes made from cellulose acetate (CA) and TFC have also been investigated for rejecting boron from wastewater and seawater (Fam *et al.*, 2014; Valladares Linares *et al.*, 2014). Further information on the FO membrane development is detailed by Suwaileh *et al.* (2020).

Indeed, boron rejection and solute flux were calculated using equations 8.5 and 8.6 (Darwish *et al.*, 2020):

$$\text{Boron rejection} (\%) = 1 - \frac{C_d \times DF}{C_f} \times 10 \tag{8.5}$$

$$\text{Boron solute flux} = \frac{m_p}{A_m t} \tag{8.6}$$

TABLE 8.3
The effect of operational parameters on boron rejection in SWRO

Parameters	Effects on boron rejection
pH	• High boron removal from seawater occurred at alkaline conditions.
	• For instance, more than 90% could be achieved at pH 10.5 for Hydranautics CPA2, Toray SU710, Toray UTC-80-AB, FILMTEC SW30HR membranes due to the increasing amount of borate ions as the pH rises (Koseoglu *et al.*, 2008; Tang *et al.*, 2017b).
	• In addition, with increasing pH, the membrane surface goes more negative, which leads to a higher rejection due to electrostatic interactions between borate ions and the membrane (Freger *et al.*, 2015).
Temperature	• Elevated temperature results in low permeate quality as a result of the increased salt flux across RO membranes (Hilal, Kim and Somerfield, 2011).
	• For example, a study has been conducted to evaluate the effect of temperature on boron elimination performance in Izmir-Urla seawater (5.2 mg/L of boron). The boron levels in RO permeate were 0.7 and 0.9 mg/L when the temperature was raised from 9.8 to 15.4°C under 55 bar (Güler *et al.*, 2011).
Pressure	• Boron discharge increases as the water flow rises owing to the elevated operating pressure (Hilal, Kim and Somerfield, 2011).
	• A study was performed to evaluate the effect of the pressure of the second pass of RO utilizing two membranes with different manufacturing processes (KOCH-XR et GE-AK). The boron removal efficiencies were 75%–85% and 60%–75% for a feed pressure of 220 psi and 150 psi, respectively (Farhat *et al.*, 2013).
	• This is generally attributed to the higher pressure allowing high water permeation and practically fixed boron throughput.
Salinity	• Increased salinity induces charge neutralization and ion diffusion in the membrane as a result of decreased electrostatic interactions (Najid *et al.*, 2021b).
	• This is attributed to the existence of a thick electrical double layer that limits borate ion transport at low salinity (Alnouri and Linke, 2014).

where C_d and C_f are, respectively, the boron concentrations (mg/L) in the FS and in the DS, DF is the dilution factor, which is the ratio of the final DS volume and the permeate volume, m_p is the boron mass in permeate (kg), A_m is the effective area of membrane (m²), and t is the time (h).

Boron passage through FO membranes is strongly impacted by water flux, membrane orientation, type and concentration of DS, boron concentration in FS, and pH of the FS (Tang *et al.*, 2017a). It was found that by increasing the water flux from 3.6 to 28.8 L/h/m², boron rejection was augmented, respectively, by approximately 20%–60% in the AL-FS mode (when the active layer faces the FS), whereas when adopting the AL-DS mode (active layer faces the draw solution), the boron removal rate increased slightly to 15% and then decreased to less than 10% due to polarization of the boron concentration through the membrane substrate (Jin *et al.*, 2011). As for the influence of the pH solution, boron removal increases as the pH increases, as depicted in Figure 8.5 (Darwish *et al.*, 2020). The reduction in boron solute flux is because of the prevalent presence of borate ions with a larger hydrated radius than boric acid. Also, the electrostatic repulsion of the borate ions by the membrane increases boron removal by more than 92% (Fam *et al.*, 2014).

The TFC membrane has a wide range of pH 2–12 in comparison to the CA membrane, which is inoperable at pH > 8. Therefore, these findings indicate that the TFC membrane would be more beneficial for FO systems where boron removal is important. To examine the effect of DS type on boron transport, several DSs (KCl, $(NH_4)_2SO_4$, and $NH_4H_2PO_4$) at a concentration of 2 mol/L have been compared. Importantly, boron removal was noticed when the KCl DS was adopted. This is because KCl DS exhibits lower hydrated radii and greater reverse solute flux (RSF) than PO_4^{2-} and SO_4^{2-}. This is related to the BSR possibly hindering boron flux because the two fluxes occur in opposing directions. Similarly, an important content of KCl in the DS or a low content of the FS results in a high RSF along with a high water passage. The RSF interacts with the diffusion of boron (from the FS to the DS), and subsequently enhances the efficiency of boron rejection. Boron content is also an essential factor to be assessed when studying boron removal from the FO process. Unfortunately, there was an unobserved change in boron release for all membranes tested, although the boron content was raised from 10 to 100 mg/L at pH 7 and a DS content of 1 mol/L (Darwish *et al.*, 2020).

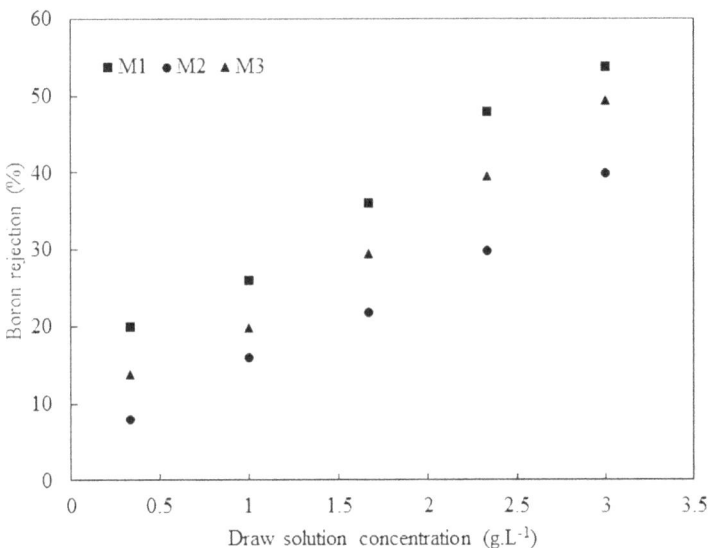

FIGURE 8.5 FS pH effect on the boron rejection using M1: FTS H_2O^{TM}, M2: Aquaporin Inside™, and M3: PSU membranes.

8.5.3 Nanofiltration

Regarding NF, limited studies have been conducted on boron rejection as a result of its low efficiency when it is used alone (Liu *et al.*, 2013; Kheriji and Hamrouni, 2016). Indeed, using NF before RO seawater desalination is a novel technology that has gained much attention (Figure 8.6).

Furthermore, the addition of many polyol compounds is becoming an efficacious approach to significantly enhance boron rejection in this system (Geffen *et al.*, 2006). Polyols such as mannitol, sorbitol, xylitol, and glucose have many hydroxyl groups that can complex with boric acid or borate ions forming many types of borate esters as shown in Figure 8.7 (Liu *et al.*, 2013).

The complexation equilibrium of boric acid with polyol elements relies on the pH. In fact, there are four mechanisms (Figure 8.8); K_1 and K_2 are the stability complexation constants of boric acid, whereas the other two involve the complexation of the borate ions (K_3 and K_4). The latter have been analyzed by various investigations (Duin *et al.*, 1984; Makkee, Kieboom and Bekkum, 2010), while the constants of the boric acid complexes are not yet found in the literature.

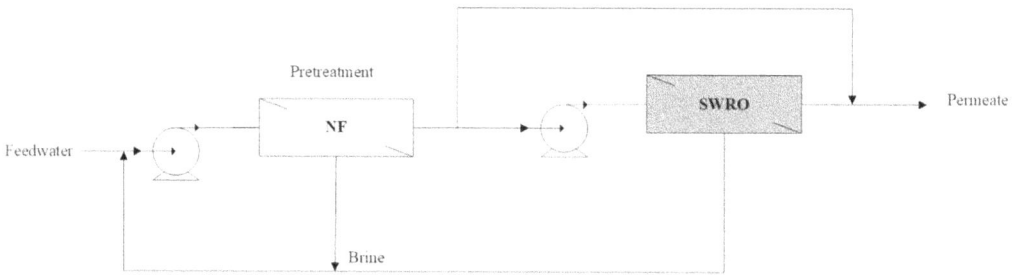

FIGURE 8.6   Schematic design of an NF-RO system (Najid *et al.*, 2021a).

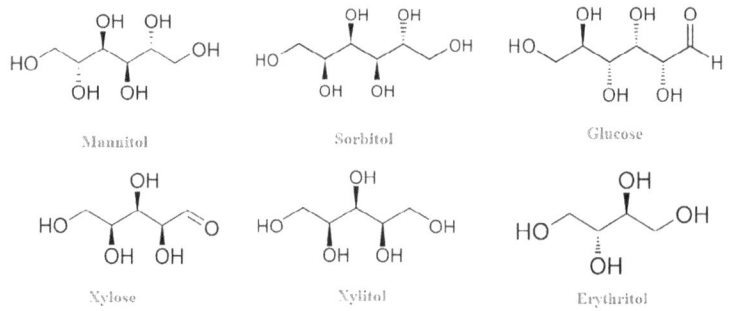

FIGURE 8.7 The different structural formulas of polyols (Liu *et al.*, 2013).

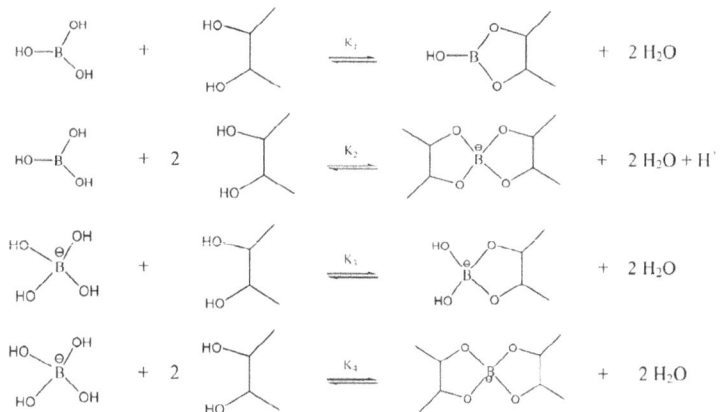

FIGURE 8.8 Complexation equilibrium of boric acid/ hydroxyborate ions with the polyol compounds (Kheriji and Hamrouni, 2016).

It would be worth noting that the type of complexes produced is mainly dependent on the selected complexing agent and the molar ratios of complexant to boron (nC/nB) and pH. It was demonstrated that the addition of complexing agents (25 mg/L of mannitol and sorbitol) enhanced the boron rejection from 34% to 45% and 52%, respectively, inducing the formation of a large size of borate complex ion species when the NF-AG membrane was used (Kheriji and Hamrouni, 2016). In this case, sorbitol exhibited the highest boron rejection rate. One reason for this may be that sorbitol complexes are more stable (K_3 (sorbitol) > K_3 (mannitol)).

Numerous polyols can react with boric acid as mentioned above and provide improved boron removal efficiencies. Nonetheless, they must satisfy several requirements: (i) high equilibrium constant and low reaction time between polyols and boric acid; (ii) safe and eco-friendly; and (iii) minimal effect on the membrane efficiency.

8.5.4 MEMBRANE DISTILLATION

MD has also attracted a lot of attention for boron rejection, as it is less affected by variations in the boron content of the raw water and needs a lower operating temperature and pressure than other membrane technologies (Tang *et al.*, 2017b). During MD, a hot salt solution is contacted with a hydrophobic membrane, which enables the diffusion of water vapor across the membrane, thus limiting the liquid flow and thus the dissolved salts throughout the pores. The mass transfer of water vapor across the membrane is assisted by the vapor pressure gap and the difference in temperature between the two sides of the hydrophobic membrane, i.e. the feeding and permeate sides, as depicted in Figure 8.9 (Alhathal Alanezi *et al.*, 2016). Some investigations have demonstrated the high efficiency of direct contact membrane distillation (DCMD) for boron rejection (Hou *et al.*, 2013; Francis *et al.*, 2014; Boubakri *et al.*, 2015). Using a hydrophobic flat sheet polyvinylidene

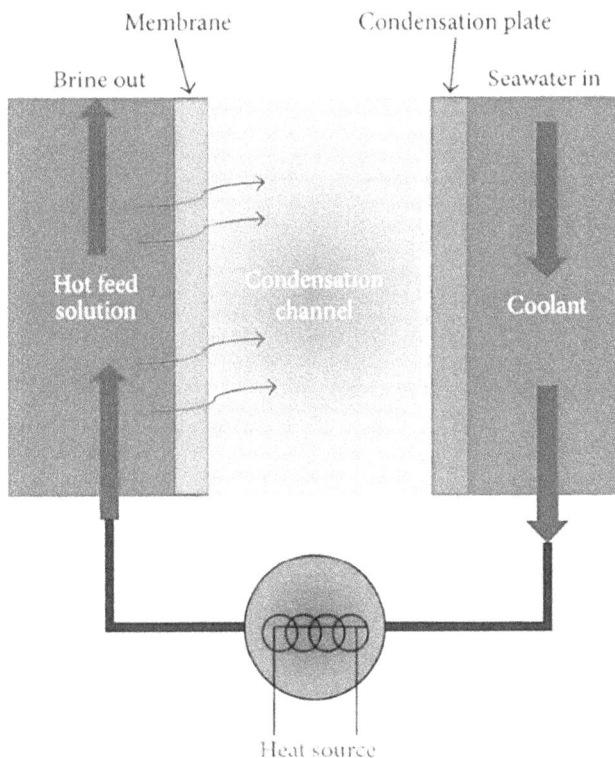

FIGURE 8.9 Principle of the MD process (Alhathal Alanezi *et al.*, 2016).

fluoride membrane, Boubakri *et al.* (2015) found that the permeate boron level was less than 0.8 mg/L over a broad concentration range of boron for a pH of 7 and a temperature of 56°C. In fact, temperature is regarded as the most relevant factor in this technique. At 74°C, significant boron removal was achieved (90.5%) with permeate flux of 27 kg/m²/h. Similar findings have been reported by Hou *et al.* (2010). Admittedly, DCMD is a potential option for removing boron from seawater. However, it remains an energy-consuming process where a high risk of inorganic fouling could be experienced due to the co-precipitation of boron and $CaSO_4$ (Hou *et al.*, 2013; Francis *et al.*, 2014; Boubakri *et al.*, 2015).

8.5.5 ELECTRODIALYSIS

ED using ion exchange membranes has also been pointed out as a potentially attractive process for eliminating boron from seawater (Dydo, 2015; Han, Liu and Chew, 2017). By using an electrical potential as a driving force, ED permits ionic species to be separated and transported across the ion-selective membrane (Figure 8.10) (Freger *et al.*, 2015). The two kinds of membranes utilized for ED separation are cation exchange membranes (CEMs) and anion exchange membranes (AEMs) (Kouzbour *et al.*, 2019). Boron concentration gradient-induced diffusion, coupled with convective ion flow resistance, is generally the primary mechanism for ED transfer of boron through the AEM (Dydo, 2015; Tang *et al.*, 2017b).

The efficiency of ED is critically reliant on a variety of crucial factors. The pH, as usual, plays an important factor in the elimination of boron. At elevated pH, $B(OH)_4{}^-$ elements prevail, resulting in higher boron transport through the membrane (Dydo, 2015). Solution salinity has also been stated to be a key parameter affecting boron transfer across the membrane (Kabay *et al.*, 2008). Unfortunately, seawater with high NaCl content (>35 g/L) can impede boron transfer as a result of competition between borate and chloride ions, resulting in decreased boron flux. Han, Liu and Chew (2017) concluded that a significant decline in the flux of borate was observed at a pH value of 10.8 and for a concentration of NaCl of 35 g/L. Conversely, boron transfer was greater at weak

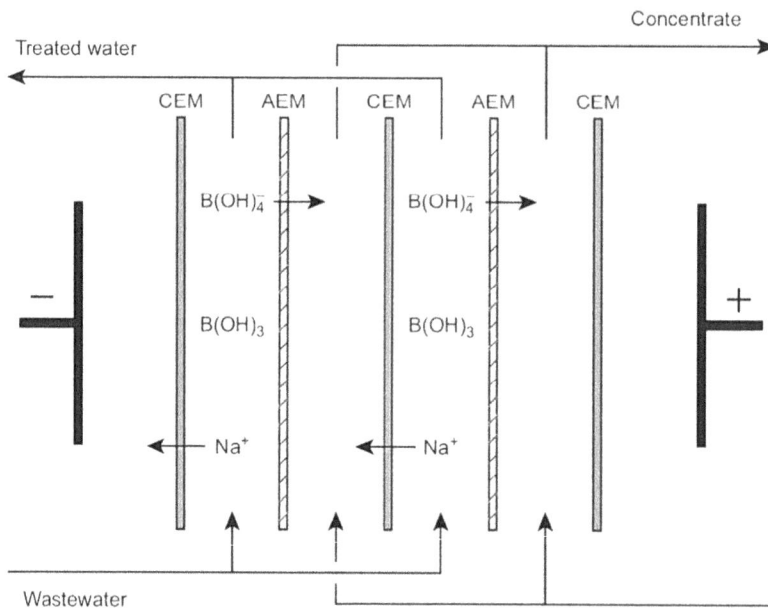

FIGURE 8.10 The electrodialysis desalination concept (Freger *et al.*, 2015).

salinity of 10 g/L. Regarding the current density, it exhibited a paramount factor in the processes of electro-chemical separation. The higher the current density, the more boron species were transferred from the cathode compartment to the anode compartment (Kabay *et al.*, 2008, p. 201). Eventually, the ED with ion exchange membranes proved to be a more appropriate technique due to the poorer permeability of boron. More than 80% boron removal could be reached (Han, Liu and Chew, 2017).

8.5.6 ELECTRODE IONIZATION

EDI has also been suggested to be effective in removing diverse ionic substances from salt water. This is a novel combination system that integrates ion exchange resin and ED technology (Hu *et al.*, 2014). The principle of EDI is shown in Figure 8.11. After the entrance of the inflow into the dilution chamber, the ionic molecules are captured by the resins moved to the surface, where cations migrate to the cathode and anions to the anode. Hence, high-charged ionic species are continuously moved into the concentration chamber owing to the high electrical current. At the same time, the water dissociates and hydrogen and hydroxide ions are released following equation (8.7). Thanks to these ions, it is possible to continuously regenerate resins without adding chemicals.

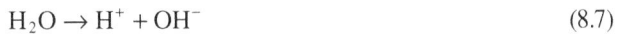

$$H_2O \rightarrow H^+ + OH^- \qquad (8.7)$$

Moreover, OH^- ions can ionize the boric acid component to be removed from the central chamber according to equation (8.8) (Hu *et al.*, 2014):

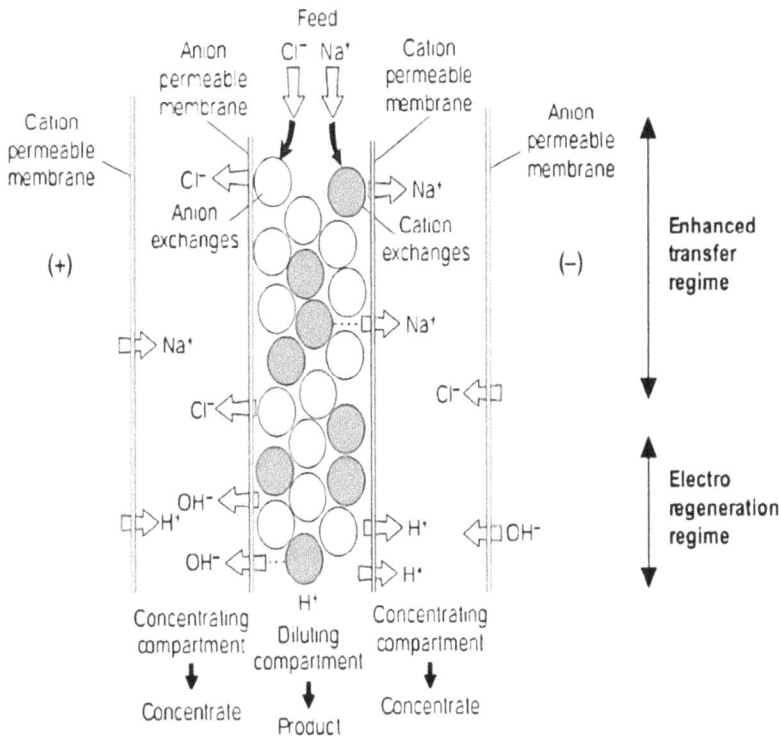

$$H_3BO_3 + OH^- \rightarrow B(OH)_4^- \qquad (8.8)$$

FIGURE 8.11 Schematic illustration of EDI process and ion transfer in EDI (Arar *et al.*, 2014).

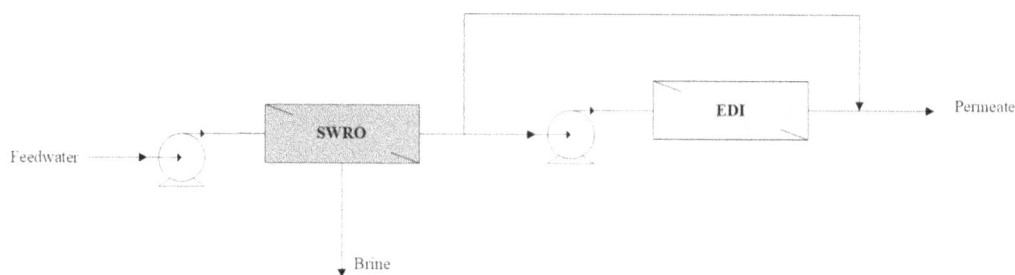

FIGURE 8.12   Schematic illustration of the RO-EDI system (Najid *et al.*, 2021a).

EDI has been shown to be an efficient desalination technology like RO, where the salt levels are decreased by more than 95% (Hu *et al.*, 2014; Liu *et al.*, 2019). However, it is recommended for aqueous solutions with poor conductivity (Jiang *et al.*, 2018). Because of this, EDI is mostly employed as a post-treatment to the RO process to effectively reject boron from saline waters as shown in Figure 8.12. In fact, Hu *et al.* (2014) examined boron removal from model seawater with 4.5 mg/L and 55,693 mg/L of boron and TDS, respectively. In this work, an important boron rejection rate of 93.33% was observed at an electric current of 2.51 Å utilizing the RO (SW30HRLE) and EDI processes as a hybrid system.

Various operation conditions, including water flow rate, electrical potential, current, and pH of the feed water, have been proved to be of importance in the EDI technique and can directly influence the rejection performance of boron. Indeed, a smaller flow rate does not yield a high level of boron elimination. Bunani, Arda and Kabay (2018) found that an optimum flow rate of 1.08 L/h was obtained, where the final resulting content was 0.647 mg/L at 20 V. Boron ions must be transported from the concentration chamber at a high flow rate. Nonetheless, lowering the flow rate in the dilute compartment appears to be more advantageous. It extends the boron residence time, whereby the ions do not leak out before ionization and transport to the concentrated compartment. Within this context, two main restriction mechanisms for controlling boron by EDI are as follows: (i) the rate of entry of borate ions into the resins, and (ii) the transport rate of such ions across the ion exchange membrane.

Electrical current and potential are also key influencing factors in EDI. Higher values can cause more water split as well as more H^+ and OH^- generation. Therefore, a suitable amount of boric acid species is regenerated under the e-field. Nevertheless, phenomena of polarization and backscattering can happen at extremely high values of current and potential (Hakim *et al.*, 2020). Influent pH is also an important factor in boron removal by EDI. The optimal pH range recommended by the majority of studies is between 7 and 10 (Jiang *et al.*, 2018). A high pH value results in an increase in the content of OH^- ions, which may prevent borate ions from being adsorbed into the anion exchange resin, resulting in a reduction in the effectiveness of boron removal. Jiang *et al.* (2018) stated that when using CEM and AEM incorporating -SO_3H and -[$N(CH_3)$]OH and IRN 160 as an ion exchange resin, boron was released at 26% at pH 10 versus just 13% at pH 11. In summary, boron removal in the EDI process was impaired by the ionization degree of H_3BO_3, borate ion transfer, and hydroxide ion exchange inhibition.

8.5.7   HYBRID PROCESS: ADSORPTION MEMBRANE FILTRATION

An efficient boron removal method from saline water by a combination of sorption (ion exchange) and membrane filtration techniques has received widespread use in the past few years (Hou *et al.*, 2010). In this instance, the formation of a boron complex with functional groups attached to the chelating ionic polymer occurred due to the adsorption of small boron particles by ion exchange.

Then, a membrane technique separates the charged boron from water. This hybrid approach is known to have a faster kinetic speed of ion exchange sorbents because of the small particle size (Güler et al., 2015). It is therefore advisable to reduce the dosage of adsorbents by utilizing relatively small particles with minimal pressure loss (Hilal, Kim and Somerfield, 2011).

As previously mentioned, a new and effective approach to improve boron removal in this system seems to be the insertion of polyols with multiple hydroxyl groups, such as sorbitol, mannitol, glycerol, etc. (Park et al., 2016; Tang et al., 2017b). In fact, the high particulate size resulting from boron polymer binding results in an efficient post-membrane filtration separation. Park et al. (2016) noticed an improvement in boron separation of around 80% with the introduction of xylitol at a molar ratio of 1:2, compared to 52.4% for seawater only without the use of polyol. These findings were achieved at a pH of 8.5, a permeation flux of 20.3 L/m^2/h, and a TMP of 50 bar. Similarly, it was stated that despite seawater containing numerous ionic substances (Na$^+$, Mg^{2+}, K$^+$, Cl$^-$, SO$_4^{2-}$, etc.), there was no meaningful inhibition of seawater reactions. Previous studies conducted on the membrane process are presented in Table 8.4.

The subsequent section will be devoted to comparison and analysis of the different membrane processes for boron removal from seawater, starting with RO, which has always been the most widely used technique in this field, but also considering other membrane processes, stand-alone or hybrid, that can become a relevant alternative.

8.6   COMPARISON AND LIMITATION OF TECHNOLOGIES

Based on the above studies, no simple and easy technique exists for the elimination of boron from seawater. The commonly applied method is two-pass RO, in which pH adjustment over 9 is performed only at the second pass to overcome the risk of inorganic fouling. In fact, various multivalent cations such as Ca^{2+} and Mg^{2+} are eliminated in the first RO step without raising the pH. It is worth noting that RO with two passes with pH adjustment was suggested in many studies to increase boron removal and meet WHO potability norms. It can also achieve the required specific boron content in irrigation water (0.5 mg/L). Nevertheless, this approach is more expensive because it requires a lot of energy and chemicals to adjust the pH. In addition, the life span of RO membranes needs special attention when considering factors that affect boron permeability. As such, more rapid diffusion of boron into the membrane can happen as a result of charge neutralization and obstruction of the membrane surface at elevated salinity. For such reasons, RO must be coupled with alternative profitable techniques such as pretreatment or post-treatment to improve mineral and boron elimination at a reduced cost.

Seawater forward osmosis (SWFO) has also been extensively tested as a promising technology because of its advantages over SWRO, including reduced energy consumption, simplicity, and effective removal with no need to adjust the pH. Nonetheless, this technique has a few drawbacks that could be the focus of various upcoming research investigations. The majority of SWFO studies are not performed under the operating parameters of an industrial facility. Hence, the practical aspect of SWFO should be conducted in large-scale operations. In addition, DS regeneration is expensive because it involves high energy consumption for the pumps, resulting in considerable operating costs. The capital expenses could be minimized if consideration is given to upgrading the fabricated membranes to increase water flow and salt and boron removal. Suwaileh et al. (2020) stated that the hydrophilicity, antifouling properties, wetting properties, and water permeability of the membranes are the major contributors to providing high-quality treatment performance at relatively inexpensive costs. Indeed, SWFO alone is unable to desalinate seawater and it would be preferable to combine SWFO with a SWRO process as a pretreatment. It was previously noted that the hybrid SWFO-SWRO system may be technically and economically preferable to the two-pass SWRO system in achieving a boron level of 0.4 mg/L in the permeate (Ban et al., 2019). This low tolerated concentration was requested by many farmers seeking secure and unrestricted irrigation for boron-sensitive crops like barley, wheat, and oats.

TABLE 8.4

Summary of research studies on boron removal from saline water/seawater by membrane-based separation processes with their pros and cons

Membrane	Feed solution	Boron (mg/L)	Removal efficiency (%)	Comments	Merits	Demerits	Refs.
				Reverse osmosis			
Filmtec™ SW30HR	Seawater from Mediterranean Sea Alanya-Kızılot shores, Turkey	5.1	>98	The Filmtec membrane is a polyamide TFC membrane pH = 10.5 Pressure = 48 bar Batch mode	Efficient with two passes of RO unit, possibility to use renewable energy	It requires high pressure and is an energy-intensive process, the use of caustic soda to ionize boric acid for pH	(Koseoglu et al., 2008)
Toray TM820A-370	Seawater from the Middle East	6.5	>94	TDS = 45 g/L Coagulation filter (FeCl₃) as pretreatment and low-pressure RO membrane for brackish water TMG20-430 as post-treatment		adjustment, brine water, and waste production	(Taniguchi et al., 2004)
				Forward osmosis			
Polyamide TFC	Model seawater		80	DS contains 35 g/L and 5 mg B/L FS contains 1 mg B/L pH = 7	More feasible if combined with RO; its operation is simple, and low energy consumption and no chemicals are needed for pH adjustment	The regeneration of its draw solutes is challenging, its combination with RO is strongly needed to obtain high quality of permeate, it suffers from internal concentration polarization, and the suitable DS and membranes choice	(Ban et al., 2019)

Nanofiltration

			Operating conditions	Advantages	Disadvantages	Reference
DOW Filmtec™ NF90-4040 (Dual-NF stage) and	Model seawater	82.94	5.698	TDS 34312 mg/L Recovery ratio of 15% Feed pressure 35 bar Flow velocity 9 cm/s Mannitol and sorbitol (0.01 mol) were added. The addition of polyols enhanced boron rejection. Sorbitol was the most effective polyol	Feasible, low operating pressure, cost-effective, the polyols are safe and environmentally friendly	The efficiency of its single-stage boron removal is often poor and the addition of polyols for enhanced complexation is highly required	(Liu et al., 2013)
				Membrane distillation			
DCMD using polyvinylidene fluoride membrane	Natural seawater from the Mediterranean coast of BorjCedria, Tunisia	90.50	5.37	TDS = 38.5 g/L pH = 7 Feed temperature = 74°C Permeate flux = 27.5 kg/m²/h	It may have low energy consumption and a high rejection	Scaling is its major problem; permeate flux is low compared to other pressure-based membrane processes, and modules are expensive, thermal polarization causes flux decrease	(Boubakri et al., 2015)
DCMD using Polyvinylidene fluoride flat sheet membrane	Natural seawater	–	4.65	Feed temperature = 53°C The permeate boron was kept 20 μg/L Even concentration factor (CF) exceeded 4 For CF > 4.7 the formation of deposits occurred et permeate flux declined sharply			(Hou et al., 2013)
				Electrodialysis			
AEM-CEM	Saline solution	–	50	NaCl = 35 g/L pH = 10.8 Current density = 2 A	Possibility to use renewable energy	Membrane scaling if it is not controlled, co-precipitation of	(Han, Liu and Chew, 2017)
AEM-CEM		80	3	NaCl = 20.3 g/L Voltage = 2 V Inverse arrangement of the spacers that are in the contact with AEM and CEM	Feasible and effective if it is combined after RO	boron with other species could occur, and cleaning and regeneration are highly required	(Oren et al., 2006)

(Continued)

TABLE 8.4 *(Continued)*

Summary of research studies on boron removal from saline water/seawater by membrane-based separation processes with their pros and cons

Membrane	Feed solution	Boron (mg/L)	Removal efficiency (%)	Comments	Merits	Demerits	Refs.
				Electrode ionization			
	Simulated seawater	5.9	93.33	A hybrid system of RO-EDI TDS = 55.693 g/L The recovery of RO is 30% and for EDI is 90% Boron in RO permeate is 1.25 mg/L	There is no need for chemicals, safety, reliability, and zero pollution generation, Production of high-quality water if combined with RO	It is not able to treat seawater directly	(Hu *et al.*, 2014)
				Adsorption membrane filtration			
Sorption microfiltration process	Seawater collected from Banyuls bay	5.083	–	Amberlite IRA 743, Purolite S108 and Diaion CRB 05 were used pH = 8.2 Resin dosage = 2 g/L Low boron was achieved in membrane filtration permeate after 180 min	It is less sensitive to species concentration, high efficiency, and low energy and cost consumption	Adsorbent recycling and waste management, difficult for large-scale application	(Alharati *et al.*, 2018)

For the NF process, it is hoped that this is a potential alternative method to RO in the future in terms of desalination. However, it is preferable to consider it as a pretreatment before the RO unit for boron removal with the addition of polyols for improved complexation. NF has several benefits like simplicity, reduced energy consumption, and the additives are environmentally friendly and reliable, which is convenient for drinking water supply. Nevertheless, further studies are required to identify and make new additives, resulting in both improved boron elimination rates and reduced costs. Although polyols improve boron removal in the NF-RO hybrid process, their regeneration from concentrate is still an issue that leads to higher costs and restricts their application.

ED can be a competitive and promising process to reject boron seawater only if fouling of the membrane is properly controlled. Indeed, scaling leads to a decrease in the permeation flux and co-precipitation of boron with other substances. Furthermore, the regeneration step is still a challenge for ED technology. Therefore, regular cleaning is necessary. Hence, ED is not a sustainable process for treating seawater. It would be preferable to apply the ED process as a post-treatment of the RO unit. Likewise, MD remains a very recent technique and requires more research to gain market acceptance and be implemented industrially. In contrast to other membrane post-treatments, the EDI process has several advantages. It produces high-quality water at a constant flow rate (continuous operation) with low energy consumption. EDI is an ecological process since it does not require the introduction of chemicals. The sole drawback is that it is not feasible as the principal deboronation process for direct treatment of seawater.

AMF has also been the topic of several patents due to its significant efficiency and reduced energy and cost consumption compared to classic fixed bed column sorption. The use of small sorbents enhances the adsorption capacity of boron because of the higher interface area. In addition, AMF has been shown to be an effective combined system for boron removal. For example, poly(GMA-co EDM) (glycidyl methacrylate-co-ethylene dimethacrylate) beads containing NMDG in a hybrid sorption-ultrafiltration (UF) unit exhibited a small residual boron level of 0.38 mg/L (Samatya, Tuncel and Kabay, 2012). This permissible boron content can be employed for irrigation applications. Besides, this combined process is constrained by the adsorbent recovery and disposal of the waste. Accordingly, AMF is preferentially proposed as a post-treatment of the RO permeate.

In summary, it should be noted that choosing the most appropriate technique for eliminating boron from seawater is not simple. Each method has its advantages and shortcomings. The choice of an effective technology depends mainly on regional decisions and conditions. In fact, a hybrid system of RO and the treatments discussed previously may be more effective in removing boron from seawater. The advantages and disadvantages of each membrane technique are summarized in Table 8.4. As a complement, the increase in the cost of deboronation processes mentioned above as well as the environmental pollution generated during the SWRO desalination process will be detailed in the following sections.

8.7   COST EVALUATION FOR BORON REMOVAL

Comparing the economics of various deboronation techniques is very complicated and relies on a number of considerations. To make an appropriate choice, the decision maker must evolve a utility function (Fu), where expenditures are kept to a minimum and returns are maximized, as shown in equation 8.9 (Bick and Oron, 2005):

$$\text{Fu} = \text{capital cost} + \text{operating cost} + \text{disposal cost} - \text{returns} \qquad (8.9),$$

where returns include improved boron abatement, enhanced salinity, environmental monitoring, automation, and continuous operation. Different economic assessments have been performed for boron elimination by multiple technologies. Regarding RO, the overall cost has been split into two major parts, namely, the capital and operating expenses, according to the type of membrane,

TABLE 8.5

CAPEX and OPEX for full-scale two-pass SWRO hybrid system with a daily capacity of 500,000 m³ using commercial polyamide membranes (Ban *et al.*, 2019)

Cost	%	Million $
Civil engineering	6.6	61.5
Equipment and materials	25.9	240.7
Pumps	11	102.8
Installations and services	5.4	50.5
Piping/high alloy	17.8	166
Intake/outfall	5	46.6
Membranes	2.2	20.3
Pretreatments	19.5	181.2
Pressure vessels	6.6	61.4
Total capital cost	100	931
Electricity	53.1	1752
Membranes	7.2	235
Parts and materials	10.4	340
Chemicals	6.7	245
Labor	22.6	740
Total operating cost	100	3312
Total water cost		4243

plant size, and operational conditions such as flow recovery and feed pressure. On the other hand, the recovery rate and pH are considered important in estimating the operating costs (Güler *et al.*, 2015). Ban *et al.* (2019) conducted an economic assessment of a two-pass SWRO system with boron discharge from model seawater that contains 5 mg/L boron. The capital (CAPEX) and operating (OPEX) expenses for this unit are shown in Table 8.5. Energy is a critical component of the overall water cost (total water cost (TWC)). It constitutes approximately 42% of the TWC, while the cost of membranes is a minimum as a result of recent improvements in membrane technologies.

Furthermore, Kayaci *et al.* (2020) performed a cost assessment for a two-pass RO plant with a 48% recovery rate to produce a permeate with a boron level of 0.5 mg/L. Similarly, electrical energy was predominant and cost $30.43 million. The chemical expenses used to raise the pH in the second RO pass and neutralize the discharge were also accounted in the OPEX, to the tune of $2.58 million. A cost study must also consider RO brine disposal, which is becoming a tricky problem. Brine retentates with high boron and other minerals may be detrimental to sea life, particularly in areas that are dependent on fishing activities (Kettani and Bandelier, 2020). It should be noted that the cost of brine disposal ranges from 5% to 33% of the TWC (Katal *et al.*, 2020).

The SWFO process, which is reported to be a viable technique for the production of good water quality, significantly reduces the OPEX compared with the SWRO process as a result of lower energy consumption, reduced chemical use, fewer membrane replacements, and longer membrane lifetimes (Güler *et al.*, 2015). It was stated that the power consumption cost was poorer because no high operating pressure was used in the SWFO technique (Osipi, Secchi and Borges, 2020). Ban *et al.* (2019) made an interesting economic comparison between the hybrid SWRO-SWRO and SWFO-SWRO systems with boron discharge. The OPEX of the SWFO-SWRO system was smaller than that of the two SWRO systems ($2,241 M and $3,312 M, respectively). On the other hand, the CAPEX of the SWFO-SWRO process was still important ($1,619 M) than that of the two-pass SWRO unit ($931 M). The high number of membranes required in the SWFO-SWRO system increases the total cost (Osipi, Secchi and Borges, 2020).

In addition, the cost of equipment and materials for piping, intake, and civil engineering resulted in high CAPEX. Therefore, further advances in SWFO membrane manufacturing are needed to make this a next-generation, environmentally friendly seawater desalination technology with substantial boron removal.

Using dual-stage NF with polyols addition exhibits lower energy needs, longer life at constant performance, and therefore better process efficiency. Vuong (2006) reported that this seawater desalination system effectively removed ions and boron from seawater with a 20% to 30% reduced energy cost than the traditional one-stage RO system. Moreover, it was found that the energy consumption and cost were as low as 0.25 kW h/m^3 for a unit cost equal to \$0.098/m^3 using dual-stage NF for boron removal from seawater. Coupling NF with the RO desalination process has also been shown to be more advantageous in eliminating divalent ions, hardness, and boron from seawater (Zhou et al., 2015). It was proven that the RO-NF process enabled the reduction of CAPEX and electrical energy (Wang et al., 2021).

For MD, important energy consumption, fouling, and chemical application for cleaning are the major problems leading to considerable OPEX (particularly the high cost of maintenance and membrane replacement) (Al-Obaidani et al., 2008; Khayet, 2013). Macedonio and Drioli (2008) reported that MD can be favorably coupled with RO as a hybrid unit to overcome the single-pass vulnerability and foster operational efficiency. Nevertheless, MD is not yet employed on a large commercial scale, and the CAPEX involving membranes and modules rises and falls irregularly. Another con is that the extremely concentrated brine discharged from the MD is also an environmental challenge and the associated costs must be considered. MD plants must first be properly engineered and designed, and then a complete and accurate comparative cost assessment will be possible.

Regarding boron elimination by ED, membrane clogging hinders the technological and cost-effectiveness of this process. The passage of borate ions through AEMs seems to be greater at pH > 11, but the possibility of fouling appears to be inevitable. Consequently, the demineralization operation is paramount, particularly for highly saline waters leading to an increase in the cost of the ED process (Dydo and Turek, 2013). Thus, an appropriate membrane must be used for reducing the investment cost (Turek, Dydo and Bandura-Zalska, 2009). Besides, it is worth mentioning that ED is more costly than the RO method because of the limited demand for ED. Dydo and Turek (2014) assessed the cost of ED for a hybrid ED-RO system with saltwater boron remediation (75 mg/L boron and 1.8 g/L TDS). The global expense of reducing the boron of this system was \$1.04/m^3. Demineralization costs prior to boron discharge are strongly accounted for when designing the ED process to remove boron from seawater with high TDS values.

The EDI method is profitable when applied in combination with RO to remove boron from seawater. The overall cost comprises investment and maintenance cost, membrane replacement, power cost covering pumps and rectifiers, and chemicals cost which are needed to clean the RO membrane. The RO-EDI system appears to be amongst the cheapest hybrid processes. Regarding AMF, boron separation is achieved at low pressure and with low energy consumption. AMF offers a simple alternative to improve boron elimination during membrane filtration. Nevertheless, the regeneration of the membrane is of great importance and must be optimized to obtain a good performance of the deboronation system and a low TWC.

8.8   LCA OF SWRO WITH BORON REJECTION

Based on ISO norms (14040, 2006; ISO 14044, 2006), LCA may be described as an evaluation tool for the environmental impacts of materials, products, processes, and activities, from beginning to end (Fernandez-Gonzalez et al., 2015). It is structured in four areas, as shown in Figure 8.13. In fact, impact assessment focuses on quantifying and storing impacts in 18 categories and 3 endpoint indicators using the ReCiPe methodology, as shown in Table 8.6.

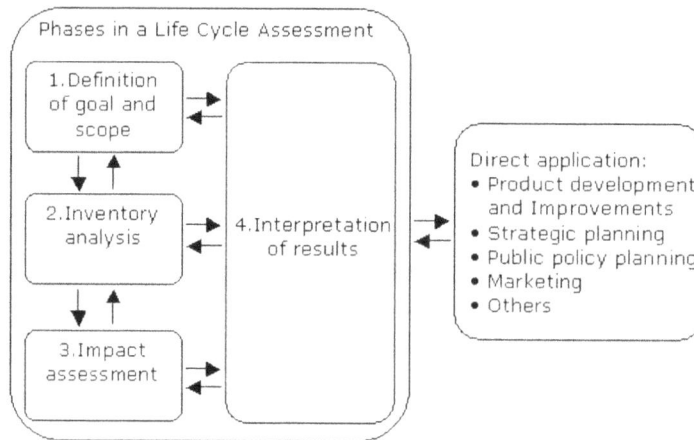

FIGURE 8.13　LCA methodology (Ismail and Hanafiah, 2017).

TABLE 8.6
List of impact categories for the characterization of the midpoint based on the ReCiPe approach (Huijbregts et al., 2017)

Impact category	Abbreviation	Unit
Ozone depletion	OD	kg CFC to air
Terrestrial acidification	TA	kg SO_2 to air
Global warming	GW	kg CO_2 to air
Fresh water eutrophication	FE	kg PO_4 to fresh water
Marine eutrophication	ME	kg N to marine water
Human toxicity	HT	kg 1,4-DCB to urban air
Fresh water toxicity	FT	kg 1,4-DCB to fresh water
Ionizing radiation	IR	kBq Co-60 to air
Marine ecotoxicity	MET	kg 1,4-DCB to marine water
Terrestrial ecotoxicity	TET	kg 1,4-DCB to indus-water
Photochemical oxidant formation	POF	kg NO_x to air
Photochemical ozone creation	POZ	kg C_2H_4 to air
Particulate matter formation	PMF	kg $PM_{2.5}$ to air
Water use	WD	m^3 water consumed
Metal depletion	MD	kg metal
Fossil resource scarcity	FD	kg oil
Land use	LU	$m^2 \times yr$ annual crop land
Abiotic depletion	AD	kg Sb

Large numbers of LCA analyses have been conducted for SWRO desalination, as shown in Table 8.7. Unfortunately, they have rarely been conducted for RO desalination with boron removal specifically. Based on the existing literature, the primary sources of significant environmental impacts may be from power consumption, intake, and brine disposal (Cherif and Belhadj, 2018; Elsaid et al., 2020; Aziz and Hanafiah, 2021). As a matter of fact, energy consumption, involving the GW category, is a main environmental impact contributing to greenhouse gases (GHGs) generation due to the energy intensity of the RO desalination system, taking into account the elimination of boron (Du et al., 2019). Indeed, fossil fuel burning including oil, gas, and coal is the principal source of energy for the majority of SWRO plants, whereby CO_2 releases outweigh those of other GHGs

TABLE 8.7
Overview of some of the LCA studies conducted for SWRO desalination

RO plant location	Scenario description	LCA tool	Indicator	Database	Assessed environmental impact	References
RO, Australia	Conventional SWRO, FO-low-pressure RO membrane, UF-low-pressure RO membrane	SimaPro	CML	Ecoinvent	AD, MET, TA, ME, GW, HT and TET	Pazouki et al. (2021)
Florianópolis RO, Brazil	SWRO, indirect potable wastewater reuse, rainwater harvesting	SimaPro	ReCiPe	Ecoinvent	MET, AD, FE, ME, GW, HT, FD, MD, OD, IR, TA, TET, POZ and PMF	Tarpani et al. (2021)
Shuwaikh RO and Zour RO, Kuwait	Two RO plants with seven multi-stage flash distillation plants (MSF) in different locations	SimaPro	CML 2001	Ecoinvent	MET, AD, ME, GW, HT, OD and POZ	Al-Shayji and Aleisa (2018)
Gulf RO, Kuwait	Two RO plants were powered with PV and nuclear energy and five MSF plants were powered with natural gas, heavy oil, crude oil, solar, gas oil-diesel	SimaPro	CML 2001	European Reference Life Cycle Database	MET, AD, ME, GW, HT, OD and POZ	Aleisa and Al-Shayji (2018)
RO, UAE	PV-RO/solar still/truck delivery	SimaPro	EI 99	Ecoinvent	MET, AD, ME, GW, HT, OD, POZ, FD and LU	Jijakli et al. (2012)
RO, Australia	MF-RO/UF-RO/fertilizer drawn FO-NF using cellulose triacetate and TFC membranes	SimaPro	Australian Indicator	Ecoinvent, Australian Database	MET, ME, GW, HT, OD and FD	Kim et al. (2017)
Fujairah RO, UAE	Sedimentation and UF pretreatment	SimaPro	EI 99	Ecoinvent	MET, AD, ME, GW, HT, OD, FD and LU	Al-Sarkal and Arafat (2013)
RO, Australia	Assessment of end-of-life options for RO: landfill, incineration, electric arc furnace, recycling, gasification, conversion to UF, direct RO reuse	SimaPro	ReCiPe	Ecoinvent, Australian Database	GW	Lawler et al. (2015)
RO, Spain	Brackish water and seawater	SimaPro	Energy demand and TDS in brine	Ecoinvent	MET, AD, ME, GW, HT, OD and POZ	Muñoz and Fernández-Alba (2008)
RO, South Africa	RO and ion exchange and softening process	Manual	CML 2 baseline	N/A	GW	Ras and von Blottnitz (2011)
RO, Greece	Wind turbine and local grid	Manual	EE	Manual	MET, AD, ME, GW, OD and POZ	Karvounis (2017)

2015

2050

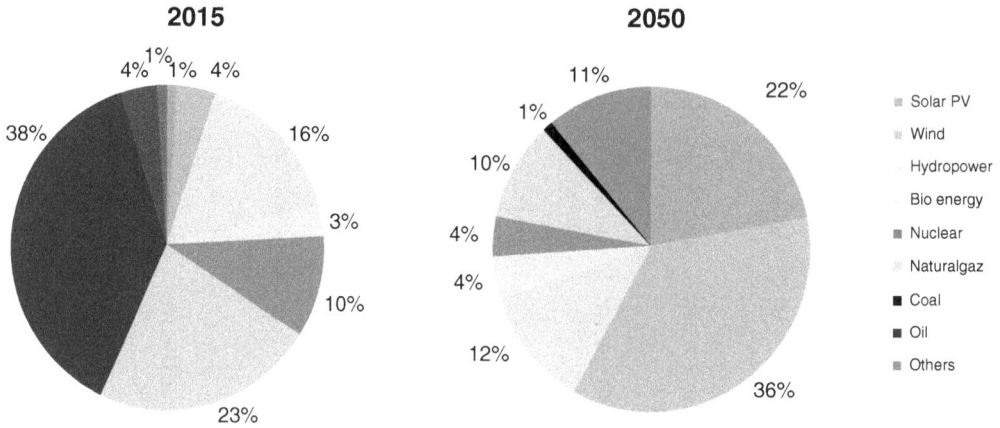

FIGURE 8.14 Growing importance of RES, particularly solar and wind, in the electricity sector between 2015 and 2050 (IRENA, 2018).

(Cherif and Belhadj, 2018). Du *et al.* (2019) carried out an LCA study (CML 2001 method) to insure this, wherein GW was the main impact category as power consumption contributes a substantial share to CO_2 emissions. Therefore, a two-pass RO design with permeate splitting may supply drinking water with specific electricity consumption of 3.17 to 3.88 kW h/m^3, and with a specific environmental impact between 1.49 and 1.81 kgCO$_2$/m^3, respectively. To minimize dependence on fossil fuels and electricity, it seems more beneficial to use green and clean renewable energy sources (RES), such as solar stills, photovoltaic (PV) panels, and wind power. The International Renewable Energy Agency (IRENA) has predicted that RES will outpace other sources of energy and account for more than 85% of the world's energy consumption by 2050, as illustrated in Figure 8.14 (IRENA, 2018).

In LCA analysis, it is worth mentioning that the power consumption over the life cycle of the RO system is generally referred to as embodied energy (EE), expressed in MJ or kWh. EE, therefore, encompasses the energy needed to run moto-pumps, pipes, tanks for water storage, and others such as connections and cables (Cherif and Belhadj, 2018). The overall EE model for the moto-pump and RO membrane is given by equations 8.10 and 8.11:

$$EE_{MP} = 5.52 \times W_{MP}^{1.47} \tag{8.10}$$

$$EE_{RO} = 6,443.5 \times S^{0.94}, \tag{8.11}$$

where W_{MP} is the motor pump weight and S is the RO membrane's surface.

For renewable power generation, EE might be described for PV and WT systems using equations 8.12 and 8.13:

$$EE_{PV} = 3,863 \times S_{PV} - 47.26 \tag{8.12}$$

$$EE_{WT} = 2,359.7 \times S_{WT} + 49.43, \tag{8.13}$$

where S_{PV} and S_{WT} are, respectively, the surface area of PV panels and the WT swept area.

GHGs emissions and EE integration are important aspects to consider when optimizing the RO desalination system with boron removal and to satisfy the drinking water requirements at reduced environmental and economic demands. Further information can be found in Mo, Wang and Jacobs (2016) and Cherif and Belhadj (2018).

Apart from energy consumption and GHGs emissions, the impacts related to water intake in desalination facilities also need to be assessed (Panagopoulos and Haralambous, 2020). Typically,

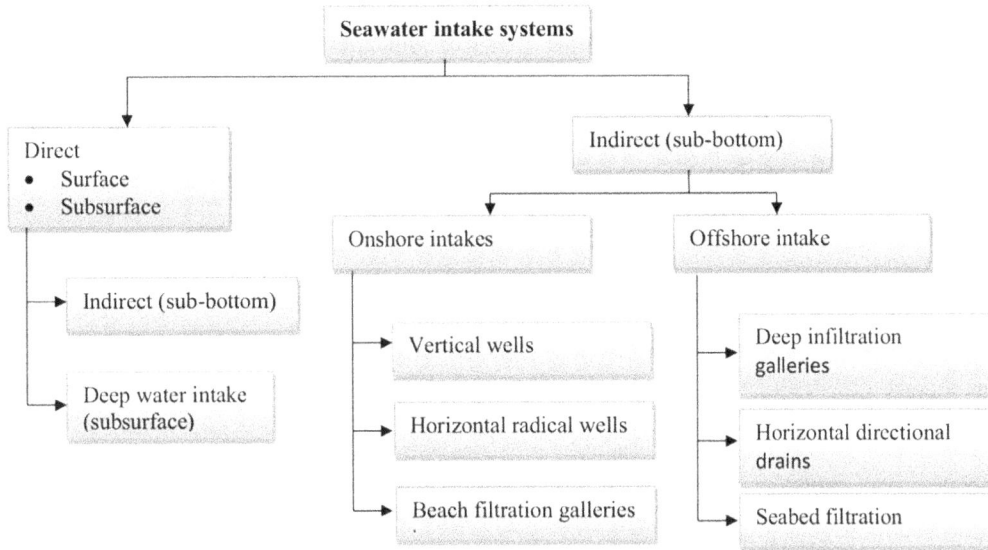

FIGURE 8.15 Seawater intake systems for desalination plants (Najid *et al.*, 2022).

SWRO plants are supplied by a direct (open intake) or indirect intake (sub-bottom intake), as shown in Figure 8.15. A direct intake causes organisms to encroach and entangle themselves in the intake screens and suction racks (Lattemann *et al.*, 2010). Even though direct intake is favored because of its simplicity of installation and high volumetric flow rate, it has significant environmental implications for the surrounding maritime environment (Elsaid *et al.*, 2020). Therefore, there is a need for a pretreatment phase to deal with high turbidities, and algal growth, and to minimize the occurrence of red tide and oil contamination. In contrast, in the case of an indirect intake, seawater is drawn from underground and no extensive pretreatment is required (Anderson, Boerlage and Dixon, 2017).

For the SWRO plant discharge, the brine is known to have a salt level 1.3- to 1.7-fold higher than in the feed seawater (Shatat, Worall and Riffat, 2013). The elevated salinity of the latter has a significant impact since it necessitates substantial energy consumption to surmount the osmotic pressure, resulting in the combustion of a large amount of fossil fuels and thus important GHGs generation (Ruso *et al.*, 2007). The inherent danger of brine concentrates is not just in their salinity, but because they can have high levels of boron and other minerals that adversely affect sea life, such as coral reefs (Mannan *et al.*, 2019). With regard to chemicals, numerous substances present environmental hazards, including surface active agents, coagulants, scavengers, caustic soda to adjust pH, and anti-scalants, which can form halogen by-products (Elsaid *et al.*, 2020). In addition, high boron retentates have to be recirculated to improve the recovery rate and prevent the toxicity of the incoming water supplies, causing harmful effects on the surrounding sea life (fish, seagrass, plankton, etc.) (Schoderboeck *et al.*, 2010).

To diminish the ecological effects of brine disposal, it is necessary to dilute the brine and set up diffusers in places where the mixture is appropriate (Panagopoulos and Haralambous, 2020) or by injecting it into deeply polluted wells between 322 and 2,575 m below ground (Younos, 2009). Thereby, Del-Pilar-Ruso *et al.* (2015) showed that once the diffusers were placed at the end of the outfall of a 5-km-long, 33-m-deep pipeline for a 65,000 m³/d SWRO plant, better mixing of the brine with the seawater reduced the salinity of the latter from about 50 g/L to approximately >35 g/L, thus allowing the recovery of the seabed community.

Therefore, managing the RO concentrate is extremely necessary to cope with the negative impacts on the environment. The choice of the most appropriate treatment depends on the composition of the RO concentrate and the extent of the treatment needed. In fact, various brine discharge operations are implemented in desalination plants, such as sewage disposal, evaporation ponds, surface water discharge, deep well injection, and land use (Panagopoulos and Haralambous, 2020). Table 8.8 provides a summary of the strengths and weaknesses of each brine discharge technique and their building costs for a SWRO desalination plant with a capacity of 40,000 m^3 per day (45% recovery −48,800 m^3/d of brine) (Charisiadis, 2018). Figure 8.16 outlines the environmental impacts related to a SWRO desalination plant with boron elimination and some suggested remedial strategies.

TABLE 8.8

The pros and cons of brine disposal methods and their construction cost (Charisiadis, 2018; Panagopoulos, Haralambous and Loizidou, 2019)

Brine disposal methods	Advantages	Limitations	Construction cost ($ mm)
Deep well injection	Appropriate for inland desalination plants	Unsuitable for countries with high seismic activity	15–25
	Reasonable construction costs	Potential marine environment pollution	
	Low energy requirement		
Evaporation ponds	Easy to construct and operate	Expensive	140–180
	Applied in inland and inshore regions	Applied only for small desalination plants	
		High footprint area	
Surface water discharge	Applied for all desalination plant sizes	Adverse impact on marine ecosystem	6.5–30
	Cost-effective		
Sewage discharge	Low construction cost	Applied only for small desalination plants	1.5–6
	Easy to construct	Potential marine environment pollution	
	Low energy requirement		
Land application	Easy to construct and implement	Expensive	30–40
	Applied in inland and inshore regions	Limited to small brine flows	

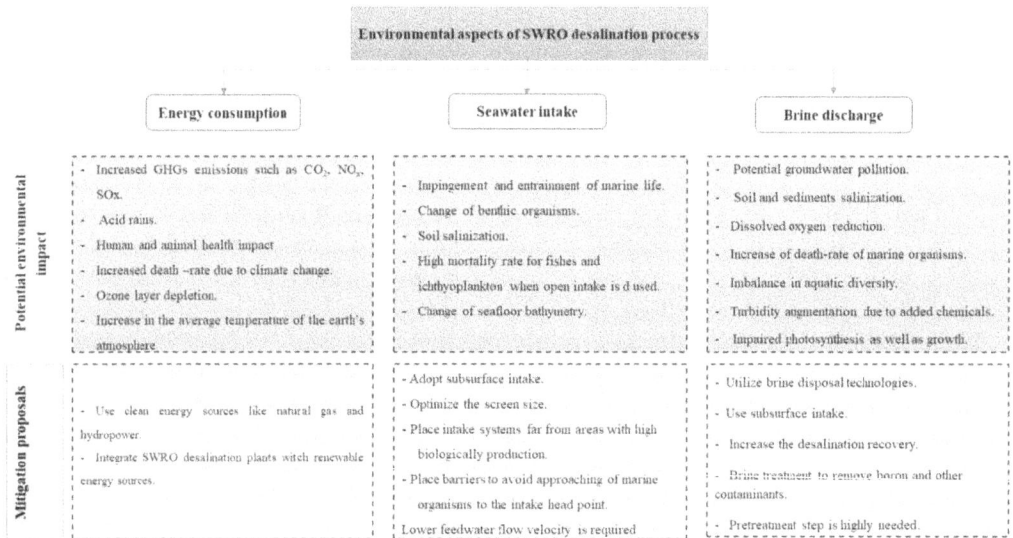

FIGURE 8.16 Potential environmental impacts and different mitigation proposals of SWRO desalination process (Missimer *et al.*, 2013; Shahabi *et al.*, 2014; Del-Pilar-Ruso *et al.*, 2015; Elsaid *et al.*, 2020).

8.9 CONCLUSIONS AND OUTLOOK

The presence of boron in seawater has attracted worldwide attention due to its toxicity and potential impact on human health and aquatic environments. Improper rejection of this component from seawater has been found to result in non-compliance with WHO standards for drinking water and irrigation water. Consequently, different membrane separation technologies have been suggested for the removal of boron from seawater, with updated information.

As indicated in the previous paragraphs, RO is a good technique for the desalination of boron-restricted seawater. However, the necessity of a multi-pass, to decrease the boron content under the recommended guideline, represents a serious limitation and leads to an increase in capital costs. Therefore, combining an RO system with other treatments, such as FO, NF, MD, ED, EDI, and AMF, seems to be more feasible and efficient. For instance, SWFO and NF with the addition of polyols coupled to SWRO have proven their ability to mitigate boron in seawater. However, these techniques need to be tested and conducted under the operating conditions of an industrial plant. For the removal of boron by the MD process, it exhibits good performance, although it is not yet practical due to membrane clogging and high power consumption, which prevent its use. Economically, this process is costly and requires frequent cleaning of the membrane. For the EDI technique, it can only be applied in a hybrid system with the RO technique to alleviate the boron content. Also, the ED process continues to be a disabled process due to the drawbacks of membrane fouling and the need for frequent cleaning and high operating costs. Overall, it is difficult to choose the most suitable treatment for boron elimination. Every process has its advantages and limitations.

It should be noted that a number of associated adverse environmental impacts need to be considered in the SWRO system, including power consumption, seawater intake, and brine disposal. In fact, this chapter has aimed to provide the current state of the art of LCA assessments for SWRO desalination considering the impact of boron and the main results being reported to assist in informing and refining public discussions; in particular, LCA analyses have rarely been conducted in this sense. Some overarching information can be considered as follows: (i) LCA studies of seawater desalination are, usually, data-intensive. It may often be better to reduce the system boundary by skipping a large number of background flows. Nevertheless, the omission of some data, including chemicals, building materials, and membranes, must be deliberate as it depends on the impact categories and the purpose of the LCA, (ii) the environmental effects of seawater desalination need to be coupled with its socio-economic dimensions, offering a broad outlook on the viability of SWRO desalination, and (iii) minimizing power consumption is a major objective in the design of a seawater desalination process. Therefore, energy based on the second law can also be applied to evaluate membrane process projects, which focus on the qualitative aspect of the consumed energy.

LIST OF ABBREVIATIONS

AL-DS	Active layer-draw solution
AL-FS	Active layer-feed solution
AMF	Adsorption membrane filtration
AEM	Anion exchange membranes
BWRO	Brackish water reverse osmosis
CAPEX	Capital expenses
CEM	Cation exchange membranes
CA	Cellulose acetate
DCMD	Direct contact membrane distillation
DS	Draw solution
ED	Electrodialysis
EDI	Electrode ionization
EE	Embodied energy

EU	European Union
FS	Feed solution
FO	Forward osmosis
GMA-co EDM	Glycidyl methacrylate-co-ethylene dimethacrylate
GHG	Greenhouse gas
IRENA	International Renewable Energy Agency
LCA	Life cycle assessment
MD	Membrane distillation
NF	Nanofiltration
NCC/GuA	Nanocrystalline cellulose/gum arabic conjugates
NMDG	N-Methyl-D-glucamine
OPEX	Operating expenses
PV	Photovoltaic panels
PF	Polarization factor
RES	Renewable energy sources
RO	Reverse osmosis
RSF	Reverse solute flux
SWFO	Seawater forward osmosis
SWRO	Seawater reverse osmosis
TDS	Total dissolved solids
TFC	Thin film composite
TWC	Total water cost
WHO	World Health Organization
WT	Wind turbine

REFERENCES

Abdelnour, S.A. *et al.* (2018) 'The vital roles of boron in animal health and production: A comprehensive review', *Journal of Trace Elements in Medicine and Biology: Organ of the Society for Minerals and Trace Elements (GMS)*, 50, pp. 296–304. Available at: https://doi.org/10.1016/j.jtemb.2018.07.018.

Al Haddabi, M. *et al.* (2014) 'Boron removal from seawater using date palm (Phoenix dactylifera) seed ash', *Desalination and Water Treatment*, 57(11), pp. 1–8. Available at: https://doi.org/10.1080/19443994.2014.1000385.

Aleisa, E. and Al-Shayji, K. (2018) 'Ecological–economic modeling to optimize a desalination policy: Case study of an arid rentier state', *Desalination*, 430, pp. 64–73. Available at: https://doi.org/10.1016/j.desal.2017.12.049.

Alharati, A.A. (2018) *Élimination du bore contenu dans l'eau de mer par un système hybride de sorption par résines échangeuses d'ions et de microfiltration*. PhD Thesis. Claude Bernard Lyon 1, Ecole Doctorale de Chimie de Lyon.

Alharati, A. *et al.* (2018) 'Boron removal from seawater using a hybrid sorption/microfiltration process without continuous addition of resin', *Chemical Engineering and Processing - Process Intensification*, 131, pp. 227–233. Available at: https://doi.org/10.1016/j.cep.2018.07.019.

Alhathal Alanezi., A. *et al.* (2016) 'Performance investigation of o-ring vacuum membrane distillation module for water desalination', *Journal of Chemistry*, 2016, p. 12. Available at: https://doi.org/10.1155/2016/9378460.

Alnouri, S.Y. and Linke, P. (2014) 'Optimal seawater reverse osmosis network design considering product water boron specifications', *Desalination*, 345, pp. 112–127. Available at: https://doi.org/10.1016/j.desal.2014.04.030.

Al-Obaidani, S. *et al.* (2008) 'Potential of membrane distillation in seawater desalination: Thermal efficiency, sensitivity study and cost estimation', *Journal of Membrane Science*, 323(1), pp. 85–98. Available at: https://doi.org/10.1016/j.memsci.2008.06.006.

Al-Sarkal, T. and Arafat, H.A. (2013) 'Ultrafiltration versus sedimentation-based pretreatment in Fujairah-1 RO plant: Environmental impact study', *Desalination*, 317, pp. 55–66. Available at: https://doi.org/10.1016/j.desal.2013.02.019.

Al-Shayji, K. and Aleisa, E. (2018) 'Characterizing the fossil fuel impacts in water desalination plants in Kuwait: A Life Cycle Assessment approach', *Energy*, 158, pp. 681–692. Available at: https://doi.org/10.1016/j.energy.2018.06.077.

Anderson, D.M., Boerlage, S.F.E. and Dixon, M.B. (2017) *Harmful Algal Blooms (HABs) and Desalination: A Guide to Impacts, Monitoring and Management*. Report. Intergovernmental Oceanographic Commission of UNESCO, p. 538. Available at: https://doi.org/10.25607/OBP-203.

Arar, Ö. *et al.* (2014) 'Various applications of electrodeionization (EDI) method for water treatment—A short review', *Desalination*, 342, pp. 16–22. Available at: https://doi.org/10.1016/j.desal.2014.01.028.

Argust, P. (1998) 'Distribution of boron in the environment', *Biological Trace Element Research*, 66(1–3), pp. 131–143. Available at: https://doi.org/10.1007/BF02783133.

Asim, S. *et al.* (2018) 'The effect of Nanocrystalline cellulose/Gum Arabic conjugates in crosslinked membrane for antibacterial, chlorine resistance and boron removal performance', *Journal of Hazardous Materials*, 343, pp. 68–77. Available at: https://doi.org/10.1016/j.jhazmat.2017.09.023.

Aziz, N.I.H.A. and Hanafiah, M.M. (2021) 'Application of life cycle assessment for desalination: Progress, challenges and future directions', *Environmental Pollution*, 268, pp. 1–14. Available at: https://doi.org/10.1016/j.envpol.2020.115948.

Ban, S.-H. *et al.* (2019) 'Comparative performance of FO-RO hybrid and two-pass SWRO desalination processes: Boron removal', *Desalination*, 471, pp. 1–10. Available at: https://doi.org/10.1016/j.desal.2019.114114.

Baransi-Karkaby, K., Bass, M. and Freger, V. (2019) 'In situ modification of reverse osmosis membrane elements for enhanced removal of multiple micropollutants', *Membranes*, 9(2). Available at: https://doi.org/10.3390/membranes9020028.

Bick, A. and Oron, G. (2005) 'Post-treatment design of seawater reverse osmosis plants: Boron removal technology selection for potable water production and environmental control', *Desalination*, 178(1–3), pp. 233–246. Available at: https://doi.org/10.1016/j.desal.2005.01.001.

Bint El Hassan, S. *et al.* (2018) *Water Around the Mediterranean*. Available at: https://www.cmimarseille.org/knowledge-library/water-around-mediterranean (Accessed: 21 March 2020).

Boretti, A. and Rosa, L. (2019) 'Reassessing the projections of the World Water Development Report', *npj Clean Water*, 2(1), p. 15. Available at: https://doi.org/10.1038/s41545-019-0039-9.

Boubakri, A. *et al.* (2015) 'Effect of operating parameters on boron removal from seawater using membrane distillation process', *Desalination*, 373, pp. 86–93. Available at: https://doi.org/10.1016/j.desal.2015.06.025.

Bunani, S., Arda, M. and Kabay, N. (2018) 'Effect of operational conditions on post-treatment of RO permeate of geothermal water by using electrodeionization (EDI) method', *Desalination*, 431, pp. 100–105. Available at: https://doi.org/10.1016/j.desal.2017.10.032.

Butterwick, L., de Oude, N. and Raymond, K. (1989) 'Safety assessment of boron in aquatic and terrestrial environments', *Ecotoxicology and Environmental Safety*, 17(3), pp. 339–371. Available at: https://doi.org/10.1016/0147-6513(89)90055-9.

Charisiadis, C. (2018) *Brine Zero Liquid Discharge (ZLD) Fundamentals and Design; A Guide to the Basic Conceptualization of the ZLD/MLD Process Design and the Relative Technologies Involved*. Lenntech Water Treatment Solutions. Available at: https://doi.org/10.13140/RG.2.2.19645.31205.

Cherif, H. and Belhadj, J. (2018) 'Chapter 15 - Environmental life cycle analysis of water desalination processes', in V.G. Gude (ed.) *Sustainable Desalination Handbook*. Butterworth-Heinemann, pp. 527–559. Available at: https://doi.org/10.1016/B978-0-12-809240-8.00015-0.

Cho, B.-Y. *et al.* (2015) 'A study on boron removal for seawater desalination using the combination process of mineral cluster and RO membrane system', *Environmental Engineering Research*, 20(3), pp. 285–289. Available at: https://doi.org/10.4491/eer.2014.0083.

Conseil canadien des ministres de l'environnement (2009) 'Recommandations canadiennes pour la qualité des eaux : protection de la vie aquatique - bore.' Canadian Government.

Darwish, N.B. *et al.* (2020) 'Experimental investigation of forward osmosis process for boron removal from water', *Journal of Water Process Engineering*, 38, p. 101570. Available at: https://doi.org/10.1016/j.jwpe.2020.101570.

Degryse, F. (2017) 'Boron fertilizers: Use, challenges and the benefit of slow-release sources – A review', *Journal of Boron*, 2(3), pp. 111–122.

Del-Pilar-Ruso, Y. *et al.* (2015) 'Benthic community recovery from brine impact after the implementation of mitigation measures', *Water Research*, 70, pp. 325–336. Available at: https://doi.org/10.1016/j.watres.2014.11.036.

Dolati, M. *et al.* (2017) 'Boron removal from aqueous solutions by electrocoagulation at low concentrations', *Journal of Environmental Chemical Engineering*, 5(5), pp. 5150–5156. Available at: https://doi.org/10.1016/j.jece.2017.09.055.

DOW Filmtec Reverse Osmosis Membranes (2021) *Filter Water*. Available at: https://secure.livechatinc.com/.

Du, Y. *et al.* (2019) 'Economic, energy, exergo-economic, and environmental analyses and multiobjective optimization of seawater reverse osmosis desalination systems with boron removal', *Industrial & Engineering Chemistry Research*, 58(31), pp. 14193–14208. Available at: https://doi.org/10.1021/acs.iecr.9b01933.

Duin, M. *et al.* (1984) 'Studies on borate esters 1: The ph dependence of the stability of esters of boric acid and borate in aqueous medium as studied by 11B NMR', *Tetrahedron*, 40, pp. 2901–2911. Available at: https://doi.org/10.1016/S0040-4020(01)91300-6.

Dydo, P. (2015) 'Boron removal by electrodialysis', in E. Drioli and L. Giorno (eds) *Encyclopedia of Membranes*. Berlin, Heidelberg: Springer, pp. 1–3. Available at: https://doi.org/10.1007/978-3-642-40 872-4_75-4.

Dydo, P. and Turek, M. (2013) 'Boron transport and removal using ion-exchange membranes: A critical review', *Desalination*, 310, pp. 2–8. Available at: https://doi.org/10.1016/j.desal.2012.08.024.

Dydo, P. and Turek, M. (2014) 'The concept for an ED–RO integrated system for boron removal with simultaneous boron recovery in the form of boric acid', *Desalination*, 342, pp. 35–42. Available at: https://doi.org/10.1016/j.desal.2013.09.020.

Elevli, B., Yaman, İ. and Laratte, B. (2022) 'Estimation of the Turkish boron exportation to Europe', *Mining*, 2(2), pp. 155–169. Available at: https://doi.org/10.3390/mining2020009.

Elsaid, K. *et al.* (2020) 'Environmental impact of desalination processes: Mitigation and control strategies', *Science of The Total Environment*, 740, p. 140125. Available at: https://doi.org/10.1016/j.scitotenv.2020.140125.

Ezechi, E.H. *et al.* (2014) 'Boron removal from produced water using electrocoagulation', *Process Safety and Environmental Protection*, 92(6), pp. 509–514. Available at: https://doi.org/10.1016/j.psep.2014.08.003.

Fam, W. *et al.* (2014) 'Boron transport through polyamide-based thin film composite forward osmosis membranes', *Desalination*, 340, pp. 11–17. Available at: https://doi.org/10.1016/j.desal.2014.02.010.

Farhat, A. *et al.* (2013) 'Boron removal in new generation reverse osmosis (RO) membranes using two-pass RO without pH adjustment', *Desalination*, 310, pp. 50–59. Available at: https://doi.org/10.1016/j.desal.2012.10.003.

Fernandez-Gonzalez, C. *et al.* (2015) 'Sustainability assessment of electrodialysis powered by photovoltaic solar energy for freshwater production', *Renewable and Sustainable Energy Reviews*, 47, pp. 604–615. Available at: https://doi.org/10.1016/j.rser.2015.03.018.

Filippini, G. *et al.* (2019) 'Design and economic evaluation of solar-powered hybrid multi effect and reverse osmosis system for seawater desalination', *Desalination*, 465, pp. 114–125. Available at: https://doi.org/10.1016/j.desal.2019.04.016.

Francis, L. *et al.* (2014) 'Performance evaluation of the DCMD desalination process under bench scale and large scale module operating conditions', *Journal of Membrane Science*, 455, pp. 103–112. Available at: https://doi.org/10.1016/j.memsci.2013.12.033.

Freger, V. *et al.* (2015) 'Boron removal using membranes', in *Boron Separation Processes*. Elsevier, pp. 199–217. Available at: https://doi.org/10.1016/B978-0-444-63454-2.00008-3.

Geffen, N. *et al.* (2006) 'Boron removal from water by complexation to polyol compounds', *Journal of Membrane Science*, 286(1–2), pp. 45–51. Available at: https://doi.org/10.1016/j.memsci.2006.09.019.

Ghaseminezhad, S.M., Barikani, M. and Salehirad, M. (2019) 'Development of graphene oxide-cellulose acetate nanocomposite reverse osmosis membrane for seawater desalination', *Composites Part B: Engineering*, 161, pp. 320–327. Available at: https://doi.org/10.1016/j.compositesb.2018.10.079.

Güler, E. *et al.* (2011) 'Integrated solution for boron removal from seawater using RO process and sorption-membrane filtration hybrid method', *Journal of Membrane Science*, 375(1), pp. 249–257. Available at: https://doi.org/10.1016/j.memsci.2011.03.050.

Güler, E. *et al.* (2015) 'Boron removal from seawater: State-of-the-art review', *Desalination*, 356, pp. 85–93. Available at: https://doi.org/10.1016/j.desal.2014.10.009.

Hakim, A.N. *et al.* (2020) 'Ionic separation in electrodeionization system: Mass transfer mechanism and factor affecting separation performance', *Separation & Purification Reviews*, 49(4), pp. 294–316. Available at: https://doi.org/10.1080/15422119.2019.1608562.

Han, L., Liu, Y. and Chew, J.W. (2017) 'Boron transfer during desalination by electrodialysis', *Journal of Membrane Science*, 547, pp. 64–72. Available at: https://doi.org/10.1016/j.memsci.2017.10.036.

Health and Ecological Criteria Division, Office of Science and Technology (2008) *Drinking Water Health Advisory for Boron*. Washington DC: U.S. Environmental Protection Agency, pp. 1–65.

Helvaci, C. (2017) 'Borate deposits: An overview and future forecast with regard to mineral deposits', *Journal of Boron*, 2(2), pp. 59–70.

Helvacı, C. and Palmer, M.R. (2017) 'Origin and distribution of evaporite borates: The primary economic sources of boron', *Elements: An International Magazine of Mineralogy, Geochemistry, and Petrology*, 13(4), pp. 249–254. Available at: https://doi.org/10.2138/gselements.13.4.249.

Hilal, N., Kim, G.J. and Somerfield, C. (2011) 'Boron removal from saline water: A comprehensive review', *Desalination*, 273(1), pp. 23–35. Available at: https://doi.org/10.1016/j.desal.2010.05.012.

Hou, D. *et al.* (2010) 'Boron removal from aqueous solution by direct contact membrane distillation', *Journal of Hazardous Materials*, 177(1–3), pp. 613–619. Available at: https://doi.org/10.1016/j.jhazmat.2009.12.076.

Hou, D. *et al.* (2013) 'Boron removal and desalination from seawater by PVDF flat-sheet membrane through direct contact membrane distillation', *Desalination*, 326, pp. 115–124. Available at: https://doi.org/10.1016/j.desal.2013.07.023.

Hu, Z.J. *et al.* (2014) 'The study on the effect on seawater desalination and boron removal by RO-EDI desalination system', *Advanced Materials Research*, 955–959, pp. 3211–3215. Available at: https://doi.org/10.4028/www.scientific.net/AMR.955-959.3211.

Huijbregts, M.A.J. *et al.* (2017) 'ReCiPe2016: A harmonised life cycle impact assessment method at midpoint and endpoint level', *The International Journal of Life Cycle Assessment*, 22(2), pp. 138–147. Available at: https://doi.org/10.1007/s11367-016-1246-y.

Hunt, J.D. *et al.* (2021) 'Deep seawater cooling and desalination: Combining seawater air conditioning and desalination', *Sustainable Cities and Society*, 74, p. 103257. Available at: https://doi.org/10.1016/j.scs.2021.103257.

Hydranautics – A Nitto Group Company (2021). Available at: https://membranes.com/.

Hyung, H. and Kim, J.-H. (2006) 'A mechanistic study on boron rejection by sea water reverse osmosis membranes', *Journal of Membrane Science*, 286(1–2), pp. 269–278. Available at: https://doi.org/10.1016/j.memsci.2006.09.043.

Ibrahim, G.P.S., Isloor, A.M. and Farnood, R. (2020) '6 - Fundamentals and basics of reverse osmosis', in A. Basile, A. Cassano, and N.K. Rastogi (eds) *Current Trends and Future Developments on (Bio-) Membranes*. Elsevier, pp. 141–163. Available at: https://doi.org/10.1016/B978-0-12-816777-9.00006-X.

IRENA (2018) *Global Energy Transformation: A Roadmap to 2050*. Abu Dhabi: International Renewable Energy Agency, p. 76.

Isa, M.H. *et al.* (2014) 'Boron removal by electrocoagulation and recovery', *Water Research*, 51, pp. 113–123. Available at: https://doi.org/10.1016/j.watres.2013.12.024.

Ismail, H. and Hanafiah, M. (2017) 'Management of end-of-life electrical and electronic products: The challenges and the potential solutions for management enhancement in developing countries context', *Acta Scientifica Malaysia*, 1, pp. 05–08. Available at: https://doi.org/10.26480/asm.02.2017.05.08.

Jiang, B. *et al.* (2018) 'Removal of high level boron in aqueous solutions using continuous electrodeionization (CEDI)', *Separation and Purification Technology*, 192, pp. 297–301. Available at: https://doi.org/10.1016/j.seppur.2017.10.012.

Jijakli, K. *et al.* (2012) 'How green solar desalination really is? Environmental assessment using life-cycle analysis (LCA) approach', *Desalination*, 287, pp. 123–131. Available at: https://doi.org/10.1016/j.desal.2011.09.038.

Jin, X. *et al.* (2011) 'Boric acid permeation in forward osmosis membrane processes: Modeling, experiments, and implications', *Environmental Science & Technology*, 45, pp. 2323–2330.

Kabay, N. *et al.* (2008) 'Removal of boron from water by electrodialysis: Effect of feed characteristics and interfering ions', *Desalination*, 223(1–3), pp. 63–72. Available at: https://doi.org/10.1016/j.desal.2007.01.207.

Kabay, N., Güler, E. and Bryjak, M. (2010) 'Boron in seawater and methods for its separation — A review', *Desalination*, 261(3), pp. 212–217. Available at: https://doi.org/10.1016/j.desal.2010.05.033.

Karvounis, P. (2017) 'A review of desalination potential in Greek islands using renewable energy sources, a life cycle assessment of different units', *European Journal of Sustainable Development*, 6, pp. 19–32. Available at: https://doi.org/10.14207/ejsd.2017.v6n2p19.

Katal, R. *et al.* (2020) 'An overview on the treatment and management of the desalination brine solution', in *Desalination - Challenges and Opportunities*. Singapore: IntechOpen, p. 28. Available at: https://www.intechopen.com/books/desalination-challenges-and-opportunities/an-overview-on-the-treatment-and-management-of-the-desalination-brine-solution (Accessed: 2 September 2020).

Kayaci, S. *et al.* (2020) 'Technical and economic feasibility of the concurrent desalination and boron removal (CDBR) process', *Desalination*, 486, pp. 1–13. Available at: https://doi.org/10.1016/j.desal.2020.114474.

Kettani, M. and Bandelier, P. (2020) 'Techno-economic assessment of solar energy coupling with large-scale desalination plant: The case of Morocco', *Desalination*, 494, pp. 1–18. Available at: https://doi.org/10.1016/j.desal.2020.114627.

Khayet, M. (2013) 'Solar desalination by membrane distillation: Dispersion in energy consumption analysis and water production costs (a review)', *Desalination*, 308, pp. 89–101. Available at: https://doi.org/10.1016/j.desal.2012.07.010.

Kheriji, J. and Hamrouni, B. (2016) 'Boron removal from brackish water by reverse osmosis and nanofiltration membranes: Application of Spiegler–Kedem model and optimization', *Water Science and Technology: Water Supply*, 16(3), pp. 684–694. Available at: https://doi.org/10.2166/ws.2015.178.

Kim, J.E. *et al.* (2017) 'Environmental and economic impacts of fertilizer drawn forward osmosis and nanofiltration hybrid system', *Desalination*, 416, pp. 76–85. Available at: https://doi.org/10.1016/j.desal.2017.05.001.

Kochkodan, V., Darwish, N.B. and Hilal, N. (2015) 'The chemistry of boron in water', in *Boron Separation Processes*. Amsterdam: Elsevier, pp. 35–63. Available at: https://doi.org/10.1016/B978-0-444-63454-2.00002-2.

Koseoglu, H. *et al.* (2008) 'Boron removal from seawater using high rejection SWRO membranes — Impact of pH, feed concentration, pressure, and cross-flow velocity', *Desalination*, 227(1–3), pp. 253–263. Available at: https://doi.org/10.1016/j.desal.2007.06.029.

Kouzbour, S. *et al.* (2019) 'Comparative analysis of industrial processes for cadmium removal from phosphoric acid: A review', *Hydrometallurgy*, 188, pp. 222–247. Available at: https://doi.org/10.1016/j.hydromet.2019.06.014.

Kucera, J. (2015) 'Reverse osmosis: Industrial processes and applications', in *Reverse Osmosis: Design, Processes and Applications for Engineers*. John Wiley & Sons: Hoboken, p. 472. Available at: https://www.wiley.com/en-us/Reverse+Osmosis%3A+Industrial+Processes+and+Applications%2C+2nd+Edition-p-9781118639740 (Accessed: 25 September 2020).

Lattemann, S. *et al.* (2010) 'Chapter 2 Global desalination situation', in *Sustainability Science and Engineering*. Elsevier, pp. 7–39. Available at: https://doi.org/10.1016/S1871-2711(09)00202-5.

Lawler, W. *et al.* (2015) 'Comparative life cycle assessment of end-of-life options for reverse osmosis membranes', *Desalination*, 357, pp. 45–54. Available at: https://doi.org/10.1016/j.desal.2014.10.013.

Liu, J. *et al.* (2013) 'Complexation-enhanced boron removal in a dual-stage nanofiltration seawater desalination process', *Separation Science and Technology*, 48(11), pp. 1648–1656. Available at: https://doi.org/10.1080/01496395.2012.752010.

Liu, Y. *et al.* (2019) 'A deep desalination and anti-scaling electrodeionization (EDI) process for high purity water preparation', *Desalination*, 468, pp. 1–10. Available at: https://doi.org/10.1016/j.desal.2019.114075.

Macedonio, F. and Drioli, E. (2008) 'Pressure-driven membrane operations and membrane distillation technology integration for water purification', *Desalination*, 223(1–3), pp. 396–409. Available at: https://doi.org/10.1016/j.desal.2007.01.200.

Maddah, H.A. *et al.* (2018) 'Evaluation of various membrane filtration modules for the treatment of seawater', *Applied Water Science*, 8(6), p. 13. Available at: https://doi.org/10.1007/s13201-018-0793-8.

Makkee, M., Kieboom, A. and Bekkum, H. (2010) 'Studies on borate esters III. Borate esters of D-mannitol, D-glucitol, D-fructose and D-glucose in water', *Recueil des Travaux Chimiques des Pays-Bas*, 104, pp. 230–235. Available at: https://doi.org/10.1002/recl.19851040905.

Mannan, M. *et al.* (2019) 'Examining the life-cycle environmental impacts of desalination: A case study in the State of Qatar', *Desalination*, 452, pp. 238–246. Available at: https://doi.org/10.1016/j.desal.2018.11.017.

Missimer, T.M. *et al.* (2013) 'Subsurface intakes for seawater reverse osmosis facilities: Capacity limitation, water quality improvement, and economics', *Desalination*, 322, pp. 37–51. Available at: https://doi.org/10.1016/j.desal.2013.04.021.

Mo, W., Wang, H. and Jacobs, J.M. (2016) 'Understanding the influence of climate change on the embodied energy of water supply', *Water Research*, 95, pp. 220–229. Available at: https://doi.org/10.1016/j.watres.2016.03.022.

Moseman, R.F. (1994) 'Chemical disposition of boron in animals and humans', *Environmental Health Perspectives*, 102(Suppl 7), pp. 113–117. Available at: https://doi.org/10.1289/ehp.94102s7113.

Muñoz, I. and Fernández-Alba, A.R. (2008) 'Reducing the environmental impacts of reverse osmosis desalination by using brackish groundwater resources', *Water Research*, 42(3), pp. 801–811. Available at: https://doi.org/10.1016/j.watres.2007.08.021.

Najid, N. *et al.* (2021a) 'Energy and environmental issues of seawater reverse osmosis desalination considering boron rejection: A comprehensive review and a case study of exergy analysis', *Process Safety and Environmental Protection*, 156, pp. 373–390. Available at: https://doi.org/10.1016/j.psep.2021.10.014.

Najid, N. *et al.* (2021b) 'Comparison analysis of different technologies for the removal of boron from seawater: A review', *Journal of Environmental Chemical Engineering*, 9(2), p. 105133. Available at: https://doi.org/10.1016/j.jece.2021.105133.

Najid, N. *et al.* (2022) 'Fouling control and Modeling in reverse osmosis for seawater desalination: A review', *Computers & Chemical Engineering*, p. 107794. Available at: https://doi.org/10.1016/j.compchemeng.2022.107794.

Oren, Y. *et al.* (2006) 'Boron removal from desalinated seawater and brackish water by improved electrodialysis', *Desalination*, 199(1–3), pp. 52–54. Available at: https://doi.org/10.1016/j.desal.2006.03.141.

Osipi, S.R., Secchi, A.R. and Borges, C.P. (2020) '13 - Cost analysis of forward osmosis and reverse osmosis in a case study', in A. Basile, A. Cassano, and N.K. Rastogi (eds) *Current Trends and Future Developments on (Bio-) Membranes*. Elsevier, pp. 305–324. Available at: https://doi.org/10.1016/B978-0-12-816777-9.00013-7.

Panagopoulos, A. and Haralambous, K.-J. (2020) 'Environmental impacts of desalination and brine treatment - Challenges and mitigation measures', *Marine Pollution Bulletin*, 161, pp. 1–12. Available at: https://doi.org/10.1016/j.marpolbul.2020.111773.

Panagopoulos, A., Haralambous, K.-J. and Loizidou, M. (2019) 'Desalination brine disposal methods and treatment technologies - A review', *Science of The Total Environment*, 693, pp. 1–23. Available at: https://doi.org/10.1016/j.scitotenv.2019.07.351.

Park, B. *et al.* (2016) 'Enhanced boron removal using polyol compounds in seawater reverse osmosis processes', *Desalination and Water Treatment*, 57(17), pp. 7910–7917. Available at: https://doi.org/10.1080/19443994.2015.1038596.

Parks, J. and Edwards, M. (2005) 'Boron in the environment', *Critical Reviews in Environmental Science and Technology*, 35, pp. 81–114. Available at: https://doi.org/10.1080/10643380590900200.

Pazouki, P. *et al.* (2021) 'Comparative environmental life cycle assessment of alternative osmotic and mixing dilution desalination system configurations', *Desalination*, 504, p. 114963. Available at: https://doi.org/10.1016/j.desal.2021.114963.

Prats, D., Chillon-Arias, M.F. and Rodriguez-Pastor, M. (2000) 'Analysis of the influence of pH and pressure on the elimination of boron in reverse osmosis', *Desalination*, 128(3), pp. 269–273. Available at: https://doi.org/10.1016/S0011-9164(00)00041-2.

Rahmawati, K. (2011) *Boron Removal in Seawater Reverse Osmosis System*. Thuwal: King Abdullah University of Science and Technology.

Ras, C. and von Blottnitz, H. (2011) 'A comparative life cycle assessment of process water treatment technologies at the Secunda industrial complex, South Africa', *Water SA*, 38, pp. 549–554. Available at: https://doi.org/10.4314/wsa.v38i4.10.

Redondo, J., Busch, M. and De Witte, J.-P. (2003) 'Boron removal from seawater using FILMTECTM high rejection SWRO membranes', *Desalination*, 156(1–3), pp. 229–238. Available at: https://doi.org/10.1016/S0011-9164(03)00345-X.

Rioyo, J. *et al.* (2018) '"High-pH softening pretreatment" for boron removal in inland desalination systems', *Separation and Purification Technology*, 205, pp. 308–316. Available at: https://doi.org/10.1016/j.seppur.2018.05.030.

Ruiz-García, A., León, F.A. and Ramos-Martín, A. (2019a) 'Different boron rejection behavior in two RO membranes installed in the same full-scale SWRO desalination plant', *Desalination*, 449, pp. 131–138. Available at: https://doi.org/10.1016/j.desal.2018.07.012.

Ruiz-García, A., León, F.A. and Ramos-Martín, A. (2019b) 'Different boron rejection behavior in two RO membranes installed in the same full-scale SWRO desalination plant', *Desalination*, 449, pp. 131–138. Available at: https://doi.org/10.1016/j.desal.2018.07.012.

Ruso, Y.D.P. *et al.* (2007) 'Spatial and temporal changes in infaunal communities inhabiting soft-bottoms affected by brine discharge', *Marine Environmental Research*, 64(4), pp. 492–503. Available at: https://doi.org/10.1016/j.marenvres.2007.04.003.

Samatya, S., Tuncel, A. and Kabay, N. (2012) 'Boron removal from geothermal water by a novel monodisperse porous poly(GMA-co-EDM) resin containing N-methyl-D-glucamine functional group', *Solvent Extraction and Ion Exchange*, 30(4), pp. 341–349. Available at: https://doi.org/10.1080/07366299.2012.686857.

Sassi, K.M. and Mujtaba, I.M. (2013) 'MINLP based superstructure optimization for boron removal during desalination by reverse osmosis', *Journal of Membrane Science*, 440, pp. 29–39. Available at: https://doi.org/10.1016/j.memsci.2013.03.012.

Schlesinger, W.H. and Vengosh, A. (2016) 'Global boron cycle in the anthropocene', *Global Biogeochemical Cycles*, 30(2), pp. 219–230. Available at: https://doi.org/10.1002/2015GB005266.

Schoderboeck, L. *et al.* (2010) 'Effects assessment: Boron compounds in the aquatic environment', *Chemosphere*, 82, pp. 483–487. Available at: https://doi.org/10.1016/j.chemosphere.2010.10.031.

Shahabi, M.P. *et al.* (2014) 'Environmental life cycle assessment of seawater reverse osmosis desalination plant powered by renewable energy', *Renewable Energy*, 67, pp. 53–58. Available at: https://doi.org/10.1016/j.renene.2013.11.050.

Shatat, M., Worall, M. and Riffat, S. (2013) 'Opportunities for solar water desalination worldwide: Review', *Sustainable Cities and Society*, 9, pp. 67–80. Available at: https://doi.org/10.1016/j.scs.2013.03.004.

Suwaileh, W. *et al.* (2020) 'Forward osmosis membranes and processes: A comprehensive review of research trends and future outlook', *Desalination*, 485, pp. 1–21. Available at: https://doi.org/10.1016/j.desal.2020.114455.

Tang, Y.P. *et al.* (2017a) 'Development of novel diol-functionalized silica particles toward fast and efficient boron removal', *Industrial & Engineering Chemistry Research*, 40(56), pp. 1–9.

Tang, Y.P. *et al.* (2017b) 'Recent advances in membrane materials and technologies for boron removal', *Journal of Membrane Science*, 541, pp. 434–446. Available at: https://doi.org/10.1016/j.memsci.2017.07.015.

Taniguchi, M. *et al.* (2004) 'Boron removal in RO seawater desalination', *Desalination*, 167, pp. 419–426. Available at: https://doi.org/10.1016/j.desal.2004.06.157.

Tarpani, R.R.Z. *et al.* (2021) 'Comparative life cycle assessment of three alternative techniques for increasing potable water supply in cities in the Global South', *Journal of Cleaner Production*, 290, p. 125871. Available at: https://doi.org/10.1016/j.jclepro.2021.125871.

Toray Membrane: Innovation by Chemistry (2021). Available at: https://www.water.toray/search_result.html?ajaxUrl=%2F%2Fss.marsflag.com%2Ftoray__membrane__membrane%2Fx_search.x&ct=&d=&doctype=all&htmlLang=en&imgsize=3&page=1&pagemax=10&q=boron%20rejection&sort=0.

Tu, K.L., Nghiem, L.D. and Chivas, A.R. (2010) 'Boron removal by reverse osmosis membranes in seawater desalination applications', *Separation and Purification Technology*, 75(2), pp. 87–101. Available at: https://doi.org/10.1016/j.seppur.2010.07.021.

Turek, M., Dydo, P. and Bandura-Zalska, B. (2009) 'Boron removal from dual-staged seawater nanofiltration permeate by electrodialysis', *Desalination and Water Treatment*, 10(1–3), pp. 60–63. Available at: https://doi.org/10.5004/dwt.2009.782.

Turkbay, T. *et al.* (2022) 'Prior knowledge of the data on the production capacity of boron facilities in Turkey', *Cleaner Engineering and Technology*, p. 100539. Available at: https://doi.org/10.1016/j.clet.2022.100539.

Valladares Linares, R. *et al.* (2014) 'Higher boron rejection with a new TFC forward osmosis membrane', *Desalination and Water Treatment*, 55(10), pp. 1–7.

Vuong, D.X. (2006) 'Two stage nanofiltration seawater desalination system'. Available at: https://patents.google.com/patent/US7144511B2/en (Accessed: 15 March 2022).

Wang, H. *et al.* (2021) 'Comprehensive analysis of a hybrid FO-NF-RO process for seawater desalination: With an NF-like FO membrane', *Desalination*, 515, p. 115203. Available at: https://doi.org/10.1016/j.desal.2021.115203.

Wijmans, J.G. and Baker, R.W. (1995) 'The solution-diffusion model: A review', *Journal of Membrane Science*, 107(1), pp. 1–21. Available at: https://doi.org/10.1016/0376-7388(95)00102-I.

Wolska, J. and Bryjak, M. (2013) 'Methods for boron removal from aqueous solutions — A review', *Desalination*, 310, pp. 18–24. Available at: https://doi.org/10.1016/j.desal.2012.08.003.

Xu, F. *et al.* (2007) *Advances in Plant and Animal Boron Nutrition: Proceedings of the 3rd International Symposium on all Aspects of Plant and Animal Boron Nutrition.* Springer. Available at: https://doi.org/10.1007/978-1-4020-5382-5.

Younos, T. (2009) 'Environmental issues of desalination: Environmental issues', *Journal of Contemporary Water Research & Education*, 132(1), pp. 11–18. Available at: https://doi.org/10.1111/j.1936-704X.2005.mp132001003.x.

Zhou, D. *et al.* (2015) 'Development of lower cost seawater desalination processes using nanofiltration technologies — A review', *Desalination*, 376, pp. 109–116. Available at: https://doi.org/10.1016/j.desal.2015.08.020.

9 Development and Characterization of Metal-Chelating Membranes Fabricated Using Semi-Interpenetrating Polymer Networks for Water Treatment Applications

Haad Bessbousse, Hamid Barkouch, Meryem Amar, Ouafa Tahiri Alaoui, Asma Zouitine, Fatna Zaakour, Marion Bellier, François Perreault, Laurent Lebrun and Jean-François Verchère

9.1 INTRODUCTION

Toxic heavy metals are a result of natural or industrial pollution, and they pose a rising hazard to the environment. Heavy metals are not biodegradable; therefore, the health hazards associated with their buildup throughout the food chain and in the human body are becoming an increasing source of worry (K. Renu, 2021; S. Sun, 2022). Mercury is of special concern because of its high toxicity and volatility (U. Forstner, 1979; S. Kumari, 2020). The removal of heavy metals contained in liquid effluents can be technologically challenging due to strict national and international regulations. Therefore, there is a clear need for novel, more efficient techniques to achieve the removal of metal ions from wastewater.

As for other toxic ions, numerous methods exist for the removal of mercury from liquid effluents. Ordinary treatment techniques used for the retention of toxic ions include sorption, ion exchange, coagulation, photoreduction, or membrane separation (S.A. Razzak, 2022; T.A. Saleh, 2022). More recently, combining sorption and membrane filtration into a single metal removal process has been highlighted as a potential high-efficiency method for the elimination of metal ions as high through-put removal can be performed at low cost. However, to achieve high metal sorption during membrane filtration, new membrane materials tailored for both filtration and sorption remain necessary.

Toxic metals ions can be eliminated by sorption metal oxides, activated carbon, or ion-exchange materials. Sorbent designs often include a support matrix, which can be made of inorganic (as silica) or organic (as cellulose) materials on which are found complexing agents that enable the interaction of the material with a specific ion or contaminant. The sorption interaction may occur either on a granular material (e.g. a resin) or on a membrane. Compared to granular media, membranes have the advantage of having faster filtration times.

Membranes having pores of larger diameter achieve higher permeation performance; however, typical porous membranes such as ultrafiltration (UF) membranes typically cannot keep ions in the feed caused by the large pore diameter and the inactive membrane surface. Studies have shown a noticeable improvement in metal-chelating UF membranes by using soluble polymers in water that

DOI: 10.1201/9781003326281-11

can bind to the target ions and form large polymer-ion complexes that are then easily retained by UF membranes (named polymer-enhanced ultrafiltration (PEUF) (M.A. Barakat, 2010; Y. Huang, 2019)). A similar method for eliminating toxic ions is to utilize ion-complexing membranes (B. Lam, 2018; S. Xu, 2020). Several polymers were tested, such as poly(acrylic acid) (V. Beaugeard, 2020), chitosan (B. Lam, 2018; Y. Huang, 2019), poly(ethyleneimine) (Y. Huang, 2016), and alginate (N. Fatin-Rouge, 2006). These sorptive membranes can provide high removal efficiency for heavy metal ions with low energy consumption and high filtration rate.

The objective of this chapter is to achieve ion-sorptive membranes that have a complexing water-soluble polymer immobilized on the membrane to combine chelation and filtration in a single step, which will complex metal ions during filtration (L. Lebrun, 2002; H. Bessbousse, 2012). The ion-sorptive membranes differ from the PEUF process by having the complexed metal ions rejected in the retentate, which results in lower polymer consumption compared to PEUF without compromising the selectivity of the process. In addition, the number of procedural steps is decreased by integrating ion complexation with the filtration step. As PEUF, the complexing membranes cannot be used ceaselessly, since the membrane accumulates ions over time. Once saturated, the membranes must be regenerated at acidic pH.

The chelating polymer used in ion-sorptive membranes is soluble in water; therefore, it should be fixed on the membrane. It can be accomplished by grafting a chelating polymer on a support polymer (E.A. Hegazy, 2000). In comparison, the semi-interpenetrating polymer network (s-IPN) approach is a relatively simple method for immobilizing the chelating polymer in the matrix. In this method, chains of a chelating polymer are stabilized in the film without chemical bonding (L. Lebrun, 2007). Due to its outstanding film-formation capabilities and high density of reticulable OH groups (L. Lebrun, 2002), polyvinyl alcohol (PVA) was selected as the matrix (H. Bessbousse, 2008a). Several reagents can be used to crosslink PVA, including glutaraldehyde, 1,2-dibromoethane (DBE), or hydrochloric acid (H. Bessbousse, 2009; O. Farid, 2016). In this work, a PVA matrix crosslinked with gaseous DBE is used.

A complexing membrane that can filter heavy metal-contaminated water quantitatively (selective) and quickly (porous) will be a useful devise for filtration-contaminated water. When selectivity is determined by the nature of the chelating polymer, permeability is determined by the membrane's physical features, such as thickness or the number and size of pores. Controlling the porosity of organic membranes might result in membranes with high permeability. Several chemical and physical methods have been proposed in the literature to modify pore size distribution in membranes (X. Tan 2019; S.R. Ravichandran, 2022), such as stretching (S.R. Ravichandran, 2022), track etching (H. Bessbousse, 2016; S.R. Ravichandran, 2022), electrospinning (X. Tan 2019; S.R. Ravichandran, 2022), and the use of sacrificial grains which may be removed after the solid material is formed, resulting in a macroporous membrane (X. Zeng, 1996; H. Bessbousse, 2010a), and the development of porous membranes via phase inversion, which are examples of these procedures (M. Mulder, 1991; H. Bessbousse, 2010a; S. Mondal, 2020; C.C. Ho, 2021).

Novel PVA-based chelating membranes that integrate different complexing polymers, such as poly(vinylimidazole) or poly(4-vinylpyridine) (P_4VP), have been characterized in previous studies (H. Bessbousse, 2008a; H. Bessbousse, 2008b; H. Bessbousse, 2009; H. Bessbousse, 2010b; H. Bessbousse, 2012). In the present chapter, the design of a complexing membrane, by the s-IPN technique, containing P_4VP (Figure 9.1) as a good chelating polymer for Hg(II) is reported. The performance of this membrane was reviewed under static (sorption) and dynamic (filtration) conditions for the elimination of mercury ions from water under several conditions. For an industrial treatment and to filter high volumes rapidly, two methods were chosen to increase the PVA/P_4VP membrane porosity. For the first method called impregnation, the membrane is incorporated with silica granules which are subsequently removed to form pores in the membrane (X. Zeng, 1996; G.C. Steenkamp, 2002). In the other one, the membrane is fabricated by inversion of phases by submersion in a nonsolvent bath until precipitation (M. Mulder, 1991; C.O. M'Bareck, 2006; H. Zhao, 2022). All membranes were prepared by s-IPN. The structural and morphology characteristics of the membranes, their swelling, water permeability, sorption performance, and their ability to treat water contaminated with mercury were examined.

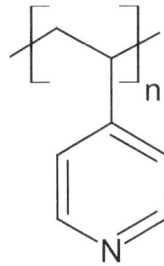

FIGURE 9.1 Poly(4-VP), P_4VP molecule.

9.2 EXPERIMENTAL

9.2.1 Fabrication of Ion-Complexing Membranes by s-IPN

9.2.1.1 Fabrication of the Initial or Unmodified Membrane

The PVA (M_w = 124,000–186,000 g/mol, 99% hydrolys) was solubilized in deionized water at 10 g/L. The P_4VP (M_w = 160,000 g/mol) was solubilized in an aqueous solution of acetic acid at 2% vol. The solutions were then combined in a 60:40 ratio. After five days of solvent evaporation of 122 cm³ of PVA/P_4VP solution (into a 13.5-cm-diameter petri dish) at room temperature, a film of ~140 cm² of surface area and 70 μm of thickness was elaborated. DBE (99%) was used for crosslinking of the film, which was done at 140°C (H. Bessbousse, 2012).

9.2.1.2 Fabrication of Different Porous Membranes

The porous membrane was elaborated precisely as the initial or unmodified membrane. The cross-linking was performed with DBE.

9.2.1.3 Modification by Impregnation with Silica Particles

Silica particles were used as a pore-forming agent. The silica particles were included during the blending step by adding 10 mg of silica (granulometry 0.045–0.065 mm) in the polymer solution, which was blended until homogeneous (~6 h). To produce pores within the crosslinked membrane, the silica particles were eliminated by submerging the membrane for 2 h at 80°C in a NaOH solution (5% wt). The permeable membrane was washed with deionized water. Removing the silica particles did not visibly alter the mechanical properties of the membrane.

9.2.1.4 Modification by Phase Inversion

Porous PVA/P_4VP membranes were obtained using nonsolvent-induced phase inversion with water as the solvent and acetone as the nonsolvent. The PVA and P_4VP solution was 100 g/L in water. The PVA/P_4VP blend was first deaerated before casting on a pre-cleaned glass plate using a custom-made Gardner cut (265 μm). The film was left to evaporate at room temperature for less than 1 h before immersion in acetone, which precipitated the membrane. The obtained membrane was then crosslinked with DBE as previous described.

9.2.2 Characterization Experiments

Swelling was measured by weighting the dry membranes (weight, m_D) and then immersing them at ambient temperature in deionized water for 24 h. The samples were then collected, wiped carefully to absorb the surplus water from the surface, and weighed (m_S). S_w (swelling ratio) was determined by H. Bessbousse (2012) and X. Jin (2022).

$$S_W = \frac{m_S - m_D}{m_D} \tag{9.1}$$

Fourier transform infrared spectroscopy (FTIR) (Nicolet Avatar 360) was also used for checking the crosslinking treatment efficiency. The spectra of the PVA/P$_4$VP film (before crosslinking) was compared to the PVA/P$_4$VP membrane (after crosslinking) spectra.

Scanning electron microscopy (SEM) images were obtained by a Jeol JEM-1200EX II (France) for studying the films' and membranes' structure and morphology. Before observation, the material samples were frozen in nitrogen at −196°C and covered with a gold layer.

9.2.3 SORPTION AND DESORPTION EXPERIMENTS

The sorption of metal ions (i.e. Hg(II)) from water by the PVA/P$_4$VP membranes was characterized. For this, the mercury concentrations were measured by an atomic absorption flame spectrometer (AAS, from Varian). The calibration of the AAS was performed with a Hg(II) ions standard (1% wt/v in HNO$_3$). All AAS measurements were reproducible within 0.4% accuracy and duplicated. A wavelength value of 253.5 nm was used for Hg measurements.

Mercury nitrate was used for preparing all Hg(II) ion solutions. A Tacussel LPH-230T pH meter with a combined glass electrode was used for pH measurements.

For sorption experiments, membrane samples were weighed (m$_D$) dry and placed into stirred aqueous solutions of Hg(II) (volume V, initial concentration c$_0$, at desired pH). The solutions were stirred using a mechanical shaker at 10.4 rad/s at room temperature (T = 21 ± 3°C) for 24 h.

For kinetic studies, samples of about 3 cm^3 of solution were withdrawn at known times. For the other sorption experiments, samples of about 3 cm^3 of solution were withdrawn before introduction of the membrane and at equilibrium (at 24 h of sorption).

The retention ratio R (%) of mercury is defined as

$$R = \frac{\left(c_o - c_{eq}\right)}{c_o} \times 100, \tag{9.2}$$

where c_0 and c_{eq} (mg/L) are, respectively, the starting and equilibrium Hg(II) concentrations. V is the solution volume (cm^3).

f (%) is designed as the membrane's ion retention efficiency, calculated as

$$f = \frac{n_r}{n_{VP}} \times 100 \tag{9.3}$$

- n$_r$ is the number (mol) of mercury ions fixed by the membrane
- n$_{VP}$ is the number (mol) of 4-vinylpyridine monomer repetition units in the membrane. n$_r$ and n$_{VP}$ are determined using equations (9.4) and (9.5) as:

$$n_r = \frac{\left(c_o - c_{eq}\right)V}{M \times 10^3} \tag{9.4}$$

$$n_{VP} = \frac{0.4 m_D}{105}, \tag{9.5}$$

where V is the volume of the solution (L) in contact with the membrane, M is the molar mass of mercury ions in g/mol, 105 g/mol is the molar weight of vinylpyridine (VP), with 10^3 for the conversion from c$_0$ in mg/L to g/L, and 0.4 due to the membrane includes 40% (w/w) P$_4$VP.

The mass q of mercury sorbed at equilibrium per gram of membrane (mg/g) was calculated as follows (H. Bessbousse, 2008a; F. Wang, 2022):

$$q = \frac{\left(c_0 - c_{eq}\right)V}{m_D} \tag{9.6}$$

In the desorption (or regeneration) experiments, membranes were first loaded with mercury ions in a 24 h sorption step using the same conditions described above. Then, desorption was realized by putting the membranes in 50 cm^3 of HNO$_3$ at 0.5 M, which was used as a desorption medium with continuous stirring at 100 rpm at ambient temperature for 24 h. The desorption ratio D was determined as:

$$D = \frac{n_d}{n_r} \times 100 \qquad (9.7)$$

- n_r is the ion quantity (mol) fixed into the membrane.
- n_d is the quantity (mol) of ions desorbed. It was determined from the final ion concentration in the desorption medium.

9.2.4 FILTRATION AND PERMEABILITY EXPERIMENTS

The filtration experiments were carried out at constant temperature ($25 \pm 0.5°C$) using a two-chamber filtration cell (V = 450 cm^3), with the membrane area of about 28 cm^2, which was described in previous studies (L. Lebrun, 2007; H. Bessbousse 2010b). The membrane was first weighed dry (m_D) and then swollen in water for a minimum of 4 h before placing it in the membrane cell. Then, 400 cm^3 of mercury ions with different initial concentrations to be treated were placed in the upstream feed chamber, with continuous stirring at 200 rpm.

Differential pressure was applied to generate permeate flow, which was collected at various time points during the filtration process. Aliquots were also taken in the upstream chamber at the beginning of each experiment. At ~370 cm^3 of filtrate volume, the remaining upstream volume (~30 mL) was collected to measure the final Hg(II) concentration.

The performance of PVA/P$_4$VP membranes was determined by the elimination ratio E, calculated as:

$$E = \frac{(c_0 - c_F)}{c_0} \times 100 \qquad (9.8)$$

The filtration flux (or permeability) is the volume of filtrate produced to the area of the membrane. Determination of the pure flux of water was realized with the same cell used for filtration experiments. Before flux measurement, membranes were allowed to swell in deionized water for more than 3 h. The filtration flux of distilled water, J_v, was determined using equation (9.9). The measurement of flux was carried out at a pressure ΔP of 2.5 bar by:

$$J_v = \frac{\Delta V}{(S \cdot \Delta t)} \qquad (9.9)$$

ΔV is the volume of cumulated permeate for a given period of time Δt and S is the membrane active surface area.

9.3 RESULTS AND DISCUSSION

9.3.1 CHARACTERIZATION OF CHELATING MEMBRANES

9.3.1.1 Film Crosslinking and Membrane Swelling

A film containing a matrix, made from a water-soluble polymer such as PVA, reacts in aqueous solutions by two processes: dissolution and swelling. It may dissolve in water. Therefore, the film needs to be crosslinked at 140°C with DBE vapour. In a crosslinked matrix, the polymer will swell, due to the affinity of the polymer chains with water molecules, but not be dissolved. S_w (swelling ratio) is used as an indicator of the efficiency of crosslinking, with an optimal S_w below 0.7.

The reaction time t of DBE crosslinking was studied using a change in the S_w. After 60 min, S_w reached a constant value (0.66), indicating full crosslinking at 1 h for PVA/P$_4$VP membranes.

9.3.1.2   FTIR Measurements

FTIR spectra (Figure 9.2) reveal the crosslinking reaction in the membrane. They show the ether links formation from the HO groups of the PVA chains (E. Da Silva, 2002) through a decrease in the intensity of the stretching vibration band of OH groups (3050–3650/cm) and the appearance of new peaks $\cong$ 1180/cm from the vibrations of ether groups.

9.3.1.3   Morphology

To study the morphology of initial and porous PVA/P$_4$VP membranes, SEM images were obtained (Figures 9.3–9.5). Figure 9.3 shows SEM images of the initial PVA/P$_4$VP. It was found to be homogeneous and dense (Figure 9.3a). However, when the surface is visualized, Figure 9.3b shows small folds and a wrinkled texture. This surface morphology is due to crosslinking by DBE, since, when analysed before crosslinking, Figure 9.3c shows a smooth surface. Thus, PVA/P$_4$VP is completely homogeneous with a heterogeneous surface before the crosslinking reaction.

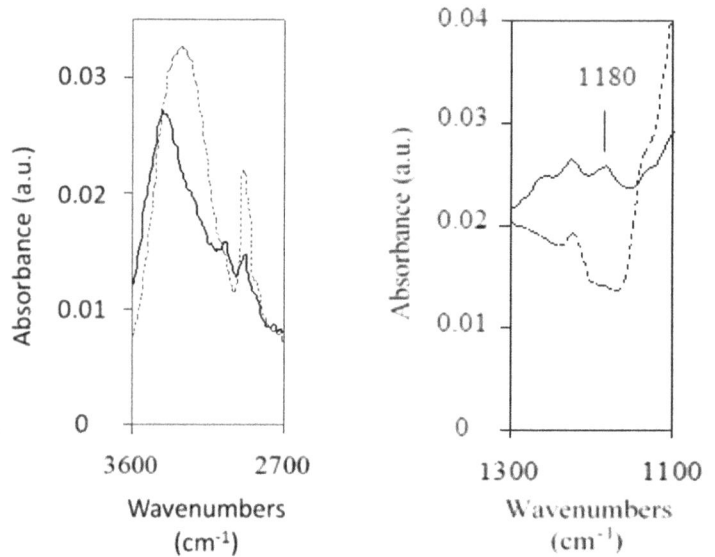

FIGURE 9.2 FTIR analysis of the crosslinking of PVA/P$_4$VP films with gaseous DBE, before (---) and after (—) DBE exposure for 1 h at 140°C.

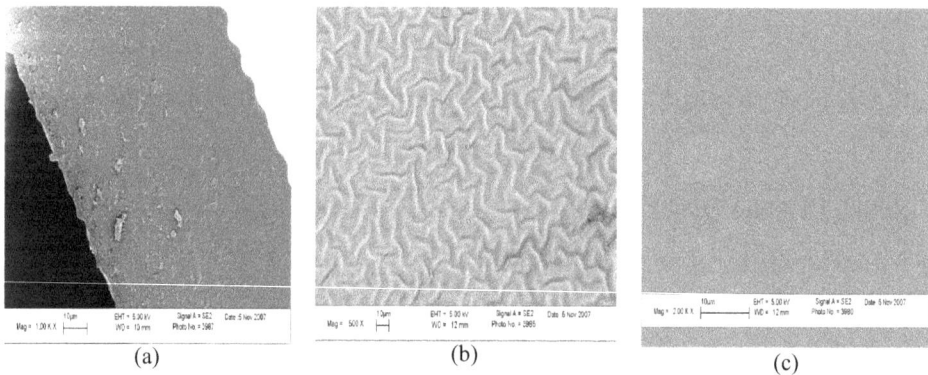

(a)   (b)   (c)

FIGURE 9.3 SEM micrographs of the cross-section (a, 1,000×) and surface (b, 500×) of the crosslinked initial PVA/P$_4$VP membrane, as well as the membrane surface before (c, 1,000×).

Figures 9.4 and 9.5 concern the crosslinked membranes with a porosity enhanced by silica impregnation and phase inversion, respectively. Both membranes prominently show the existence of pores when compared to the classical dense membrane of Figure 9.3. Figure 9.4 shows that the membrane is microporous with pores that are approximately 15–25 µm in diameter. For the phase inversion technique, Figure 9.5a and b illustrates two kinds of pores.

(i) homogeneous distribution of finger-like macropores of approximately 50 to 90 µm diameter (Figure 9.5a).
(ii) micropores of around 0.3–2.5 µm diameter, representing a large part of the membrane porosity (Figure 9.5b).

All pores are disposed haphazardly but homogeneously through the porous membrane.

(a)

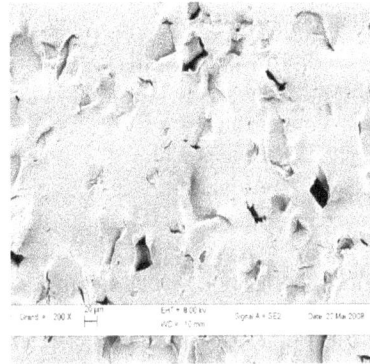

(b)

FIGURE 9.4 SEM micrographs of the surface of the PVA/P$_4$VP membrane where porosity was enhanced by impregnating the membrane with silica particles (a, 5,000×) and subsequently removing them by dissolution (b, 200×).

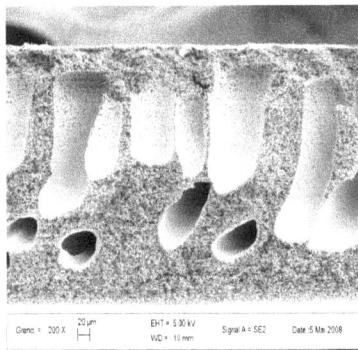

(a)

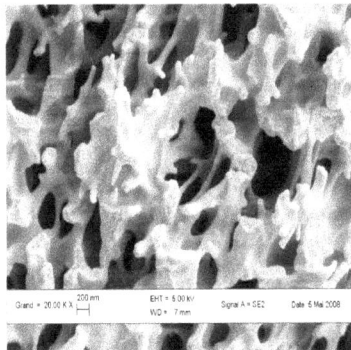

(b)

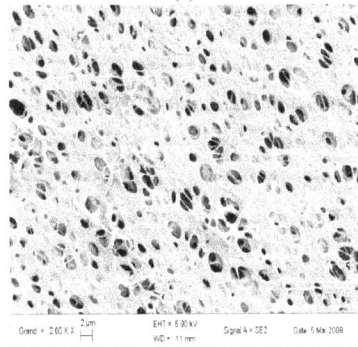

(c)

(d)

FIGURE 9.5 SEM micrographs of PVA/P$_4$VP membranes with their porosity enhanced by phase inversion, visualized as cross-section (a, 200×) (b, 20,000×) and from the top surface (c, 2,000×) and bottom surface (d, 2,000×).

Figure 9.5c and d shows a difference between the top and the bottom of the surface of the membrane. The first one is more porous with pore size between 0.5 and 2 μm, while the diameter of the pores on the bottom side is ~0.1–0.5 μm. Thus, the surface is microporous. This difference in the morphology of the two sides is a characteristic of the phase inversion technique, where one surface interfaces directly with the nonsolvent liquid while the other surface is in contact with the solid support, here the glass plate, which slows down its precipitation.

The mechanical resistance of the porous membrane was verified under pressure by water flux determination and filtration experiments.

9.3.2   APPLICATION OF CHELATING MEMBRANES TO WATER TREATMENT

9.3.2.1   Sorption Study in Static Conditions

9.3.2.1.1   Effect of pH

The pH has a strong influence on the speciation of metals in an aqueous solution. Therefore, pH may have a strong impact on the extent of Hg(II) sorption. In addition, the protonation of the P_4VP ligand will change in function of pH. Table 9.1 reports the influence of pH on the retention ratio R of mercury (c_0 = 100 mg/L; V = 50 cm^3) by a PVA/P_4VP membrane (m_D = 100 mg).

Since the solubility of Hg(II) changes with pH, experimental conditions need to be adjusted to take potential precipitation into consideration. For example, at pH 2.5, Hg(II) ions are stable in solution at concentrations below $<2 \times 10^3$ mg/L. However, since precipitation is not immediate, sorption studies can be realized at pH 2.5 and c_0 about 4×10^3 mg/L without significant precipitation being observed for the duration of the filtration. When the pH is changed from 1 to 3, the retention ratio increases to a value of 99% at pH 3. This trend agrees with the increasing protonation of atoms of nitrogen in a P_4VP polymer at low pH, which decreases the number of linking sites for mercury. Previous studies have shown that P_4VP has a pK_a of 5.0 ± 0.3 (M. Satoh, 1989; U. Pinaeva, 2019). The authors (N. Oyama, 1980; M. Ito, 2018) confirmed that the polymer is strongly protonated at pH less than 3. Based on the high performance of the P_4VP membrane at higher pH and the need to limit precipitation, the sorption of mercury was further studied at pH 2.5. For the treatment of waste streams in the neutral range, it should be noted that acidification may be required for the chelating membrane technology. However, this may be not practical with highly buffered waters; therefore, consideration needs to be given to the types of water treated.

9.3.2.1.2   Sorption Kinetics

The sorption kinetic by a membrane (m_D around 300 mg) was studied using a Hg(II) solution at pH 2.5 (c_0 around 250 mg/L; V = 230 cm^3). The change of mercury ions concentration, c_t, as a function of time t, is illustrated in Figure 9.6. Results show a slow retention of mercury by the membrane with a plateau reached at ~1,260 min. Based on these findings, the time for further sorption experiments was established as 24 h, at which a stable sorption condition is reached.

TABLE 9.1

Retention ratio of mercury ions by PVA/P_4VP membranes at different pH. The solubility limit is determined using the product of solubility (pK_s = 25.0) of Hg(OH)$_2$ (A. Ringbom, 1967; H. Bessbousse, 2012)

pH	1	2	3
R (%)	41	58	99
Limit of solubility of Hg(II) (mg/L)	2×10^6	2×10^4	2×10^2

Ambient temperature; c_0 = 100 mg/L; m_D = 100 mg; V = 50 cm^3.

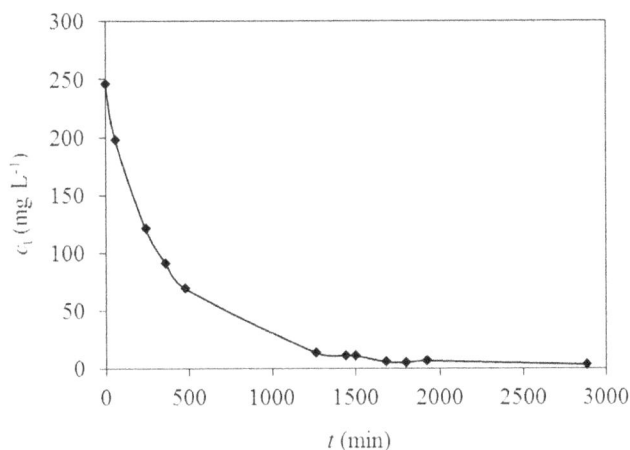

FIGURE 9.6 Change in the concentration of mercury ions over time in contact with the PVA/P_4VP chelating membrane.

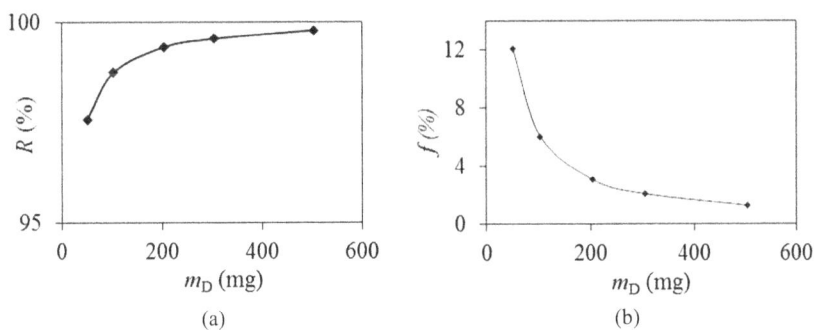

(a) (b)

FIGURE 9.7 Influence of m_D on Hg(II) sorption measured by the change in the rate R (a) and the rate f (b).

9.3.2.1.3 Effect of Membrane Mass

Hg(II) sorption was investigated at different membrane mass (m_D from 50 to 500 mg) for 100 mg/L of mercury ions to understand the effect of the PVA/P_4VP membrane mass on Hg(II) sorption. The change in Hg(II) was evaluated after 24 h contact time at a pH of 2.5. Figure 9.7 illustrates the change in R(a) and f(b).

Figure 9.7a shows that the retention ratio R increased slightly with an increase in membrane mass, reaching a maximum value of R = 99 ± 2% at m_D = 500 mg. This increase in R can be assigned to an increase in the number of complexing sites as m_D increases. At equilibrium, most of the Hg(II) ions were found to be fixed by the membrane at the highest membrane mass used. In contrast, Figure 9.7b shows that f consistently diminished to 1%, with an increase of the mass of the membrane. The decrease in f is explained by the fact that the same quantity of mercury ions was interacting with an increasing number of VP chelating sites, which leads to a reduced ratio of engaged VP sites to the total number of VP sites. Based on these experiments results, an m_D of 100 mg was selected for the subsequent experiments.

9.3.2.1.4 Effect of Water Hardness

Divalent ions like magnesium and calcium ions can interact with the sorption of mercury ions. To understand the performance of the chelating membranes in hard water, the effect of calcium ions on the establishment of the mercury-P_4VP complex was investigated. This experiment was

TABLE 9.2

Impact of the addition of Ca(II) on the complexation of mercury ions by the PVA/P$_4$VP membrane

[Ca(II)] mg/L	R (%)	q (mg/g)
0	99	44
20	98	43
40	99	43
100	97	41
200	97	41

Ambient temperature; $m_D \approx 100$ mg; pH 2.5; V = 50 cm³; $c_0 = 100$ mg/L.

realized for calcium ion concentrations up to 200 mg/L, using $m_D = 100$ mg and a fixed concentration of mercury (V = 50 cm³, $c_0 = 100$ mg/L) at pH 2.5. Table 9.2 shows the results for q and the retention ratio R for increasing concentrations of Ca(II). Ca(II), at a concentration of 40 mg/L and lower, did not affect the R value. As the Ca(II) concentration increased, R decreased by only 2.5% up to a Ca(II) concentration of 200 mg/L. This small change in R value with increasing Ca(II) concentration indicates that Ca does not compete with Hg for the VP binding sites in the membrane. Therefore, the PVA/P$_4$VP membrane is likely to be effective even in hard water streams.

9.3.2.1.5 Effect of Mercury(II) Ion Concentration

Sorption experiments were performed at increasing starting concentrations of mercury ions ($c_0 \approx 70$–3,500 mg/L). The pH of Hg(II) solution was fixed at 2.5. The changes in R and f values over the range of Hg(II) c_0 are shown in Figure 9.8. At low c_0 values, R is shown to be close to 100% due to the few Hg(II) ions in solutions, which are completely complexed by the membrane. As the concentration of Hg(II) in solution increases, R gradually decreases due to the progressive saturation of the finite quantity of chelating sites on the membrane. When all sites are engaged, any increase in c_0 results in a higher proportion of Hg(II) in solution. Concomitantly, f increased with the initial mercury ions concentration because of the number of accessible sites being gradually linked by the growing quantity of mercury ions (Figure 9.8b). The highest rate of f $\approx 60\%$ is determined for around $c_0 \approx 2,000$ mg/L. As stated in (9.3), f is the ratio between the quantity of mercury ions fixed and the quantity of nitrogen atoms into the membrane or P$_4$VP repetition, which all have one nitrogen atom. Since f = 60% corresponds to one mercury ion linked with 1/0.6 = 1.7 N atom, the complexation of Hg(II) ions is obtained with 1 and mostly 2 N atoms. Therefore, mercury ions form a bridge between two repeating units in the polymer structure.

Mercury ion uptake by the membrane was also studied by the shape of the sorption isotherms. As expected, q increased with increasing c_{eq} (Figure 9.9), with a plateau value reached for c_{eq} of about 1,000 mg/L, where the membrane reached complete saturation. The maximum mass of mercury ions fixed at equilibrium was about q ≈ 460 mg per gram of membrane. To compare with the maximum sorption capacity of P$_4$VP, the theoretical q value was determined as such: for $m_D = 1$ g, the membrane's P$_4$VP content is 40%, which corresponds to N = 3.81 mmol P$_4$VP units or N atoms based on the molar weight of P$_4$VP of 105 g/mol. As an average chelating site between Hg(II) and P$_4$VP has 1.7 N atoms, the maximum sorption of mercury ions is N/1.7 = 2.24 mmol. Therefore, the weight of mercury ions that theoretically could be fixed in 1 gram of membrane is determined, based on the molar weight of mercury. The theoretical q value of 450 mg/g is found to be very close to the 460 mg/g measured experimentally.

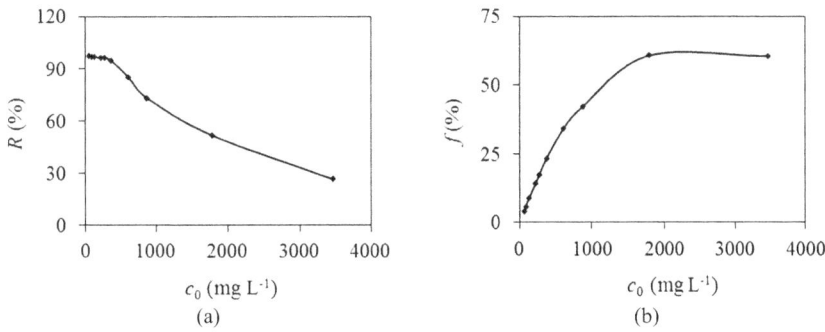

FIGURE 9.8   Effect of the initial Hg(II) concentration, c_0, on mercury retention ratio R (a) and efficiency ratio f (b) for PVA/P_4VP membranes.

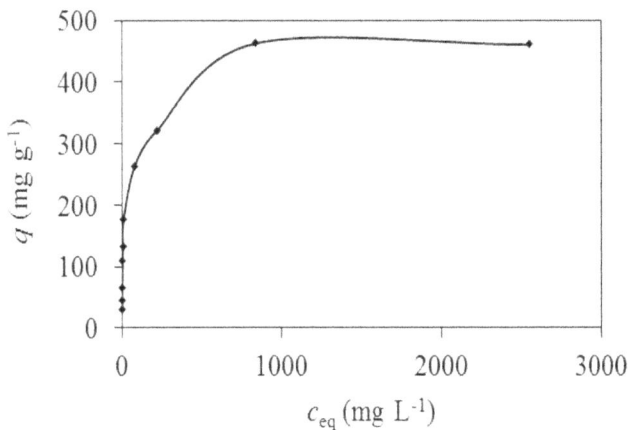

FIGURE 9.9   Sorption isotherm of mercury ions by PVA/P_4VP membranes.

9.3.2.1.6   Membrane Regeneration

For the applicability of PVA/P_4VP membranes for the elimination of heavy metal ions from an aqueous solution cost considerations need to be taken into account for the regeneration of the membrane. Therefore, the reusability of PVA/P_4VP membranes was evaluated after desorption of mercury from the membranes using nitric acid.

Multiple regeneration cycles were performed to understand the stability of the membranes through this regeneration process. The exposure to the regeneration medium (i.e. HNO_3) was repeated five times with the same membrane. The change in retention ratio R, during each individual sorption step, and the desorption ratio D, during each subsequent desorption step, are given in Table 9.3. Both R and D showed high values (ranging between 90% and 100%) and did not show any significant change over the different sorption/desorption cycles. Therefore, the membrane shows good reusability and stability for Hg(II) sorption.

9.3.2.2   Dynamic Sorption in Frontal Filtration

Beyond metal ions sorption capacity in static conditions, the performance of the metal-chelating membranes needs to be evaluated under dynamic conditions. Therefore, the Hg(II) removal capacity of the PVA/P_4VP membrane was investigated by tracking the concentration of mercury ions in the water filtrated, c_F, as a function of time according to the protocol in Section 9.2.4.

TABLE 9.3

Performance of the PVA/P$_4$VP membrane over several mercury ions sorption/desorption cycles, using HNO$_3$ at 0.5 M as the desorption medium

Cycle no.	q (mg/g)	R (%)	D (%)
1	46	100	100
2	42	90	98
3	42	91	99
4	43	93	98
5	45	96	97

Ambient temperature; $m_D \approx 100$ mg; pH of sorption 2.5; V = 50 cm^3; $c_0 = 100$ mg/L.

Two different initial concentrations to be treated ($c_0 \approx 16$ or 89 mg/L, pH = 2.5) were placed in the cell. A differential pressure of 0.3 ± 0.01 MPa was applied to generate permeate flow. The typical flow rate of the membrane at an applied pressure of 0.3 MPa was 0.53–0.56 cm^3/h, which is relatively low compared to that of non-chelating membranes. While a lower filtration rate can be desirable in chelating membranes to allow time for ions to complex with the P$_4$VP polymer, progress should still be made in the membrane design (by decreasing the thickness or changing the porosity or pore morphology, for example) before its practical use in full-scale applications can be considered. On the other side, the filtrate mercury ions' average concentration (c_F) in both experiments (Table 9.4) revealed nearly total elimination of metal ions by the PVA/P$_4$VP membrane.

Of note, it can be observed that saturation of Hg(II) complexing was not observed under dynamic conditions since the c_F concentrations in the filtrate remained very low up to the end of the filtration experiment. The theoretical chelating capacity of the membrane was previously determined in Section 9.3.2.1 as 450 mg/g. Based on this value, the theoretical volume for membrane saturation, V_{calc}, can be found. The values of V_{calc} are showed in Table 9.4 for the two filtration conditions. In these considerations, the mercury concentration in the filtrated water, which showed very low value for all samples, was considered to be negligible. Table 9.4 shows that mercury masses of about 166 mg correspond to volumes of 1,866 (for $c_0 \sim 89$ mg/L) and 10,006 cm^3 (for $c_0 \sim 16$ mg/L) of water filtrate, which is much higher than the experimental volume used for the dynamic sorption experiments. Therefore, the membranes could not reach saturation in the duration of the experiments used.

For both Hg(II) concentrations, E was close to 100% (Table 9.4). This shows that PVA/P$_4$VP can be very efficient for the retention of metal ions from wastewater streams.

In conclusion, PVA/P$_4$VP membranes show high chelating capacity for mercury ions. A membrane that is only 369 mg in mass can filtrate until 10 L of a mercury solution at 16 mg/L before saturation. This high capacity suggests good process stability over time for the removal of metals by chelating membranes.

9.3.2.3 Changing the Porosity of Chelating Membranes

The main objective was to get a permeable membrane that can quickly eliminate mercury by filtration. For this, the porosity of the PVA/P$_4$VP membranes was enhanced using two different methods: "impregnation with silica particles" and "phase inversion".

9.3.2.3.1 Water Fluxes and Permeability

For the development of economically viable filtration membranes, the filtration flux is an important parameter. The filtration flux of distilled water, J_v, was determined for both porous membranes (silica impregnation or phase inversion). The flows of water via the membranes were significantly ameliorated by augmenting the porosity of membranes, especially by impregnating the membrane with silica particles (Table 9.5).

TABLE 9.4

Filtration of mercury ion by PVA/P$_4$VP membranes and the theoretical volume of filtrate

c_0 (mg/L)	m_D (mg)	Flow rate (cm³/h)	c_F (mg/L)	E (%)	P$_4$VP (mmol)	Hg retained (mmol)	Hg capacity (mg)	V_{calc} (L)
89	369.6	0.53	0.1	99.9	1.408	0.704	166.3	1.866
16	369.0	0.56	0	100	1.406	0.703	166.1	10.006

pH 2.5; T = 25°C; V = 400 cm³.

TABLE 9.5

Flows of deionized water through the PVA/ P$_4$VP membranes

PVA/P$_4$VP membranes	Flux (cm³/h)
Initial membrane	0.5
Porosity enhanced by phase inversion	9
Porosity enhanced by silica particles	47

ΔP = 2.5 bar; membrane area = 28 cm².

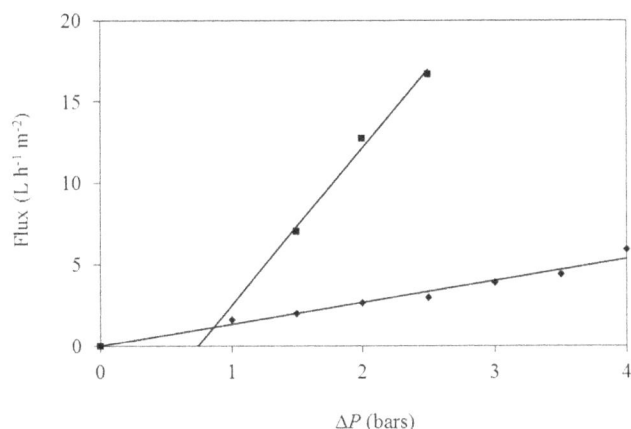

FIGURE 9.10 Effect of ΔP on J_v for the PVA/P$_4$VP membranes with improved porosity, enhanced using impregnation and dissolution of silica particles (■) or through the phase inversion process (◆).

The flow of water through the different PVA/P$_4$VP membranes was determined at different applied pressures (Figure 9.10). The plot of membranes' fluxes versus ΔP revealed linear slopes, thereby agreeing with Darcy's law (equation 9.10). From the slopes, the coefficients of permeability, L_p (L/h/m²/bar), are determined (M. Porter, 1988; C.O. M'Bareck, 2006) based on

$$J_v = L_p (\Delta P - \Delta \pi). \qquad (9.10)$$

J_v is the water flux (L/h/m²), ΔP is the transmembrane pressure (bar), and $\Delta \pi$ is the minimum ΔP producing permeation.

The effect of pressure (1–4 bar) on the flux for the porous membranes is shown in Figure 9.10. For the phase inversion modified membrane, the flux is linear with pressure. For the silica impregnation modified membrane, a linear regression through the origin shows a poor correlation. On the other hand, a regression without using the point of origin is of much better quality due to the fact that the water flux, for the membranes developed by phase inversion, increases in function of the applied pressure starting from a minimum threshold value of only ~0.7 bar. The lines drawn for both membranes have different slopes, which allow the calculation of the permeability coefficients of the modified membranes: 10 L/h/m²/bar in the porous membrane fabricated using silica impregnation and removal and only 1 L/h/m²/bar for the porous membrane obtained through inversion of phases. Therefore, the permeability coefficient of the PVA/P$_4$VP membrane developed by impregnation and removal of silica is about ten times higher than that for the membrane obtained by phase inversion. The higher permeability of the membrane prepared by silica impregnation and removal is explained by the different structure of this membrane. In the case of the membrane made by inversion of phases, the membranes show asymmetry in their pore structure, with a very porous bottom layer supporting a denser, low permeability skin layer of low thickness but low porosity. Conversely, the addition and then removal of the silica particles results in the formation of large pores across the membrane thickness. By creating higher porosity and larger pores, it is therefore normal that the silica impregnation method enables higher fluxes than the phase inversion approach. However, it seems that these higher fluxes only appear from a minimum pressure value (0.7 bar). This can be explained by the hydrophobicity of this membrane and by an incomplete dissolution of silica, known to be hydrophilic, by the alkaline treatment with NaOH. When compared to the initial PVA/P$_4$VP membrane, high-porosity membranes developed by impregnation of silica or phase inversion allow higher water fluxes. Both approaches can achieve faster treatment of water contaminated with Hg(II) compared to the unmodified membrane.

9.3.2.3.2 *Frontal Filtration Experiments*

The effect of porosity, enhanced by the two membrane modifications discussed above, on Hg(II) filtration and retention was evaluated. For this purpose, porous PVA/P$_4$VP membranes were tested in frontal filtration, according to the protocol in Section 9.2.4. Different pressures, 1.5 and 3 bars, were used to filter an aqueous solutions of mercury ions (400 cm³) at $c_0 = 100$ mg/L using the membranes prepared by inversion of phases or impregnation of silica, respectively, in order to achieve the same flow rates between membranes. The pH of the Hg(II) solution was fixed at 2.5. The filtrate flow rate was periodically measured: in general, it is, in the stationary state, of the order of 11 cm³/h and 18 cm³/h, respectively, for the membranes developed by phase inversion and by impregnation of silica, instead of 0.5 cm³/h (at P = 3 bars) for the initial membrane. Comparing the Hg(II) solution fluxes to the water flux shows no difference in the membrane flux. The change in c_F in function of the filtrate volume (V_F) is presented in Figure 9.11 for the initial and modified PVA/P$_4$VP membranes. The c_F value remained below 3 mg/L for all membranes, indicating saturation if the chelating membranes had not been reached yet. In the membrane produced by phase inversion, the c_F reached the highest value at ~2.5 mg/L, suggesting that this membrane has a lower retention capacity for Hg(II) compared to the other membranes, which both showed low c_F values. The different measured parameters associated with Hg(II) removal during filtration are given in Table 9.6 for the different membranes. The c_F values show good retention of Hg(II) for all membranes. The best results are obtained with the initial PVA/P$_4$VP membrane and the membrane changed by impregnation and dissolution of silica, which both achieve close to complete Hg(II) removal. The membrane changed by inversion of phases, despite its lower performance compared to the other two membranes, still achieves ~98% elimination of mercury ions. Therefore, enhancing the porosity of the membrane, particularly with the silica impregnation and dissolution approach, is able to increase the water flux without compromising the Hg(II) retention properties.

TABLE 9.6

Filtration of mercury ions by the initial and high-porosity modified PVA/P$_4$VP membranes

Membrane	ΔP (bar)	m_D (mg)	c_0 (mg/L)	c_F (mg/L)	E (%)
Initial membrane	3	369	89	0.1	99.9
Porosity enhanced by phase inversion	3	405	100	2.5	97.5
Porosity enhanced by silica particles	1.5	390	98	0	100

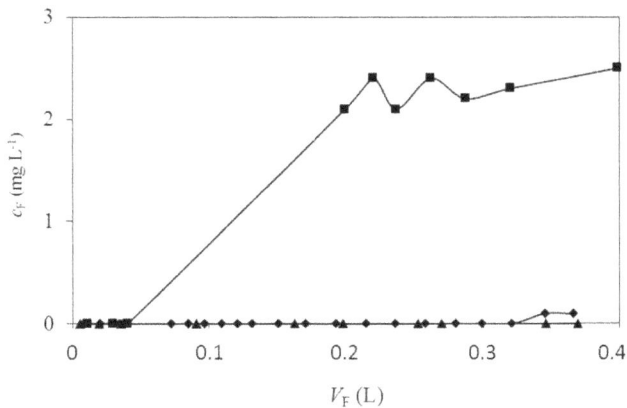

FIGURE 9.11   Change in the mercury concentration in the filtrate (c_F) as a function of the cumulative filtrate volume (V_F) for the initial ($\Delta P = 3$ bar) (♦) PVA/P$_4$VP membrane or the membranes changed by inversion of phases ($\Delta P = 3$ bar) (■) or silica impregnation and removal ($\Delta P = 1.5$ bar) (▲).

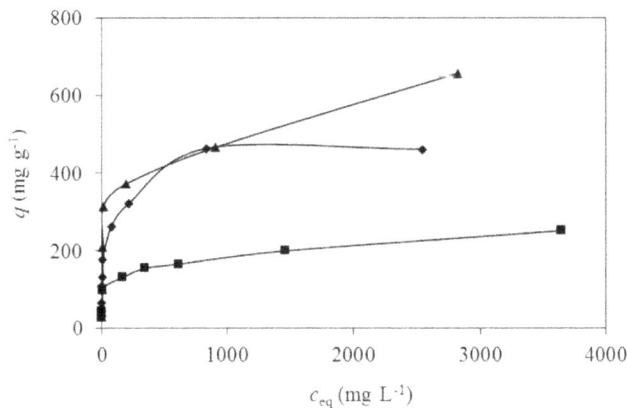

FIGURE 9.12   The isotherm sorption for mercury ions by the initial (♦) PVA/P$_4$VP membrane and the high-porosity membranes obtained by phase inversion (■) or silica impregnation (▲).

9.3.2.3.3   Sorption Isotherm Analysis

The maximum amount of Hg(II) that can be retained by membrane mass q_{max} was characterized for the modified membranes using isotherm analysis. Figure 9.12 presents the plots of q as a function of c_{eq} for an initial concentration c_0 ranging between 20 and 4,000 mg/L ($m_D \approx 100$ mg, V = 50 cm^3, room temperature, pH = 2.5). From these plots, a Langmuir isotherm could be established and the values of q_{max} identified by extrapolation to the value of q at infinite concentration (Table 9.7).

As shown in Table 9.7, the phase inversion process resulted in a decrease in the Hg(II) sorption capacity, while impregnation of silica particles and subsequent dissolution to increase the porosity also increased Hg(II) sorption capacity. This higher sorption capacity can explain how the more permeable PVA/P$_4$VP membrane changed by impregnation of silica was able to achieve a higher flux without reducing Hg(II) removal. The higher porosity of the membranes obtained by silica impregnation most likely results in a higher number of sorption sites per unit of mass, with the additional sites located on the inner pore surface. Similarly, the dense, low-porosity skin layer of the membranes obtained by phase inversion can decrease the number of sorption sites per unit of mass, explaining the lower q_{max} value for this membrane. This lower porosity explains both the lower water flux and the lower Hg(II) retention by the phase inversion membranes.

The theoretical maximum sorption capacity was calculated as in Section 9.3.2.1. Considering that a complexing site would possess, on average, a single N atom, a membrane would be able to sorb a maximum of $3.81 \times 200.7 = 764$ mg of Hg(II) per g of membrane. This maximum theoretical sorption capacity was compared with experimentally measured sorption capacities for the initial membrane and the high-porosity modified membranes (Table 9.7). The modified membrane obtained by silica impregnation and dissolution has a maximum sorption capacity ($q_{max} = 656$ mg/g) closer to the theoretical value than the other two membranes. The similarity between the measured and theoretical values for this membrane indicates that a majority of the complexing sites in P_4VP are available for Hg(II) sorption. In comparison, the modified membrane obtained by phase inversion had a much lower maximum sorption capacity, suggesting that a significant portion of the complexing P_4VP polymer was not accessible to the Hg(II) ions during filtration.

The sorption equilibrium of a sorbate on its sorbent can be described using isotherm models such as those of Langmuir and Freundlich to obtain a thermodynamical interpretation of the sorption process. The Langmuir model is described as

$$q = \frac{q_{max} b c_{eq}}{1 + b c_{eq}}, \tag{9.11}$$

where q_{max} and b are the saturated fixed mass (mmol g^{-1}) and the coefficient of sorption (L mmol^{-1}), respectively. The constants q_{max} and b are obtained from a modification of (11) based on the Hanes-Woolf treatment to obtain the linearized equation:

$$\frac{c_{eq}}{q} = \frac{c_{eq}}{q_{max}} + \frac{1}{b q_{max}}. \tag{9.12}$$

TABLE 9.7
q_{max} of mercury ions by the PVA/P_4VP membranes

PVA/P_4VP membranes	q_{max} (mg/g)	q_{max} (mmol/g)
SInitial membrane	459	2.29
Porosity enhanced by phase inversion	251	1.25
Porosity enhanced by silica particle	656	3.27
Theoretical	764	3.81

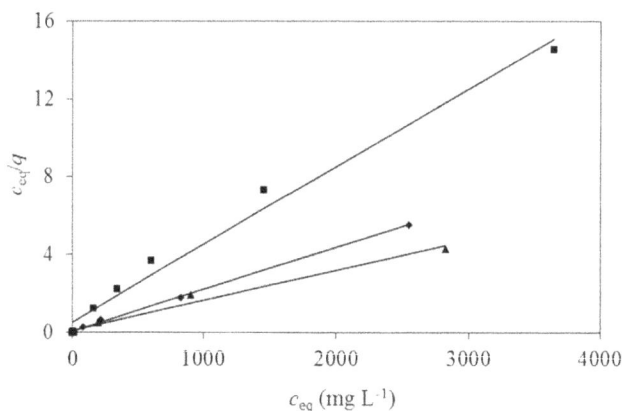

FIGURE 9.13 Plot of the linear form of the Langmuir isotherm model for sorption of mercury ions by the initial (♦) PVA/P_4VP membranes and the high-porosity membranes obtained by phase inversion (■) or silica impregnation (▲).

TABLE 9.8

Sorption of mercury ions by PVA/P₄VP membranes described by the different parameters of the Langmuir isotherm model

PVA/P$_4$VP membrane	Langmuir		
	b (L/mol)	q$_{max}$ (mmol/g)	r^2
Initial membrane	8861	2.32	0.9986
Porosity enhanced by phase inversion	1511	1.25	0.9852
Porosity enhanced by silica particle	3038	3.24	0.9837

The obtained linear plots are shown in Figure 9.13, with the associated parameters presented in Table 9.8 for the different membranes. The q_{max} calculated using isotherm analysis are comparable to the experimentally measured sorption capacity of Table 9.7. Based on a linear correlation coefficient close to 1 and the similar q_{max} values in Tables 9.7 and 9.8, the Langmuir model appears to accurately describe the mercury ions' sorption approach by the chelating membranes. In addition, based on the Langmuir model description, it can be concluded that the sorption of mercury ions by the PVA/P$_4$VP membranes is limited by the availability of complexation sites on the membrane.

9.4 CONCLUSION

The s-IPN technique was used to develop chelating membranes based on a complexing polymer, P$_4$VP, and a support polymer, PVA. The chelating membranes were evaluated for their potential applications in industrial wastewater treatment. A strong affinity for the complexation of mercury was obtained by using the chelating polymer P$_4$VP in a PVA matrix crosslinked by DBE vapour. The prepared PVA/P$_4$VP membrane showed excellent Hg(II) retention, reaching 100% for 100 mg of membrane and 100 mg/L of mercury ions at pH 2.5 after 24 h of equilibrium. The Hg(II) sorption process was found to be stable even under varying pH or in the presence of competing Ca(II) ions. Further testing of the PVA/P$_4$VP membrane showed that, in frontal filtration mode, a large volume of contaminated water can be treated without saturation of the sorption sites on the membrane; however, the filtration flux (approximately 0.5 cm³/h) was found to be very low and unapplicable for industrial applications. To increase the water flux of the PVA/P$_4$VP membrane, two simple techniques were used to obtain porous and permeable PVA/P$_4$VP membranes: phase inversion or impregnation and dissolution of silica. Analysis of the modified membranes by SEM and water flux measurements demonstrated the formation of pores inside the membranes, which resulted in an increase in water permeation, particularly for the membranes obtained using impregnation and dissolution of silica. The increase in filtration flow rates allows the use of these membranes for low-pressure filtration (3 bar) of contaminated water. The high-porosity membrane obtained by this technique was able to achieve close to 100 times higher flux than the initial PVA/P$_4$VP membrane without affecting the Hg(II) removal capacity, maintaining close to 100% Hg(II) retention. Sorption isotherm analysis provided evidence of the increased sorption capacity for the membrane obtained by impregnation and dissolution of silica. The order of the sorption capacities of the three fabricated membranes, for Hg(II) ions, was PVA/P$_4$VP (silica), 656 mg/g > PVA/P$_4$VP (initial membrane), 460 mg/g > PVA/P$_4$VP (phase inversion), 251 mg/g. The sorption isotherms data agree with the Langmuir model and validate that ions are fixed by a chelation equilibrium. The better performance of the membrane changed by silica impregnation was attributed to a better accessibility of the complexation sites, which are distributed over the entire internal surface of the pores. On the contrary, the membrane changed by phase inversion has a dense skin, low porosity, and low permeability, in which the accessibility of the sites may be decreased compared to the classical membrane. It may also have lost a fraction of the complexing polymer by dissolution in acetone. Ultimately, these permeable membranes open an interesting perspective to obtain really usable filtration systems.

NOMENCLATURE

ABBREVIATIONS

AAS	Spectroscopy of atomic absorption
DBE	1,2-Dibromoethane
PEUF	Polymer-enhanced ultrafiltration
PVA	Polyvinyl alcohol
P_4VP	Poly(4-vinylpyridine)
SEM	Microscopy of scanning electron
s-IPN	Semi-interpenetrating polymer network
UF	Ultrafiltration

SYMBOLS

b	Langmuir isotherm's coefficient of sorption (L/mg)
c_{eq}	Ion concentration at equilibrium (mg/L)
c_F	Filtrate ion concentration (mg/L)
c_0	Starting ion concentration (mg/L)
c_t	Ion concentration at time t
D	Rate of desorption (%)
E	Rate of elimination (%)
f	Rate of retention efficiency (%)
L_p	Permeability coefficient (L/h/m²/bar)
M	Molar weight of Hg(II) ions (g/mol)
m_D	Weight of dry membrane (g or mg)
m_S	Weight of humid membrane (g or mg)
M_w	Polymer molecular mass (g/mol)
N	Number of nitrogen atoms in a chelating site
n_d	Desorbed ion amount (mol)
n_r	Ion retained amount in the film (mol)
n_{VP}	4-VP repeat units in the film (mol)
P	Pressure (bar)
q	Retained ion mass at equilibrium per gram of membrane (mg/g)
q_{max}	Maximum weight of retained ion per gram of membrane (mg/g)
R	Rate of retention (%)
r^2	Correlation coefficient
S	Area membrane (m⁻²)
S_W	Rate of membrane swelling
t	Time (min or h)
V	Solution volume (L or cm³)
V_{calc}	Calculated filtrate volume at membrane saturation (L or cm³)
V_F	Filtrate volume (L or cm³)
$\Delta\pi$	The pressure for which the minimal flow shows: osmotic pressure (bar)
J_v	Flux of water via the membrane (L/h/m²)

REFERENCES

A. Ringbom. 1967. *Complexation in analytical chemistry* (French translation). Paris: Dunod.

B. Lam, S. Déon, N. Morin-Crini, G. Crini, P. Fievet. 2018. Polymer-enhanced ultrafiltration for heavy metal removal: Influence of chitosan and carboxymethyl cellulose on filtration performances. *Journal of Cleaner Production*, 171, 927–933.

C.C. Ho, J.F. Su, L.P Cheng. 2021. Fabrication of high-flux asymmetric polyethersulfone (PES) ultrafiltration membranes by nonsolvent induced phase separation process: Effects of H_2O contents in the dope. *Polymer*, 217, 123451.

C.O. M'Bareck, Q.T. Nguyen, S. Alexandre, I. Zimmerlin. 2006. Fabrication of ion-exchange ultrafiltration membranes for water treatment I. Semi-interpenetrating polymer networks of polysulfone and poly(acrylic acid). *Journal of Membrane Science*, 278, 10–18.

E. Da Silva, L. Lebrun, M. Métayer. 2002. Elaboration of a membrane with bipolar behaviour using the semi-interpenetrating polymer networks technique. *Polymer*, 43, 5311–5320.

E.A. Hegazy, H.A. Abd El-Rehim, H.A Shawky. 2000. Investigations and characterization of radiation grafted copolymers for possible practical use in waste water treatment. *Radiation Physics and Chemistry*, 57, 85–95.

F. Wang, B. Pang, T. Yang, J. Liu. 2022. Fabrication of nitrogen-doped graphene quantum dots hybrid membranes and its sorption for Cu(II), Co(II) and Pb(II) in mixed polymetallic solution. *Journal of Molecular Liquids*, 362, 119690.

G.C. Steenkamp, K. Keizer, H.W.J.P. Neomagus, H.M. Krieg. 2002. Copper(II) removal from polluted water with alumina/chitosan composite membranes. *Journal of Membrane Science*, 197, 147–156.

H. Bessbousse, J.F. Verchère, L. Lebrun. 2010a. Increase in permeate flux by porosity enhancement of a sorptive UF membrane designed for the removal of mercury(II). *Journal of Membrane Science*, 364, 167–176.

H. Bessbousse, J.F. Verchère, L. Lebrun. 2012. Characterisation of metal-complexing membranes prepared by the semi-interpenetrating polymer networks technique. Application to the removal of heavy metal ions from aqueous solutions. *Chemical Engineering Journal*, 187, 16–28.

H. Bessbousse, N. Zran, J. Fauléau, B.Godin, V. Lemée, T.L. Wade, M.C. Clochard. 2016. Poly(4-vinyl pyridine) radiografted PVDF track etched membranes as sensors for monitoring trace mercury in water. *Radiation Physics and Chemistry*, 118, 48–54.

H. Bessbousse, T. Rhlalou, J.F. Verchère, L. Lebrun. 2008a. Removal of heavy metal ions from aqueous solutions by filtration with a novel complexing membrane containing poly(ethyleneimine) in a poly(vinyl alcohol) matrix. *Journal of Membrane Science*, 307, 249–259.

H. Bessbousse, T. Rhlalou, J.F. Verchère, L. Lebrun. 2008b. Sorption and filtration of Hg(II) ions from aqueous solutions with a membrane containing poly(ethyleneimine) as a complexing polymer. *Journal of Membrane Science*, 325, 997–1006.

H. Bessbousse, T. Rhlalou, J.F. Verchère, L. Lebrun. 2009. Novel metal-complexing membrane containing poly(4-vinylpyridine) for removal of Hg(II) from aqueous solution. *Journal of Physical Chemistry B*, 113, 8588–8598.

H. Bessbousse, T. Rhlalou, J.F. Verchère, L. Lebrun. 2010b. Mercury removal from wastewater using a poly(vinyl alcohol)\poly(vinylimidazole) complexing membrane. *Chemical Engineering Journal*, 164, 37–48.

H. Zhao, D. Zhang, H. Sun, Y. Zhao, M. Xie. 2022. Adsorption and detection of heavy metals from aqueous water by PVDF/ATP-CDs composite membrane. *Colloids and Surfaces A: Physicochemical and Engineering Aspects*, 641, 128573.

K. Renu, R. Chakraborty, H. Myakala, R. Koti, A.C. Famurewa, H. Madhyastha, B. Vellingiri, A. George, A.V. Gopalakrishnan. 2021. Molecular mechanism of heavy metals (lead, chromium, arsenic, mercury, nickel and cadmium) - Induced hepatotoxicity - A review. *Chemosphere*, 271, 129735.

L. Lebrun, E. Da Silva, M. Métayer. 2002. Elaboration of ion-exchange membranes with semi-interpenetrating polymer networks containing poly(vinyl alcohol) as polymer matrix. *Journal of Applied Polymer Science*, 84, 1572–1580.

L. Lebrun, F. Vallée, B. Alexandre, Q.T. Nguyen. 2007. Preparation of chelating membranes to re-move metal cations from aqueous solutions. *Desalination*, 207, 9–23.

M. Ito, K. Takano, H. Hanochi, Y. Asaumi, S.I Yusa, Y. Nakamura, S. Fujii. 2018. pH-responsive aqueous bubbles stabilized with polymer particles carrying poly(4-vinylpyridine) colloidal stabilizer. *Frontiers in Chemistry*, 6, 269.

M. Mulder. 1991. *Basic Principles of Membrane Technology*. Dordrecht: Kluwer Academic Publishers.

M.C. Porter (Ed.). 1988. *Handbook of industrial membrane technology*. Park Ridge, NJ: Noyes Publications, 264–265.

M. Satoh, E. Yoda, T. Hayashi, J. Komiyama. 1989. Potentiometric titration of poly(vinylpyridines) and hydrophobic interaction in the counterion binding. *Macromolecules*, 22, 1808–1812.

M.A. Barakat, E. Schmidt. 2010. Polymer-enhanced ultrafiltration process for heavy metals removal from industrial wastewater. *Desalination*, 256, 90–93.

N. Fatin-Rouge, A. Dupont, A. Vidonne, J. Dejeu, P. Fievet, A. Foissy. 2006. Removal of some divalent cations from water by membrane-filtration assisted with alginate. *Water Research*, 40, 1303–1309.

N. Oyama, F.C. Anson. 1980. Electrostatic binding of metal complexes to electrode surfaces coated with highly charged polymeric films. *Journal of The Electrochemical Society*, 127, 247–250.

O. Farid, F. Mansour, M. Habib, J. Robinson, S. Tarleton. 2016. Investigating the sorption influence of poly(vinyl alcohol) (PVA) at different crosslinking content. *Journal of Environmental Chemical Engineering*, 4, 293–298.

S. Kumari, A.R. Jamwal, N. Mishra, D.K. Singh. 2020. Recent developments in environmental mercury bioremediation and its toxicity: A review. *Environmental Nanotechnology, Monitoring & Management*, 13, 100283.

S. Mondal, S.K. Majumder. 2020. Fabrication of the polysulfone-based composite ultrafiltration membranes for the adsorptive removal of heavy metal ions from their contaminated aqueous solutions. *Chemical Engineering Journal*, 401, 126036.

S. Sun, H. Zhang, Y. Luo, C. Guo, X. Ma, J. Fan, J. Chen, N. Geng. 2022. Occurrence, accumulation, and health risks of heavy metals in Chinese market baskets. *Science of the Total Environment*, 829, 154597.

S. Xu, Y. Liu, Y. Yu, X. Zhang, J. Zhang, Y. Li. 2020. PAN/PVDF chelating membrane for simultaneous removal of heavy metal and organic pollutants from mimic industrial wastewater. *Separation and Purification Technology*, 235, 116185.

S.A. Razzak, M.O. Faruque, Z. Alsheikh, L. Alsheikhmohamad, D. Alkuroud, A. Alfayez, S.M.Z. Hossaine, M.M. Hossain. 2022. A comprehensive review on conventional and biological-driven heavy metals removal from industrial wastewater. *Environmental Advances*, 7, 100168.

S.R. Ravichandran, C.D. Venkatachalam, M. Sengottian, S. Sekar, B.S.S. Ramasamy, M. Narayanan, A.V. Gopalakrishnan, S. Kandasamy, R. Raja. 2022. A review on fabrication, characterization of membrane and the influence of various parameters on contaminant separation process. *Chemosphere*, 306, 135629.

T.A. Saleh, M. Mustaqeem, M. Khaled. 2022. Water treatment technologies in removing heavy metal ions from wastewater: A review. *Environmental Nanotechnology, Monitoring & Management*, 17, 100617.

U. Forstner, G.T.W. Wittmann. 1979. *Metal pollution in the aquatic environment.* New York: Springer-Verlag.

U. Pinaeva, D. Lairez, O. Oral, A. Faber, M.C. Clochard, T.L. Wade, P. Moreau, J.P. Ghestem, M. Vivier, S. Ammor, R. Nocua, A. Soulé. 2019. Early warning sensors for monitoring mercury in water. *Journal of Hazardous Materials*, 376, 37–47.

V. Beaugeard, J. Muller, A. Graillot, X. Ding, J-J. Robin, S. Monge. 2020. Acidic polymeric sorbents for the removal of metallic pollution in water: A review. *Reactive and Functional Polymers*, 152, 104599.

X. Jin, C. Wei, C. Wu, W. Zhang. 2022. Gastric fluid-induced double network hydrogel with high swelling ratio and long-term mechanical stability. *Composites Part B*, 236, 109816.

X. Tan, D. Rodrigue. 2019. A review on porous polymeric membrane preparation. Part I: Production techniques with polysulfone and poly(vinylidene fluoride). *Polymers*, 11, 1160.

X. Zeng, E. Ruckenstein. 1996. Control of pore sizes in macroporous chitosan and chitin membranes. *Industrial & Engineering Chemistry Research*, 35, 4169–4175.

Y. Huang, J. Du, Y. Zhang, D. Lawless, X. Feng. 2016. Batch process of polymer-enhanced ultrafiltration to recover mercury(II) from wastewater. *Journal of Membrane Science*, 514, 229–240.

Y. Huang, X. Feng. 2019. Polymer-enhanced ultrafiltration: Fundamentals, applications and recent developments. *Journal of Membrane Science*, 586, 53–83.

10 Graphene and Graphene Oxide-Based Nanomaterials as a Promising Tool for the Mitigation of Heavy Metals from Water and Wastewater

Zohaib Abbas, Iqra Shahid, Shafaqat Ali,
Muhammad Nihal Naseer, Muhammad Awais Khalid,
Shamas Tabraiz, Ihsan Elahi Zaheer and Muhammad Zeeshan

10.1 INTRODUCTION

The emission of toxic metal ions has emerged among the most severe ecological issues in recent years, posing a global threat to human beings (Wu et al. 2018; Wu et al. 2019). In recent years, rapid urbanization and industrialization have increased industrial effluents (Smaali et al. 2021; Vasseghian et al. 2021a). Industrial sectors such as smelting, mechatronics, mining, batteries, and chemical processes (e.g., making plastic products) produce large quantities of toxic metals, which are often discharged without adequate treatment, directly threatening marine and coastal existence (Zahoor et al. 2018). Inorganic toxic substances and additives pollute lakes and rivers severely (Yang et al. 2018c). As compared to organic pollutants, dissolved heavy metals (HMs) cannot be broken down into less or non-toxic forms and can therefore cause serious health problems when they come into contact with live species throughout the food web (Godiya et al. 2019). The environmental impacts from these industries, particularly in developing countries, are severe (Velusamy et al. 2021). Toxic substances are even discharged into the air through agricultural practices and industrial effluents.

HM ions are currently being removed from the environment using several treatment techniques, including coagulation/flocculation, solvent extraction, electrodialysis, and thermal decomposition (RoyChoudhury et al. 2019; Saleh 2020, 2021; Yu et al. 2018; Zhao et al. 2018). Even though some techniques have socioeconomic and environmental drawbacks, which include large quantities of sludge production, high operational expenses, toxic by-products, or the cleanup system's complication, however, they offer many benefits as well (Fang et al. 2018a). Adsorption is amongst the most popular and commonly applied approaches for removing HMs (Al-Jabari 2016). It is simple, easy to use, low-cost, highly efficient, widely adaptable, and neutral to dangerous substances and contaminants (Huang et al. 2019; Makertihartha et al. 2017; Tan et al. 2016; Zhang et al. 2017). Therefore, new adsorbent materials like nano adsorbents have been developed based on nanotechnology (Wang et al. 2017a; Zhou et al. 2019). Adsorbents like chitosan (CS), activated carbon, zeolite, nanomaterials, SiO_2, or fibers are being used to remove contaminants (Nongbe et al. 2018; Pap et al. 2017; Radi et al. 2019). Unfortunately, most of the above-mentioned materials are not stable at higher or lower pH levels or have poor sorption capacities (Nicomel et al. 2016). Graphene-based substances, among all these adsorption processes, appear to have become an efficient technique for overcoming the drawbacks.

DOI: 10.1201/9781003326281-12

Graphene was made in 2004 through mechanical exfoliation by Novoselve et al. (2004). Graphene is a sp_2-linked single-atom-thick tetrahedra-organized carbon plate. With its distinct biological, thermal, structural, and functional properties, graphene has received considerable importance (De Silva et al. 2017; Perreault et al. 2015). By eliminating interference factors (vulnerable functional groups), graphene membranes demonstrate more robust separation properties (Dai et al. 2020; Liu et al. 2019a). Graphene-based material is particularly promising for making new technological breakthroughs than commercial membranes currently available (Nagarajan et al. 2018; Nie et al. 2021; Zhang et al. 2021). The graphene-modified version, graphene oxide (GO), has an oxygen-functionalized honeycomb-like structure which is obtained using an exfoliating process and by chemical oxidation of graphene (Makertihartha et al. 2017; Peng et al. 2017; Zunita et al. 2018). Compared to other graphene-based materials, GO has a better membrane and antifouling properties. Environmental applications of graphene and graphene-based materials include innovative bioadsorbents, membranes, solar cells, electrode materials, and antimicrobial agents for pollutant removal or monitoring (Chen et al. 2018; Perreault et al. 2015; Yu et al. 2018). Another major advantage of GO is its ability to eliminate toxic metal pollutants from water, including copper, chromium, lead, cobalt (Zheng et al. 2019). Previous research has shown that hydrophilic oxygen-containing amino groups on GO plates can trap HMs via surface inclusions, π-π interaction, and electrostatic repulsion in both the amine groups and metal oxides (Ahmad and Liu 2021). All these properties make GO an attractive candidate for large-scale HM ion recovery from aqueous solutions (Ahmad et al. 2021; Ahmad and Liu 2021; Alosaimi 2021; Zhang et al. 2019a).

Graphene-based membranes have been studied, but few studies have been conducted on their use directly for HM removal as well as their application as modifiers for other membranes (Ahmad et al. 2020; Huang et al. 2019; Piumie et al. 2018). Development of GO nanocomposite membranes for pollutant separation from water has been studied by Zhang et al. (2015). They fabricated the membranes by using a blending method to create specific adsorption characteristics. Mostly, such types of designed membranes were used as adsorbents for chelating anionic and cationic dyes. Recent research studies suggest that GO composite insulation materials have become a reasonable alternative for purifying effluents.

The effectiveness of nanofiltration membranes using GO nanoparticles was evaluated by Zinadini *et al.* (2014). They analyzed the effect of GO nanomaterials on membranes' antifouling properties and on morphology. The results demonstrated that the amount of flux increased by incorporating GO to the membrane structure. Furthermore, by using this type of membrane, exquisite removal of dye from industrial effluents was further achieved. Chen *et al.* (2020) investigated the use of GO with recycled newspapers to suspend HM ions from industrial effluents. According to the study findings, GO paper as an environmentally friendly, low-cost, and renewable material can adsorb toxic metals along with industrial wastewater (Chen et al. 2020). Another study reported different types of metals; zinc, lead, selenium, mercury, copper, nickel, zinc, and arsenic were discovered in Shandong Province's Dawen River basin (Liu et al. 2021). HMs can bioaccumulate in aquatic life and can enter our food chain. For example, shellfish from the Pearl Coast of Southern China was reported to have high levels of As (Yu et al. 2021).

Duru et al. (2016) reviewed GO's applications for removing precious and HMs. In the review, they highlighted the adsorption efficiency of GO nanocomposites as well as their mechanisms for Cr (Zunitaa et al. 2020), Pb (Zhang et al. 2017), Cu (Zhang et al. 2017), and Au(III) adsorption. Peng et al. (2017) reviewed HM absorption from water by GO and its composites. The mechanism of adsorption and operational conditions were highlighted in the review.

Yusuf et al. (2015) discussed ways to remove dyes and HMs using graphene and its derivatives. In addition to historical overviews, synthesis methods, and toxicity concerns, this review discussed techniques for the elimination of additives as well as industrial wastes by using graphene and its modified versions. Khan et al. (2016) studied methods for fabricating GO and nanoparticles, including suction evaporation, beam centrifugal casting, adhesive percolation, convection cooling method.

This chapter primarily focused on graphene applications for HMs and radionuclides removal rather than characterization or fabrication techniques. These results demonstrate that GO and its derived products can be effective environmental adsorbents, particularly for water treatment. This research was conducted to understand the adsorption mechanism as well as the practical benefits of graphene materials and GO membranes mostly for the removal of toxic metals from the environment.

10.2 TOXICITY OF HMS

Excessive HM ions in groundwater sources end up causing adverse health and environmental challenges globally (Beroigui et al. 2020; Islam et al. 2020; Jaiswal and Pandey 2019; Li et al. 2021a). HMs are magnetic materials of densities greater than 5 g/cm^3 (El-Kady and Abdel-Wahhab 2018; Talukdar et al. 2020; Zamora-Ledezma et al. 2021). HMs' primary source in an aquatic environment are batteries, mining, electroplating, refineries, fertilizers, textiles, pulp, and paper industries. Huge quantities of heavy metal-contaminated wastewater are generated from these industries (Joseph et al. 2020; Ngueagni et al. 2020). These industrial effluents can cause severe environmental impact (Velusamy et al. 2021). HMs that are toxic make their way into the environment, endangering long-term ecological safety and adversely affecting animal and plant growth and development (Dubey et al. 2018; Ismail et al. 2019; Li et al. 2021c; Rai et al. 2019). Although HMs are essential

TABLE 10.1
Effects of toxic HMs

Heavy metals	EPA regulatory limit (USEPA	Toxicity	References
Mercury	1 μg/L	Damages the gastrointestinal tract and causes kidney failure, has an impact on cognition, brain, adulation, speech, fine and gross motor skills, and visual-spatial abilities.	Aram et al. (2021)
Cadmium	5 μg/L	Damages the kidneys and has a negative impact on the respiratory, cardiorespiratory, and muscular systems, like rheumatoid arthritis, osteoporosis, and osteoarthritis.	Hu et al. (2016) and Reyes-Hinojosa et al. (2019)
Copper	1000 μg/L	Causes gastrointestinal (Saleh 2021) constipation; hematuria, chronic diarrhea, liver disease, lack of appetite, or nausea are some of the symptoms.	Alengebawy et al. (2021)
Lead	10 μg/L	Causes neurological problems, high blood pressure, digestive problems, recuperative medical problems, termination of pregnancy in women, sexual dysfunction in men, behavioral problems in children, kidney damage, muscle and joint pain, and anemia.	Ahmad et al. (2020)
Chromium	50 μg/L	Kidney dysfunction, GI disorders, dermal diseases, cancers of the bronchi, laryngeal, urinary system, lymphatic system, testicles, bone, and thyroid are becoming more common.	Odhong et al. (2019) and Pavesi and Moreira (2020)
Zinc	3000 μg/L	Skin irritation, retching, constipation, pigmentation, and anxiousness are all possible side effects.	Abdolmohammad-Zadeh et al. (2020) and Asadi et al. (2020)
Aluminum	100 μg/L	Alzheimer's disease (Ahmad and Liu), dialysis(Alzheimer's disease, neurodegenerative disorders, and motor neurone disease are all conditions that can occur as a result of dialysis). It also has an impact on the skeletal system, nerve cells, spine, clotting factors, liver, and kidneys.	Berihu (2015) and Ghosh et al. (2021)

for plant growth and for human health, their concentrations over the prescribed amount are toxic. HMs that are toxic, such as aluminum, nickel, bronze, tin, lead, iron oxide, copper, chromium, and magnesium, are a major global concern. Their presence in the food chain can adversely affect living organisms including microbes, plants, animals, and humans (Martínez-Cortijo and Ruiz-Canales 2018; Paschoalini et al. 2019; Rai et al. 2019; Selvi et al. 2019). However, even at low concentrations, lead, chromium, arsenic, and mercury are extremely hazardous (Beroigui et al. 2020; Hsini et al. 2020; Kim et al. 2019; Zhang et al. 2020a). In aquatic ecosystems, fish are a key bioindicator of HM contamination as they are the top predators and can accumulate extensive amounts of HMs in their tissues (Maury-Brachet et al. 2020; Muhammad and Ahmad 2020; Zhong et al. 2018). HMs have a negative impact on crop production, chlorophyll content, organic matter, and production efficiency of soil. HMs in soil also limit plants' ability to absorb and translocate essential nutrients to their tissues (Awan et al. 2020; Khan et al. 2021; Li et al. 2021c; Liu et al. 2017b). Furthermore, when compared to other vegetables, leafy vegetables accumulate as well as absorb more HMs in their leaves, placing them at a much greater risk of ruination. Metals could be approaching the food supply chain through fruits and veggies (Fatemi et al. 2020). After being released into biological systems through the food supply chain, HMs probably end up threatening the health of people, causing chronic diseases, deformities, and cancers (Achary et al. 2017; Lian et al. 2019; Sall et al. 2020; Yan et al. 2020; Yang and Massey 2019). According to the U.S. Environmental Protection Agency and the Global Medical Council, toxic metals such as zinc, chromium, cadmium, lead, mercury, or nickel are noxious substances that rank among major contaminants and are known human carcinogens (Jaskulak et al. 2019; O'Connor et al. 2019; Selvi et al. 2019). Table 10.1 illustrates the detrimental effects of various toxic metals on human health.

10.3 GRAPHENE-BASED MATERIALS FOR HM REMOVAL

HMs can be removed from wastewater by using different methods. Treatment strategies are primarily determined by economy and initial HM concentration (Qalyoubi et al. 2021). According to recent research, the most popular methods for removing toxic substances are flocculation, ultrafiltration, environmental remediation, photocatalytic degradation, ozonation, solvent extraction, and adsorption (Al Ketife et al. 2020; Almomani et al. 2020; Almomani and Bohsale 2021; Bashir et al. 2019; Ding et al. 2016; Khraisheh et al. 2021; Kyzas et al. 2019; Mansoorianfar et al. 2020). Membrane-based techniques have become popular in recent years for removal of various HM pollutants from waterbodies because of the higher binding affinity, minimum space restriction, and good stability (Moradi et al. 2020). In recent years, researchers have preferred nanocomposites to create effective membranes for treating wastewater. Several nanocomposites are being used to eliminate toxic metals, including MoS_2, aluminum oxide, polydopamine, Polyvinylidene fluoride (PVDF)-based nanoparticles, Ethylenediaminetetraacetic acid (EDTA), GO, as well as teflon-based nanoparticles (Fang et al. 2017; Mishra et al. 2020; Rathour et al. 2020; Sunil et al. 2018; Wang et al. 2018c; Xia et al. 2017). The membranes synthesized from all these materials are not efficient for removing HMs. So, a lot of research is being carried out to develop innovative membrane to remove HMs.

Graphene-based nanomaterials such as graphene, graphene oxide, reduced graphene oxide, as well as graphene quantum dots (GQDs) are seen as effective in removing HMs from membranes (Liu et al. 2019b; Ma et al. 2018b; Nagaraj et al. 2018) and are depicted in Figure 10.1.

10.3.1 PRISTINE GRAPHENE OR REDUCED GO

In comparison to graphene, GO is a derivative of pristine graphite, which can be synthesized by mechanical oxidation following Hummers', Brodie's, Staudenmaier's, or some modified methods (Peng et al. 2017; Sherlala et al. 2018; Zaaba et al. 2017). Considerable amount of HM ions can be adsorbed by GO in water because of its -C-O, -COH, -COOH groups, and certain chemical bonding which can chelate toxic HMs (Kong et al. 2018a; Kong et al. 2018b). GO is considered to be a promising biosorbent for the elimination of HMs like Au(III) and Pt, Pb, Cu, Zn, Cd, Co, etc. (Alosaimi 2021; Asadi et al. 2020; Joseph et al. 2020; Ngueagni et al. 2020; Qalyoubi et al. 2021; Yu et al. 2021; Zhou et al. 2021b). Many of the HM ions discussed above exist as cations, while others exist as anions. As an example, Cr(VI) exists as $HCrO_4^-$ and $Cr_2O_7^{2-}$ and $H_2AsO_4^-$ and $HAsO_4^-$ are metalloid compounds (Kong et al. 2019). In aqueous solution, GO and reduced GO can accumulate Na^+, Co^{2+}, Zn^{2+}, Mg^{2+}, Ni^{2+}, Cu^{2+}, Pb^{2+}, Cs, Mn^{2+}, Cd^{2+}, U(VI) (Eltayeb and Khan 2020; Kammoun et al. 2017; Zunitaa et al. 2020). Table 10.2 summarizes the detection limits of targeted HMs. During the adsorption of HM ions, chemical or physical bonds are created among toxic HMs and the adsorbent to remove and separate them. The toxic metal ion adsorption by GO is significantly determined by electrostatic interactions, intercalation (coordination chelation), reverse osmosis, and hydrophobic interactions, as shown in Figure 10.1. Adsorbents based on graphene can be improved by introducing

TABLE 10.2
HM adsorption by GO

Adsorbents	Targeted HMs	Maximum adsorption capacity (mmol/g)	References
GO	Cu	0.73	Abdi et al. (2018) and Rao et al. (2017)
	Cd	0.32	Liu et al. (2019b)
GO	Mg^{2+}, Na^+, Cd^{2+}, Co^{2+}, Ni^{2+}, Cu^{2+}, Zn^{2+}, Pb^{2+}	Na^+ (0.41), Mg^{2+} (0.95), Ni^{2+} (1.92), Cd^{2+} (1.37), Cu^{2+} (2.02), Co^{2+} (1.28), Zn^{2+} (1.71), Pb^{2+} (2.32)	Kong et al. (2020)
GO membranes	Cu^{2+}, Cd^{2+}, Ni^{2+}	Cd^{2+} (0.75), Cu^{2+} (1.14), Ni^{2+} (1.06)	Tan et al. (2015)
Reduced GO	Th(IV), Cs(I)	0.21, 0.30	Tan et al. (2016)

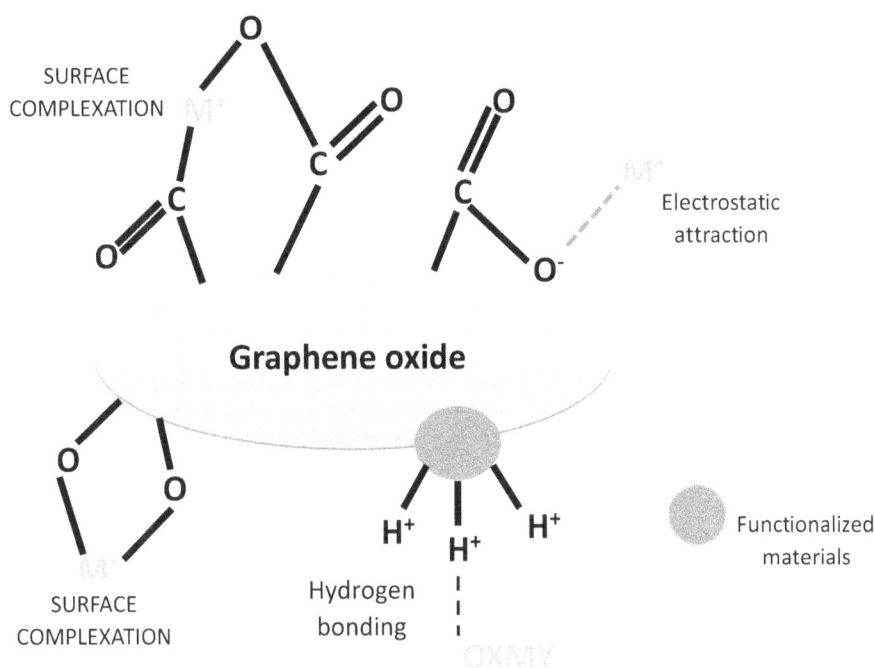

FIGURE 10.1 A diagram depicting toxic HM adsorptive structures over GO adsorbents (M^+ denotes HM cations).

interaction groups including coordination atoms. These groups include -OH, -COOH, -C-O, -C-N, -NH$_2$, -NH, -C-S, -S-S, etc. This allows the adsorbent to be improved with excellent adsorption capacity (Ahmad et al. 2020). A ceramic-supported GO-attapulgite membrane was used in a study to remove HMs from organic solvents (Liu et al. 2019b). A long sequence of adenosine triphosphate (ATP) nanorods were integrated in strands of GO laminar content, creating more water transfer channels and maximum surface permeability and permeate flux. Lead, chromite, nickel, and copper ions are almost completely removed from an aqueous medium using this structure. A few studies have investigated the elimination of As, Zn, Cr, Cd, as well as Pb ions from aqueous solutions using carboxylated GO-incorporated polyphenylsulfone nanofiltration membranes (Shukla et al. 2018). Based on the findings of this research, it is possible to conclude that the widely researched membrane has a substantial potential to dissolve HM ions under the volumetric flux dimensions used. As a result of the easy syntheses and low costs of graphene-based membranes, this study shows extremely attractive and supportive economic performance for the treatment of toxic metals ions. GO or rGO have been reported to adsorb certain HMs, but a unified standard is not yet available for evaluating their ability to degrade these toxic metals (Kong et al. 2021).

10.4 SYNTHESIS AND CHARACTERIZATION OF GRAPHENE AND GO NANOMATERIALS

Graphene is a 2D hexagonal carbon network and has gained considerable attention for its exemplary biological, electrical, thermal, and physical characteristics (De Silva et al. 2017; Sherlala et al. 2018). The synthesis of graphene can be achieved in a variety of ways. Graphene synthesis techniques are classified into two broad categories: top-down and bottom-up approaches (Coroş et al. 2019; Saeed et al. 2020). For top-down approaches, stacked graphite layers are separated to yield graphene sheets, while bottom-up approaches involve techniques such as chemical vapor deposition and synthesis of graphene from alternative carbon sources (Coroş et al. 2019; Saeedi-Jurkuyeh et al. 2020). Methods for synthesizing graphene can be divided into three groups: (1) liquid phase exfoliation, (2) mechanical exfoliation, and (3) chemical vaporization method (Chouhan et al. 2020). Despite the existence of several methods of graphene production, the current challenge is to produce good quality graphene with no or minimal contamination. Liquid phase exfoliation is a large-scale manufacturing technique, but it has poor electrical characteristics (Li et al. 2020b). Chemical vapor deposition is a cost-effective and renewable technique of industrial production (Fauzi et al. 2018; Karthik et al. 2021; Saeed et al. 2020). In research and comprehensive studies, mechanical exfoliation is commonly used for generating graphene (Karthik et al. 2021).

GO, including graphene, are among the most important derivatives. These compounds primarily contain oxygen-containing functional components (e.g., carboxyl, hydroxyl, epoxy, as well as carbonyl groups) and they can change their physiochemical properties. For example, a hydrophilic group present in GO sheets allows it to disperse in water, allowing it to be processed and stacked easily (Cai et al. 2019; Papageorgiou et al. 2017). By linking oxygen to carbon atoms, oxygen promotes molecular transport in two ways: A first step would be to enhance the permeation properties through hydrogen bonding with water and electrostatic interactions with ions (Modi and Bellare 2019). Second, they act as spacers, keeping the graphene planes separated for transporting small molecules (Vasseghian et al. 2021b). A fire tube is submerged in ZIF-8@GO/PEI and interfaced each side with a vacuum siphon to achieve a nanofiltration GO-ceramic membrane with the results showing an improvement in penetrability due to the formation of extensive water channels (Yang et al. 2018a). GO is synthesized using Brodie's, Staudenmaier's, and Hummers' methods (Vasseghian et al. 2021b; Zunitaa et al. 2020). The hydroxyl, carboxyl, epoxy, and carbonyl contents of GO nanosheets result from strong oxidation processes. Due to the dominance of carboxyl ions, the edges of the GO lamellar sheet are hydrophilic, while the base of the sheet has both hydrophilic and hydrophobic regions. As a result of its amphiphilic nature, GO is highly ion-absorbent but there are still some complications in incorporating the GO membrane (Li et al. 2019b; Peng et al. 2017).

Aside from their high cost, GO nanoparticles are susceptible to leaching from the membrane due to their affinity for water. Consequently, it is important to consider certain factors when selecting the proper fabrication method for GO membranes.

10.5 MODIFICATION OF GRAPHENE NANOMATERIALS FOR HM REMOVAL

GO has excellent mechanical and electrical properties; it has several drawbacks, such as its high hydrophilicity, extensive agglomeration, and problems in isolation from recycled water (Sherlala et al. 2018; Zhao et al. 2016). As a result, modifying GO by adding new organic compounds is the latest research project. GO is designed and synthesized to improve its good characteristics and potential for toxic metal removal from untreated sewage (Gul et al. 2016). A promising way to modify GO is by adding polymers and multidentate chelating agents, which are multifunctional organic molecules (Zare-Dorabei et al. 2016). To enhance GO for toxic metal reduction, a wide range of different organic compounds, like hydroxyl, methyl, thiol, and amino groups, are being used. Table 10.3 contains a list of some of the organic chemicals which are frequently designed to optimize GO for removal of various pollutants, as well as the functional groups that they introduce.

By binding organic compounds to GO, we can enhance its adsorption capacity. The compounds consist mainly of -SH, $-NH_2$, -COOH groups (Hosseinzadeh and Ramin 2018). The physiochemical characteristics of GO are improved with polymeric organic materials (such as CS, dendrimer, etc.), but most organic materials are not polymeric adsorbents. Other agents such as chelating and complexing agents like dipyridylamine (DPA) or diaminocyclohexanetetraacetic acid (DCTA) modify GO surfaces to facilitate the binding of metal ions to them.

GO was modified by using a dendrimer miniature structure terminated with an amine to enhance its ability to adsorb anionic molecules. In addition to having a maximum adsorption efficiency because of their functional terminal groups or empty spaces between their own branches, dendrimer molecules are highly branched nanoscale structures that are not toxic or exhibit low toxicity at low concentrations. With the combination of dendritic polymers as well as GO, it is possible to produce a highly functional composite, which is used to absorb HMs (Ghasempour et al. 2017; Gonzalo et al. 2015). Dendrimer-functionalized GO was prepared by Xiao et al. (2016) for removing Se(VI) (Xiao

TABLE 10.3

Modification of GO with organic compounds for the adsorption of HMs

Materials	Functional groups	References
Chitin	$-NH_2$	Ou et al. (2019) and Ikram et al. (2021)
CS	-NHCO, $-NH_2$	Zhou et al. (2021a)
Aminothiophenol	$-NH_2$, -SH	Zhang et al. (2020c)
Cyclodextrin	$-NH_2$	Ma et al. (2018c)
EDTA	$-NH_2$, -COOH, -OH	Mahmoudian et al. (2018)
Dendrimer	$-NH_2$	Zarei and Saedi (2018)
Alginate	-OH, -COOH	Fadillah et al. (2019)
DCTA	-COOH, $-NH_2$	Liu et al. (2016)
Amine	$-NH_2$	Ma et al. (2018a)
Glycol	-OH	Liu et al. (2019b)
Polyaniline	$-NH_2$	El-Sharkaway et al. (2019)
TETA	$-NH_2$	Yao et al. (2019)
Polypyrrole	$-NH_2$	Fang et al. (2018b)
Xanthate	-SH	Li et al. (2021b)
2-PTSC	$-NH_2$	Sherlala et al. (2018)

et al. 2016). The dendrimer introduced abundant -NH$_2$ groups, resulting in synthesized GO with higher adsorption efficiency than pristine GO.

β-cyclodextrin (β-CD), as a member of the long chain oligosaccharide's family, may also form spherical cylinders that seem to be immiscible with aqueous solution (Liu et al. 2019c).

The β-CD-enhanced GO (β-CD-GO) were used like a bioadsorbent to remove Co(II) from wastewater (Song et al. 2019).

Ceramic materials like cadmium sulfide (CdS) have provided a way to modify GO. A liquid insertion process was tested for the production of GO/CdS composites that increased Cu(II) ion degradation efficiency onto GO/CdS by incorporating ethylenediamine (en) into the carbon fiber activation method (Jiang et al. 2015).

CS, yet another excellent organic material, is usually coupled to GO to develop chitosan-graphene oxide (CS-GO) composite materials (Moradi et al. 2020). These composites have excellent characteristics as well as increased adsorption capacities (Subedi et al. 2019). CS, which is formed by deacetylating chitin from crustacean shells, is an exceptional chelating agent and HM reaction site, as it contains amines and hydroxyl groups which act as chelating agents (Chandra 2010; Le et al. 2019; Zhang et al. 2021). As highlighted by Samuel *et al.* (2018a) CS-modified graphene (MGO) has been used for Pb(II) adsorption from leather-polluted graywater (Samuel et al. 2018a). Since actual wastewater contains very few Pb(II) ions, Pb(NO$_3$)$_2$ was added till the Pb(II) concentration reached 50 mg/L. The results of the experiment were deemed successful because the Pb(II) removal efficiency from wastewater is 80%.

The GO structure can be modified for HM removal using chelating agents like 2,2′-DPA because of its importance in HM adsorption. Chelation enhances the binding capacity of strong complexing agents with HM ions. Such molecules also produce compounds with ions, enhancing GO adsorption capacity (Zare-Dorabei et al. 2016). Modified GO by Tadjarodi *et al.* (2016) employing 2-pyridinecarboxaldehyde thiosemicarbazone (2-PTSC) was used for removal of Hg(II) from liquid medium (Tadjarodi et al. 2016). The removal efficiency of modified GO was considerably higher (302 mg/g) than that of standard GO (130 mg/g). When pH conditions are optimal, Hg(II) is removed from the adsorbent via chelation among both inhibitors and non-covalent interactions with O$_2$-containing groups upon that adsorbent's edges. A similar study used 2,20-DPA to modify GO to retain Cu(II), Cd(II), and Ni(II) from a liquid medium. DPA-modified GO had higher removal efficiencies for HMs (Zare-Dorabei et al. 2016).

10.6 GO NANOCOMPOSITES FOR HM REMOVAL

Although numerous types of GO composites have been studied for their ability to remove HMs, here emphasis will be on the five best-known GO composites: GO-alginate, GO-SiO$_2$, GO-CS, magnetic GO composites, as well as nanoscale zero-valent iron (NZVI)-rGO composites.

10.6.1 GO-Chitosan

CS, a naturally occurring biodegradable polymer formed by degradation of crustacean shell chitin, is an effective chelating agent for HMs. It contains amines, which act as chelating agents for HMs, as well as hydrophilic groups, which act as reaction sites (Chandra 2010; Chokradjaroen et al. 2018; Le et al. 2019; Saheed et al. 2021; Samuel et al. 2018b; Samuel et al. 2018c; Zhang et al. 2021). CS is economical to prepare, undergoes easier polymerization and functionalization processes, and has good stability, which make it ideal for adsorption purposes and for environmental remediation (Saheed et al. 2021; Yang et al. 2019). CS, an organic material, is commonly used in combination with GO to make CS-GO composites (Moradi et al. 2020). CS-GO composites have unique physical and chemical properties as well as a high level of adsorption (Subedi et al. 2019). The amino group of CS and the carboxyl groups of GO form a chemical bond through the use of EDC (N-(3-dimethylaminopropyl)-N′-ethylcarbodiimide hydrochloride)/NHS (N-hydroxysuccinimide) or by

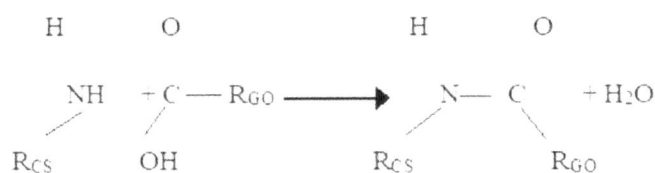

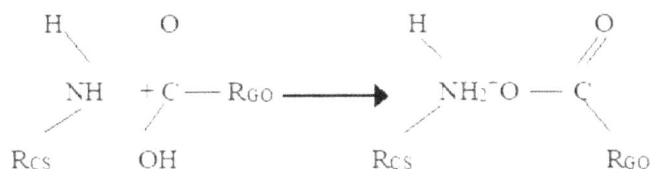

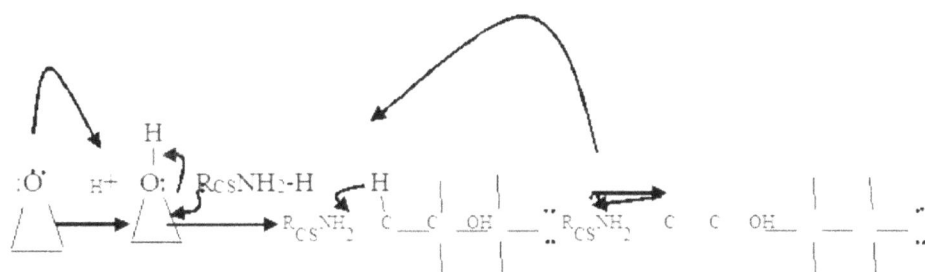

FIGURE 10.2 Interaction of GO and CS, wherein R_{CS} denotes a residual part of CS and R_{GO} denotes a residual part of GO.

physical attachment of GO with CS via hydrogen bonding (Kolanthai et al. 2018; Li et al. 2016). The reactions provided in Figure 10.2 illustrate the interactions between CS and GO.

According to Verma *et al.* (2022), EDTA-functionalized GO nanocomposites (GO-EDTA-CS) are capable of concurrently removing both organic and inorganic pollutants from polluted waters (Verma et al. 2022). The model inorganic pollutants are Hg(II) and Cu(II), while the organic pollutants are Methylene Blue and crystal violet (CV), which occur frequently in polluted water (Fauzi et al. 2018). The results show that GO-EDTA-CS is a promising adsorbent to treat wastewater containing a complex mixture of inorganic and organic pollutants. In another approach, the adsorption of As(III) ions from liquid solution is also possible through the use of ά-FeO(OH)/GO/CS nanocomposites (Shan et al. 2020). The study results indicate that the incorporation of ά-FeO(OH) into the adsorption matrix greatly enhances the adsorption capacity of the CS-GO composite toward As(III). Composites based on CS-GO have excellent adsorption capacities and can remove a variety of HMs from aquatic environments, including Cr(IV), Pb(II), and Cu(II). The GO-CS composites, for Au(III) removal, are not only effective adsorbents, but also extractants for isolating and reprocessing Au(III) from underground mining wastewater.

10.6.2 GO-Alginate Composites

The monosaccharide alginate consists of homopolymeric blocks of monomerized *-D-mannuronate (M) and C-5 epimer α-L-guluronate (G) (Nie et al. 2019). Alginate is recognized as a promising biosorption material due to its low cost, bioavailability, permeability, and high binding affinity for HMs (Zhang et al. 2020b). There are numerous environmental uses of alginate, including the capture of ions such as HMs by the cross-linked cations and ion exchange among alginate and pollutants, such as hydroxyl and carboxyl groups on its edges (Wang et al. 2019). For removing HMs from graywater, GO as well as alginate biopolymers offer great potential (Wang et al. 2019; Wang et al. 2016). Although alginate exhibits excellent elasticity, it has several drawbacks, including its high rigidity and fragility (Jiao et al. 2016; Ren et al. 2016; Thakur et al. 2016). GO sheets are a suitable reinforcement material in composites for alleviating the unsatisfying characteristics of alginate and preventing the collapse of a highly porous structure (Fei et al. 2016; Gan et al. 2018; Pan et al. 2018). In several studies, toxic HMs were removed from water by using GO-based alginate composites. It has been found that increase in GO content increases the pore size of PVA/GO-SA, and a relatively loose network structure forms because GO and PVA have a negative correlation, and the PVA content decreases the cross-linking degree (Yu et al. 2019). Additionally, once the GO concentration reaches 5%, the Pb(II) adsorption rate increases sharply, and an increase in the pore size leads to higher Pb(II) adsorption (Jiao et al. 2016). A recent study combined GO and sodium alginate directly to form double network hydrogel beads (Yang et al. 2018d). Based on the theoretical maximum adsorption capacity for Mn^{2+}, the GO-sodium alginate hydrogel beads prepared demonstrated good affinity for cationic metals. Although GO-alginate composites usually have marginally lower adsorption efficiency toward HMs than GO-CS composites, the gelling characteristic of alginate ascertained that such GO-alginate composites and water-soluble media could be kept separate.

10.6.3 GO-SiO$_2$ Composites

For a variety of reasons, HMs are removed from wastewater using silica nanomaterials (Siddeeg et al. 2021). Because of the formation of the silanol group (Si-OH) on their substrates, they behave as anthropogenic as well as natural chelating agents (Ahmed et al. 2020). Because of their 3D highly permeable lattice structure, these materials display high mechanical strength, water and thermal stability, as well as high density (Li et al. 2020a). Because of these properties, HMs can be chelated rapidly by the inner surface (Palos-Barba et al. 2020). Additionally, the surface functionalization of accessible silica is also easily modified with GO to prevent aggregation. Several studies have used electrostatic attraction (**Figure 10.3**) to prepare composites of GO and SiO$_2$ containing amino groups when amino groups of SiO$_2$ and carboxyl groups of GO are added (Bulbula et al. 2018; Li et al. 2019a). Aside from that, it must also be noted that GO-based SiO$_2$ composites can selectively adsorb HMs because they have excellent thermal stability and controllable surface coverage (Zhang et al. 2020b). For example, GO-SiO$_2$ and reversible-addition-fragmentaion chain-transfer (RAFT) polymerization were used to prepare Ni(II) ion-imprinted polymers. A higher selectivity was observed for Ni(II) in systematic samples and in the presence of Cd(II), Cu(II), Zn(II), and Co(II) (Liu et al. 2015).

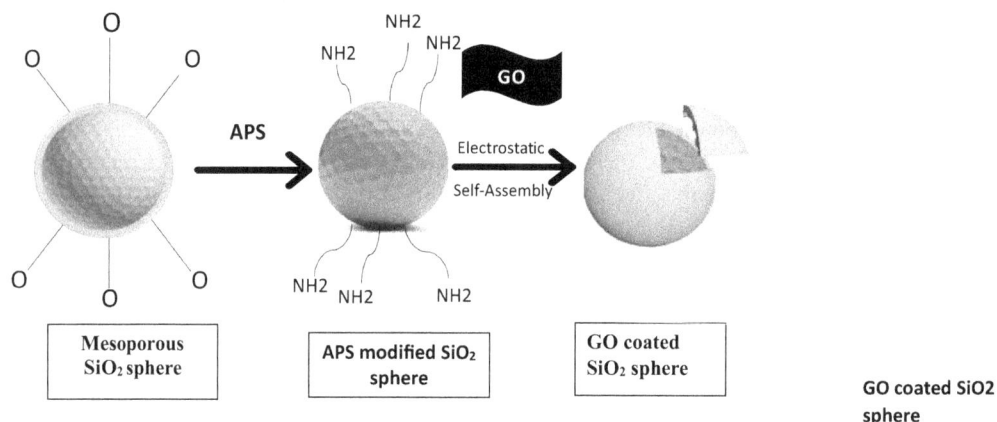

FIGURE 10.3   Schematic diagram of SiO$_2$@GO.

10.6.4 MAGNETIC GO

Magnetic nanoparticles are composed of d-block elements and their oxides that possess magnetic properties (Wang et al. 2017b; Williams et al. 2018). Magnetite (Fe_3O_4), maghemite, and hematite are just a few of the magnetic nanoparticles that have been studied extensively for ecological, technological, and biological applications (de Mendonça et al. 2019). Chemical precipitation is the principal process for synthesizing magnetic GO (Hu et al. 2017). The removal of HMs (Cu^{2+}, Cr^{3+}, Zn^{2+}, Pb^{2+}, and Ni^{2+}) is the primary requirement of magnetic GO, indicating maximum elimination capacity (Farooq and Jalees 2020). A recent study has successfully removed arsenic ions (As(III) and As(V)) from water for the first time using functionalized magnetic GO organic solvents (MGO-IL) (Zhang et al. 2019b). Their adsorption efficiencies for As(III) and As(V) were 160.65 mg/g and 104.13 mg/g, respectively. Polypyrrole nanoparticles and Fe_3O_4 were successfully used to remove Al^{3+} metal ions from the water surface. About 100.7 mg of Al^{3+} was removed per g of polypyrrole/Fe_3O_4 (Mollahosseini et al. 2019). Similarly, a magnetic γ-PGA-Fe_3O_4-GO-(oMWCNTs) composite demonstrated the effective adsorption of several HMs and it can effectively work in a wide range of pH (2–10) (Wang et al. 2018a). The removal efficiency of Cu(II), Ni(II), and Cd(II) were 574, 625, 384, 62, and 71 mg/g, respectively, according to the Langmuir model. Furthermore, GO, diethylenetriamine, and Fe_3O_4 nanoparticles (AMGO) were used to develop different magnetic nanocomposites (Zhao et al. 2016). Amino groups in diethylenetriamine formed stable chelates with HMs, enabling AMGO to bind Cr(IV) very efficiently, and the maximum removal efficiency was 123.4 mg/g; MGO, on the other hand, had an elimination capacity of 257.5 mg/g.

10.6.5 NZVI-rGO COMPOSITES

NZVI is one of several engineered nanomaterials that can be used to remediate the environment (Lv et al. 2019; Ren et al. 2018). NZVI has a very low bulk density and a low buffering capacity, as well as good injectability into water sources. NZVI is typically made up of approximately one Fe core that serves as an electrophile and with at least one iron oxide shell that serves as a contaminant adsorbent, for example, for eliminating HMs and phosphates, nitrates, and organic contaminants that have been reductively degraded or oxidatively degraded by molecular oxygen (Dong et al. 2018; Eljamal et al. 2018; Liu et al. 2017a; Maamoun et al. 2018; Mu et al. 2017). NZVI removes both metallic ions (Ni(II), U(VI)), Cd(II), Cu(II), and Zn(II) (Liu et al. 2017c; Tsarev et al. 2017)) and oxyanions (Se(IV)/Se(VI) or Cr(VI), As(V))) through direct or coordinated reactions, converting itself to biologically beneficial iron oxide or hydroxide (Lv et al. 2019; Song et al. 2021). For efficient HM removal from water, nZVI-rGO was synthesized without the use of toxic reductant agents or surfactants through a gaseous hydrogen reduction method (Khurana et al. 2018). Studies have reported that once NZVI nanocrystals are supported with GO, their concentration decreases, while their scattering tendency and specific surface area continue to improve. The removal of Cu(II) using NZVI-imprinted GO/bentonite (GO-B-nZVI) was evaluated. The study concludes that applying GO can increase the extraction efficiency of Cu(II); in contrast, the removal efficiency through bentonite is 26%, through B-nZVI it is 71%, and through GO-B-nZVI it is 82% for 16 hours. However, the Langmuir removal efficiency of GO-B-nZVI and B-nZVI was 184.5 mg/g and 130.7 mg/g, respectively (Shao et al. 2018).

10.7 ADSORPTION MECHANISM OF GO

The adsorption technique involves a surface interaction of molecules that is superficial; it is commonly used to remove metals and non-metals from water (Thakur and Kandasubramanian 2019; Yang et al. 2018b). Three basic mechanisms of adsorption of GO-based nanomaterials are described below: adsorption isotherms, adsorption kinetics, and adsorption thermodynamics.

10.7.1 Adsorption Isotherm

An adsorption isotherm relates the concentration of the pollutant at a particular temperature to the limit of contaminants dissolved for each unit of mass of the adsorbate process (Ahmad et al. 2020). The adsorption isotherm is a useful method for assessing hypothetical effects of pollutants on GO composites, exploring the mechanism of adsorption, and interpreting thermodynamic parameters (Kyzas et al. 2018; Velusamy et al. 2021; Zhang et al. 2020b). There seem to be two noteworthy adsorption isotherms that can be used for HMs: Freundlich and Langmuir isotherms were being used, as also the Temkin and Dubinin-Radushkevich (D-R) models. HM ions are studied extensively using Langmuir and Freundlich models (Ahmad et al. 2020; Huang et al. 2018; Velusamy et al. 2021; White et al. 2018; Zhang et al. 2020b). Freundlich's work is not limited to monolayer adsorption on homogeneous surfaces, but the Langmuir model is used to analyze this problem. Therefore, the Langmuir model is found to be a better fit for most studies.

Thus, the Langmuir equation's non-linear form is as follows:

$$q_e = \frac{q_{max} K_L C_e}{1 + K_L C_e} \tag{10.1}$$

C denotes the equilibrium concentration (mg/L), q denotes the number of adsorbed metals at equilibrium (mg/g), K_L denotes the equilibrium constant (L/mg), and q_{max} denotes high removal efficiency (mg/g) of the adsorbent. In contrast, the linear equation is as follows:

$$\frac{C_e}{q_e} = \frac{C_e}{q_m} + \frac{1}{bq_m} \tag{10.2}$$

q_m represents the saturated single layer adsorption efficiency, equilibrium sorbate concentration is represented by C_e, and the equilibrium adsorption constant is measured by b.

The Freundlich isotherm's non-linear equations and linear equations are as follows:

$$q_e = K_F \ C_e^{\frac{1}{n}} \tag{10.3}$$

$$\log q_e = \log k_F + \frac{\log C_e}{n} \tag{10.4}$$

Q_e denotes the equilibrium sorbate load on the sorbent (mg/g); K_F denotes the adsorption capacity ($mg^{1-n}L^n g^{-1}$); n is an adsorptive energy indicator; and C_e denotes equilibrium fluid composition of the sorbate (mg/L) (Qiao et al. 2020; Velusamy et al. 2021).

10.7.2 Adsorption Kinetics

An analysis of environmental data using adsorption kinetic models was undertaken to examine HMs' adsorption mechanism on GO composites. Different kinetic models have been used to explain the adsorption of toxic substances with the help of GO composites. Pseudo-first-order (PFO), pseudo-second-order (PSO), as well as intraparticle diffusion are some of most popular kinetic methods (Sherlala et al. 2018). Sharif *et al.* (2018) studied the adsorption mechanisms of Pb(II) and Cu(II) on Alg-MGO by using PFO as well as PSO, and researchers discovered that this technique is effective with PFO (Sharif et al. 2018). As a result, adsorbates (Pb(II) and Cu(II)) were found to accumulate on the Alg-MGO surface during the degradation mechanism. According to Peer *et al.* (2018), PSO fit all of the experimental data better than PFO due to a higher R^2 (>0.9939), but PFO was found to be consistent at low metal ion concentrations (Peer et al. 2018). Due to the presence of multiple amino groups in mGO1st- and mGO2nd-PAMAM, both coordination and electrostatic attraction were major contributors to adsorption. Several studies have used the intraparticle diffusion model, even though it is not used as frequently as PSO and PFO. For example, it was noticed that intra-particle diffusion lines did not cross the origin in adsorption of Cd(II) onto á-PGA-Fe$_3$O$_4$-GO-(o-MWCNTs), suggesting that intraparticle diffusion isn't the only rate-limiting point in the adsorption mechanism (Wang et al. 2018a).

10.7.3 ADSORPTION THERMODYNAMICS

To evaluate adsorbate-adsorbent systems using thermodynamic factors, change in enthalpy (ΔH), Gibbs free energy (ΔG), as well as change in entropy (ΔS) are commonly used. Negative ΔG is indicative of adsorption performance, while (H) identifies adsorption characteristics (Ahmad *et al.* 2020; Zhang *et al.* 2020b). When ΔH becomes positive, the reaction is endothermic; however, when ΔH is negative, it indicates that the response has been expressed as:

$$\Delta G = -RT \ln K_L \tag{10.5}$$

According to the Langmuir equilibrium constant, enthalpy and entropy changes can be represented as:

$$\ln \quad K_L = \frac{\rho S}{R} - \frac{\rho H}{RT} \tag{10.6}$$

Hence, if ln K$_L$ is presented against 1/T, a straight line should result. The curve and intercept of this line will yield the ΔH and ΔS values. Based on the equation, the following thermodynamic parameters were determined: entropy (S), standard free energy (G) and enthalpy (H):

$$\Delta G = \Delta H - T\Delta S \tag{10.7}$$

ΔG versus T plotted with slope and intercept were used to calculate ΔS and ΔH values (Xu et al. 2018a).

10.8 FACTOR EFFECTING THE ABSORPTION OF HMS BY GO

Several factors influence the HM removal efficiencies of GO-based composites from wastewater. Temperature, adsorbent dose, pH, and contact time are the most important variables.

10.8.1 EFFECT OF PH

Adsorbents for HMs are primarily affected by pH value, which determines their adsorption efficiency (Wang et al. 2018b). This is because of the reactive species within target metallic materials and the surface charge of adsorbents (Ahmad et al. 2020; Parastar et al. 2021; Zhang et al. 2020b). HM removal efficiency is frequently governed not only by the pH of the solution, but also by point of zero charge (pH$_{pzc}$) of the adsorbents. pH$_{pzc}$ denotes the pH at which particles have no net charges on their surfaces (Ahmad et al. 2020; Farooq and Jalees 2020; Zhang et al. 2020b). Therefore, the surface of an adsorbent is neutral because the rate of accumulation of opposite charges on it is approximately equivalent. Ions do not interact with this neutral surface via electrostatic interactions. Adsorbents for cationic HMs should have a pH$_{pzc}$ that is more acidic than the pH$_{pzc}$ for anionic HMs to allow cationic HMs to be immobilized across a broader pH range (Ahmad et al. 2020; Liu et al. 2019c). Adsorbents become negatively charged and cationic HMs are attracted to them by electrostatic attraction whenever the solution's pH exceeds pH$_{pzc}$. Conversely, when the solution's pH is lower than pH$_{pzc}$, adsorbents become positively charged and can be used to adsorb metals which are anionic in nature (Ahmad et al. 2020).

Hadadian et al. (2018) created innovative zinc oxide-graphene (ZnO-Gr) nanoparticles for Ni removal with pH_{pzc} of 3.26. The results showed that at pH values under 3.26 (pH_{pzc}), Ni(II) could not adsorb on ZnO-Gr surfaces due to massive electrostatic repulsion among the ionic species of Ni(II) or the proactive surfaces of the adsorbent. A pH of 3.26–5.6, however, brought very low Ni(II) adsorption because Ni(II) and H^+ competed for adsorbed sites (Hadadian et al. 2018). However, because H^+ ions have a higher charge density, they win. Ni(II) removal increases only when the pH reaches 5.6–8.2, and the optimal pH for adsorption is reached at pH levels greater than 6. Whether the surface sites are negatively or positively charged, at pH greater than 8.2, Ni(II) doesn't exist as Ni^{2+}, but rather as $Ni(OH)^+$, $Ni(OH)_3^-$, or $Ni(OH)_2$. Hence, it is assumed that electrostatic interaction is ineffective, which leads to a decrease in adsorption. The structure and properties of GO-based nanostructures are largely determined by the solution's pH, as is the behavior of various types of metal complexes on various types of GO-modified nanomaterials (Lim et al. 2018a; Lim et al. 2018b). Whereas the efficiency of ionic species extraction is highly dependent on solution pH, when applying GO-based nanomaterials to eliminate toxic metals, the solution's pH should be considered first.

10.8.2 ABSORBENT DOSAGE

Toxic metals should be eliminated from alkaline media by applying appropriate amounts of adsorbent. It is essential to choose the most appropriate amount to reduce excessive usage and increase the utilization rate (De Beni et al. 2022; Peng et al. 2017). Increasing doses of adsorbent reduce HM removal percentages, which is associated with greater surface areas and more surface functional groups available. After the saturation point, increasing the adsorbent dose does not help in enhancing the effectiveness of pollutants removal, and the adsorbent capacity decreases because a relatively high concentration of the substance cannot be used efficiently (Xu et al. 2018b). In addition to component dosage of GO composites, different component concentrations also ought to be considered as an internal adsorbent parameter. The effects of β-CD and GO concentration on the adsorption qualities of Cu(II) and Pb(II) were evaluated. Results showed that Pb(II) and Cu(II) reached maximum adsorption efficiencies at 2.0 g of β-CD, because of the effective diffusion of β-CD and Fe_3O_4. Similarly, high dose of Go effect, 2.0 g of -CD was placed on a plate and a plate value of 50 mg was measured. This could be due to GO sheet aggregation at high concentrations (Ma et al. 2018c).

10.8.3 CONTACT TIME

Toxic metal removal efficiency is affected by contact time as well, so determining the highest removal rate and optimum time is critical (Zhang et al. 2020b). Treatment methods are categorized into two steps: The adsorption rate rises rapidly during the preliminary stages of the adsorption mechanism because of the high number of active metal-binding areas, or as the number of active metal-binding sites decreases gradually, the adsorption rate decreases gradually till equilibrium is reached (Zhou et al. 2017). A half of the Cu optimum concentration was reached within five minutes after 120-minute adsorption on CS- ion-imprinted polymers (Kong et al. 2017). Similarly, the TEPAGO/$MnFe_2O_4$ adsorption capacity increased strongly for the first 20 minutes, before deteriorating slowly until it reached equilibrium at 90 minutes (Xu et al. 2018b). The duration of adsorbent equilibrium is determined by a variety of factors, including composite materials and metal complexes, actual adsorbate concentrations, and so on. Moreover, it is reported that the equilibrium removal rates of Cu(II) as well as Pb(II) were reached after 240 and 120 minutes, respectively (Ma et al. 2018c). Another study results showed that as immediate accumulation of HMs increased from 50 to 330 mg/L, adsorption capacity increased, due to a greater driving force produced by different concentration gradient pressures in contrast to a lower initial concentration (Wang et al. 2018a).

10.8.4 TEMPERATURE

Molecular interaction and solubility are greatly influenced by temperature, which affects adsorption rates (Fadlalla et al. 2020). As temperature increases, the number of O_2 functional groups in the GO substrate surface increases, accelerating the process of metal ion diffusion. As a result, the adsorption rate increases as temperature rises, implying that the process is temperature dependent and spontaneous (Yang et al. 2017). Many isothermal models, including the Langmuir, Dubinin-Radushkevich (D-R), Freundlich, and SIPS models, have already been applied to predict the adsorption of toxic metals ions on GO-modified nanomaterials (Chen et al. 2017; Liu et al. 2019c). Investigation the impacts of temperature (318, 298, and 308 K) on Cr(IV) removal efficiency on modified graphene oxide/chitosan composite with disodium ethylenediaminetetraacetate (EDTA-2Na) (GEC) showed that the sorption process was dependent on temperature resulting in positive relation of $\Delta H°$ and the negative relation of $\Delta G°$ (Zhang et al. 2016). Additionally, the positive factor of $\Delta S°$ during Cr(IV) adsorption onto GEC showed an increasing alignment at the solid-liquid interface. Hg(II) immobilization applying GO/mCS and GO/CS was also highlighted in a similar manner. The negative coefficient of $\Delta G°$ decreases as temperature rises from 298 to 338 K, indicating one of most reliable adsorption mechanisms. Because controlling the temperature of effluents is typically more complex and power intensive it is not necessary to analyze the impact of temperature when developing GO or GO-modified nanoparticles for removing HMs. In general, it is preferable to preconcentrate HMs from an aqueous medium to GO or GO-based nanocomposites at room temperature and conditions.

10.9 CONCLUSIONS AND FUTURE OUTLOOKS

This review summarizes research on the management and efficiency of GO-based nanomaterials for the adsorption of HM ions in wastewater. GO-based composite structures are commonly studied as novel biosorbents because of their minimum maintenance requirements and high affinity for different metal ions. There are still some key problems to solve in terms of composites-based GO and their wastewater management. (1) Several different approaches to design GO-based composite products with improved adsorption capacities are required. Both engineering expertise and theoretical guidance are required to be incorporated. (2) A thorough investigation of the biosorption and sampling of metal complexes under difficult circumstances should be undertaken. (3) The formation of GO-modified nanocomposites should be enhanced and the planning process modernized. (4) Use of composites with different structural groups for binding HMs is crucial in synthesizing nanocomposites which can be used to remove many different types of metal ions under complicated conditions. (5) A nanocomposite should have the ability to remain stable in extreme conditions. (6) Simple and fast methods to separate nanocomposites from aqueous solutions are required to be investigated. (7) The recyclability, ease of maintenance, and even the irreversibility of some HMs are examples of other parameters which need to be investigated and validated before the full-scale application of GO-based composites. In the future, such issues should be considered when preparing and applying GO-based nanomaterials. GO-based composites are expected to be used in treating wastewater in the coming years, even though many issues remain in the formation of GO-based nanoparticles and their incorporation in pollution cleanup.

REFERENCES

Abdi G, Alizadeh A, Zinadini S, Moradi G. 2018. Removal of dye and heavy metal ion using a novel synthetic polyethersulfone nanofiltration membrane modified by magnetic graphene oxide/metformin hybrid. *Journal of Membrane Science*. 552: 326–335.

Abdolmohammad-Zadeh H, Ayazi Z, Hosseinzadeh S. 2020. Application of Co_3O_4 nanoparticles as an efficient nano-sorbent for solid-phase extraction of zinc(II) ions. *Microchemical Journal*. 153: 104268.

Achary MS, Satpathy K, Panigrahi S, Mohanty A, Padhi R, Biswas S, Prabhu R, Vijayalakshmi S, Panigrahy R. 2017. Concentration of heavy metals in the food chain components of the nearshore coastal waters of Kalpakkam, southeast coast of India. *Food Control*. 72: 232–243.

Ahmad H, Husain FM, Khan RA. 2021. Graphene oxide lamellar membrane with enlarged inter-layer spacing for fast preconcentration and determination of trace metal ions. *RSC Advances*. 11(20): 11889–11899.

Ahmad H, Liu C. 2021. Ultra-thin graphene oxide membrane deposited on highly porous anodized aluminum oxide surface for heavy metal ions preconcentration. *Journal of Hazardous Materials*. 415: 125661.

Ahmad SZN, Salleh WNW, Ismail AF, Yusof N, Yusop MZM, Aziz F. 2020. Adsorptive removal of heavy metal ions using graphene-based nanomaterials: Toxicity, roles of functional groups and mechanisms. *Chemosphere*. 248: 126008.

Ahmed MO, Shrpip A, Mansoor M. 2020. Synthesis and characterization of new Schiff base/thiol-functionalized mesoporous silica: An efficient sorbent for the removal of Pb(II) from aqueous solutions. *Processes*. 8(2): 246.

Al-Jabari M. 2016. Kinetic models for adsorption on mineral particles comparison between Langmuir kinetics and mass transfer. *Environmental Technology & Innovation*. 6: 27–37.

Al Ketife AM, Al Momani F, Judd S. 2020. A bioassimilation and bioaccumulation model for the removal of heavy metals from wastewater using algae: New strategy. *Process Safety and Environmental Protection*. 144: 52–64.

Alengebawy A, Abdelkhalek ST, Qureshi SR, Wang M-Q. 2021. Heavy metals and pesticides toxicity in agricultural soil and plants: Ecological risks and human health implications. *Toxics*. 9(3): 42.

Almomani F, Bhosale R, Khraisheh M, Almomani T. 2020. Heavy metal ions removal from industrial wastewater using magnetic nanoparticles (MNP). *Applied Surface Science*. 506: 144924.

Almomani F, Bohsale RR. 2021. Bio-sorption of toxic metals from industrial wastewater by algae strains Spirulina platensis and Chlorella vulgaris: Application of isotherm, kinetic models and process optimization. *Science of the Total Environment*. 755: 142654.

Alosaimi AM. 2021. Polysulfone membranes based hybrid nanocomposites for the adsorptive removal of Hg(II) ions. *Polymers*. 13(16): 2792.

Aram SA, Lartey PO, Amoah SK, Appiah A. 2021. Gold eco-toxicology: Assessment of the knowledge gap on the environmental and health effects of mercury between artisanal small scale and medium scale gold miners in Ghana. *Resources Policy*. 72: 102108.

Asadi R, Abdollahi H, Gharabaghi M, Boroumand Z. 2020. Effective removal of Zn(II) ions from aqueous solution by the magnetic $MnFe_2O_4$ and $CoFe_2O_4$ spinel ferrite nanoparticles with focuses on synthesis, characterization, adsorption, and desorption. *Advanced Powder Technology*. 31(4): 1480–1489.

Awan SA, Ilyas N, Khan I, Raza MA, Rehman AU, Rizwan M, Rastogi A, Tariq R, Brestic M. 2020. Bacillus siamensis reduces cadmium accumulation and improves growth and antioxidant defense system in two wheat (Triticum aestivum L.) varieties. *Plants*. 9(7): 878.

Bashir A, Malik LA, Ahad S, Manzoor T, Bhat MA, Dar G, Pandith AH. 2019. Removal of heavy metal ions from aqueous system by ion-exchange and biosorption methods. *Environmental Chemistry Letters*. 17(2): 729–754.

Berihu BA. 2015. Histological and functional effect of aluminium on male reproductive system. *International Journal of Pharmaceutical Sciences and Research*. 6(8): 1122–1132.

Beroigui M, Naylo A, Walczak M, Hafidi M, Charzyński P, Świtoniak M, Różański S, Boularbah A. 2020. Physicochemical and microbial properties of urban park soils of the cities of Marrakech, Morocco and Toruń, Poland: Human health risk assessment of fecal coliforms and trace elements. *Catena*. 194: 104673.

Bulbula ST, Dong Y, Lu Y, Yang X-Y. 2018. Graphene oxide coating enhances adsorption of lead ions on mesoporous SiO_2 spheres. *Chemistry Letters*. 47(2): 210–212.

Cai J, Chen J, Zeng P, Pang Z, Kong X. 2019. Molecular mechanisms of CO_2 adsorption in diamine-cross-linked graphene oxide. *Chemistry of Materials*. 31(10): 3729–3735.

Chandra V. 2010. Park. Jaesung, Y. Chun, JW Lee, IC Hwang, KS Kim. *ACS Nano*. 7: 3979–3986.

Chen H, Meng Y, Jia S, Hua W, Cheng Y, Lu J, Wang H. 2020. Graphene oxide modified waste newspaper for removal of heavy metal ions and its application in industrial wastewater. *Materials Chemistry and Physics*. 244: 122692.

Chen L, Feng S, Zhao D, Chen S, Li F, Chen C. 2017. Efficient sorption and reduction of U(VI) on zero-valent iron-polyaniline-graphene aerogel ternary composite. *Journal of Colloid and Interface Science*. 490: 197–206.

Chen L, Li Y, Chen L, Li N, Dong C, Chen Q, Liu B, Ai Q, Si P, Feng J. 2018. A large-area free-standing graphene oxide multilayer membrane with high stability for nanofiltration applications. *Chemical Engineering Journal*. 345: 536–544.

Chokradjaroen C, Theeramunkong S, Yui H, Saito N, Rujiravanit R. 2018. Cytotoxicity against cancer cells of chitosan oligosaccharides prepared from chitosan powder degraded by electrical discharge plasma. *Carbohydrate Polymers*. 201: 20–30.

Chouhan A, Mungse HP, Khatri OP. 2020. Surface chemistry of graphene and graphene oxide: A versatile route for their dispersion and tribological applications. *Advances in Colloid and Interface Science*. 283: 102215.

Coroş M, Pogăcean F, Măgeruşan L, Socaci C, Pruneanu S. 2019. A brief overview on synthesis and applications of graphene and graphene-based nanomaterials. *Frontiers of Materials Science*. 13(1): 23–32.

Dai F, Yu R, Yi R, Lan J, Yang R, Wang Z, Chen J, Chen L. 2020. Ultrahigh water permeance of a reduced graphene oxide nanofiltration membrane for multivalent metal ion rejection. *Chemical Communications*. 56(95): 15068–15071.

De Beni E, Giurlani W, Fabbri L, Emanuele R, Santini S, Sarti C, Martellini T, Piciollo E, Cincinelli A, Innocenti M. 2022. Graphene-based nanomaterials in the electroplating industry: A suitable choice for heavy metal removal from wastewater. *Chemosphere*. 292: 133448.

de Mendonça ESDT, de Faria ACB, Dias SCL, Aragon FF, Mantilla JC, Coaquira JA, Dias JA. 2019. Effects of silica coating on the magnetic properties of magnetite nanoparticles. *Surfaces and Interfaces*. 14: 34–43.

De Silva K, Huang H-H, Joshi R, Yoshimura M. 2017. Chemical reduction of graphene oxide using green reductants. *Carbon*. 119: 190–199.

Ding S, Yang Y, Li C, Huang H, Hou L-a. 2016. The effects of organic fouling on the removal of radionuclides by reverse osmosis membranes. *Water Research*. 95: 174–184.

Dong H, Qiang Z, Lian J, Li J, Yu J, Qu J. 2018. Deiodination of iopamidol by zero valent iron (ZVI) enhances formation of iodinated disinfection by-products during chloramination. *Water Research*. 129: 319–326.

Dubey S, Shri M, Gupta A, Rani V, Chakrabarty D. 2018. Toxicity and detoxification of heavy metals during plant growth and metabolism. *Environmental Chemistry Letters*. 16(4): 1169–1192.

Duru İ, Ege D, Kamali AR. 2016. Graphene oxides for removal of heavy and precious metals from wastewater. *Journal of Materials Science*. 51(13): 6097–6116.

El-Kady AA, Abdel-Wahhab MA. 2018. Occurrence of trace metals in foodstuffs and their health impact. *Trends in Food Science & Technology*. 75: 36–45.

El-Sharkaway E, Kamel RM, El-Sherbiny IM, Gharib SS. 2019. Removal of methylene blue from aqueous solutions using polyaniline/graphene oxide or polyaniline/reduced graphene oxide composites. *Environmental Technology*.

Eljamal O, Mokete R, Matsunaga N, Sugihara Y. 2018. Chemical pathways of nanoscale zero-valent iron (NZVI) during its transformation in aqueous solutions. *Journal of Environmental Chemical Engineering*. 6(5): 6207–6220.

Eltayeb NE, Khan A. 2020. Preparation and properties of newly synthesized polyaniline@ graphene oxide/Ag nanocomposite for highly selective sensor application. *Journal of Materials Research and Technology*. 9(5): 10459–10467.

Fadillah G, Saleh TA, Wahyuningsih S, Putri ENK, Febrianastuti S. 2019. Electrochemical removal of methylene blue using alginate-modified graphene adsorbents. *Chemical Engineering Journal*. 378: 122140.

Fadlalla MI, Kumar PS, Selvam V, Babu SG. 2020. Emerging energy and environmental application of graphene and their composites: A review. *Journal of Materials Science*. 55(17): 7156–7183.

Fang L, Li L, Qu Z, Xu H, Xu J, Yan N. 2018a. A novel method for the sequential removal and separation of multiple heavy metals from wastewater. *Journal of Hazardous Materials*. 342: 617–624.

Fang W, Jiang X, Luo H, Geng J. 2018b. Synthesis of graphene/SiO$_2$@polypyrrole nanocomposites and their application for Cr(VI) removal in aqueous solution. *Chemosphere*. 197: 594–602.

Fang X, Li J, Li X, Pan S, Zhang X, Sun X, Shen J, Han W, Wang L. 2017. Internal pore decoration with polydopamine nanoparticle on polymeric ultrafiltration membrane for enhanced heavy metal removal. *Chemical Engineering Journal*. 314: 38–49.

Farooq MU, Jalees MI. 2020. Application of magnetic graphene oxide for water purification: Heavy metals removal and disinfection. *Journal of Water Process Engineering*. 33: 101044.

Fatemi H, Pour BE, Rizwan M. 2020. Isolation and characterization of lead (Pb) resistant microbes and their combined use with silicon nanoparticles improved the growth, photosynthesis and antioxidant capacity of coriander (Coriandrum sativum L.) under Pb stress. *Environmental Pollution*. 266: 114982.

Fauzi FB, Ismail E, Ani MH, Bakar SNSA, Mohamed MA, Majlis BY, Din MFM, Abid MAAM. 2018. A critical review of the effects of fluid dynamics on graphene growth in atmospheric pressure chemical vapor deposition. *Journal of Materials Research*. 33(9): 1088–1108.

Fei Y, Li Y, Han S, Ma J. 2016. Adsorptive removal of ciprofloxacin by sodium alginate/graphene oxide composite beads from aqueous solution. *Journal of Colloid and Interface Science*. 484: 196–204.

Gan L, Li H, Chen L, Xu L, Liu J, Geng A, Mei C, Shang S. 2018. Graphene oxide incorporated alginate hydrogel beads for the removal of various organic dyes and bisphenol A in water. *Colloid and Polymer Science*. 296(3): 607–615.

Ghasempour A, Pajootan E, Bahrami H, Arami M. 2017. Introduction of amine terminated dendritic structure to graphene oxide using poly(propylene Imine) dendrimer to evaluate its organic contaminant removal. *Journal of the Taiwan Institute of Chemical Engineers*. 71: 285–297.

Ghosh B, Sharma RK, Yadav S. 2021. Toxic effects of aluminium on testis in presence of ethanol coexposure. *European Journal of Anatomy*. 25(3): 359–367.

Godiya CB, Liang M, Sayed SM, Li D, Lu X. 2019. Novel alginate/polyethyleneimine hydrogel adsorbent for cascaded removal and utilization of Cu^{2+} and Pb^{2+} ions. *Journal of Environmental Management*. 232: 829–841.

Gonzalo S, Rodea-Palomares I, Leganés F, García-Calvo E, Rosal R, Fernández-Piñas F. 2015. First evidences of PAMAM dendrimer internalization in microorganisms of environmental relevance: A linkage with toxicity and oxidative stress. *Nanotoxicology*. 9(6): 706–718.

Gul K, Sohni S, Waqar M, Ahmad F, Norulaini NN, Mohd Omar AK. 2016. Functionalization of magnetic chitosan with graphene oxide for removal of cationic and anionic dyes from aqueous solution. *Carbohydrate Polymers*. 152: 520–531.

Hadadian M, Goharshadi EK, Fard MM, Ahmadzadeh H. 2018. Synergistic effect of graphene nanosheets and zinc oxide nanoparticles for effective adsorption of Ni(II) ions from aqueous solutions. *Applied Physics A*. 124(3): 1–10.

Hosseinzadeh H, Ramin S. 2018. Effective removal of copper from aqueous solutions by modified magnetic chitosan/graphene oxide nanocomposites. *International Journal of Biological Macromolecules*. 113: 859–868.

Hsini A, Naciri Y, Laabd M, El Ouardi M, Ajmal Z, Lakhmiri R, Boukherroub R, Albourine A. 2020. Synthesis and characterization of arginine-doped polyaniline/walnut shell hybrid composite with superior clean-up ability for chromium (VI) from aqueous media: Equilibrium, reusability and process optimization. *Journal of Molecular Liquids*. 316: 113832.

Hu X, Zhao Y, Wang H, Tan X, Yang Y, Liu Y. 2017. Efficient removal of tetracycline from aqueous media with a Fe_3O_4 nanoparticles@graphene oxide nanosheets assembly. *International Journal of Environmental Research and Public Health*. 14(12): 1495.

Hu Y, Cheng H, Tao S. 2016. The challenges and solutions for cadmium-contaminated rice in China: A critical review. *Environment International*. 92: 515–532.

Huang D, Li B, Wu M, Kuga S, Huang Y. 2018. Graphene oxide-based Fe–Mg (hydr)oxide nanocomposite as heavy metals adsorbent. *Journal of Chemical & Engineering Data*. 63(6): 2097–2105.

Huang D, Wu J, Wang L, Liu X, Meng J, Tang X, Tang C, Xu J. 2019. Novel insight into adsorption and co-adsorption of heavy metal ions and an organic pollutant by magnetic graphene nanomaterials in water. *Chemical Engineering Journal*. 358: 1399–1409.

Ikram R, Mohamed Jan B, Abdul Qadir M, Sidek A, Stylianakis MM, Kenanakis G. 2021. Recent advances in chitin and chitosan/graphene-based bio-nanocomposites for energetic applications. *Polymers*. 13(19): 3266.

Islam ARMT, Siddiqua MT, Zahid A, Tasnim SS, Rahman MM. 2020. Drinking appraisal of coastal groundwater in Bangladesh: An approach of multi-hazards towards water security and health safety. *Chemosphere*. 255: 126933.

Ismail A, Riaz M, Akhtar S, Goodwill JE, Sun J. 2019. Heavy metals in milk: Global prevalence and health risk assessment. *Toxin Reviews*. 38(1): 1–12.

Jaiswal D, Pandey J. 2019. Investigations on peculiarities of land-water interface and its use as a stable testbed for accurately predicting changes in ecosystem responses to human perturbations: A sub-watershed scale study with the Ganga River. *Journal of Environmental Management*. 238: 178–193.

Jaskulak M, Grobelak A, Grosser A, Vandenbulcke F. 2019. Gene expression, DNA damage and other stress markers in Sinapis alba L. exposed to heavy metals with special reference to sewage sludge application on contaminated sites. *Ecotoxicology and Environmental Safety*. 181: 508–517.

Jiang T, Liu W, Mao Y, Zhang L, Cheng J, Gong M, Zhao H, Dai L, Zhang S, Zhao Q. 2015. Adsorption behavior of copper ions from aqueous solution onto graphene oxide–CdS composite. *Chemical Engineering Journal*. 259: 603–610.

Jiao C, Xiong J, Tao J, Xu S, Zhang D, Lin H, Chen Y. 2016. Sodium alginate/graphene oxide aerogel with enhanced strength–toughness and its heavy metal adsorption study. *International Journal of Biological Macromolecules*. 83: 133–141.

Joseph IV, Tosheva L, Doyle AM. 2020. Simultaneous removal of Cd(II), Co(II), Cu(II), Pb(II), and Zn(II) ions from aqueous solutions via adsorption on FAU-type zeolites prepared from coal fly ash. *Journal of Environmental Chemical Engineering*. 8(4): 103895.

Kammoun M, Ghorbel I, Charfeddine S, Kamoun L, Gargouri-Bouzid R, Nouri-Ellouz O. 2017. The positive effect of phosphogypsum-supplemented composts on potato plant growth in the field and tuber yield. *Journal of Environmental Management*. 200: 475–483.

Karthik V, Selvakumar P, Senthil Kumar P, Vo D-VN, Gokulakrishnan M, Keerthana P, Tamil Elakkiya V, Rajeswari R. 2021. Graphene-based materials for environmental applications: A review. *Environmental Chemistry Letters*. 19(5): 3631–3644.

Khan I, Awan SA, Rizwan M, Ali S, Hassan MJ, Brestic M, Zhang X, Huang L. 2021. Effects of silicon on heavy metal uptake at the soil-plant interphase: A review. *Ecotoxicology and Environmental Safety*. 222: 112510.

Khan ZU, Kausar A, Ullah H, Badshah A, Khan WU. 2016. A review of graphene oxide, graphene buckypaper, and polymer/graphene composites: Properties and fabrication techniques. *Journal of Plastic Film & Sheeting*. 32(4): 336–379.

Khraisheh M, AlMomani F, Al-Ghouti M. 2021. Electrospun Al_2O_3 hydrophobic functionalized membranes for heavy metal recovery using direct contact membrane distillation. *International Journal of Energy Research*. 45(6): 8151–8167.

Khurana I, Shaw AK, Saxena A, Khurana JM, Rai PK. 2018. Removal of trinitrotoluene with nano zerovalent iron impregnated graphene oxide. *Water, Air, & Soil Pollution*. 229(1): 1–16.

Kim L, Catrina GA, Stanescu B, Pascu LF, Tanase G, Manolache D. 2019. The chemical fractions and leaching of heavy metals in ash from medical waste incineration using two different sequential extraction procedures.

Kolanthai E, Sindu PA, Khajuria DK, Veerla SC, Kuppuswamy D, Catalani LH, Mahapatra DR. 2018. Graphene oxide—A tool for the preparation of chemically crosslinking free alginate–chitosan–collagen scaffolds for bone tissue engineering. *ACS Applied Materials & Interfaces*. 10(15): 12441–12452.

Kong D, Wang N, Qiao N, Wang Q, Wang Z, Zhou Z, Ren Z. 2017. Facile preparation of ion-imprinted chitosan microspheres enwrapping Fe_3O_4 and graphene oxide by inverse suspension cross-linking for highly selective removal of copper (II). *ACS Sustainable Chemistry & Engineering*. 5(8): 7401–7409.

Kong Q, Preis S, Li L, Luo P, Wei C, Li Z, Hu Y, Wei C. 2020. Relations between metal ion characteristics and adsorption performance of graphene oxide: A comprehensive experimental and theoretical study. *Separation and Purification Technology*. 232: 115956.

Kong Q, Shi X, Ma W, Zhang F, Yu T, Zhao F, Zhao D, Wei C. 2021. Strategies to improve the adsorption properties of graphene-based adsorbent towards heavy metal ions and their compound pollutants: A review. *Journal of Hazardous Materials*. 415: 125690.

Kong Q, Wei C, Preis S, Hu Y, Wang F. 2018a. Facile preparation of nitrogen and sulfur co-doped graphene-based aerogel for simultaneous removal of Cd^{2+} and organic dyes. *Environmental Science and Pollution Research*. 25(21): 21164–21175.

Kong Q, Wei J, Hu Y, Wei C. 2019. Fabrication of terminal amino hyperbranched polymer modified graphene oxide and its prominent adsorption performance towards Cr(VI). *Journal of Hazardous Materials*. 363: 161–169.

Kong Q, Xie B, Preis S, Hu Y, Wu H, Wei C. 2018b. Adsorption of Cd^{2+} by an ion-imprinted thiol-functionalized polymer in competition with heavy metal ions and organic acids. *RSC Advances*. 8(16): 8950–8960.

Kyzas GZ, Bomis G, Kosheleva RI, Efthimiadou EK, Favvas EP, Kostoglou M, Mitropoulos AC. 2019. Nanobubbles effect on heavy metal ions adsorption by activated carbon. *Chemical Engineering Journal*. 356: 91–97.

Kyzas GZ, Deliyanni EA, Bikiaris DN, Mitropoulos AC. 2018. Graphene composites as dye adsorbents. *Chemical Engineering Research and Design*. 129: 75–88.

Le TTN, Le VT, Dao MU, Nguyen QV, Vu TT, Nguyen MH, Tran DL, Le HS. 2019. Preparation of magnetic graphene oxide/chitosan composite beads for effective removal of heavy metals and dyes from aqueous solutions. *Chemical Engineering Communications*. 206(10): 1337–1352.

Li J, Chen Y, Lu H, Zhai W. 2021a. Spatial distribution of heavy metal contamination and uncertainty-based human health risk in the aquatic environment using multivariate statistical method. *Environmental Science and Pollution Research*. 28(18): 22804–22822.

Li L, He M, Feng Y, Wei H, You X, Yu H, Wang Q, Wang J. 2021b. Adsorption of xanthate from aqueous solution by multilayer graphene oxide: An experimental and molecular dynamics simulation study. *Advanced Composites and Hybrid Materials*. 4(3): 725–732.

Li L, Liu F, Duan H, Wang X, Li J, Wang Y, Luo C. 2016. The preparation of novel adsorbent materials with efficient adsorption performance for both chromium and methylene blue. *Biointerfaces*.141: 253–259.

Li M, Feng J, Huang K, Tang S, Liu R, Li H, Ma F, Meng X. 2019a. Amino group functionalized SiO_2@graphene oxide for efficient removal of Cu(II) from aqueous solutions. *Chemical Engineering Research and Design*. 145: 235–244.

Li X, Yang Q, Ye Y, Zhang L, Hong S, Ning N, Tian M. 2020a. Quantifying 3D-nanosized dispersion of SiO_2 in elastomer nanocomposites by 3D-scanning transmission electron microscope (STEM). *Composites Part A: Applied Science and Manufacturing*. 131: 105778.

Li Y, Yu H, Liu L, Yu H. 2021c. Application of co-pyrolysis biochar for the adsorption and immobilization of heavy metals in contaminated environmental substrates. *Journal of Hazardous Materials*. 420: 126655.

Li Y, Zhao W, Weyland M, Yuan S, Xia Y, Liu H, Jian M, Yang J, Easton CD, Selomulya C. 2019b. Thermally reduced nanoporous graphene oxide membrane for desalination. *Environmental Science & Technology*. 53(14): 8314–8323.

Li Z, Young RJ, Backes C, Zhao W, Zhang X, Zhukov AA, Tillotson E, Conlan AP, Ding F, Haigh SJ. 2020b. Mechanisms of liquid-phase exfoliation for the production of graphene. *ACS Nano*. 14(9): 10976–10985.

Lian M, Wang J, Sun L, Xu Z, Tang J, Yan J, Zeng X. 2019. Profiles and potential health risks of heavy metals in soil and crops from the watershed of Xi River in Northeast China. *Ecotoxicology and Environmental Safety*. 169: 442–448.

Lim JY, Mubarak N, Abdullah E, Nizamuddin S, Khalid M. 2018a. Recent trends in the synthesis of graphene and graphene oxide based nanomaterials for removal of heavy metals—A review. *Journal of Industrial and Engineering Chemistry*. 66: 29–44.

Lim JY, Mubarak N, Khalid M, Abdullah E, Arshid N. 2018b. Novel fabrication of functionalized graphene oxide via magnetite and 1-butyl-3-methylimidazolium tetrafluoroborate. *Nano-Structures & Nano-Objects*. 16: 403–411.

Liu A, Liu J, Han J, Zhang W-x. 2017a. Evolution of nanoscale zero-valent iron (nZVI) in water: Microscopic and spectroscopic evidence on the formation of nano-and micro-structured iron oxides. *Journal of Hazardous Materials*. 322: 129–135.

Liu B, Ai S, Zhang W, Huang D, Zhang Y. 2017b. Assessment of the bioavailability, bioaccessibility and transfer of heavy metals in the soil-grain-human systems near a mining and smelting area in NW China. *Science of the Total Environment*. 609: 822–829.

Liu C, Liu Y, Feng C, Wang P, Yu L, Liu D, Sun S, Wang F. 2021. Distribution characteristics and potential risks of heavy metals and antimicrobial resistant Escherichia coli in dairy farm wastewater in Tai'an, China. *Chemosphere*. 262: 127768.

Liu F, Shan C, Zhang X, Zhang Y, Zhang W, Pan B. 2017c. Enhanced removal of EDTA-chelated Cu(II) by polymeric anion-exchanger supported nanoscale zero-valent iron. *Journal of Hazardous Materials*. 321: 290–298.

Liu M-L, Guo J-L, Japip S, Jia T-Z, Shao D-D, Zhang S, Li W-J, Wang J, Cao X-L, Sun S-P. 2019a. One-step enhancement of solvent transport, stability and photocatalytic properties of graphene oxide/polyimide membranes with multifunctional cross-linkers. *Journal of Materials Chemistry A*. 7(7): 3170–3178.

Liu S, Wang H, Chai L, Li M. 2016. Effects of single-and multi-organic acid ligands on adsorption of copper by Fe_3O_4/graphene oxide-supported DCTA. *Journal of Colloid and Interface Science*. 478: 288–295.

Liu W, Wang D, Soomro RA, Fu F, Qiao N, Yu Y, Wang R, Xu B. 2019b. Ceramic supported attapulgite-graphene oxide composite membrane for efficient removal of heavy metal contamination. *Journal of Membrane Science*. 591: 117323.

Liu X, Ma R, Wang X, Ma Y, Yang Y, Zhuang L, Zhang S, Jehan R, Chen J, Wang X. 2019c. Graphene oxide-based materials for efficient removal of heavy metal ions from aqueous solution: A review. *Environmental Pollution*. 252: 62–73.

Liu Y, Meng X, Liu Z, Meng M, Jiang F, Luo M, Ni L, Qiu J, Liu F, Zhong G. 2015. Preparation of a two-dimensional ion-imprinted polymer based on a graphene oxide/SiO_2 composite for the selective adsorption of nickel ions. *Langmuir*. 31(32): 8841–8851.

Lv X, Qin X, Wang K, Peng Y, Wang P, Jiang G. 2019. Nanoscale zero valent iron supported on MgAl-LDH-decorated reduced graphene oxide: Enhanced performance in Cr(VI) removal, mechanism and regeneration. *Journal of Hazardous Materials*. 373: 176–186.

Ma G, Zhang M, Zhu L, Chen H, Liu X, Lu C. 2018a. Facile synthesis of amine-functional reduced graphene oxides as modified quick, easy, cheap, effective, rugged and safe adsorbent for multi-pesticide residues analysis of tea. *Journal of Chromatography A*. 1531: 22–31.

Ma J, He Y, Zeng G, Li F, Li Y, Xiao J, Yang S. 2018b. Bio-inspired method to fabricate poly-dopamine/reduced graphene oxide composite membranes for dyes and heavy metal ion removal. *Polymers for Advanced Technologies*. 29(2): 941–950.

Ma Y-X, Shao W-J, Sun W, Kou Y-L, Li X, Yang H-P. 2018c. One-step fabrication of β-cyclodextrin modified magnetic graphene oxide nanohybrids for adsorption of Pb(II), Cu(II) and methylene blue in aqueous solutions. *Applied Surface Science*. 459: 544–553.

Maamoun I, Eljamal O, Khalil AM, Sugihara Y, Matsunaga N. 2018. Phosphate removal through nano-zerovalent iron permeable reactive barrier; column experiment and reactive solute transport modeling. *Transport in Porous Media*. 125(2): 395–412.

Mahmoudian M, Nozad E, Hosseinzadeh M. 2018. Characterization of EDTA functionalized graphene oxide/ polyethersulfone (FGO/PES) nanocomposite membrane and using for elimination of heavy metal and dye contaminations. 폴리머. 42(3): 434–445.

Makertihartha I, Rizki Z, Zunita M, Dharmawijaya PT. 2017. Graphene based nanofiltration for mercury removal from aqueous solutions. *Advanced Science Letters*. 23(6): 5684–5686.

Mansoorianfar M, Khataee A, Riahi Z, Shahin K, Asadnia M, Razmjou A, Hojjati-Najafabadi A, Mei C, Orooji Y, Li D. 2020. Scalable fabrication of tunable titanium nanotubes via sonoelectrochemical process for biomedical applications. *Ultrasonics Sonochemistry*. 64: 104783.

Martínez-Cortijo J, Ruiz-Canales A. 2018. Effect of heavy metals on rice irrigated fields with waste water in high pH Mediterranean soils: The particular case of the Valencia area in Spain. *Agricultural Water Management*. 210: 108–123.

Maury-Brachet R, Gentes S, Dassié EP, Feurtet-Mazel A, Vigouroux R, Laperche V, Gonzalez P, Hanquiez V, Mesmer-Dudons N, Durrieu G. 2020. Mercury contamination levels in the bioindicator piscivorous fish Hoplias aïmara in French Guiana rivers: Mapping for risk assessment. *Environmental Science and Pollution Research*. 27(4): 3624–3636.

Mishra S, Singh AK, Singh JK. 2020. Ferrous sulfide and carboxyl-functionalized ferroferric oxide incorporated PVDF-based nanocomposite membranes for simultaneous removal of highly toxic heavy-metal ions from industrial ground water. *Journal of Membrane Science*. 593: 117422.

Modi A, Bellare J. 2019. Efficiently improved oil/water separation using high flux and superior antifouling polysulfone hollow fiber membranes modified with functionalized carbon nanotubes/graphene oxide nanohybrid. *Journal of Environmental Chemical Engineering*. 7(2): 102944.

Mollahosseini A, Khadir A, Saeidian J. 2019. Core–shell polypyrrole/Fe_3O_4 nanocomposite as sorbent for magnetic dispersive solid-phase extraction of Al +3 ions from solutions: Investigation of the operational parameters. *Journal of Water Process Engineering*. 29: 100795.

Moradi G, Zinadini S, Rajabi L, Derakhshan AA. 2020. Removal of heavy metal ions using a new high performance nanofiltration membrane modified with curcumin boehmite nanoparticles. *Chemical Engineering Journal*. 390: 124546.

Mu Y, Jia F, Ai Z, Zhang L. 2017. Iron oxide shell mediated environmental remediation properties of nano zerovalent iron. *Environmental Science: Nano*. 4(1): 27–45.

Muhammad S, Ahmad K. 2020. Heavy metal contamination in water and fish of the Hunza River and its tributaries in Gilgit–Baltistan: Evaluation of potential risks and provenance. *Environmental Technology & Innovation*. 20: 101159.

Nagaraj A, Munusamy MA, Al-Arfaj AA, Rajan M. 2018. Functional ionic liquid-capped graphene quantum dots for chromium removal from chromium contaminated water. *Journal of Chemical & Engineering Data*. 64(2): 651–667.

Nagarajan S, Abessolo Ondo D, Gassara S, Bechelany M, Balme S, Miele P, Kalkura N, Pochat-Bohatier C. 2018. Porous gelatin membrane obtained from pickering emulsions stabilized by graphene oxide. *Langmuir*. 34(4): 1542–1549.

Ngueagni PT, Woumfo ED, Kumar PS, Siéwé M, Vieillard J, Brun N, Nkuigue PF. 2020. Adsorption of Cu(II) ions by modified horn core: Effect of temperature on adsorbent preparation and extended application in river water. *Journal of Molecular Liquids*. 298: 112023.

Nicomel NR, Leus K, Folens K, Van Der Voort P, Du Laing G. 2016. Technologies for arsenic removal from water: Current status and future perspectives. *International Journal of Environmental Research and Public Health*. 13(1): 62.

Nie L, Chuah CY, Bae TH, Lee JM. 2021. Graphene-based advanced membrane applications in organic solvent nanofiltration. *Advanced Functional Materials*. 31(6): 2006949.

Nie L, Wang C, Hou R, Li X, Sun M, Suo J, Wang Z, Cai R, Yin B, Fang L. 2019. Preparation and characterization of dithiol-modified graphene oxide nanosheets reinforced alginate nanocomposite as bone scaffold. *Sn Applied Sciences*. 1(6): 1–16.

Nongbe MC, Bretel G, Ekou T, Ekou L, Yao BK, Le Grognec E, Felpin F-X. 2018. Cellulose paper grafted with polyamines as powerful adsorbent for heavy metals. *Cellulose*. 25(7): 4043–4055.

Novoselov KS, Geim AK, Morozov SV, Jiang D-e, Zhang Y, Dubonos SV, Grigorieva IV, Firsov AA. 2004. Electric field effect in atomically thin carbon films. *Science*. 306(5696): 666–669.

O'Connor D, Zheng X, Hou D, Shen Z, Li G, Miao G, O'Connell S, Guo M. 2019. Phytoremediation: Climate change resilience and sustainability assessment at a coastal brownfield redevelopment. *Environment International*. 130: 104945.

Odhong C, Wilkes A, van Dijk S, Vorlaufer M, Ndonga S, Sing'ora B, Kenyanito L. 2019. Financing large-scale mitigation by smallholder farmers: What roles for public climate finance? *Frontiers in Sustainable Food Systems*. 3: 3.

Ou X, Yang X, Zheng J, Liu M. 2019. Free-standing graphene oxide–chitin nanocrystal composite membrane for dye adsorption and oil/water separation. *ACS Sustainable Chemistry & Engineering*. 7(15): 13379–13390.

Palos-Barba V, Moreno-Martell A, Hernández-Morales V, Peza-Ledesma CL, Rivera-Muñoz EM, Nava R, Pawelec B. 2020. SBA-16 cage-like porous material modified with APTES as an adsorbent for Pb^{2+} ions removal from aqueous solution. *Materials*. 13(4): 927.

Pan L, Wang Z, Yang Q, Huang R. 2018. Efficient removal of lead, copper and cadmium ions from water by a porous calcium alginate/graphene oxide composite aerogel. *Nanomaterials*. 8(11): 957.

Pap S, Knudsen TŠ, Radonić J, Maletić S, Igić SM, Sekulić MT. 2017. Utilization of fruit processing industry waste as green activated carbon for the treatment of heavy metals and chlorophenols contaminated water. *Journal of Cleaner Production*. 162: 958–972.

Papageorgiou DG, Kinloch IA, Young RJ. 2017. Mechanical properties of graphene and graphene-based nanocomposites. *Progress in Materials Science*. 90: 75–127.

Parastar M, Sheshmani S, Shokrollahzadeh S. 2021. Cross-linked chitosan into graphene oxide-iron(III) oxide hydroxide as nano-biosorbent for Pd(II) and Cd(II) removal. International *Journal of Biological Macromolecules*. 166: 229–237.

Paschoalini A, Savassi L, Arantes F, Rizzo E, Bazzoli N. 2019. Heavy metals accumulation and endocrine disruption in Prochilodus argenteus from a polluted neotropical river. *Ecotoxicology and Environmental Safety*. 169: 539–550.

Pavesi T, Moreira JC. 2020. Mechanisms and individuality in chromium toxicity in humans. *Journal of Applied Toxicology*. 40(9): 1183–1197.

Peer FE, Bahramifar N, Younesi H. 2018. Removal of Cd(II), Pb(II) and Cu(II) ions from aqueous solution by polyamidoamine dendrimer grafted magnetic graphene oxide nanosheets. *Journal of the Taiwan Institute of Chemical Engineers*. 87: 225–240.

Peng W, Li H, Liu Y, Song S. 2017. A review on heavy metal ions adsorption from water by graphene oxide and its composites. *Journal of Molecular Liquids*. 230: 496–504.

Perreault F, De Faria AF, Elimelech M. 2015. Environmental applications of graphene-based nanomaterials. *Chemical Society Reviews*. 44(16): 5861–5896.

Piumie R, Power A, Chandra S, Chapman J. 2018. Graphene, electrospun membranes and granular activated carbon for eliminating heavy metals, pesticides and bacteria in water and wastewater treatment processes. *Analyst*. 143(23): 5629–5645.

Qalyoubi L, Al-Othman A, Al-Asheh S. 2021. Recent progress and challenges of adsorptive membranes for the removal of pollutants from wastewater. Part II: Environmental applications. *Case Studies in Chemical and Environmental Engineering*. 3: 100102.

Qiao D, Li Z, Duan J, He X. 2020. Adsorption and photocatalytic degradation mechanism of magnetic graphene oxide/ZnO nanocomposites for tetracycline contaminants. *Chemical Engineering Journal*. 400: 125952.

Radi S, El Abiad C, Moura NM, Faustino MA, Neves MGP. 2019. New hybrid adsorbent based on porphyrin functionalized silica for heavy metals removal: Synthesis, characterization, isotherms, kinetics and thermodynamics studies. *Journal of Hazardous Materials*. 370: 80–90.

Rai PK, Lee SS, Zhang M, Tsang YF, Kim K-H. 2019. Heavy metals in food crops: Health risks, fate, mechanisms, and management. *Environment International*. 125: 365–385.

Rao Z, Feng K, Tang B, Wu P. 2017. Surface decoration of amino-functionalized metal–organic framework/graphene oxide composite onto polydopamine-coated membrane substrate for highly efficient heavy metal removal. *ACS Applied Materials & Interfaces*. 9(3): 2594–2605.

Rathour RKS, Bhattacharya J, Mukherjee A. 2020. Selective and multicycle removal of Cr(VI) by graphene oxide–EDTA composite: Insight into the removal mechanism and ionic interference in binary and ternary associations. *Environmental Technology & Innovation*. 19: 100851.

Ren H, Gao Z, Wu D, Jiang J, Sun Y, Luo C. 2016. Efficient Pb(II) removal using sodium alginate–carboxymethyl cellulose gel beads: Preparation, characterization, and adsorption mechanism. *Carbohydrate Polymers*. 137: 402–409.

Ren L, Dong J, Chi Z, Huang H. 2018. Reduced graphene oxide-nano zero value iron (rGO-nZVI) microelectrolysis accelerating Cr(VI) removal in aquifer. *Journal of Environmental Sciences*. 73: 96–106.

Reyes-Hinojosa D, Lozada-Pérez C, Cuevas YZ, López-Reyes A, Martínez-Nava G, Fernández-Torres J, Olivos-Meza A, Landa-Solis C, Gutiérrez-Ruiz M, Del Castillo ER. 2019. Toxicity of cadmium in musculoskeletal diseases. *Environmental Toxicology and Pharmacology*. 72: 103219.

RoyChoudhury P, Majumdar S, Sarkar S, Kundu B, Sahoo GC. 2019. Performance investigation of Pb(II) removal by synthesized hydroxyapatite based ceramic ultrafiltration membrane: Bench scale study. *Chemical Engineering Journal*. 355: 510–519.

Saeed M, Alshammari Y, Majeed SA, Al-Nasrallah E. 2020. Chemical vapour deposition of graphene—Synthesis, characterisation, and applications: A review. *Molecules*. 25(17): 3856.

Saeedi-Jurkuyeh A, Jafari AJ, Kalantary RR, Esrafili A. 2020. A novel synthetic thin-film nanocomposite forward osmosis membrane modified by graphene oxide and polyethylene glycol for heavy metals removal from aqueous solutions. *Reactive and Functional Polymers*. 146: 104397.

Saheed IO, Da Oh W, Suah FBM. 2021. Chitosan modifications for adsorption of pollutants–A review. *Journal of Hazardous Materials*. 408: 124889.

Saleh TA. 2020. Characterization, determination and elimination technologies for sulfur from petroleum: Toward cleaner fuel and a safe environment. *Trends in Environmental Analytical Chemistry*. 25: e00080.

Saleh TA. 2021. Protocols for synthesis of nanomaterials, polymers, and green materials as adsorbents for water treatment technologies. *Environmental Technology & Innovation*. 24: 101821.

Sall ML, Diaw AKD, Gningue-Sall D, Efremova Aaron S, Aaron J-J. 2020. Toxic heavy metals: Impact on the environment and human health, and treatment with conducting organic polymers, a review. *Environmental Science and Pollution Research*. 27: 29927–29942.

Samuel MS, Shah SS, Bhattacharya J, Subramaniam K, Singh NP. 2018a. Adsorption of Pb(II) from aqueous solution using a magnetic chitosan/graphene oxide composite and its toxicity studies. *International Journal of Biological Macromolecules*. 115: 1142–1150.

Samuel MS, Shah SS, Subramaniyan V, Qureshi T, Bhattacharya J, Singh NP. 2018b. Preparation of graphene oxide/chitosan/ferrite nanocomposite for chromium (VI) removal from aqueous solution. *International Journal of Biological Macromolecules*. 119: 540–547.

Samuel MS, Subramaniyan V, Bhattacharya J, Parthiban C, Chand S, Singh NP. 2018c. A GO-CS@MOF [Zn(BDC)(DMF)] material for the adsorption of chromium (VI) ions from aqueous solution. *Composites Part B: Engineering*. 152: 116–125.

Selvi A, Rajasekar A, Theerthagiri J, Ananthaselvam A, Sathishkumar K, Madhavan J, Rahman PK. 2019. Integrated remediation processes toward heavy metal removal/recovery from various environments-A review. *Frontiers in Environmental Science*. 7: 66.

Shan H, Peng S, Zhao C, Zhan H, Zeng C. 2020. Highly efficient removal of As(III) from aqueous solutions using goethite/graphene oxide/chitosan nanocomposite. *International Journal of Biological Macromolecules*. 164: 13–26.

Shao J, Yu X, Zhou M, Cai X, Yu C. 2018. Nanoscale zero-valent iron decorated on bentonite/graphene oxide for removal of copper ions from aqueous solution. *Materials*. 11(6): 945.

Sharif A, Khorasani M, Shemirani F. 2018. Nanocomposite bead (NCB) based on bio-polymer alginate caged magnetic graphene oxide synthesized for adsorption and preconcentration of lead (II) and copper (II) ions from urine, saliva and water samples. *Journal of Inorganic and Organometallic Polymers and Materials*. 28(6): 2375–2387.

Sherlala A, Raman A, Bello M, Asghar A. 2018. A review of the applications of organo-functionalized magnetic graphene oxide nanocomposites for heavy metal adsorption. *Chemosphere*. 193: 1004–1017.

Shukla AK, Alam J, Alhoshan M, Dass LA, Ali FAA, Mishra U, Ansari MA. 2018. Removal of heavy metal ions using a carboxylated graphene oxide-incorporated polyphenylsulfone nanofiltration membrane. *Environmental Science: Water Research & Technology*. 4(3): 438–448.

Siddeeg SM, Tahoon MA, Alsaiari NS, Shabbir M, Rebah FB. 2021. Application of functionalized nanomaterials as effective adsorbents for the removal of heavy metals from wastewater: A review. *Current Analytical Chemistry*. 17(1): 4–22.

Smaali A, Berkani M, Merouane F, Vasseghian Y, Rahim N, Kouachi M. 2021. Photocatalytic-persulfate-oxidation for diclofenac removal from aqueous solutions: Modeling, optimization and biotoxicity test assessment. *Chemosphere*. 266: 129158.

Song H, Liu W, Meng F, Yang Q, Guo N. 2021. Efficient sequestration of hexavalent chromium by graphene-based nanoscale zero-valent iron composite coupled with ultrasonic pretreatment. *International Journal of Environmental Research and Public Health*. 18(11): 5921.

Song W, Hu J, Zhao Y, Shao D, Li J. 2019. Correction: Efficient removal of cobalt from aqueous solution using β-cyclodextrin modified graphene oxide. *RSC Advances*. 9(70): 40975–40976.

Subedi N, Lähde A, Abu-Danso E, Iqbal J, Bhatnagar A. 2019. A comparative study of magnetic chitosan (Chi@Fe$_3$O$_4$) and graphene oxide modified magnetic chitosan (Chi@Fe$_3$O$_4$GO) nanocomposites for efficient removal of Cr(VI) from water. *International Journal of Biological Macromolecules*. 137: 948–959.

Sunil K, Karunakaran G, Yadav S, Padaki M, Zadorozhnyy V, Pai RK. 2018. Al-Ti$_2$O$_6$ a mixed metal oxide based composite membrane: A unique membrane for removal of heavy metals. *Chemical Engineering Journal*. 348: 678–684.

Tadjarodi A, Ferdowsi SM, Zare-Dorabei R, Barzin A. 2016. Highly efficient ultrasonic-assisted removal of Hg(II) ions on graphene oxide modified with 2-pyridinecarboxaldehyde thiosemicarbazone: Adsorption isotherms and kinetics studies. *Ultrasonics Sonochemistry*. 33: 118–128.

Talukdar D, Jasrotia T, Sharma R, Jaglan S, Kumar R, Vats R, Kumar R, Mahnashi MH, Umar A. 2020. Evaluation of novel indigenous fungal consortium for enhanced bioremediation of heavy metals from contaminated sites. *Environmental Technology & Innovation*. 20: 101050.

Tan L, Wang S, Du W, Hu T. 2016. Effect of water chemistries on adsorption of Cs(I) onto graphene oxide investigated by batch and modeling techniques. *Chemical Engineering Journal*. 292: 92–97.

Tan P, Sun J, Hu Y, Fang Z, Bi Q, Chen Y, Cheng J. 2015. Adsorption of Cu^{2+}, Cd^{2+} and Ni^{2+} from aqueous single metal solutions on graphene oxide membranes. *Journal of Hazardous Materials*. 297: 251–260.

Thakur K, Kandasubramanian B. 2019. Graphene and graphene oxide-based composites for removal of organic pollutants: A review. *Journal of Chemical & Engineering Data*. 64(3): 833–867.

Thakur S, Pandey S, Arotiba OA. 2016. Development of a sodium alginate-based organic/inorganic superabsorbent composite hydrogel for adsorption of methylene blue. *Carbohydrate Polymers*. 153: 34–46.

Tsarev S, Collins RN, Ilton ES, Fahy A, Waite TD. 2017. The short-term reduction of uranium by nanoscale zero-valent iron (nZVI): Role of oxide shell, reduction mechanism and the formation of U(V)-carbonate phases. *Environmental Science: Nano*. 4(6): 1304–1313.

Vasseghian Y, Berkani M, Almomani F, Dragoi E-N. 2021a. Data mining for pesticide decontamination using heterogeneous photocatalytic processes. *Chemosphere*. 270: 129449.

Vasseghian Y, Dragoi E-N, Almomani F, Berkani M. 2021b. Graphene-based membrane techniques for heavy metal removal: A critical review. *Environmental Technology & Innovation*. 24: 101863.

Velusamy S, Roy A, Sundaram S, Kumar Mallick T. 2021. A review on heavy metal ions and containing dyes removal through graphene oxide-based adsorption strategies for textile wastewater treatment. *The Chemical Record*. 21(7): 1570–1610.

Verma M, Lee I, Oh J, Kumar V, Kim H. 2022. Synthesis of EDTA-functionalized graphene oxide-chitosan nanocomposite for simultaneous removal of inorganic and organic pollutants from complex wastewater. *Chemosphere*. 287: 132385.

Wang B, Wan Y, Zheng Y, Lee X, Liu T, Yu Z, Huang J, Ok YS, Chen J, Gao B. 2019. Alginate-based composites for environmental applications: A critical review. *Critical Reviews in Environmental Science and Technology*. 49(4): 318–356.

Wang F, Lu X, Li X-y. 2016. Selective removals of heavy metals (Pb^{2+}, Cu^{2+}, and Cd^{2+}) from wastewater by gelation with alginate for effective metal recovery. *Journal of Hazardous Materials*. 308: 75–83.

Wang L, Hu D, Kong X, Liu J, Li X, Zhou K, Zhao H, Zhou C. 2018a. Anionic polypeptide poly (γ-glutamic acid)-functionalized magnetic Fe$_3$O$_4$-GO-(o-MWCNTs) hybrid nanocomposite for high-efficiency removal of Cd(II), Cu(II) and Ni(II) heavy metal ions. *Chemical Engineering Journal*. 346: 38–49.

Wang S, Li X, Liu Y, Zhang C, Tan X, Zeng G, Song B, Jiang L. 2018b. Nitrogen-containing amino compounds functionalized graphene oxide: Synthesis, characterization and application for the removal of pollutants from wastewater: A review. *Journal of Hazardous Materials*. 342: 177–191.

Wang X, Wang A, Ma J. 2017a. Visible-light-driven photocatalytic removal of antibiotics by newly designed C$_3$N$_4$@MnFe$_2$O$_4$-graphene nanocomposites. *Journal of Hazardous Materials*. 336: 81–92.

Wang Y, Tseng L-T, Murmu PP, Bao N, Kennedy J, Ionesc M, Ding J, Suzuki K, Li S, Yi J. 2017b. Defects engineering induced room temperature ferromagnetism in transition metal doped MoS$_2$. *Materials & Design*. 121: 77–84.

Wang Z, Sim A, Urban JJ, Mi B. 2018c. Removal and recovery of heavy metal ions by two-dimensional MoS$_2$ nanosheets: Performance and mechanisms. *Environmental Science & Technology*. 52(17): 9741–9748.

White RL, White CM, Turgut H, Massoud A, Tian ZR. 2018. Comparative studies on copper adsorption by graphene oxide and functionalized graphene oxide nanoparticles. *Journal of the Taiwan Institute of Chemical Engineers*. 85: 18–28.

Williams GV, Prakash T, Kennedy J, Chong SV, Rubanov S. 2018. Spin-dependent tunnelling in magnetite nanoparticles. *Journal of Magnetism and Magnetic Materials*. 460: 229–233.

Wu Y, Pang H, Liu Y, Wang X, Yu S, Fu D, Chen J, Wang X. 2019. Environmental remediation of heavy metal ions by novel-nanomaterials: A review. *Environmental Pollution*. 246: 608–620.

Wu Y, Pang H, Yao W, Wang X, Yu S, Yu Z, Wang X. 2018. Synthesis of rod-like metal-organic framework (MOF-5) nanomaterial for efficient removal of U(VI): Batch experiments and spectroscopy study. *Science Bulletin*. 63(13): 831–839.

Xia Z, Baird L, Zimmerman N, Yeager M. 2017. Heavy metal ion removal by thiol functionalized aluminum oxide hydroxide nanowhiskers. *Applied Surface Science*. 416: 565–573.

Xiao W, Yan B, Zeng H, Liu Q. 2016. Dendrimer functionalized graphene oxide for selenium removal. *Carbon*. 105: 655–664.

Xu J, Cao Z, Zhang Y, Yuan Z, Lou Z, Xu X, Wang X. 2018a. A review of functionalized carbon nanotubes and graphene for heavy metal adsorption from water: Preparation, application, and mechanism. *Chemosphere*. 195: 351–364.

Xu W, Song Y, Dai K, Sun S, Liu G, Yao J. 2018b. Novel ternary nanohybrids of tetraethylenepentamine and graphene oxide decorated with $MnFe_2O_4$ magnetic nanoparticles for the adsorption of Pb(II). *Journal of Hazardous Materials*. 358: 337–345.

Yan A, Wang Y, Tan SN, Mohd Yusof ML, Ghosh S, Chen Z. 2020. Phytoremediation: A promising approach for revegetation of heavy metal-polluted land. *Frontiers in Plant Science*. 11: 359.

Yang F, Massey IY. 2019. Exposure routes and health effects of heavy metals on children. *Biometals*. 32(4): 563–573.

Yang H, Wang N, Wang L, Liu H-X, An Q-F, Ji S. 2018a. Vacuum-assisted assembly of ZIF-8@GO composite membranes on ceramic tube with enhanced organic solvent nanofiltration performance. *Journal of Membrane Science*. 545: 158–166.

Yang S, Li Y, Wang S, Wang M, Chu M, Xia B. 2018b. Advances in the use of carbonaceous materials for the electrochemical determination of persistent organic pollutants. A review. *Microchimica Acta*. 185(2): 1–14.

Yang X, Wang Z, Shao L. 2018c. Construction of oil-unidirectional membrane for integrated oil collection with lossless transportation and oil-in-water emulsion purification. *Journal of Membrane Science*. 549: 67–74.

Yang X, Xia L, Song S. 2017. Arsenic adsorption from water using graphene-based materials as adsorbents: A critical review. *Surface Review and Letters*. 24(01): 1730001.

Yang X, Zhou T, Ren B, Hursthouse A, Zhang Y. 2018d. Removal of Mn(II) by sodium alginate/graphene oxide composite double-network hydrogel beads from aqueous solutions. *Scientific Reports*. 8(1): 1–16.

Yang Z, Miao H, Rui Z, Ji H. 2019. Enhanced formaldehyde removal from air using fully biodegradable chitosan grafted β-cyclodextrin adsorbent with weak chemical interaction. *Polymers*. 11(2): 276.

Yao M, Wang Z, Liu Y, Yang G, Chen J. 2019. Preparation of dialdehyde cellulose graftead graphene oxide composite and its adsorption behavior for heavy metals from aqueous solution. *Carbohydrate Polymers*. 212: 345–351.

Yu S, Liu Y, Ai Y, Wang X, Zhang R, Chen Z, Chen Z, Zhao G, Wang X. 2018. Rational design of carbonaceous nanofiber/Ni-Al layered double hydroxide nanocomposites for high-efficiency removal of heavy metals from aqueous solutions. *Environmental Pollution*. 242: 1–11.

Yu Y, Liu L, Chen X, Xiang M, Li Z, Liu Y, Zeng Y, Han Y, Yu Z. 2021. Brominated flame retardants and heavy metals in common aquatic products from the pearl river delta, south china: Bioaccessibility assessment and human health implications. *Journal of Hazardous Materials*. 403: 124036.

Yu Y, Zhang G, Ye L. 2019. Preparation and adsorption mechanism of polyvinyl alcohol/graphene oxide-sodium alginate nanocomposite hydrogel with high Pb(II) adsorption capacity. *Journal of Applied Polymer Science*. 136(14): 47318.

Yusuf M, Elfghi F, Zaidi SA, Abdullah E, Khan MA. 2015. Applications of graphene and its derivatives as an adsorbent for heavy metal and dye removal: A systematic and comprehensive overview. *RSC Advances*. 5(62): 50392–50420.

Zaaba N, Foo K, Hashim U, Tan S, Liu W-W, Voon C. 2017. Synthesis of graphene oxide using modified hummers method: Solvent influence. *Procedia Engineering*. 184: 469–477.

Zahoor M, Arshad A, Khan Y, Iqbal M, Bajwa SZ, Soomro RA, Ahmad I, Butt FK, Iqbal MZ, Wu A. 2018. Enhanced photocatalytic performance of CeO_2–TiO_2 nanocomposite for degradation of crystal violet dye and industrial waste effluent. *Applied Nanoscience*. 8(5): 1091–1099.

Zamora-Ledezma C, Negrete-Bolagay D, Figueroa F, Zamora-Ledezma E, Ni M, Alexis F, Guerrero VH. 2021. Heavy metal water pollution: A fresh look about hazards, novel and conventional remediation methods. *Environmental Technology & Innovation*. 22: 101504.

Zare-Dorabei R, Ferdowsi SM, Barzin A, Tadjarodi A. 2016. Highly efficient simultaneous ultrasonic-assisted adsorption of Pb(II), Cd(II), Ni(II) and Cu(II) ions from aqueous solutions by graphene oxide modified with 2,2′-dipyridylamine: Central composite design optimization. *Ultrasonics Sonochemistry*. 32: 265–276.

Zarei A, Saedi S. 2018. Synthesis and application of Fe_3O_4@SiO_2@carboxyl-terminated PAMAM dendrimer nanocomposite for heavy metal removal. *Journal of Inorganic and Organometallic Polymers and Materials.* 28(6): 2835–2843.

Zhang G, Zhou M, Xu Z, Jiang C, Shen C, Meng Q. 2019a. Guanidyl-functionalized graphene/polysulfone mixed matrix ultrafiltration membrane with superior permselective, antifouling and antibacterial properties for water treatment. *Journal of Colloid and Interface Science.* 540: 295–305.

Zhang L, Luo H, Liu P, Fang W, Geng J. 2016. A novel modified graphene oxide/chitosan composite used as an adsorbent for Cr(VI) in aqueous solutions. *International Journal of Biological Macromolecules.* 87: 586–596.

Zhang M, Cui J, Lu T, Tang G, Wu S, Ma W, Huang C. 2021. Robust, functionalized reduced graphene-based nanofibrous membrane for contaminated water purification. *Chemical Engineering Journal.* 404: 126347.

Zhang M, Ma X, Li J, Huang R, Guo L, Zhang X, Fan Y, Xie X, Zeng G. 2019b. Enhanced removal of As(III) and As(V) from aqueous solution using ionic liquid-modified magnetic graphene oxide. *Chemosphere.* 234: 196–203.

Zhang M, Sun M, Wang J, Yan X, Hu X, Zhong J, Zhu X, Liu X. 2020a. Geographical distribution and risk assessment of heavy metals: A case study of mine tailings pond. *Chemistry and Ecology.* 36(1): 1–15.

Zhang Q, Hou Q, Huang G, Fan Q. 2020b. Removal of heavy metals in aquatic environment by graphene oxide composites: A review. *Environmental Science and Pollution Research.* 27(1): 190–209.

Zhang X, Liu Y, Sun C, Ji H, Zhao W, Sun S, Zhao C. 2015. Graphene oxide-based polymeric membranes for broad water pollutant removal. *RSC Advances.* 5(122): 100651–100662.

Zhang Y, Peng W, Xia L, Song S. 2017. Adsorption of Cd(II) at the interface of water and graphene oxide prepared from flaky graphite and amorphous graphite. *Journal of Environmental Chemical Engineering.* 5(4): 4157–4164.

Zhang Y, Ruan H, Guo C, Liao J, Shen J, Gao C. 2020c. Thin-film nanocomposite reverse osmosis membranes with enhanced antibacterial resistance by incorporating p-aminophenol-modified graphene oxide. *Separation and Purification Technology.* 234: 116017.

Zhao D, Gao X, Chen S, Xie F, Feng S, Alsaedi A, Hayat T, Chen C. 2018. Interaction between U(VI) with sulfhydryl groups functionalized graphene oxides investigated by batch and spectroscopic techniques. *Journal of Colloid and Interface Science.* 524: 129–138.

Zhao D, Gao X, Wu C, Xie R, Feng S, Chen C. 2016. Facile preparation of amino functionalized graphene oxide decorated with Fe_3O_4 nanoparticles for the adsorption of Cr(VI). *Applied Surface Science.* 384: 1–9.

Zheng Y, Cheng B, You W, Yu J, Ho W. 2019. 3D hierarchical graphene oxide-NiFe LDH composite with enhanced adsorption affinity to Congo red, methyl orange and Cr(VI) ions. *Journal of Hazardous Materials.* 369: 214–225.

Zhong W, Zhang Y, Wu Z, Yang R, Chen X, Yang J, Zhu L. 2018. Health risk assessment of heavy metals in freshwater fish in the central and eastern North China. *Ecotoxicology and Environmental Safety.* 157: 343–349.

Zhou C, Xu P, Lai C, Zhang C, Zeng G, Huang D, Cheng M, Hu L, Xiong W, Wen X. 2019. Rational design of graphic carbon nitride copolymers by molecular doping for visible-light-driven degradation of aqueous sulfamethazine and hydrogen evolution. *Chemical Engineering Journal.* 359: 186–196.

Zhou C, Zhu H, Wang Q, Wang J, Cheng J, Guo Y, Zhou X, Bai R. 2017. Adsorption of mercury (II) with an Fe_3O_4 magnetic polypyrrole–graphene oxide nanocomposite. *RSC Advances.* 7(30): 18466–18479.

Zhou S, Xie Y, Zhu F, Gao Y, Liu Y, Tang Z, Duan Y. 2021a. Amidoxime modified chitosan/graphene oxide composite for efficient adsorption of U(VI) from aqueous solutions. *Journal of Environmental Chemical Engineering.* 9(6): 106363.

Zhou X-Q, Carbo-Bague I, Siegler MA, Hilgendorf J, Basu U, Ott I, Liu R, Zhang L, Ramu V, IJzerman AP. 2021b. Rollover cyclometalation vs. nitrogen coordination in tetrapyridyl anticancer gold(III) complexes: Effect on protein interaction and toxicity. *JACS Au.* 1(4): 380–395.

Zinadini S, Zinatizadeh AA, Rahimi M, Vatanpour V, Zangeneh H. 2014. Preparation of a novel antifouling mixed matrix PES membrane by embedding graphene oxide nanoplates. *Journal of Membrane Science.* 453: 292–301.

Zunita M, Makertiharta I, Irawanti R, Prasetya N, Wenten I, editors. 2018. *IOP Conference Series: Materials Science and Engineering.* 395: 012021.

Zunitaa M, Irawantia R, Koesmawatib TA, Lugitoa G, Wentena IG. 2020. Graphene oxide (GO) membrane in removing heavy metals from wastewater: A review. *Chemical Engineering.* 82: 415–420.

11 Heavy Metal Removal from Wastewater via Photocatalytic Membrane

Shahnaz Ghasemi

11.1 INTRODUCTION

A large number of toxic pollutants with low biodegradability have been found in surface water, groundwater, and municipal-treated wastewater [1]. Such complex contaminants cannot be degraded and removed by existing treatment processes. One of the efficient ways to destroy organic pollutants is through advanced oxidation processes (AOPs) [2]. Hydroxyl radicals (OH) generated through AOPs as non-selective oxidizing agents can break down all contaminations depending on their concentration. Among AOPs, heterogeneous photocatalysis appears as a promising solution for low-level concentrations of pollutants. However, the short lifespan, low reusability, and poor separation capability of heterogeneous photocatalysis systems have limited their practical application in environmental remediation [3, 4]. To overcome these challenges, implementing photocatalytic membrane reactors (PMRs) would be a tremendously successful method for separating photocatalysts and photodecomposition products from reaction mixtures [5]. PMRs are a combination of photocatalysis with pressure-driven membrane separation, including nanofiltration (NF), ultrafiltration (UF), and microfiltration (MF) [6].

11.1.1 PHOTOCATALYSIS

In 1972, Fujishima and Honda developed TiO_2 electrodes for water splitting through photoelectrochemicals. Since then, photocatalysis, one of the earlier forms of AOPs, has evolved rapidly for various applications [7]. Technological advancements have made photocatalysis an alternative to existing water treatment methods. The principle of photocatalysis has relied on the absorption of light by a substrate and the generation of electron-hole pairs. The substrate that absorbs light is known as a photocatalyst, and the most widely employed photocatalysts are semiconductors [8]. Based on their physical state, photocatalytic reactions can be classified as homogeneous or heterogeneous reactions. Homogeneous photocatalysis happens when both semiconductors and reactants are in the same phase, while heterogeneous photocatalysts are those that employ semiconductors and reactants with different phases [9].

11.1.2 HETEROGENEOUS PHOTOCATALYSTS

A variety of hazardous materials have been successfully degraded using heterogeneous photocatalysts. Heterogeneous photocatalysts offer several benefits for wastewater remediation. Some of the significant advantages of such photocatalysts are that they are environment-friendly and safe; also, there is an option to carry out reactions at room temperature and under normal pressure without the utilization of dangerous and potent chemical oxidant/reducing agents [10]. Furthermore, these materials can convert contaminants to non-toxic by-products in various liquid, solid, and gaseous phases [11]. Heterogeneous photocatalysts are also able to couple with other technologies such as

DOI: 10.1201/9781003326281-13

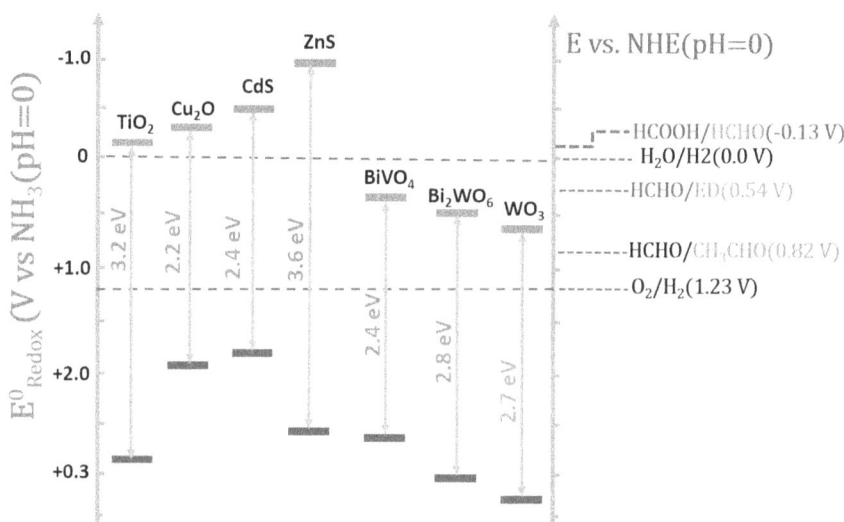

FIGURE 11.1   The energy level positions and bandgap of some semiconductors [21].

membrane separation and membrane reactors [12]. Energy band theory explains the physicochemical processes of heterogeneous photocatalysis. Based on this theory, when two atomic orbitals combine to form molecules, atomic orbitals split into molecular orbitals with lower and higher energy levels, so-called bonding orbitals and anti-bonding orbitals, respectively.

When many atoms combine to form a crystal lattice, the number of bonding and anti-bonding orbitals becomes larger. Consequently, the energy difference between them becomes minimal. Therefore, the continuous energy levels of bonding and anti-bonding orbitals in a crystal lattice build up the valence band (VB) and the conduction band (CB), respectively. An interval of energy containing no orbitals is called the bandgap [13, 14]. An electron can occupy the highest energy level at an absolute zero temperature, known as the Fermi level. Accordingly, the Fermi level measures electron occupancy probability between VBs and CBs [15]. The two main types of semiconductors are intrinsic semiconductors and extrinsic semiconductors. The number of electrons and holes in intrinsic semiconductors is equal, and the Fermi level lies between VB and CB. In contrast, extrinsic semiconductors possess an unequal number of electrons and holes [16]. During light irradiation, electrons and holes generated by excitations from the VB to the CB are considered as charge carriers in semiconductors [17]. Considering irradiation sources, most photocatalysts, such as TiO_2, ZnO, Nb_2O_5, and others, are typical semiconductors that can be activated under UV irradiation. On exposure to UV light, the aforementioned photocatalysts can completely degrade organic and inorganic contamination in the liquid phase through selective reactions. Since the solar spectrum contains only about 5% UV light, and, on the other hand, because of the extreme energy input, high-energy UV light increases by-product formation [18]. Recently, a great deal of interest has been shown in applying photocatalysts that can be activated through visible light as an energy source [19]. In contrast, CdS, Bi_2MoO_6, Bi_2WO_6, WO_3, Ag_2O, Cu_2O, $BiVO_4$, CdSe, $RbPb_2Nb_3O_{10}$, etc., are examples of photocatalysts possessing visible-light photoactivity [20]. Figure 11.1 demonstrates the energy levels and band-edge positions of some semiconductors.

11.1.3.   MECHANISM OF PHOTOCATALYTIC OXIDATION

The illumination of the photocatalyst using the photon energy larger than its bandgap causes electrons at the VB of the photocatalyst to be excited to the CB, creating holes at the VB, producing electron (e^-) and hole (h^+) pairs in the bulk phase [22]. When charge carriers move to the surface of the photocatalyst, h^+ in the VB oxidizes adsorbed H_2O to hydroxyl radicals (OH). In contrast,

e^- in the CB reduces the adsorbed O_2 to superoxide radicals ($^\bullet O_2^-$). The photogenerated radicals can degrade pollutants adsorbed on the photocatalyst surface (Figure 11.2) from the surrounding environment to small molecules of H_2O and CO_2. Briefly, photogeneration of radical species in TiO_2 as a widely accepted photocatalyst can be described as follows [23]:

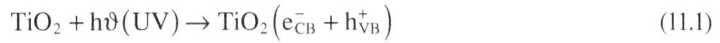

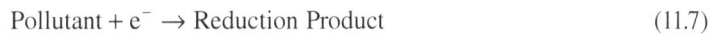

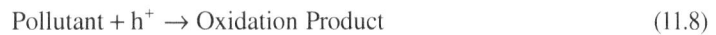

$$TiO_2 + h\vartheta\,(UV) \rightarrow TiO_2\left(e_{CB}^- + h_{VB}^+\right) \tag{11.1}$$

$$O_2\left(h_{VB}^+\right) + OH^- \rightarrow TiO_2 + OH^\bullet \tag{11.2}$$

$$TiO_2\left(e_{CB}^-\right) + O_2 \rightarrow TiO_2 + O_2^{\bullet -} \tag{11.3}$$

$$O_2^{\bullet -} + H^+ \rightarrow HO_2^\bullet \tag{11.4}$$

$$\text{Pollutant} + OH^\bullet \rightarrow H_2O + CO_2 \tag{11.5}$$

$$\text{Pollutant} + O_2^{\bullet -} \rightarrow H_2O + CO_2 \tag{11.6}$$

$$\text{Pollutant} + e^- \rightarrow \text{Reduction Product} \tag{11.7}$$

$$\text{Pollutant} + h^+ \rightarrow \text{Oxidation Product} \tag{11.8}$$

However, a fraction of photogenerated carriers can also be recombined on the surface or in the bulk of the photocatalyst, releasing the accumulated energy in heat or photon forms [24]. Therefore, the efficiency of photocatalysis highly depends on the recombination rate of photogenerated charge carriers and the bandgap of the photocatalyst [25].

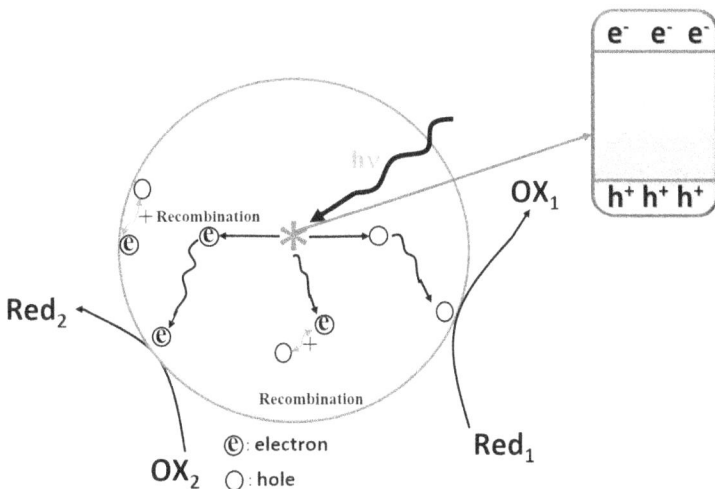

FIGURE 11.2   Mechanism of photocatalytic oxidation/reduction processes.

11.1.4 Factors Affecting the Photocatalytic Process

Suitable photocatalyst material must possess high photoactivity toward UV or visible light, resistance to photocorrosion, stability in the reaction media, and be affordable [26]. Photocatalytic efficiency depends on several operational parameters, which are classified into intrinsic and extrinsic. Parameters include surface area, bandgap, particle size, surface defects and surface charge, functional groups, the recombination rate of charge carriers in bulk and surface, and interfacial charge transfer related to the photocatalyst itself, known as intrinsic parameters. Due to the higher surface-to-volume ratio and large surface area, photocatalysts with nanoscale particle size possess superior light absorption and minimum light reflection, facilitate charge transfer, and therefore decrease the e/h^+ recombination rate [23, 27]. However, a large bandgap leads to insufficient charge separation due to a higher e/h^+ recombination rate and poor light utilization. Chemical and physical doping can introduce defects into photocatalysts structures and induce oxygen vacancies and lattice distortions, creating new energy levels that can trap charge carriers or narrow bandgaps to improve light absorption [28]. In contrast, extrinsic parameters are external factors that include photocatalyst loading, pollutant concentration, solution pH, light intensity, and temperature, governed by the photocatalytic reaction. Some studies have shown that reaction temperature affects photocatalytic reactions rate and photocatalyst loading. The increasing temperature causes higher diffusion and mass transfer, enhancing the reaction rate. A higher quantity of photocatalysts is required at lower temperatures for the same mass transfer. With an increase in temperature, the catalyst requirement also decreases [29].

11.2 MEMBRANE PROCESSES

Membrane technology has received considerable attention for wastewater treatment and has been developed rapidly in practical applications worldwide. Membranes have existed since the 18th century, and, since then, they have undergone many improvements to meet the demands of a wide range of applications. Besides being used for filtration, extraction, and distillation, membranes can also be used as storage systems for gas in biogas plants or as catalysts in syntheses. Depending on the application, membranes can be applied individually or in combination with other techniques. Besides being used for filtration, extraction, and distillation, membranes can also be used as storage systems for gas in biogas plants or as catalysts in synthesis processes. Depending on the application, membranes can be implemented individually or in conjunction with other techniques. However, to increase the efficiency of the separation process and achieve superior performance, hybrid membrane processes have been considered, which combine membrane technology with other processes [30, 31].

11.2.1 Membrane Bioreactors

Membrane bioreactors (MBRs) have attracted significant industrial attention worldwide for treating various wastewater. Biological processes such as activated sludge and membrane filtration have been combined in this technology. The MBR process produces highly purified effluents while eliminating most pathogenic bacteria and viruses [32]. The MBR system provides longer sludge retention time, reduces waste-activated sludge production, and controls sludge retention time. Combining biological and filtration processes allows the treatment of high concentrations of organic compounds with high loading rates of contamination. Although MBR technology has the aforementioned benefits, it also faces several challenges, such as increased operating and capital costs, membrane fouling, high cleaning and replacement costs, and bioreactor foaming [33]. There are two types of MBRs regarding configuration, including external and submerged reactors. The external MBR is categorized into recirculated and side-stream types, while the submerged one is classified into immersed and integrated reactors. External MBR has a stirred tank as the main reactor coupled with a membrane unit for water purification. The primary liquid is pumped to the membrane module at a high flow rate (2–4 m/s), and the concentrated liquid is returned to the tank.

In submerged MBRs, membranes are immersed directly in the primary suspension or in a separate tank, in which the permeation rate is controlled by low transmembrane pressure [34, 35]. Submerged MBRs, the most common configuration of MBR for wastewater treatment, use less energy to operate than external MBRs because there is no pumping involved, while a larger membrane surface area is required compared to cross-flow MBRs [36]. Furthermore, MBRs can be classified into biomass separation MBR, enzymatic MBR, and ion-exchange MBR categories. In the former, the cell units can be microbial cells, animal cells, or plant cells. In contrast, the second involves membrane separation and enzymatic reaction. Water streams contaminated with ionic pollutants can be cleaned using membrane-supported biofilm reactors [33, 37].

11.2.2 ADSORPTIVE MEMBRANE REACTORS

Adsorption processes, commonly used as pretreatment steps to eliminate inorganic and organic contaminants, can effectively be implemented to improve contaminant removal [38]. Various emerging pollutants can be removed from wastewater using adsorptive membranes that cannot be eliminated conventionally [39]. Adsorptive membrane technology utilizes a variety of adsorbents and membranes, including NF, UF, and MF membranes, to remove contaminants by dual adsorption/filtration mechanisms [40]. Adsorptive membranes require less space and possess advantages such as high removal rates and efficiencies due to the large surface area, low operating pressures, high permeabilities, and reusability character [41]. In order to remove contamination, adsorbent membranes utilize a two-step process, sieving and adsorption mechanism. The sieving process involves the contact of the water with the membrane's active layer, resulting in particles larger than the membrane pores being collected on the membrane surface. The water containing smaller particles passes through the active layer, and smaller particles can adsorb to the walls of pores [42]. To enhance adsorption capacity, efficiency, and selectivity, adsorptive membranes can be modified by creating and increasing reactive functional groups such as $-NH_2$ and $-COOH$ on the active layer that can interact with particles through ion exchange or surface complexation [43, 44]. Due to critical factors such as high permeability and low-pressure requirement, and affordable cost, MF and UF membranes are widely implemented as adsorptive membranes among different membrane technologies [42]. There are several types of adsorptive membranes depending on the polymer or adsorbent used in the membrane. Adsorption membranes are classified according to polymer type into natural polymers [45] like chitosan and synthetic polymer types like polyacrylonitrile (PAN) [46], polyurethane with cellulose acetate [47], and polyethersulfone (PES) [48]. According to the adsorbent materials used in the membrane, adsorptive membranes can also be divided into four main groups inorganic fillers, organic fillers, biomaterials, and hybrid filler-based membranes [41]. In addition, mixed matrix membranes (MMMs) are a hybrid membrane type that combines the merits of the polymer matrix and inorganic/organic fillers [49]. According to the physical properties of the polymer, MMMs can be classified as solid polymer, liquid polymer, and solid-liquid polymer. In addition, three major types of MMMs, according to the type of filler and its location within the hybrid membrane structure, are inorganic fillers, organic fillers, and hybrid fillers [50].

11.2.3 ACTIVATED CARBON-BASED MEMBRANE REACTORS

Activated carbon (powdered activated carbon (PAC)) is commonly used as an absorbent in water treatment. Activated carbon-based membrane combines PAC adsorption and filtration processes, demonstrating general applicability with superior permeate water quality. There are different membrane technologies, such as those using MF, UF, and MBR membranes [51, 52]. For instance, in addition to drinking water treatment applications, PAC/UF was explored as a possible pretreatment alternative for applications such as reverse osmosis (RO) in seawater desalination [53]. PAC/MBR offers the advantage of combining adsorption and biodegradation in the filtration reactor, further enhancing permeate quality [54]. Several studies show the effect of PAC on filterability improvement and a reduction of the membrane. PAC affects the permeate quality by interacting with dissolved and suspended solids, biomass, and metal ions and forming a continuous cake layer on the membrane. The cake layer that contains PAC retains adsorption capacity and controls membrane fouling by changing the structure, composition, and, thereby, the resistance of the cake layer [55].

11.2.4 Photocatalytic Membrane Reactor

Recently, PMRs, coupling membrane separation techniques with AOPs, specifically heterogeneous photocatalysis, have emerged as a promising alternative for water decontamination [56]. Degradation of organic materials, separation of ions, heavy metal removal, and virus destruction can be fulfilled using PMRs [Figure 11.3].

They can also minimize membrane fouling and reduce the cost and energy consumption of wastewater treatment. A photocatalyst reactor consists of two major types depending on how the catalyst is applied: a slurry reactor, a fixed bed reactor with suspended photocatalysts fixed in a solid support structure, or a membrane with immobilized photocatalysts. One of the significant advantages of immobilized PMRs is photocatalyst recovery and recyclability; however, the active surface area is reduced, resulting in a decrease in photocatalytic activity. The photocatalytic degradation efficiency is usually higher when a suspended photocatalyst has been used in slurry reactors [58]. PMRs with suspended photocatalysts have been upgraded with a membrane process that separates the photocatalysts from the reaction medium after treatment. In this regard, the membrane creates a barrier to maintain the photocatalyst particles and allows for their recyclability [59]. Considering membrane technology, pressure-driven membranes such as MF, UF, or NF are the most typical membranes used in the separation process in PMRs [60]. However, these membranes have been manufactured with hydrophobic materials such as polypropylene, polyvinylidene fluoride, or polytetrafluoroethylene, which leads to membrane fouling [61]. Fouling is a phenomenon that impairs the practical application of the membrane. Membrane pores can be blocked, leading to low flow, decreasing membrane life, and eventually increasing operating costs. Fouling is greatly dependent on organic pollutants and photocatalyst reactions, as well as their interaction with membrane surfaces. The photocatalytic reactions are initiated by producing hydroxyl radicals under UV or visible irradiation. The produced hydroxyl radicals can oxidize and degrade most organic pollutants; consequently, fouling is significantly decreased, and the flux behavior is dramatically improved [62].

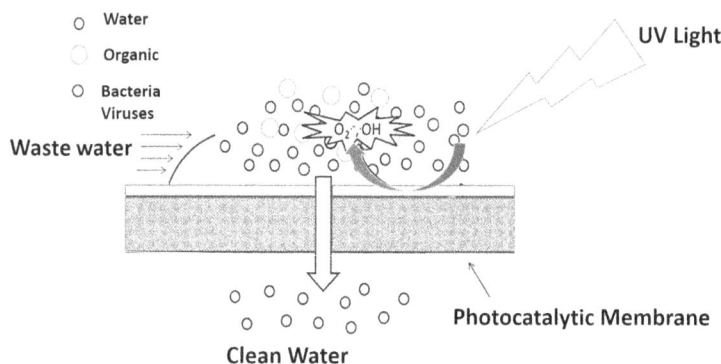

FIGURE 11.3 Scheme of PMR for eliminating various pollutants [57].

11.3 FABRICATION AND CHARACTERIZATION TECHNIQUES OF PHOTOCATALYTIC MEMBRANES

In PMRs, the membrane supporting the photocatalyst can be MF, UF, or NF. The photocatalyst can be incorporated into the membrane surface using several strategies, such as blending, dip-coating, magnetron sputtering, gas-phase deposition, and vacuum filtration [63]. In all these approaches, controlling the position and dispersion of photocatalyst particles in the membrane matrix are the key factors that can be conducted by either chemical addition or external force such as magnetic or electromagnetic sources [64]. Magnetic field-induced alignment is one of the effective methods that can control the dispersal and alignment of diamagnetic or ferromagnetic photocatalysts in a polymer texture through a magnetic field [65]. Dip-coating is one of the most influential and extensively applied immobilization methods to fabricate photocatalytic ceramic membranes. Wang et al. fabricated TiO_2-Al_2O_3 membranes by dipping the α-Al_2O_3 disk into a TiO_2 solution [66]. Deposition of TiO_2 nanoparticles on the surface increases the light illumination to the membrane surface, thereby improving photocatalytic activity.

Another method – flame spray pyrolysis (FSP) – directly deposits the photocatalyst nanoparticles onto the membrane through a one-step deposition process with suitable control of membrane properties [67, 68]. Foglia et al. fabricated photoactive membranes coated with TiO_2 and Pt/TiO_2 nanoparticles through FSP and deposited them on glass fiber filters by the flame beam method [68]. Magnetron sputtering, a physical vapor deposition sputtering technique, has been widely utilized to deposit a broad range of industrial coatings due to its potential to deposit high-quality and high-purity films on large surfaces with excellent adhesion [69]. Using magnetron sputtering followed by anodizing, Fischer et al. synthesized TiO_2 nanotubes on the surface of a polyethersulfone (PES) polymer membrane. Crystallized TiO_2 with an anatase phase obtained at low temperatures demonstrated high photocatalytic activity due to the increased surface area and the ability to absorb light [70]. Chemical vapor deposition (CVD) is another method that can deposit precursor compounds in the vapor phase on to a membrane surface. Chemical reactions like oxidation, hydrolysis, thermal decomposition, and compound formation can occur once the chamber temperature is raised appropriately. A layer of solid deposits can be formed by adsorbing reaction intermediates through surface reaction on the membrane surface. CVD, however, can damage the membrane surface because of the high heating temperature. Thus, plasma-assisted CVD (PACVD) or plasma-enhanced CVD (PECVD) is proposed to address this problem and reduce the heating temperature. The coating layer generated by these techniques is of high quality and has a controlled pore-size distribution. Sol-gel is considered a viable option to modify membrane surfaces and provides higher surface areas and more stable surfaces. A sol-gel reaction involves hydrolyzing a precursor and polycondensing the hydrolyzed products. Sol-gel reactions proceed through either the gelation of colloidal particles or polymer net growth based on the solvent chosen.

Water favors the colloidal route, while alcohol favors the polymer pathway. It depends on the membrane application whether to use colloidal or polymeric routes. In order to create a coating layer inside the membrane support, it is necessary to combine the sol-gel process with other coating techniques, such as dip-coating or spin-coating. Electrospinning can also be used to fabricate nanofiber membranes containing photocatalysts or other nanoparticles. Two or more polymers are blended with photocatalysts in this process, depending on their final use and properties. Y. Xu et al. fabricated TiO_2-polyethersulfone (PES) composite nanofibrous membranes through modified blending and electrospinning techniques [71]. The fabricated membrane composite adsorbed and photodegraded (Figure 11.4) various dye pollutions with a remarkable removal efficiency of 95% even after five cycles. Through the synergistic impact of adsorption and photocatalysis, it will be possible to design a highly sufficient membrane for photocatalytic environmental remediation.

After implementation of all mentioned techniques, the obtained coated layer on the membrane can be evaluated by different techniques such as X-ray diffraction (XRD), X-ray photoelectron spectroscopy (XPS), Fourier transform infrared spectroscopy (FTIR), transmission electron microscopy

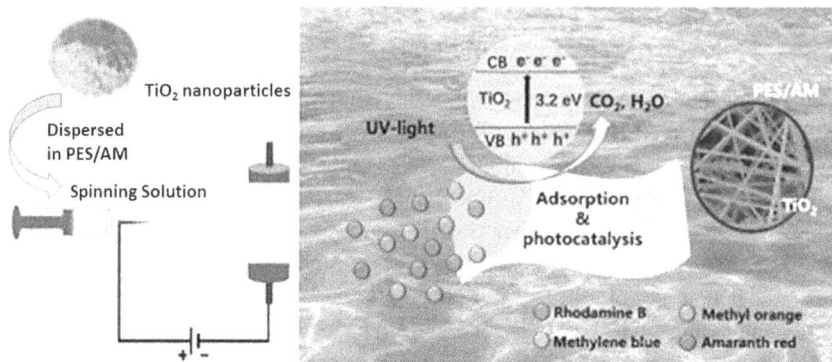

FIGURE 11.4 Photodegradation mechanism of various dye pollutions over TiO_2-polyethersulfone membrane [71].

(TEM), photocorrelation spectroscopy, nitrogen adsorption/desorption, scanning electron microscopy (SEM), scanning electron microscopy wave dispersive X-ray analysis (SEM-WDX), and gas permeability measurement, which are used to characterize the preparation of the photocatalytic membrane.

11.4 SPECIFICATION OF PMRS

11.4.1 PMR Configurations and Designs

A PMR configuration has a crucial role in minimizing membrane fouling, improving photocatalytic performance, as well as optimizing PMR operation and maintenance. Factors such as the process mode (batch or continuous flow), implemented membrane (MF, UF, NF), applied membrane modules (flat sheet, hollow fiber, submerged), and the kinds of light source are determining factors in membrane design [72]. As mentioned before, based on photocatalyst arrangement, PMRs can be divided into two main configurations: immobilized and suspended or slurry photocatalyst PMRs. Considering photocatalyst and membrane interactions, slurry PMRs are classified into two types: integrative and split [73]. In the integrative type, the membrane separation and photocatalytic reaction processes are integrated into the medium. In contrast, photocatalytic reactions and membrane separation are implemented separately in split-type PMRs [74]. Depending on the location of the light source, there are three other types of PMRs such as above the feed tank, above the membrane unit, or even above an additional vessel [73]. Basically, PMR configurations are pressure-driven or non-pressure-driven. PMRs driven by hydraulic pressure employ MF, UF, NF, or RO separation processes with a suspended or immobilized photocatalyst [75]. Figure 11.5 illustrates the typical diagram of PMRs. Non-hydraulic, pressure-driven PMRs are new PMRs in which photocatalytic processes are coupled with forward osmosis (FO), membrane distillation (MD), pervaporation (PV), or membrane crystallization (MC).

11.4.2 Mechanism of Photocatalysts on Membrane Fouling Control

Membrane fouling is a significant problem in membrane separation processes that reduces separation efficiency, elevates cleaning frequency, shortens the membrane lifespan, and increases maintenance and operational costs [76]. Membrane fouling can occur in reversible fouling and irreversible fouling. Basically, there are two kinds of fouling: organic and inorganic. There are many organic pollutants, such as alginate gels, methyl oranges, humic acid (HA), rhodamine B, azo dyes such as acid orange 7, phenolic compounds, algae, fungi, herbicides, pesticides, etc. Adsorption of organic contaminants on photocatalyst particles creates a dense cake layer on the membrane surface.

Meanwhile, organic pollutants can block membrane pores leading to pore-blocking fouling [62]. The membrane fouling caused by organic pollutants is reversible [77]. Accumulation of a less compact

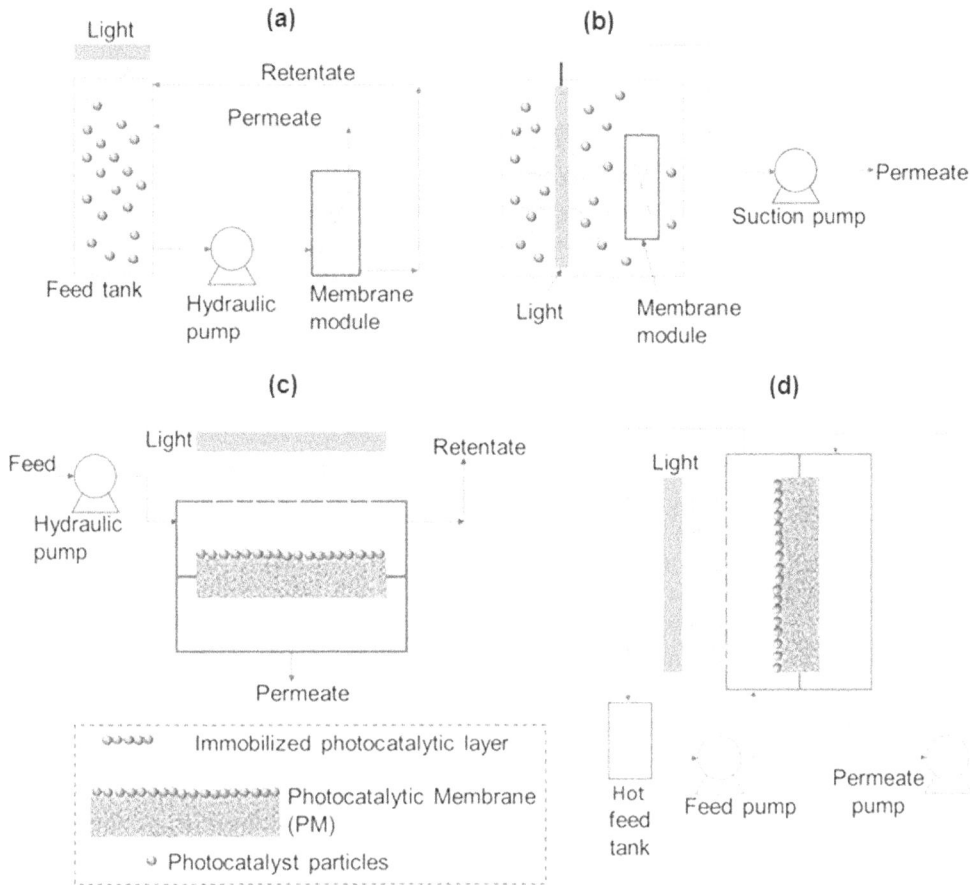

FIGURE 11.5 A schematic diagram of PMRs: (a) pressure-driven PMR with split slurry, (b) pressure-driven (submerged) PMR with integrative slurry, (c) pressure-driven immobilized PMR, and (d) non-pressure-driven immobilized PMR [75].

cake layer and polarization of the membrane surface likely cause reversible fouling, and physical cleaning is sufficient to remove it [78]. Inorganic contaminants (such as anionic and cationic salts, silica, calcium, chloride, carbonate-based compounds, and heavy metals) are able to pack densely and form crystalline particles in water [79]. Usually, inorganic pollutants lead to irreversible fouling. The adsorption of crystals inside membrane channels and stronger bonding forces with membrane materials lead to irreversible fouling that cannot be reversed physically or chemically [80]. The phenomenon of membrane fouling increases capital costs significantly. It is, therefore, necessary to identify low-cost alternatives to alleviate membrane fouling. Among the pioneer alternative techniques to mitigate membrane fouling, PMRs have a great potential to lessen the tendency of membrane fouling.

In PMRs, organic pollutants and other contaminants have complicated interactions with photocatalysts and membrane surfaces. On the one hand, the interaction between photocatalysts and organic pollution leads to the adsorption of contamination on the photocatalyst surface, making a porous and thick cake layer on the membrane surface containing photocatalyst particles and organic pollutants. On the other hand, during continuous UV/Vis irradiation of the membrane, the interaction of light source and photocatalyst generates oxidizing agents such as HO and $O_2^{\bullet-}$ radicals that can oxidize organic pollutants throughout photocatalytic oxidation; therefore, decomposition of organic pollutants and reduction of foulants occur, which significantly improves the flux behavior.

Since the micropollutants adsorbed on the membrane surface are degraded during photocatalytic oxidation, membrane fouling is minimized.

11.4.3 Effect of Photocatalyst Loading on Membrane Fouling

Membrane efficiency and fouling can be affected by photocatalyst loading. Increasing photocatalyst concentration causes more surface area for pollutant adsorption and, ultimately, more photocatalytic oxidation and removal efficiency, improving permeate flux and reducing membrane fouling. With increasing photocatalyst loading, particle interaction and particle agglomeration are increased; therefore, the active surface area decreases above certain loading levels, resulting in a constant degradation rate or a reduction with increasing photocatalyst concentrations [81]. Additionally, increasing photocatalyst concentration, on the one hand, causes more sedimentation, and it can lead to pore blocking, increased thickness, and change the texture of the cake layer [82]. Besides loading, photocatalyst particle size, and morphology, the surface area can influence membrane fouling.

11.4.4 The Remediation of Toxic Metals and Heavy Metals

Contaminations such as Cd, Cr, Cu, Ni, As, Pb, and Zn, known as heavy metals, are among the most common hazardous species found in industrial sewage [83]. The high level of toxicity, high solubility, persistence in the environment, and bioaccumulative nature of these chemicals cause significant concern for the environment, human health, and other organisms [84]. In fact, the food chain and drinking water are two possible ways for these compounds to enter metabolic pathways. As a result, heavy metal-polluted wastewater must be treated before being released into the environment. So far, typical approaches, including chemical precipitation, coagulation-flocculation, flotation, adsorption, ion exchange, electrochemical deposition, and membrane filtration, have been commonly implemented to eliminate heavy metal ions from wastewater. Among these approaches, chemical precipitation is the most common method; however, this method needs a high amount of chemicals to eradicate metal content to a standard level. In addition, the sludge produced in large amounts needs further treatment [85]. Additionally, heavy metals can be removed from effluents by the ion-exchange method. This technique typically uses synthetic organic ion-exchange resins.

Nevertheless, it can't process concentrated metal solutions because the wastewater saturates the matrix with solids [86]. Heavy metal ions can also be eliminated from wastewater through electrodeposition. The process is fulfilled in a cell containing anode and cathode electrodes in an aqueous electrolyte containing heavy metal ions. Corrosion of electrodes necessitate frequent electrode replacements, which is highly costly [87].

Due to their convenient operation and efficient separation, membrane techniques such as NF, UF, MF, and RO have been attracting considerable attention to remove inorganic pollutants such as heavy metals from wastewater. Figure 11.6 demonstrates different types of pressure-driven membranes applicable in water and wastewater technology considering membrane pore-size regimes [88].

UF membranes are among the best candidates for heavy metal removal because of their low operating pressures and low capital and energy costs. Heavy metals, however, cannot be eliminated directly through UF membranes because of their relatively large pore sizes [89]. Due to its pore size within the range of micrometers, the MF membrane can only eliminate particles and biological species within the range of micrometers and cannot remove heavy metals. The mechanism of the NF membrane is based on the rejection of multivalent anion and cation species through electrical interaction between metal ions and the membrane surface. Therefore, heavy metal elimination can be significantly impacted by the pH of media, pore diameter, and the existence of surface charge on the NF membrane. NF membranes with smaller pores and a highly charged surface demonstrate desirable efficiency for metal removal. The pH change can affect the structure of functional groups on the NF surface and the electrostatic interaction of metal ions with the membrane [90]. The reported results indicate that RO is superior to UF and NF for heavy metal removal, with a rejection

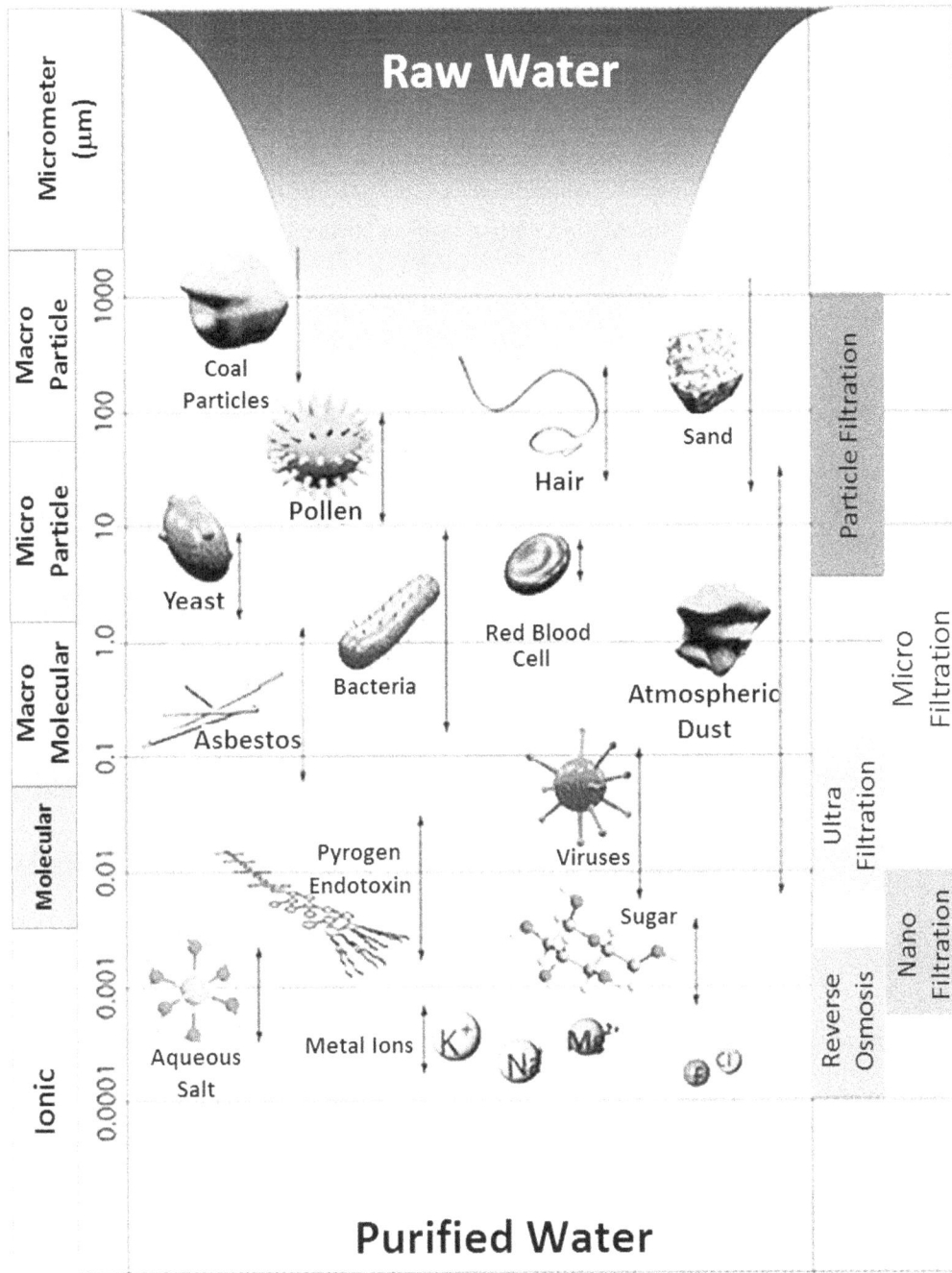

FIGURE 11.6 Schematic illustration of membrane filtration considering membrane size regimes [88].

efficiency of over 97% for metal concentrations ranging from 21 to 200 mg/L. Other benefits of RO include operating at wide pH ranges of 3–11 and pressure ranges of 4.5–15 bar, high salt rejection, biological degradation resistance, high degree of mechanical and chemical stability, and a high-temperature operating range [91, 92]. Nevertheless, RO has some drawbacks which restrict its performance. Wastewater with high turbidity or oxidized compounds such as chlorine can induce fouling due to the tiny pores of the membrane. This phenomenon can be accelerated in the presence of certain cations in wastewater, such as Cd(II) and Cu(II), resulting in a decrease in permeate flow rate, a reduction in membrane performance over time, and eventually irreversible fouling. As a result, the membrane must be replaced, which increases operational costs [89]. However, the reported results demonstrated that there is no guarantee for the complete removal of heavy metals during membrane separation. Therefore, membrane modification or implementation of hybrid systems may be required. In this regard, the PMR can be viewed as a modified membrane that takes advantage of both photocatalyst and membrane processes for heavy metal removal.

Photocatalytic elimination has recently become a popular research topic for removing heavy metals from wastewater. The photocatalytic method can be operated under room temperature and atmospheric pressure, providing a potent oxidizing agent, such as hydroxyl radicals, which can degrade a wide range of biodegradable and non-biodegradable contaminants without polluting intermediates. In addition to being compatible with other physical and chemical technologies and being powered by renewable solar energy, photocatalytic processes due to the no (or less) chemical consumption and less sludge production are known as eco-friendly methods [93]. Photocatalysts can destroy heavy metal ions through reduction or oxidation via photogenerated species such as electrons (e_{CB}^-) and holes (h_{VB}^+). In general, metal ions with a higher redox potential value than the potential level of e_{CB}^- can be directly reduced, and metal ions with a lower redox potential value than the potential of h_{VB}^+ can be oxidized (Figure 11.4) [94] based on the following equations:

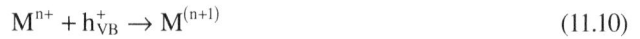

$$M^{n+} + e_{CB}^- \rightarrow M^{(n-1)+} \tag{11.9}$$

$$M^{n+} + h_{VB}^+ \rightarrow M^{(n+1)} \tag{11.10}$$

Based on the following equations, indirect reductive/oxidative removal of heavy metals promoted by electron, and hydroxyl radicals, respectively, during the photocatalytic process can be achieved through the attack of formed HO radicals [95]:

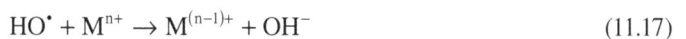

$$e_{CB}^- + O_2 \rightarrow O_2^{\cdot-} \tag{11.11}$$

$$O_2 + e_{CB}^- + H^+ \rightarrow HO_2^{\cdot} \tag{11.12}$$

$$2HO_2^{\cdot} \rightarrow H_2O_2 + O_2 \tag{11.13}$$

$$H_2O_2 + O_2^{\cdot-} \rightarrow HO^{\cdot} + HO^- + O_2 \tag{11.14}$$

$$H_2O_2 + e^- \rightarrow HO^{\cdot} + HO^- \tag{11.15}$$

$$HO^{\cdot} + M^{n+} \rightarrow M^{n+1} + OH^- \tag{11.16}$$

$$HO^{\cdot} + M^{n+} \rightarrow M^{(n-1)+} + OH^- \tag{11.17}$$

TiO_2 is a well-known and commercially available photocatalyst due to its remarkable optical and electronic properties, physical and chemical stability, non-toxicity, and low cost. Considering these benefits, most PMRs have taken advantage of TiO_2 [96]. However, the fast recombination rate of

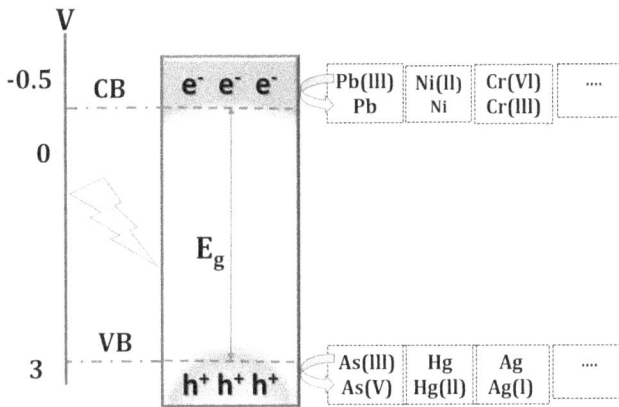

FIGURE 11.7 Mechanism of photocatalytic treatment of heavy metal.

electrons/holes and the triggering photocatalytic reactions using only UV light restrict the practical utilization of TiO_2 [97]. In addition, TiO_2 nanoparticles aggregate severely during the photocatalytic process, decreasing their active sites and light-harvesting ability. The recovery of fine particles of TiO_2 from suspension systems is another major challenge. Before the technology can be applied in an industrial setting, TiO_2 must be recoverable and reusable [98].

However, results showed that neither the photocatalysis technology nor the membrane techniques can individually remove heavy metal contaminants to the desired levels, regardless of how well the system is operated. As a result, hybrid PMR systems are required to capture the benefits of these two technologies while minimizing their drawbacks as separate technologies. Thus, in light of the benefits of PMRs, they can be implemented as an affordable, efficient, and low energy-consuming method (utilizing light irradiation) for photo-detoxication of heavy metals. Nowadays, extensive research is being conducted to develop effective photocatalysts for heavy metal removal through PMRs. G.A. Mulungulungu et al. employed graphitic carbon nitride, $g-C_3N_4$, as an effective photocatalyst due to its significant specific surface area, sufficient hydrophilicity, and surface functional group to fabricate a photocatalytic membrane. The fabricated membrane was utilized to eliminate Cu^{2+} ions from wastewater through membrane filtration method [99]. Following 15 hours of filtration, the fabricated membrane containing $g-C_3N_4$ as photocatalyst showed 47.82 $L/m^2/h/bar$ water permeance and a 98.03% rejection rate for copper ions at a concentration of about 2,500 mg/L. The synergistic effect of amino functional groups facilitates complex formation among amino groups, and heavy metal ions, surface charges, and the photocatalytic activity of $g-C_3N_4$ lead to a remarkable 98.03% removal efficiency of copper ions. In another study, T. Rathna et al. described improved PMRs for removing Cr(VI) from an aqueous medium using hydrothermally synthesized TiO_2-WO_3 nanoparticles embedded in polyaniline (PANI) membranes. Due to suitable hydrophilicity, improved antifouling activity, higher water adsorption, effective contact angle, and sufficient porosity, TiO_2-WO_3@PANI membranes revealed increased heavy metal removal efficiency. The highest rejection percentage of 98.50% in a period of 60 min and transmembrane pressure of 0.5 MPa was obtained with the 5 wt% TiO_2-WO_3 incorporated in the PANI membrane [100]. C.J. Chang et al. fabricated a polymer/BiOBr-modified gauze as a novel dual-functional membrane for Ni^{2+} from wastewater [101]. The surface of the gauze was modified using crosslinked polymer containing hydrophilic diepoxy (PEGDE), a crosslinking agent (PEI), and diamine (TETA), followed by spray coating of a BiOBr photocatalyst. The fabricated membrane was able to remove Ni^{2+} through adsorption and photocatalytic phenomena. In other research, Ferreira-Neto et al. obtained stable hybrid aerogels that possessed outstanding properties through hydrothermal growth of MoS_2 nanostructures on bacterial nanocellulose (BC)-based organic macro/mesoporous scaffolds [102]. Fabricated PMRs with developed photoactive/adsorptive BC/MoS_2 hybrid membranes (Figure 11.7) demonstrated remarkable removal efficiency of 88% through adsorptive and photocatalytic mechanisms for eliminating Cr(VI) (Figure 11.8).

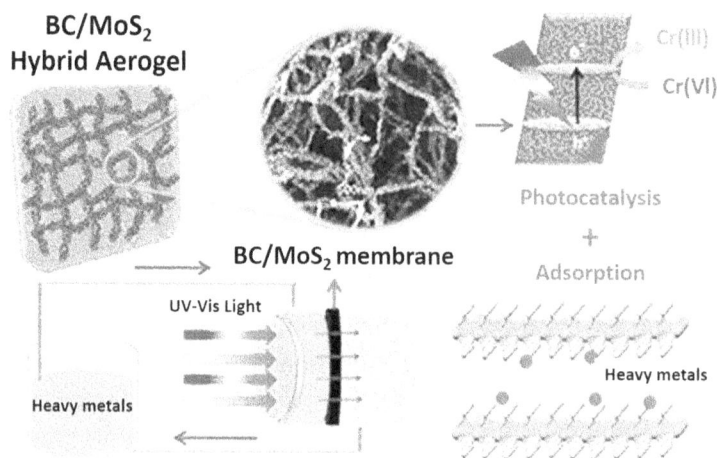

FIGURE 11.8 BC/MoS_2 as bifunctional adsorbent/photocatalyst membranes for Cr(VI) removal [102].

J.T. Jung et al. investigated the effects of operational parameters like UV intensity and TiO_2 dosage on the performance of photocatalytic microfiltration systems. The obtained results revealed that the maximum reaction efficiency of 90% for heavy metal removal was obtained in a TiO_2 dosage of 0.3 g/L and with UV intensity of 64 W [103].

11.5 CHALLENGES AND FUTURE PERSPECTIVES OF PHOTOCATALYTIC MEMBRANE TECHNOLOGY

Due to the synergistic impact of both photocatalysis and membrane separation, considerable attention has been paid to the application of PMRs to remove organic and inorganic pollutants, including heavy metals. Integration not only increases pollutant removal efficiency but also improves the performance of each of the two technologies. In one respect, photocatalysis reduces membrane fouling. Alternatively, a membrane provides efficient separation of photocatalysts, by-products, and photodegradation products. Therefore, PMRs are among the most promising water reuse and recycling technologies. Although PMRs offer many advantages over existing systems for water purification, there are still many technical obstacles to overcome to apply this technique in a practical setting. As such, a variety of factors should be considered in future commercial and large-scale applications, including operational expenses, feasibility studies, and economic analyses, removal efficiency, metal ion concentrations, chemical requirements, and environmental standards and legislation. In this context, low-cost, chemically and mechanically stable, and environmentally friendly photocatalysts that can benefit from solar light instead of ultraviolet light are needed. On the other hand, recovery and reuse of photocatalyst during water purification require highly stable and flexible membrane support materials. The designed PMR should be capable of operating in actual wastewater conditions such as acidic, neutral, and basic media. Another crucial factor is the fundamental understanding beyond the photocatalytic process to direct the reaction pathway toward producing non-toxic or less toxic by-products during wastewater treatment. Overall, the challenge of developing a practical photocatalytic system for water treatment remains substantial, and there is still a significant amount of work to be done to develop a practical photocatalytic water treatment system.

For the membrane reactor, reactor design and membrane materials need to be considered. The attempt should focus on developing low energy demand and low fouling systems; in this case, submerged systems can be considered as the most promising configurations. With regard to membrane materials, ceramic and polymer materials are both prime candidates and the key concern remains the resistance of these membranes to photogenerated radicals and light irradiation. Research should consider the membrane damage caused by UV or visible-light irradiation and abrasion by photo-catalyst particles or light sources. Therefore, comprehensive studies are required to investigate the aging, lifespan, and mechanism by which reactive oxygen species shorten the stability of membranes over a long period of time. Energy consumption is another major factor that increases capital expenditures and operational expenses. A shift toward renewable solar energy might be the best way to implement this technology in the water industry. UV light can be eliminated and energy consumption reduced by utilizing visible-light-induced photocatalysts. The utilization of energy storage systems such as batteries, in this case, is an intelligent way to maximize solar power.

11.6 CONCLUSIONS

Here, the benefits of integrating membrane technology with photocatalyst processing, as well as the commitment to the development of photocatalyst membranes for water treatment technologies, have been discussed. The mechanism of the photocatalytic process and the effect of photocatalysts to reduce membrane fouling and increase its efficiency have been reviewed. Altogether, integrating heterogeneous photocatalysis and membrane separation processes as PMRs could be one of the novel solutions in the water and wastewater industry in the near future. Efficient photocatalyst separation and recycling as well as less membrane fouling are the main advantages of PMRs. Overall, the antifouling behavior of the composite membrane is due to the hydrophilic thin layer created by the photocatalyst, preventing the attachment of foulants on the membrane surface. However, photocatalyst immobilization on membrane surfaces reduces the surface area of the photocatalyst per unit volume, resulting in a decreased efficiency of the photocatalytic reaction. In this regard, PMRs featuring suspended photocatalyst configurations are more affordable for large-scale applications. Suspended PMRs are categorized into integrated and split kinds; reduced pipe length and less occupation area requirements are the advantages of integrated PMRs, while the split PMR is easy to install and maintain. Several factors, including operational expenses, feasibility studies, economic analyses, removal efficiency, metal ion concentrations, and chemical requirements, affect the large-scale and practical application of PMRs in the wastewater industry. These influencing factors should be considered comprehensively in practical industrial applications. Although there is great interest and increasing published literature in PMRs for water treatment, there are still serious challenges regarding membrane stability and operation cost, limiting the large-scale production and practical applications of PMRs.

REFERENCES

1. Lee, C.M., P. Palaniandy, and I. Dahlan, Pharmaceutical residues in aquatic environment and water remediation by TiO$_2$ heterogeneous photocatalysis: a review. *Environmental Earth Sciences*, 2017. 76(17): pp. 1–19.
2. Chuang, Y.-H. and H.-J. Shi, UV/chlorinated cyanurates as an emerging advanced oxidation process for drinking water and potable reuse treatments. *Water Research*, 2022. 211: p. 118075.
3. Al-Maqdi, K.A., et al., Challenges and recent advances in enzyme-mediated wastewater remediation—a review. *Nanomaterials*, 2021. 11(11): p. 3124.
4. Pan, X., et al., TiO$_2$/chlorophyll S-scheme composite photocatalyst with improved photocatalytic bactericidal performance. *ACS Applied Materials & Interfaces*, 2021. 13(33): pp. 39446–39457.
5. Molinari, R., C. Lavorato, and P. Argurio, The evolution of photocatalytic membrane reactors over the last 20 years: a state of the art perspective. *Catalysts*, 2021. 11(7): p. 775.
6. Mozia, S., M. Tomaszewska, and A.W. Morawski, A new photocatalytic membrane reactor (PMR) for removal of azo-dye Acid Red 18 from water. *Applied Catalysis B: Environmental*, 2005. 59(1): pp. 131–137.
7. Guo, Y., P. Qi, and Y. Liu, A review on advanced treatment of pharmaceutical wastewater. *IOP Conference Series: Earth and Environmental Science*, 2017. 63: p. 012025.
8. Mohd Kaus, N.H., et al., Effective strategies, mechanisms, and photocatalytic efficiency of semiconductor nanomaterials incorporating rGO for environmental contaminant degradation. *Catalysts*, 2021. 11(3): p. 302.

9. Li, Y., et al., Recent advances in waste water treatment through transition metal sulfides-based advanced oxidation processes. *Water Research*, 2021. 192: p. 116850.

10. Barba-Nieto, I., et al., Sunlight-operated TiO_2-based photocatalysts. *Molecules*, 2020. 25(17): p. 4008.

11. Palmisano, G., et al., Photocatalysis: a promising route for 21st century organic chemistry. *Chemical Communications*, 2007. 38 (33): pp. 3425–3437.

12. Molinari, R., et al., Photocatalytic processes in membrane reactors, 2017. 3: pp. 101–138.

13. Ibhadon, A.O. and P. Fitzpatrick, Heterogeneous photocatalysis: recent advances and applications. *Catalysts*, 2013. 3(1): pp. 189–218.

14. Fox, M.A. and M.T. Dulay, Heterogeneous photocatalysis. *Chemical Reviews*, 1993. 93(1): pp. 341–357.

15. Ishii, H., et al., Kelvin probe study of band bending at organic semiconductor/metal interfaces: examination of Fermi level alignment. *Physica Status Solidi (a)*, 2004. 201(6): pp. 1075–1094.

16. Weller, P.F., An introduction to principles of the solid state. Extrinsic semiconductors. *Journal of Chemical Education*, 1971. 48(12): p. 831.

17. Graetzel, M. and A.J. Frank, Interfacial electron-transfer reactions in colloidal semiconductor dispersions. Kinetic analysis. *The Journal of Physical Chemistry*, 1982. 86(15): pp. 2964–2967.

18. Molinari, R., C. Lavorato, and P. Argurio, Visible-light photocatalysts and their perspectives for building photocatalytic membrane reactors for various liquid phase chemical conversions. *Catalysts*, 2020. 10(11): p. 1334.

19. Meng, X., Z. Li, and Z. Zhang, Pd-nanoparticle-decorated peanut-shaped $BiVO_4$ with improved visible light-driven photocatalytic activity comparable to that of TiO_2 under UV light. *Journal of Catalysis*, 2017. 356: pp. 53–64.

20. Sang, Y., H. Liu, and A. Umar, Photocatalysis from UV/Vis to near-infrared light: towards full solar-light spectrum activity. *ChemCatChem*, 2015. 7(4): pp. 559–573.

21. Shen, Z., et al., Direct conversion of formaldehyde to ethylene glycol via photocatalytic carbon–carbon coupling over bismuth vanadate. *Catalysis Science & Technology*, 2016. 6(17): pp. 6485–6489.

22. Albero J., García H., Photocatalytic CO2 Reduction in Heterogeneous photocatalysis. *Green Chemistry and Sustainable Technology*, J.C. Colmenares and Y.-J. Xu, Editors. 2016. Springer. pp. 1–31.

23. Kumar, A. and G. Pandey, A review on the factors affecting the photocatalytic degradation of hazardous materials. *Material Science & Engineering International Journal*, 2017. 1(3): pp. 1–10.

24. Guo, R., et al., Effects of morphology on the visible-light-driven photocatalytic and bactericidal properties of $BiVO_4$/CdS heterojunctions: a discussion on photocatalysis mechanism. *Journal of Alloys and Compounds*, 2020. 817: p. 153246.

25. Bilal Tahir, M., K. Nadeem Riaz, and A.M. Asiri, Boosting the performance of visible light-driven WO_3/g-C_3N_4 anchored with $BiVO_4$ nanoparticles for photocatalytic hydrogen evolution. *International Journal of Energy Research*, 2019. 43(11): pp. 5747–5758.

26. Molinari, R., C. Lavorato, and P. Argurio, Recent progress of photocatalytic membrane reactors in water treatment and in synthesis of organic compounds. A review. *Catalysis Today*, 2017. 281: pp. 144–164.

27. Variar, A.G., et al., Influence of various operational parameters in enhancing photocatalytic reduction efficiency of carbon dioxide in a photoreactor: a review. *Journal of Industrial and Engineering Chemistry*, 2021. 99: pp. 19–47.

28. Jeon, J.P., et al., Enhancing the photocatalytic activity of TiO_2 catalysts. *Advanced Sustainable Systems*, 2020. 4(12): p. 2000197.

29. Ye, L., et al., Noble-metal loading reverses temperature dependent photocatalytic hydrogen generation in methanol–water solutions. *Chemical Communications*, 2016. 52(78): pp. 11657–11660.

30. Obotey Ezugbe, E. and S. Rathilal, Membrane technologies in wastewater treatment: a review. *Membranes*, 2020. 10(5): p. 89.

31. Ahmad, T., C. Guria, and A. Mandal, A review of oily wastewater treatment using ultrafiltration membrane: a parametric study to enhance the membrane performance. *Journal of Water Process Engineering*, 2020. 36: p. 101289.

32. Mutamim, N.S.A., et al., Membrane bioreactor: applications and limitations in treating high strength industrial wastewater. *Chemical Engineering Journal*, 2013. 225: pp. 109–119.

33. Coutte, F., et al., Recent trends in membrane bioreactors, in *Current Developments in Biotechnology and Bioengineering*, Christian Larroche, Maria Ángeles Sanromán, Guocheng Du, Ashok Pandey, Editors. 2017, Elsevier. pp. 279–311.

34. Erkan, H.S., N.B. Turan, and G.Ö. Engin, Membrane bioreactors for wastewater treatment, in *Comprehensive Analytical Chemistry*. Dotse Selali Chormey, Sezgin Bakırdere, Nouha Bakaraki Turan, Güleda Önkal Engin, Editors. 2018, Elsevier. pp. 151–200.

35. Yamamoto, K., et al., Direct solid-liquid separation using hollow fiber membrane in an activated sludge aeration tank, in *Water Pollution Research and Control Brighton*. L. Lijklema, K.R. Imhoff, K.J. Ives, D. Jenkins, R.G. Ludwig, M. Suzuki, D.F. Toerien, A.B. Wheatland, A. Milburn, E.J. Izod, Editors. 1988, Elsevier. pp. 43–54.

36. Radjenović, J., et al., Membrane bioreactor (MBR) as an advanced wastewater treatment technology, in *Emerging Contaminants from Industrial and Municipal Waste*, Damià Barceló, Mira Petrovic, Editors. 2007, Springer. pp. 37–101.

37. Wang, Z., et al., Research and applications of membrane bioreactors in China: progress and prospect. *Separation and Purification Technology*, 2008. 62(2): pp. 249–263.

38. Vo, T.S., et al., Heavy metal removal applications using adsorptive membranes. *Nano Convergence*, 2020. 7(1): pp. 1–26.

39. Qalyoubi, L., A. Al-Othman, and S. Al-Asheh, Recent progress and challenges on adsorptive membranes for the removal of pollutants from wastewater. Part I: Fundamentals and classification of membranes. *Case Studies in Chemical and Environmental Engineering*, 2021. 3: p. 100086.

40. Chong, W.C., et al. Adsorptive membranes for heavy metal removal–a mini review, in *AIP Conference Proceedings*. 2019, 2157 (1):020005, AIP Publishing LLC. Manager, conference Proceedings: Emily Prendergast, Editorial Assistant, Conference Proceedings: Francesca Tangreti.

41. Huang, Z.Q. and Z.F. Cheng, Recent advances in adsorptive membranes for removal of harmful cations. *Journal of Applied Polymer Science*, 2020. 137(13): p. 48579.

42. Zhang, X., et al., Developing new adsorptive membrane by modification of support layer with iron oxide microspheres for arsenic removal. *Journal of Colloid and Interface Science*, 2018. 514: pp. 760–768.

43. Adam, M.R., et al., Chapter 12 - Adsorptive membranes for heavy metals removal from water, in *Membrane Separation Principles and Applications*, A.F. Ismail, et al., Editors. 2019, Elsevier. pp. 361–400.

44. Nasir, A.M., et al., Adsorptive nanocomposite membranes for heavy metal remediation: recent progresses and challenges. *Chemosphere*, 2019. 232: pp. 96–112.

45. Khademian, E., et al., A systematic review on carbohydrate biopolymers for adsorptive remediation of copper ions from aqueous environments-part A: classification and modification strategies. *Science of the Total Environment*, 2020. 738: p. 139829.

46. Kumar, M., et al., Novel adsorptive ultrafiltration membranes derived from polyvinyltetrazole-co-polyacrylonitrile for Cu(II) ions removal. *Chemical Engineering Journal*, 2016. 301: pp. 306–314.

47. Sivakumar, M., et al., Ultrafiltration application of cellulose acetate–polyurethane blend membranes. *European Polymer Journal*, 1999. 35(9): pp. 1647–1651.

48. Zhao, C., et al., Modification of polyethersulfone membranes–a review of methods. *Progress in Materials Science*, 2013. 58(1): pp. 76–150.

49. Cheng, Y., et al., Advanced porous materials in mixed matrix membranes. *Advanced Materials*, 2018. 30(47): p. 1802401.

50. Nasir, R., et al., Material advancements in fabrication of mixed-matrix membranes. *Chemical Engineering & Technology*, 2013. 36(5): pp. 717–727.

51. Zhang, S., et al., Effect of powdered activated carbon dosage on sludge properties and membrane bioreactor performance in a hybrid MBR-PAC system. *Environmental Technology*, 2019. 40(9): pp. 1156–1165.

52. Song, Y., et al., Powder activated carbon pretreatment of a microfiltration membrane for the treatment of surface water. *International Journal of Environmental Research and Public Health*, 2015. 12(9): pp. 11269–11277.

53. Valavala, R., et al., Pretreatment in reverse osmosis seawater desalination: a short review. *Environmental Engineering Research*, 2011. 16(4): pp. 205–212.

54. Satyawali, Y. and M. Balakrishnan, Performance enhancement with activated carbon (PAC) addition in a membrane bioreactor (MBR) treating distillery effluent. *Journal of Hazardous Materials*, 2009. 170(1): pp. 457–465.

55. Dong, Y., et al., Ultrafiltration enhanced with activated carbon adsorption for efficient dye removal from aqueous solution. *Chinese Journal of Chemical Engineering*, 2011. 19(5): pp. 863–869.

56. Argurio, P., et al., Photocatalytic membranes in photocatalytic membrane reactors. *Processes*, 2018. 6(9): p. 162.

57. Zhang, X., D.K. Wang, and J.C. Diniz da Costa, Recent progresses on fabrication of photocatalytic membranes for water treatment. *Catalysis Today*, 2014. 230: pp. 47–54.

58. Gogate, P.R. and A.B. Pandit, A review of imperative technologies for wastewater treatment I: oxidation technologies at ambient conditions. *Advances in Environmental Research*, 2004. 8(3): pp. 501–551.

59. Szymański, K., A.W. Morawski, and S. Mozia, Effectiveness of treatment of secondary effluent from a municipal wastewater treatment plant in a photocatalytic membrane reactor and hybrid UV/H$_2$O$_2$–ultrafiltration system. *Chemical Engineering and Processing-Process Intensification*, 2018. 125: pp. 318–324.

60. Darowna, D., et al., The influence of feed composition on fouling and stability of a polyethersulfone ultrafiltration membrane in a photocatalytic membrane reactor. *Chemical Engineering Journal*, 2017. 310: pp. 360–367.

61. Lin, Y.-R., et al., Sulfur-doped g-C$_3$N$_4$ nanosheets for photocatalysis: Z-scheme water splitting and decreased biofouling. *Journal of Colloid and Interface Science*, 2020. 567: pp. 202–212.

62. Lee, S.-A., et al., Use of ultrafiltration membranes for the separation of TiO$_2$ photocatalysts in drinking water treatment. *Industrial & Engineering Chemistry Research*, 2001. 40(7): pp. 1712–1719.

63. Mohamad Said, K.A., et al., Innovation in membrane fabrication: magnetic induced photocatalytic membrane. *Journal of the Taiwan Institute of Chemical Engineers*, 2020. 113: pp. 372–395.

64. Shi, Y.-D., et al., Low magnetic field-induced alignment of nickel particles in segregated poly(l-lactide)/poly(ε-caprolactone)/multi-walled carbon nanotube nanocomposites: towards remarkable and tunable conductive anisotropy. *Chemical Engineering Journal*, 2018. 347: pp. 472–482.

65. Erb, R.M., et al., Composites reinforced in three dimensions by using low magnetic fields. *Science*, 2012. 335(6065): pp. 199–204.

66. Wang, X., et al., Synthesis of high quality TiO$_2$ membranes on alumina supports and their photocatalytic activity. *Thin Solid Films*, 2012. 520(7): pp. 2488–2492.

67. Tricoli, A. and T.D. Elmøe, Flame spray pyrolysis synthesis and aerosol deposition of nanoparticle films. *AIChE Journal*, 2012. 58(11): pp. 3578–3588.

68. Della Foglia, F., et al., Hydrogen production by photocatalytic membranes fabricated by supersonic cluster beam deposition on glass fiber filters. *International Journal of Hydrogen Energy*, 2014. 39(25): pp. 13098–13104.

69. Kelly, P. and J. Bradley, Pulsed magnetron sputtering—process overview and applications. *Journal of Optoelectronics and Advanced Materials*, 2009. 11(9): pp. 1101–1107.

70. Fischer, K., R. Gläser, and A. Schulze, Nanoneedle and nanotubular titanium dioxide–PES mixed matrix membrane for photocatalysis. *Applied Catalysis B: Environmental*, 2014. 160: pp. 456–464.

71. Xu, Y., et al., Dual-functional polyethersulfone composite nanofibrous membranes with synergistic adsorption and photocatalytic degradation for organic dyes. *Composites Science and Technology*, 2020. 199: p. 108353.

72. Molinari, R., et al., Photocatalytic membrane reactors for wastewater treatment. *Current Trends and Future Developments on (Bio-) Membranes*, 2020: pp. 83–116.

73. Zheng, X., et al., Photocatalytic membrane reactors (PMRs) in water treatment: configurations and influencing factors. *Catalysts*, 2017. 7(8): p. 224.

74. Mozia, S., Photocatalytic membrane reactors (PMRs) in water and wastewater treatment. A review. *Separation and Purification Technology*, 2010. 73(2): pp. 71–91.

75. Chen, L., P. Xu, and H. Wang, Photocatalytic membrane reactors for produced water treatment and reuse: fundamentals, affecting factors, rational design, and evaluation metrics. *Journal of Hazardous Materials*, 2022. 424: p. 127493.

76. Zhang, W., et al., Membrane fouling in photocatalytic membrane reactors (PMRs) for water and wastewater treatment: a critical review. *Chemical Engineering Journal*, 2016. 302: pp. 446–458.

77. Mi, B. and M. Elimelech, Silica scaling and scaling reversibility in forward osmosis. *Desalination*, 2013. 312: pp. 75–81.

78. Motsa, M.M., et al., Organic fouling in forward osmosis membranes: the role of feed solution chemistry and membrane structural properties. *Journal of Membrane Science*, 2014. 460: pp. 99–109.

79. Choo, K.-H., et al., Use of an integrated photocatalysis/hollow fiber microfiltration system for the removal of trichloroethylene in water. *Journal of Hazardous Materials*, 2008. 152(1): pp. 183–190.

80. Arkhangelsky, E., et al., Effects of scaling and cleaning on the performance of forward osmosis hollow fiber membranes. *Journal of Membrane Science*, 2012. 415: pp. 101–108.

81. Wang, P., A.G. Fane, and T.-T. Lim, Evaluation of a submerged membrane vis-LED photoreactor (sMPR) for carbamazepine degradation and TiO$_2$ separation. *Chemical Engineering Journal*, 2013. 215: pp. 240–251.

82. Zhang, J., et al., Influence of azo dye-TiO$_2$ interactions on the filtration performance in a hybrid photocatalysis/ultrafiltration process. *Journal of Colloid and Interface Science*, 2013. 389(1): pp. 273–283.

83. Barakat, M.A., New trends in removing heavy metals from industrial wastewater. *Arabian Journal of Chemistry*, 2011. 4(4): pp. 361–377.

84. Tahir, M.B., H. Kiran, and T. Iqbal, The detoxification of heavy metals from aqueous environment using nano-photocatalysis approach: a review. *Environmental Science and Pollution Research*, 2019. 26(11): pp. 10515–10528.

85. Pohl, A., Removal of heavy metal ions from water and wastewaters by sulfur-containing precipitation agents. *Water, Air, & Soil Pollution*, 2020. 231(10): pp. 1–17.

86. Phetrak, A., J. Lohwacharin, and S. Takizawa, Analysis of trihalomethane precursor removal from sub-tropical reservoir waters by a magnetic ion exchange resin using a combined method of chloride concentration variation and surrogate organic molecules. *Science of the Total Environment*, 2016. 539: pp. 165–174.

87. Kuleyin, A. and H.E. Uysal, Recovery of copper ions from industrial wastewater by electrodeposition. *International Journal of Electrochemical Science*, 2020. 15: pp. 1474–1485.

88. Lee, A., J.W. Elam, and S.B. Darling, Membrane materials for water purification: design, development, and application. *Environmental Science: Water Research & Technology*, 2016. 2(1): pp. 17–42.

89. Kurniawan, T.A., et al., Physico–chemical treatment techniques for wastewater laden with heavy metals. *Chemical Engineering Journal*, 2006. 118(1): pp. 83–98.

90. Gherasim, C.-V. and P. Mikuláŝek, Influence of operating variables on the removal of heavy metal ions from aqueous solutions by nanofiltration. *Desalination*, 2014. 343: pp. 67–74.

91. Ujang, Z. and G. Anderson, Application of low-pressure reverse osmosis membrane for Zn^{2+} and Cu^{2+} removal from wastewater. *Water Science and Technology*, 1996. 34(9): pp. 247–253.

92. Madaeni, S. and Y. Mansourpanah, COD removal from concentrated wastewater using membranes. *Filtration & Separation*, 2003. 40(6): pp. 40–46.

93. Molinari, R., et al., Overview of photocatalytic membrane reactors in organic synthesis, energy storage and environmental applications. *Catalysts*, 2019. 9(3): p. 239.

94. Litter, M.I., Treatment of chromium, mercury, lead, uranium, and arsenic in water by heterogeneous photocatalysis. *Advances in Chemical Engineering*, 2009. 36: pp. 37–67.

95. Gao, X. and X. Meng, Photocatalysis for heavy metal treatment: a review. *Processes*, 2021. 9(10): p. 1729.

96. Zhao, H. and Y. Li, Removal of heavy metal ion by floatable hydrogel and reusability of its waste material in photocatalytic degradation of organic dyes. *Journal of Environmental Chemical Engineering*, 2021. 9(4): p. 105316.

97. Wang, Y., et al., Synthesis and characterization of activated carbon-coated $SiO_2/TiO_{2-x}C_x$ nanoporous composites with high adsorption capability and visible light photocatalytic activity. *Materials Chemistry and Physics*, 2012. 135(2–3): pp. 579–586.

98. Shoneye, A., et al., Recent progress in photocatalytic degradation of chlorinated phenols and reduction of heavy metal ions in water by TiO_2-based catalysts. *International Materials Reviews*, 2022. 67(1): pp. 47–64.

99. Mulungulungu, G.A., T. Mao, and K. Han, Efficient removal of high-concentration copper ions from wastewater via 2D g-C_3N_4 photocatalytic membrane filtration. *Colloids and Surfaces A: Physicochemical and Engineering Aspects*, 2021. 623: p. 126714.

100. Rathna, T., J. PonnanEttiyappan, and D. RubenSudhakar, Fabrication of visible-light assisted TiO_2-WO_3-PANI membrane for effective reduction of chromium(VI) in photocatalytic membrane reactor. *Environmental Technology & Innovation*, 2021. 24: p. 102023.

101. Chang, C.-J., et al., Polymer/BiOBr-modified gauze as a dual-functional membrane for heavy metal removal and photocatalytic dye decolorization. *Polymers*, 2020. 12(9): p. 2082.

102. Ferreira-Neto, E.P., et al., Bacterial nanocellulose/MoS_2 hybrid aerogels as bifunctional adsorbent/photocatalyst membranes for in-flow water decontamination. *ACS Applied Materials & Interfaces*, 2020. 12(37): pp. 41627–41643.

103. Jung, J.T., J.O. Kim, and W.Y. Choi. Performance of photocatalytic microfiltration with hollow fiber membrane, in *Materials Science Forum*. Hyungsun Kim, Junichi Hojo and Soo Wohn Lee, Editors. 2007. Trans Tech Publications Ltd. pp. 95–98, 544–545.

12 Ionic Liquid Membrane Technology for Heavy Metal Removal from Aqueous Effluents

María José Salar García, S. Sánchez-Segado, V.M. Ortiz-Martínez, C. Godinez-Seoane, and L.J. Lozano-Blanco

12.1 INTRODUCTION

In recent years, heavy metal pollution has increased significantly due to the growth of industrial activity in developing countries. Among them, the discharge of contaminated wastewater from industries of different natures such as the metal industry, mining, fossil fuel manufacturing, or solid waste management, among others, represents the main sources of pollution. In this context, heavy metals have garnered the attention of the scientific community due to their well-known toxicity and carcinogenic nature as well as other adverse effects on humankind's health. The main challenge of heavy metal pollution is due to most of the current methods for water treatment such as chlorination or solar disinfection, which are not useful for heavy metal removal.

Heavy metals are commonly defined as metals and metalloids whose density varies from 3.5 up to 7 g/cm^3 and their atomic weights range between 63.5 and 200.6. This group includes elements such as arsenic (As), cadmium (Cd), copper (Cu), chromium (Cr), lead (Pb), mercury (Hg), nickel (Ni), and thallium (Tl), among others. Despite all of them being naturally distributed on Earth, their toxicity even at low traces and their non-biodegradability currently pose one of the most important environmental and health concerns [1].

Table 12.1 presents the reference values of different heavy metals in drinking water according to the World Health Organization (WHO) and their potentially harmful effects on human health [2].

Ionic liquids (ILs) comprising an organic cation, such as imidazolium, ammonium, phosphonium, pyridinium, pyrrolidinium, or morpholinium, and an organic/inorganic anion, such as

TABLE 12.1

Reference values of heavy metals in drinking water [2]

Heavy metal	Maximum concentration in drinking water (mg/L)	Toxicity
Arsenic	0.01	Skin problems, carcinogenic
Copper	2	Liver damage
Lead	0.01	Nervous, excretory, and circulatory system damage
Chromium	0.05	Headache, vomiting, carcinogen
Cadmium	0.003	Renal issues, carcinogenic
Mercury	0.006 (usually < 0.0005)	Arthritis and damage to the brain, kidney, and heart
Nickel	0.07 (usually < 0.02)	Skin problems, asthma, carcinogenic
Zinc	Usually, 0.01–0.05	Depression, lethargy

DOI: 10.1201/9781003326281-14

hexafluorophosphate, tetrafluoroborate, triflate, tosylate or dicyanamide, among others, have been used because their high selectivity, polarity, high ionic conductivity, and thermal stability, near-zero vapor pressure, and non-flammability, which makes them suitable as multifaceted solvents for a wide variety of green processes [3–5]. ILs are able to interact with other molecules by different kinds of bonding, e.g., hydrogen bonding, Van der Waals forces, ion pair, and coulombic or electrostatic interactions, which can be modified by varying the IL structure as previously mentioned. Their tunable properties make it possible to adapt any IL to a specific process (task-specific ILs), which makes them ideal compounds for separation processes, electroanalytical and electrochemical applications, catalysis, synthesis, etc. [6–9] (see Figure 12.1).

According to these properties, ILs have been widely used for heavy metal removal from liquid effluents of different natures by liquid-liquid extraction, adsorption, and membrane technology [10, 11]. Conventional organic solvents have been commonly used for liquid-liquid extraction where the extraction process is based on the distribution of the target compound between two non-miscible liquid phases. Despite ILs overcoming the environmental problems caused by organic solvents, liquid-liquid extraction poses some limitations related to the loss of ILs during the separation process, which affects the number of extracting cycles and the overall cost of the process [10]. Unlike liquid-liquid extraction, adsorption techniques consist of the mass transfer of the target solute from a liquid solution to the surface of a specific material, such as activated carbon or zeolite among others, by adsorption [12]. The addition of ILs to these adsorbent materials allows for reducing the amount of absorbent required, the loss of IL, the energy demand, as well as the overall cost of the process [13, 14]. Finally, one of the most effective methods for selectively recovering solutes from liquid phases is membrane technology [15]. There are a wide variety of membranes, e.g., supported liquid membranes (SLMs), polymer inclusion membranes, bulk liquid membranes, and emulsion liquid membranes which are permeable to specific solutes [16–21]. The practical application of membrane technology for water treatment is mainly limited by the long-term stability of the membranes and the loss of ILs due to the requirements of operating conditions such as long periods running in continuous mode (months or even years). However, improvements in the immobilization method of ILs overcome these limitations by enhancing the stability of the membrane [6].

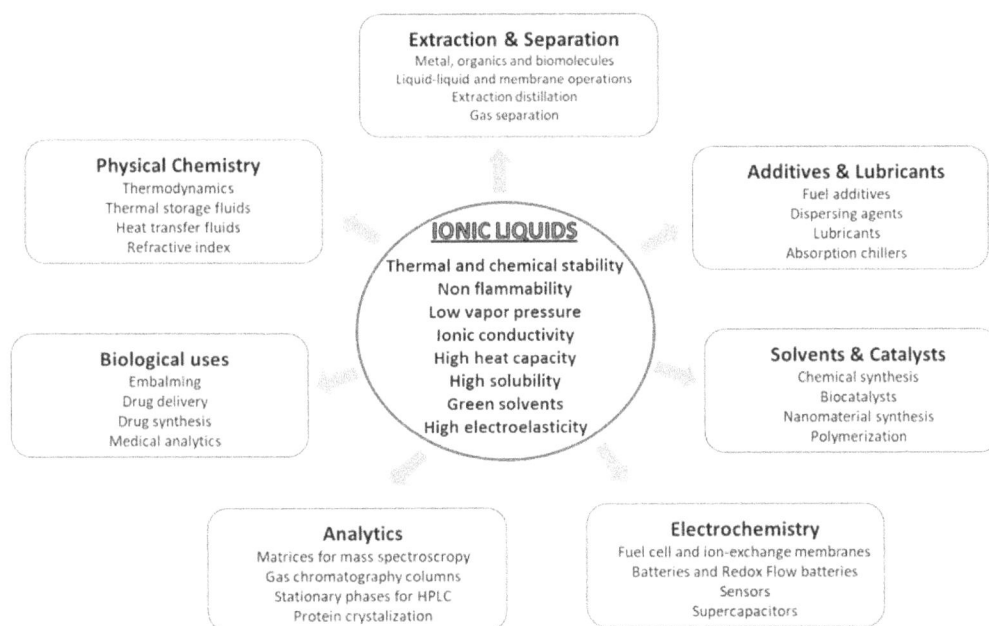

Extraction & Separation
Metal, organics and biomolecules
Liquid-liquid and membrane operations
Extraction distillation
Gas separation

Physical Chemistry
Thermodynamics
Thermal storage fluids
Heat transfer fluids
Refractive index

Additives & Lubricants
Fuel additives
Dispersing agents
Lubricants
Absorption chillers

IONIC LIQUIDS
Thermal and chemical stability
Non flammability
Low vapor pressure
Ionic conductivity
High heat capacity
High solubility
Green solvents
High electroelasticity

Biological uses
Embalming
Drug delivery
Drug synthesis
Medical analytics

Solvents & Catalysts
Chemical synthesis
Biocatalysts
Nanomaterial synthesis
Polymerization

Analytics
Matrices for mass spectroscropy
Gas chromatography columns
Stationary phases for HPLC
Protein crystalization

Electrochemistry
Fuel cell and ion-exchange membranes
Batteries and Redox Flow batteries
Sensors
Supercapacitors

FIGURE 12.1 Most common applications of ILs.

Due to the several advantages of retaining ILs into a solid support for heavy metal removal, hereafter the different types of IL-based membranes in which the IL is properly embedded into a solid matrix are described. Recent advances in the use of IL-based membranes for the extraction of metal ions are highlighted in the present chapter; also, the most commonly used ILs and polymers as well as the extraction mechanism are also addressed. Furthermore, the potential practical applications for water treatment, current limitations related to their lifetime along with the ways of improving their stability, and prospects for their scaling up are discussed.

12.2 DISCUSSION

12.2.1 SUPPORTED IONIC LIQUID MEMBRANES (SILMs)

SILMs consist of a porous matrix in whose pores an IL is retained, which is material responsible for the selectivity of the whole membrane and not the supporting material. This type of membrane brings some benefits over bulk IL membranes such as the lower amount of IL needed, easy synthesis method, and the possibility of recycling the SILM for being reused in another extraction cycle. SILMs can present different structures such as hollow fibers, flat sheets, etc. (see Figure 12.2), and in this section, the different SILMs used for heavy metal removal from aqueous solutions are grouped according to the type of cation in the IL structure.

12.2.1.1 Ammonium-Based ILs

Tricaprylmethilammonium chloride (Aliquat 336) is one of the most commonly used ILs. Güell et al. [22] used Aliquat 336 to prepare hollow fiber-supported liquid membranes (HFSLMs) for the selective removal of Cr(VI) at trace levels. The assays were performed in a hollow fiber module made of hydrophobic polypropylene hollow fibers impregnated with the ammonium-based IL. Their results show that low amounts of IL favor Cr(VI) extraction and recovery because high amounts of IL result in an increase in the viscosity of the organic solution and therefore an increase in the internal resistance of the membrane. Among the different stripping phases compositions evaluated,

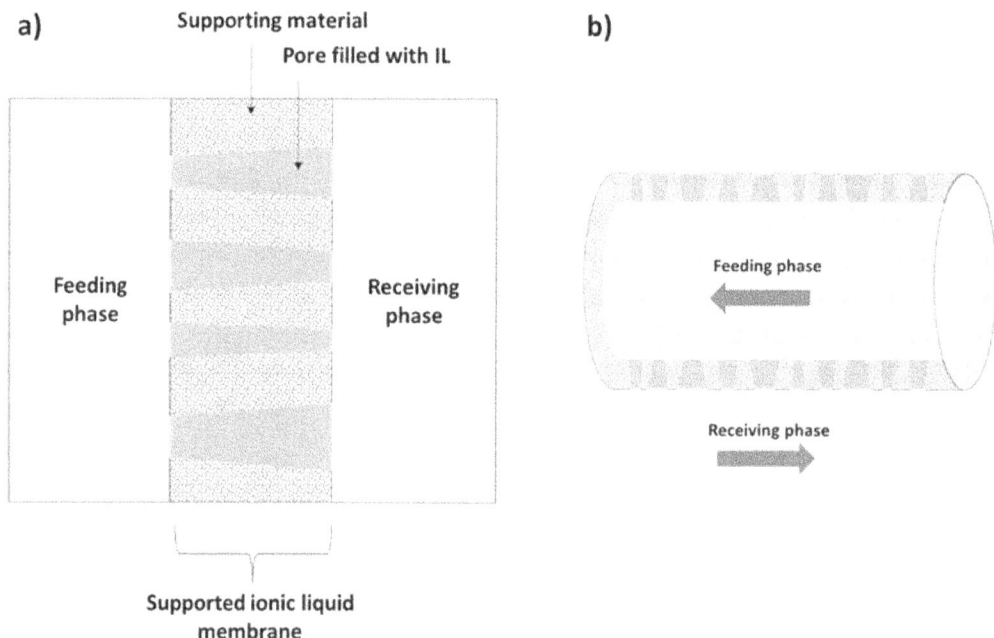

FIGURE 12.2 Scheme of different configurations of SILMs: (a) flat sheets and (b) hollow fibers.

0.5 M of NaNO$_3$ and 0.5 M of HNO$_3$ allowed the highest percentage of Cr(VI) recovery, 90.1% and 92.4%, respectively.

Aliquat 336 was also employed by Tungkijanansin et al. [23] to prepare HFSLMs for the pre-concentration of Cr(VI) and subsequent detection using a colorimetric microfluidic paper-based analytical device based on diphenylcarbazide (µPAD). The IL was retained inside of the pores of a polypropylene hollow fiber membrane (Accurel Q3/2) and the assays were performed in a U-shape set-up. Their results show the concentration yield of Cr(VI) decreases as the amount of Aliquat 336 in the membrane increases, which might be due to the high viscosity of this ammonium-based IL, which hinders metal ion transport, which is in line with those reported by [22]. For this reason, the lowest amount of Aliquat 336 (5%$_{v/v}$) was selected to perform the assays. By contrast, it was also observed that an increase of the receiving phase (NaCl) favors the extraction process, with 0.5 mol/L being the optimum concentration associated with the most intensive color caused by Cr(VI). This evidence is due to high concentrations of NaCl reducing this parameter due to the thickness of the double layer. Regarding the extraction time, the longer the time the higher the enrichment factor; so the extracting time selected was 15 min (enrichment factor of 60). The system proposed in [23] allows for the detection of very low concentrations of Cr(VI) (10 µg/L) by the naked eye thanks to the colorimetric device after 15 min in a 30 ml sample of water. The results reported are in line with those obtained by the ICP-AES technique (plasma atomic emission spectroscopy), which might be enormously useful for the detection and extraction of Cr(VI) at ppb levels in aqueous effluents.

In addition to Cr(VI), Aliquat 336 has been reported to be an excellent carrier when it is immo-bilized in a flat sheet (FSSLM) and HFSLM is used for the selective extraction of Ge(IV) [24]. Two different polymeric supports were used to prepare the FSSLM: polymeric Durapore disk polytetra-fluoroethylene (PTFE) and Millipore HVHP04700 Durapore PVDF (polyvinylidene fluoride), with the PTFE matrix being the material which allowed higher efficiencies in the transport of Ge(VI) and higher permeability coefficients. Regarding the HFSLM, a hollow fiber polypropylene/polyeth-ylene module was used. In order to optimize the membrane synthesis process, the effect of the IL concentration was evaluated. The results showed that amounts of carrier lower than 1%$_{v/v}$ are not enough to form complexes with the metal ion and the transport efficiency of the membrane is lower than 10% after 21 h. Thus, the transport efficiency of Ge(VI) increases as the concentration of IL in the membrane also increases. However, an amount of carrier higher than 5%$_{v/v}$ has a limiting effect on this parameter, and the transport efficiency is reduced by the increase in the viscosity of the Aliquat 336. Regardless of the concentration of carrier, Ge(VI) can be totally separated from Cd(II) and Co(II) whereas this metal ion can only be completely separated from Zn(II) at an IL concentration of 2.5% and 5%. Finally, in the case of a Ni(II) solution, the optimum amount of car-rier which allows the maximum separation of Ge(VI) is 2.5%$_{v/v}$ ($\alpha_{Ge/Ni}$ = 692.31). The concentration of tartaric acid needed in the feeding phase to transform Ge(VI) to anionic species and HCl used as a stripping phase were also optimized. Regarding the amount of tartaric acid, the transport of Ge(VI) increases as the concentration of tartaric acid also increases, reaching 100% of Ge(VI) transported at an acid to Ge molar ratio of 2. On the other hand, the higher the amount of HCl, the higher the Ge(VI) transport; however, the transport of this metal ion- was limited by concentration values of HCl higher than 1 M due to the decomposition of the extracted germanium species. As previously commented, one of the most important issues with regard to SILMs is their stability. For this reason, the authors evaluated the stability of the membranes during three cycles of 24 h each. The results showed that the membranes were able to transport almost 100% of Ge(VI) before 8.5 h during every cycle. As the amount of Ge(VI) transported does not decrease significantly, the stabil-ity of the membranes is not affected for at least three days. The authors concluded that under these optimum conditions, the Aliquat 336-based FSSLM system selectively transported 98% of Ge(VI) from a heavy metal mixture solution. On the other hand, the HFSLM system was able to completely transport the Ge(VI) from an aqueous solution after 240 min. Comparing both systems, the trans-port rate of Ge(VI) was faster through the HFSLM system than through the FSSLM system, with a transport ratio similar to those obtained in liquid-liquid extraction.

TABLE 12.2

Thioglycolate-based ILs synthesized by [25]

IL name	Abbreviation
Methyltrioctylammonium butylsulfanyl acetate	$[N_{1888}][C_4SAc]$
Methyltrioctylammonium pentylsulfanyl acetate	$[N_{1888}][C_5SAc]$
Methyltrioctylammonium hexylsulfanyl acetate	$[N_{1888}][C_6SAc]$
Methyltrioctylammonium benzylsulfanyl acetate	$[N_{1888}][BnSAc]$
Methyltrioctylphosphonium butylsulfanyl acetate	$[P_{1888}][C_4SAc]$
Methyltrioctylphosphonium pentylsulfanyl acetate	$[P_{1888}][C_5SAc]$
Methyltrioctylphosphonium hexylsulfanyl acetate	$[P_{1888}][C_6SAc]$
Methyltrioctylphosphonium benzylsulfanyl acetate	$[P_{1888}][BnSAc]$

Platzer et al. [25] synthesized eight task-specific ILs containing thioglycolate derivatives as an anion and different ammonium and phosphonium cations for the selective extraction of Cd(II) and Cu(II) in aqueous solutions. Table 12.2 provides the different ILs synthesized by the authors which were retained in a polypropylene Accurel PP S6/2 capillary hollow fiber membrane.

The results show that regardless of the type of cation, all ILs synthesized exhibit high affinity toward Cd(II) and Cu(II), which might be due to the affinity of the thioglycolate anion toward both metal ions. However, the extraction process is faster in the case of Cd(II), reaching 90% of the metal ion extraction after 30 min. The results showed a similar tendency when the ILs were retained in the hollow fiber membranes. Regardless of the type of IL, values of the extraction rate of Cd(II) above 95% were obtained after 30 min. of the process. By contrast, Cu(II) was not totally extracted after 6 h, yielding a better extraction rate using phosphonium-based ILs than those containing ammonium cations. After 6 h, the eight thioglycolate derivative ILs were able to extract more than 90% of Cd(II), reaching a maximum percentage of metal extracted of 94% for membranes containing $[P_{1888}][C_4SAc]$ or $[P_{1888}][C_5SAc]$. Even though membranes containing $[P_{1888}][BnSAc]$ allowed 91% extraction of Cd(II) after 2 h, the results obtained using the immobilized IL, in terms of speed, are slower than using bulk IL. However, the immobilization of ILs into hollow fiber membranes might be useful for Cd(II) and Cu(II) removal in continuous mode processes, which is necessary for practical applications in wastewater treatment.

More recently, a quaternary salt such as choline glycinate was used by Foong et al. [26] to synthesize electrospun nylon 6,6 nanofiltration membranes for Cu(II) removal after using a conductor like screening model for real solvent (COSMO-RS) analysis based on sigma analysis, sigma potential and activity coefficient at infinite dilution of Fe(III), Ni(II), Cu(II), and Pb(II). The COSMO-RS analysis allowed us to find a potential good combination of cation and anion structure of amino-based ILs. As a result, choline cation and amino anion showed potential properties to be used as extraction agents for the removal of heavy metal ions. The procedure of synthesis involves dissolving nylon 6,6 pellets in acetic acid glacial and formic acid (1:1). Then, the electrospinning method was employed by using the homogenous polymer solution and by applying a voltage of 21 kV, which allows ejection of the spiral of the polymer over the collector surface. Subsequently, the nanofiber membrane was immersed in the IL-deep eutectic solvent solutions (made of choline hydroxide and amino acid) with different concentrations for 24 h. Different molar ratios of choline hydroxide to glycine in the IL-deep eutectic solvent were evaluated. The results showed that a higher amount of IL-deep eutectic solvent in the membrane obtained by increasing the proportion of glycine over the choline hydroxide concentration results in a reduction of the porosity and the pore size distribution and, therefore, an increase in the diameter of the membrane due to the interaction between glycine and nylon 6,6. Regarding the hydrophilicity of the membranes, the water contact angle analysis showed an increase in the membrane hydrophilicity as the amount of IL-deep eutectic solvent increased. A similar effect of the amount of IL on membrane permeability was observed but values of the molar ratio of choline hydroxide and glycine above 1:3 result in a decrease in the membrane permeability because of a reduction in the pore size distribution of the membrane. The authors found 1:2 to be the optimum molar ratio of choline hydroxide to glycine to prepare the nylon 6,6 nanofiltration membrane for Cu(II) removal. This membrane showed a maximum permeability of 5,504 L/m^2/h/bar and the maximum percentage of Cu(II) rejected from synthetic wastewater containing different heavy metal ions after 3 h of operation was 80%.

12.2.1.2 Imidazolium-Based ILs

Another group of ILs very commonly used for extraction processes is those containing imidazolium cations with different lengths of the linear alkyl chain. As counterions, they are usually combined with organic or inorganic fluor-based anions such as hexafluorophosphate (PF$_6^-$) or bis(trifluoromethylsulfonyl)imide (NTf$_2^-$), resulting in hydrophobic ILs. Among the benefits of imidazolium-based ILs are the simplicity of their synthesis and their tendency to form complexes with metallic ions, e.g., cobalt, copper, or ruthenium, which make them suitable for divalent heavy metal removal [27].

Jean et al. [16] investigated the transport of Cd(II), Cr(III), and Hg(II) ions from acidic aqueous solutions through SILMs containing isooctylmethylimidazolium bis-2-ethylhexylphosphate as a carrier in a microporous matrix of PVDF. Parameters such as permeability coefficient P and initial mass flux J_0 were determined and related to the association constant K_{ass} and apparent diffusion coefficient D* in order to evaluate the transport of these ions across the SILMs at different ions concentrations, the amount of IL in the SLM, temperature, and pH. All experiments were repeated using both hydrophilic and hydrophobic PVDF supports and performed in a double cylindrical cell (V = 220 ml). Among the two different matrices, the results showed higher performances for all conditions studied when the IL is retained in the pores of the hydrophilic PVDF, which might be due to the reduction of the thickness of the diffusion boundary layers caused by the hydrophilicity of the support, which modifies the interface membrane/aqueous solution. Regarding the effect of the initial concentration of metal ions, the results show that the speed of the transport strongly depends on the hardness of the cation. Thus, harder cations, e.g., Cr(III), tend to form a more stable complex with the ligand than softer cations, e.g., Cd (II), which hinder the dissociation of the complex at the membrane/receiving liquid phase interface. On the contrary, the transport of weaker complexes through the membrane is faster since the formation and dissociation of the complexes are easier. Regarding the pH effect, it was shown that the transport of protons in acidic media might compete with the transport of metal ions and inhibit their transport. According to these results, the optimum extraction ratio for each metal ion was obtained at different pH values, following the trend Hg(II)$_{(pH2/2)}$ 31%, Cd(II)$_{(pH3/3)}$ 35%, and Cr(III)$_{(pH3/3)}$ 24%. The same trend was observed in a multicomponent system where all metal ions were simultaneously present. However, these values were significantly reduced when all metal ions were together in the same solution due to all of them competing at the same complexing site in the membrane surface. Regarding the selectivity of the membrane when metal ions were mixed, the following values were obtained: αCd(II)/Hg(II) = 1.2 and αCd(II)/Cr(III) = 2.6. Regarding the temperature effect, the transport of all metal ions was favored by increasing temperature, which enhances molecular mobility and facilitates complex dissociation. Finally, the stability analysis of the membranes showed that 81% of the IL was retained in the membrane after four cycles while the rest of the carrier was released to the feeding/receiving phase. These results support the potential use of SLMs based on isooctylmethylimidazolium bis-2-ethylhexylphosphate for the treatment of real wastewater containing heavy metals.

In 2020 Malas et al. [28] used 1-butyl-3-methylimidazolium dicyanamide ([BMIM][DCA]) to prepare SLMs using hydrophobic polyethylene as a polymeric matrix. Three types of membranes were prepared by soaking the polyethylene membrane in solutions containing different ethanol to IL ratios, 15:5, 12:8, and 10:10 (v/v), for 24 h and letting them dry for 24 h. Gravimetric analysis showed that the percentage of impregnation was higher in those membranes submerged in the solution containing an ethanol to IL ratio of 10:10$_{(v/v)}$), reaching up to 58.3%, whereas this value was 40.85%

lower in membranes submerged in the solutions with the lowest volume ratio of IL ($15:5_{(v/v)}$). The morphological changes in the membrane surface caused by ILs were analyzed by scanning electron microscopy (SEM). The results showed that the fibers and pores observed in the bare polyethylene membrane were covered as the volume ratio of IL to ethanol increased whereas mechanical stability did not increase significantly. Regarding the pure water flux and the porosity, both parameters decreased as the volume ratio of IL to ethanol increased. However, the stability tests reported that after being submerged in water for 2 h the IL was totally released from the membrane. As previously mentioned, this is an important limitation of SILMs which hinders their practical application. For this reason, it is recommended that a transmembrane pressure lower than 1 bar be used [29]. The membranes prepared were employed for heavy metal removal. The results showed that membranes containing the highest amount of IL were able to reject up to 68.05% of Zn(II), 52.87% of Cd(II), and 51.47% of Ni(II). The rejection mechanism of the membrane is based on the ion exchange between the dicyanamide anion and the heavy metal cations, metallic hydroxides precipitation, and electrostatic attraction between the IL anion and the heavy metal cations. Despite the stability of SILMs, this is one of the most challenging issues [10, 30] and further work is needed to overcome this limitation; the results in terms of metal removal are promising for their potential application for treating wastewater contaminated with heavy metals.

More recently, Zheng et al. [31] grafted 1-vinyl-3-butyl imidazolium tetrafluoroborate ($VBImBF_4$) over the surface of polyethersulfone (PES) by electrospinning. The result was a PES-g-IL nanofiltration membrane with heavy metal ion absorption properties. The synthesis method consists of grafting the IL on PES and irradiating the blend with 30 kGy gamma rays at room temperature for 17 h. Then, the ungrafted IL is removed using methanol and the grafting degree determined (DG). PES-g-IL with DG of 0.9%, 3.1%, and 3.6% were used to prepare the nanofiltration membranes by electrospinning by applying a voltage of 15 kV. All PES-g-IL membranes showed good mechanical properties, e.g., tensile strength ranged between 6 and 7 MPa and elongation at break was in the range of 10%–15%. Regarding the morphology of the membranes, all membranes were formed by nanofibers, but their size reduces as the DG increases reaching a minimum value of 170 nm when the DG is 3.6%, more than 1.6 times smaller than for the ungrafted PES. The reduction in size of the nanofibers is mainly due to the enhanced conductivity of the PES-g-IL solutions, which allow a better formation of the fibers during the electrospinning process. Smaller nanofibers' size results in an increase in the surface area of the membrane up to 23.5 m^2/g when the DG is 3.6%, higher than for the ungrafted PES (20.5 m^2/g). Regarding the water contact angle analysis, this parameter did not change that much over time for the ungrafted PES (130°) whereas in the cases of the PES-g-IL this value decreased after 60 s from 130° to 106° (DG: 0.9%), from 110° to 59° (DG: 3.1%), and from 84° to 0° (DG: 3.6%), the latter one being the most hydrophilic membrane. Regarding the absorption capacity of Cd(II), all PES-g-IL membranes exhibit higher values than the ungrafted PES due to the higher surface area and the presence of more adsorption sites in the grafted membranes. For this reason, PES-g-IL membrane with DG of 3.6% exhibits the highest adsorption capacity (72.5 mg/g). Regarding the pH, contact time, and temperature, all these parameters show a similar effect on the adsorption capacity of Cd(II). An increase in the pH or temperature results in an increase in the adsorption capacity. Despite the contact time initially also favoring Cd(II) absorption capacity, it shows a limiting effect once the system reaches equilibrium and therefore the absorption capacity achieves a steady state. Finally, the desorption of the metal ion and the recyclability of the membranes were evaluated after three cycles using ethylenediaminetetraacetic acid (EDTA) as a desorbing agent for Cd(II). The desorption rate of the PES-g-IL membrane (DG: 3.6%) remained close to 90% during three cycles whereas the adsorption capacity ranged between 60% and 75% for the same period. Good adsorption/desorption capacity allows the reusability of the membranes for heavy metal removal, which is crucial for their practical application in the field of wastewater treatment.

Also, in 2022, an imidazolium-based IL was used by Rosli et al. [32] to functionalize crosslinked chitosan/poly(vinyl alcohol) nanofibers with 1-allyl-3-methylimidazolium chloride (AMIMCl) for heavy metal removal. The electrospun nanofiber membranes are obtained by applying an electrical potential of 23 kV between a needle tip in contact with the polymer solution, containing hydrolyzed chitosan (HC) and polyvinyl alcohol (PVA), and the collector made of aluminum foil, where the nanofibers are deposited. Then, the nanofibers are crosslinked by glutaraldehyde and functionalized by AMIMCl. The adsorption capacity of the IL-based nanofiber synthesized was evaluated in pure heavy metal solutions of $Pb(NO_3)_2$, $Mn(NO_3)_2$, and $Mn(NO_3)_2$. The results show that the absorption capacity of the IL-modified nanofibers is higher than the values obtained with HC/PVA/GLA nanofibers, reaching a maximum value of 82.5% in the case of Pb(II) ions. The increase in the absorption capacity might be due to the electrostatic interaction between the positive charges (Pb(II) and imidazolium cations) and the negative charges (chloride anion and the pair of electrons over the nitrogen in the chitosan structure). The absorption capacity of metal ions for IL-modified HC/PVA/GLA nanofibers and non-modified nanofibers follows the same trend: Pb(II) $\gg$ Mn(II) > Cu(II). Regarding the effect of pH on the adsorption capacity of modified nanofibers, the results show that acidic pH (lower than 7) significantly reduces this parameter, which might be due to electrostatic repulsion between the metal cations and the protonated chitosan. On the other hand, basic pH values promote the precipitation of Pb(II). Among the different pH values evaluated, the maximum adsorption capacity of Pb(II) (39.0516 mg/g) was obtained at pH 9. Regarding the initial concentration and the contact time, both parameters positively affect the adsorption capacity. However, the adsorption capacity exhibits a plateau for contact time higher than 120 min due to the system reaching equilibrium. After 270 min, the percentage of Pb(II) removed increased up to 87.7%. The different tests performed to better understand the high adsorption capacity of Pb(II) compared to Mn(II) and Cu(II) showed a maximum adsorption capacity of 166.34 mg/g related to an 82.5% of Pb(II) removal.

12.2.1.3 Phosphonium-Based ILs

An alternative to ammonium and imidazolium-based ILs are those containing phosphonium cations such as trihexyltetradecylphosphonium bromide (Cyphos IL 102). Alguacil et al. [33] impregnated the pores of a polymer matrix of Durapore HVHP1400 with Cyphos IL 102 in order to synthesize FSSLM for Cr(VI) removal in aqueous solutions. The results show a maximum percentage of metal ion recovery of 58.3%, more than 14% higher than using trihexyltetradecylphophonium chloride (Cyphos IL 101). The IL-based FSSLM allowed selective removal of Cr(VI) from binary aqueous synthetic solutions containing other metal ions, e.g., Cu(II), F(III), and Cr(III). The effects of other parameters such as metal ion concentration, stirring speed, or amount of IL in the membrane, among others, were also investigated. The maximum value of mass transfer coefficient (K) was obtained for solutions with 0.01 g/L of Cr(VI) stirred at 1,000/min when the FSSLM contains 0.015 M of Cyphos IL 102. The results were compared to those obtained with a pseudo-emulsion membrane strip dispersion (PEMSD) prepared with Cyphos IL 102 (PEMSD). Despite the use of a PEMSD, the overall mass transfer coefficient did not improve, and the percentage of Cr(VI) removed was significantly higher (95.9%). These results demonstrate that Cyphos IK 102 allows the selective transport of Cr(VI) in aqueous solutions by using different SILM configurations.

The lifetime of SILM is limited by several degradation mechanisms which take place in the separation system and can be grouped according to the process which causes the degradation into: (i) the solubility of the organic phase into the surrounding aqueous phase or vice versa and (ii) the formation of emulsions. For these reasons, the use of hydrophobic ILs as an organic phase for synthesizing SILMs is one of the simplest ways to improve the stability of this type of liquid membrane. However, the hydrophobicity of the IL is not enough to avoid all degradation mechanisms. Another factor involved in membrane stability is the interaction between the IL and the polymeric matrix, which might be affected by the damage to the polymer structure. According to the literature, quaternary ammonium- and phosphonium-based ILs favor SILM stability. Membranes containing some ILs belonging to these groups involve an extraction mechanism based on the formation of an anionic metal complex between the metal cation and the counterion present in the aqueous phase (e.g., Cl⁻). The anionic metal complex is then extracted by SILM via exchange with the IL counterion present in the membrane. An alternative mechanism is the extraction of a positively charged compound formed by the binding of a bulky anion added to the IL. Both options make it possible to extract different metal ions regardless of their aqueous speciation mechanism [34]. In addition to the type of IL and the polymeric matrix selected, membrane stability also depends on the nature of the stripping solution [35]. These findings might help to select a proper IL to be used as an organic phase in SILMs, which could improve membrane stability and favor its application for heavy metal removal.

12.2.2 POLYMER IONIC LIQUID INCLUSION MEMBRANES (PILIMs)

12.2.2.1 Ammonium-Based ILs

Elias et al. [36] employed trioctylmethylammonium thiosalicylate (TOMATS) to prepare PILIMs (see Figure 12.3) for mercury (Hg) concentration from natural waters. The membranes were synthesized by the casting method using cellulose triacetate (CTA) as a polymeric matrix and chloroform as a solvent with an IL to polymer ratio of 1:1. Their results showed that the Hg(II) extraction capacity of IL-based PILIMs increases over time reaching almost 100% of extraction efficiency after 24 h. The authors combined PILIMs technology with energy dispersive X-ray fluorescence (EDXRF) for Hg(II) concentration and subsequent determination, obtaining a Hg(II) limit of detection of 0.2 μg/L. The metal ion extracted by this system was stable for at least six months, significantly longer than the stability period of the results previously discussed when using SILMs.

In addition to SILMs, Aliquat 336 has also been used to synthesize PILIMs for Cr(VI) removal by Sellami et al. [37]. In that work, PVDF was used as a polymeric matrix and 2-nitrophenyl octyl ether (2NPOE) as a plasticizer for PILIMs, which were prepared by the casting method. Their results reveal that both the IL and 2NPOE greatly modify the mechanical properties of the membrane. The higher the amount of Aliquat 336 in the membrane the lower the tensile strength and, therefore, the higher the elongation at break due to the plasticizing effect of the IL. Moreover, the addition of 2NPOE allows doubling the percentage of elongation at breakup to 238.3%, higher than using polyvinyl chloride (PVC) or cellulose triacetate (CTA) as a polymer matrix (212.9% and 13.9%, respectively). Regarding the hydrophilicity/hydrophobicity of the membranes, a balance between both parameters is needed to ensure metal extraction and membrane stability. The water

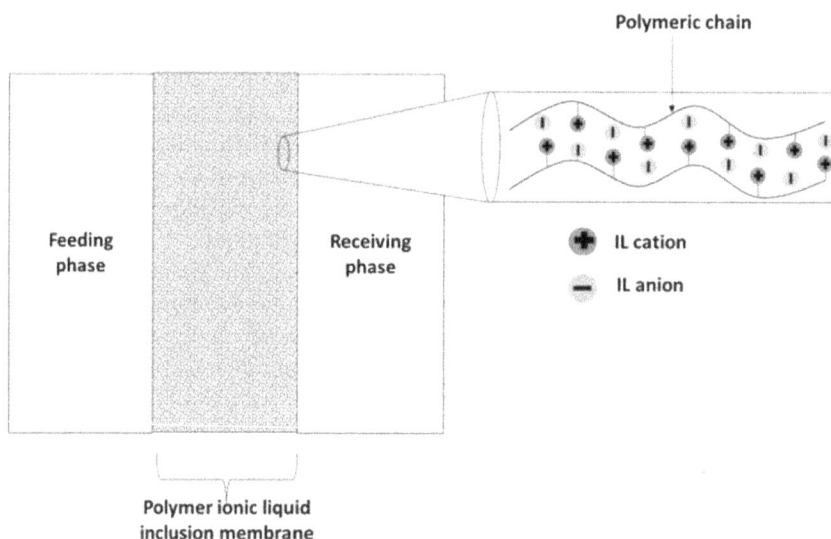

FIGURE 12.3 Generic scheme of the structure of PILIMs.

contact angle data show an increase in the hydrophilicity of the PVDF-based membranes as the amount of IL also increases. The values of the water contact angle are lower for PVDF membranes containing Aliquat 336 and 2NPOE than those using another polymeric matrix, e.g., CTA or PVC. Among the different membrane compositions, the maximum percentage of Cr(VI) extracted from an aqueous solution containing a mixture of metal ions was obtained by using a membrane with 75% of PVDF, 20% of Aliquat 336, and 5% of 2NPOE (96.9%). Other metal ions such as Cd(II) or Fe(III) were also transported through the membrane but in much lower proportion (2.7% and 0.3%, respectively) whereas the PILIMs were not permeable to Cu(II), Zn(II), Ni(II), Co(II), and Pb(II). These results demonstrate the selectivity of the Aliquat 336-based PILIMs for Cr(VI) extraction. Finally, the stability of the membrane was evaluated in terms of Cr(VI) extraction capacity after 24 cycles of 8 h each. During the first 16 cycles, the value of the recovery factor remained very close to the initial value ($\approx$98%). Then, this factor was gradually reduced to 54% after 24 cycles. Both metal ion selectivity and the stability of PILIMs make them a suitable material for practical applications such as for heavy metal removal from wastewater.

12.2.2.2 Imidazolium-Based ILs

Turgut et al. [38] evaluated the effect of the alkyl chain length of symmetric imidazolium bromide-based ILs on the extraction capacity of Cr(VI). To prepare the membranes, the different ILs synthesized were mixed with a polymer solution containing polyvinylidene fluoride-co-hexafluoropropylene (PVDF-co-HFP), the plasticizer, and acetone and the solvent was left to evaporate overnight. The micrographs of the membranes show an increase in the porosity caused by the presence of ILs, which is not observed in IL-free membranes. An increase in the porosity results in an increase in the active surface, which favors the mobility of the carrier inside of the membrane. Regarding the effect of alkyl chain length, the shorter the alkyl chain length the lower the transport values for Cr(VI). Thus, the highest transport of Cr(VI) was observed when PILIM containing 1,3-didecyl-1H-imidazol-3-ium bromide reached an initial mass transfer coefficient of 1.74×10^{-6} mol/m^2/s (J_0). It was also observed that J_0 increased as the amount of IL in PILIM also increased, regardless of the alkyl chain length. Among the different plasticizers tested, 2NPOE allowed the maximum value of J_0 when combined with 1,3-didecyl-1H-imidazol-3-ium bromide at a percentage of 11.4%$_{w/w}$. The membranes also exhibited high selectivity toward Cr(VI) when it was mixed with other metal ions such as Fe(III), Co(II), Cu(II), Ni(II) Cd(II), and Zn(II) in an aqueous solution, with Fe(II) being the most interfering cation followed by Zn(II) and Cu(II). By contrast, the transport of other metals through the membranes was limited. As the best results were obtained with PILIM containing the longest alkyl chain, the maximum percentage of Cr(VI) extracted was 85.0%, and the stability of this membrane was evaluated in-depth. The stability tests showed that the extraction percentage of the metal ion was reduced to 66.2% after the second cycle. Five cycles later, the extraction efficiency was in the range of 66%–59% whereas, during the last three cycles, the extraction efficiency of Cr(VI) was reduced by 10%, reaching a final value of 50.8%. In line with this work, the authors also investigated the effect of different additives on Cr(VI) extraction by using PVC-based PILIMs containing symmetric ILs with different alkyl chain lengths [39].

Zheng et al. [40] used 1-vinyl-3-butyl imidazole chloride (VBIM) to graft polyamide 6 (IL-g-PA6) to obtain a porous material for Cr(VI) removal. The membranes were synthesized by the immersion-precipitation technique. The morphological characterization of the membranes showed that the presence of the grafted IL in PA6 reduces the pore size distribution and therefore increases the porosity of the membranes. Despite the reduction in the pore size distribution, the water flux and Cr(VI) adsorption are also favored by the increase of IL concentration in IL-g-PA6 due to its hydrophilicity. The best results in terms of Cr(VI) adsorption were obtained with the IL-g-PA6 membrane containing 8% of IL, which allowed 2.7 times higher Cr(VI) being absorbed than with the bare PA6 membrane.

More recently, Shen et al. [41] employed 1-allyl-3-methyl imidazolium hexafluorophosphate, [AMIm][PF$_6$], to prepare silica-poly(ionic liquid) nanoparticles to be deposited over a nanofiltration membrane for heavy metal removal. The interfacial polymerization of different amounts of SiO_2-[AMIm][PF_6] nanoparticle solutions over a polyamide (PA) layer support allows several thin-film nanocomposite (TFN) membranes to be obtained using PES as a membrane polymer base. The characterization of the particles shows a reduction in the pore size distribution as a result of the grafting of poly([AMIm]$^+$) on the microporous surface of SiO_2. The mean value of the pore size distribution of the IL-based particles (0.62 nm) allows the transport of hydrated metal ions as well as water molecules [41]. The permeability tests showed that the incorporation of SiO_2-[AMIm][PF_6]/PA into the membranes allowed an increase in water permeability up to 34% and $MgCl_2$ rejection up to 41.6%. Regarding the heavy metal removal tests, it was observed that the addition of SiO_2-[AMIm][PF_6]/PA enhanced the separation of all metal ions evaluated (Cu(II), Ni(II), Cd(II), and Zn(II)), compared to the bare PA membranes, reaching an average improvement of 20.4% compared to the non-modified membrane. The maximum rejection percentage was obtained for Zn(II) (90.3%) whereas the minimum value was obtained for Cd(II), which was almost 20% lower but still higher than with the bare PA membrane. The separation process is favored by low pH or SO_4^{2-} concentration whereas a high concentration of NaCl reduces the separation rate. In essence, the incorporation of silica-poly(ionic liquid) nanoparticles in the TFN membrane allows us to not only construct water channels but also improve heavy metal separation due to the electropositivity of the nanoparticles because of the presence of ILs, which makes them suitable for heavy metal removal in wastewater treatment.

12.2.2.3   Phosphonium-Based ILs

Kogelnig et al. [42] used trihexyl(tetradecyl)phosphonium chloride (Cyphos IL 101) to prepare polymer inclusion membranes using PVC as a polymeric matrix and tetrahydrofuran (THF) as a solvent by the casting method. Different amounts of Cyphos IL 101 were used to prepare the PIMs and the results showed that an increase in the IL concentration results in an increase in membrane flexibility. Among the different membranes synthesized, those containing 20 wt% of Cyphos IL 101 were selected to perform the tests since a higher content of Cyphos IL 101 results in a sticky membrane. The stability tests in air reported a loss of weight of 0.5% after five days at room temperature, which remained at a very similar value after 40 days. However, the unplasticized membrane lost seven times more weight for the same period. In line with this, the stability tests performed in ultrapure water at room temperature showed a similar weight loss (3%) of the unplasticized membrane in air but in a shorter time (27 days), which is related to the loss of THF. By contrast, the weight loss of the plasticized membrane with 30% of Cyphos IL 101 reaches a maximum value of 7% after 27 days in water, around 6.5 higher than the maximum value obtained when the membranes were exposed to the air. On the other hand, the stability of the membranes was also investigated in hydrochloric acid solution (5 M) as a suitable extracting phase for Zn(II) ions. After 26 days, the weight of the plasticized membrane did not substantially change, which means that Cyphos IL 101 remains inside the PILIM. Comparing the extraction capacity of both membranes, that containing 30 wt% of Cyphos IL 101 reached a maximum amount of Zn(II) removal of 5 mg/g of the membrane after 145 h ($\approx$30%) whereas the unplasticized membrane was not able to extract the metal ion. Then, these results were compared with those obtained with a plasticized membrane after being submerged in an HCl (5 M) solution for 40 days, with similar values obtained for Zn(II) removal. By contrast, when the Cyphos IL 101 membrane is submerged in pure water during the same period, the extraction efficiency of the metal ion decreases significantly (more than three times lower) due to the leaching of ILs. The authors also reported that an increase in the membrane area containing the same percentage of Cyphos IL 101 allowed an increase in the extraction efficiency by up to 90% within 50 h and more than 80% of the extracted metal ion could be back-extracted by using an H_2SO_4 (1 M) solution. PILIMs exhibit degradation mechanisms similar to SILMs; however, in this case, the interactions between the liquid phase of the membrane and the polymeric matrix are stronger, which results in a higher lifetime of PILIMs compared to SILMs [34].

12.3   CONCLUSIONS AND FUTURE PROSPECTS

The unique properties of ILs have promoted their application in many research fields such as heavy metal removal due to their selectivity and adsorption capacity. ILs are being used for this purpose in the bulk or emulsion state but, in this case, the reusability of ILs is limited, which is crucial for their practical applications in wastewater treatment. An alternative to liquid membranes is IL solid membranes such as SILMs where the IL fills the pores of a porous matrix. However, the stability of these types of membranes is still limited due to the leakage of ILs from the pores because of different degradation mechanisms, as previously discussed. A more interesting alternative, from a durability and recycling point of view, is the polymeric inclusion membranes in which the IL is entrapped into the polymeric matrix. This type of membrane exhibits more stability than the liquid membranes and SILMs due to the stronger interaction between the IL and the polymeric matrix but sometimes the extraction capacity is slightly lower in the solid/liquid extraction than in the liquid/liquid process. However, due to the high cost of ILs, it is necessary to recycle them and the simplest way to do it is to immobilize the IL in a membrane.

As previously commented, there are a wide variety of ILs whose structure can be adapted to a specific process (task-specific ILs) by varying the anion and cation of their structure. However, so far, a reduced number of ILs have been used for heavy metal removal in the form of a solid membrane. Most of them contain ammonium, imidazolium, or phosphonium groups while other groups such as pyrrolidinium, pyridinium, or morpholinium have been less exploited so far for heavy metal removal. For this reason, the main potential of these organic salts is the possibility of adapting the IL structure for a specific heavy metal extraction, which results in an unlimited range of applications.

A key factor for the practical application of IL-based membranes for water treatment is the lifetime of the membranes. So far, SILMs have exhibited limited stability in aqueous solutions due to the loss of IL, which reduces the selectivity of the membranes over time. This limitation might be addressed by varying the hydrophobicity/hydrophilicity of the IL selected as well as its affinity to the polymeric matrix and the properties of the feeding and receiving phase. The emergence of PILIMs has improved the durability of the IL-based membranes since the interaction between the polymeric matrix and the IL in this type of membrane is stronger than that in SILMs. For this reason, the loss of IL is lower in PILIMs than in SILMs. According to the results discussed above, despite IL-based membranes being far from their industrial application due to their reduced durability, it seems that PILIMs are closer to their industrial application than SILMs. Future research lines might be focused on using less exploited ILs such as morpholinium-, pyrrolidinium-, or pyridinium-based ILs, whose properties might promote heavy metal extraction, when properly combined with a polymeric matrix which might enhance the stability of the membranes and therefore promote their industrial application for the treatment of real aqueous effluents.

ACKNOWLEDGMENTS

The authors wish to acknowledge the financial support of the Ministry of Science, Innovation, and Universities (MICINN) ref. RTI2018-099011-B-I00 and the Seneca Foundation Science and Technology Agency of the Region of Murcia ref. 20957/PI/18. Dr. S. Sánchez-Segado wishes to acknowledge the Ministry of Science, Innovation, and Universities of Spain its support through the "Beatriz Galindo" fellowship BEAGAL18/00079.

REFERENCES

[1] A. Bashir, L.A. Malik, S. Ahad, T. Manzoor, M.A. Bhat, G.N. Dar, A.H. Pandith, Removal of heavy metal ions from aqueous system by ion-exchange and biosorption methods, *Environ. Chem. Lett.* 17 (2019) 729–754. https://doi.org/10.1007/s10311-018-00828-y.

[2] W.H. Organization, Guias para la calidad del agua de consumo humano: Cuarta edición que incorpora la primera adenda, Ginebra, 2011. http://apps.who.int/ (accessed May 19, 2022).

[3] T. Welton, Room-temperature ionic liquids. Solvents for synthesis and catalysis, *Chem. Rev.* 99 (1999) 2071–2083. https://doi.org/10.1021/cr980032t.

[4] N. V. Plechkova, K.R. Seddon, Ionic liquids: "designer" solvents for green chemistry. In: Pietro Tundo, Alvise Perosa and Fulvio Zecchini (Eds.), *Methods and Reagents for Green Chemistry: An Introduction.* 2007, pp. 103–130. John Wiley & Sons Inc. https://doi.org/10.1002/9780470124086.CH5.

[5] N. V Plechkova, K.R. Seddon, Applications of ionic liquids in the chemical industry, *Chem. Soc. Rev.* 37 (2008) 123–150. https://doi.org/10.1039/b006677j.

[6] J. Wang, J. Luo, S. Feng, H. Li, Y. Wan, X. Zhang, Recent development of ionic liquid membranes, *Green Energy Environ.* 1 (2016) 43–61. https://doi.org/10.1016/j.gee.2016.05.002.

[7] A.D. Sawant, D.G. Raut, N.B. Darvatkar, M.M. Salunkhe, Recent developments of task-specific ionic liquids in organic synthesis, *Green Chem. Lett. Rev.* 4 (2011) 41–54. https://doi.org/10.1080/17518253.2010.500622.

[8] J. H. Davis, Jr., Task-specific ionic liquids, *Chem. Lett.* 33 (2004) 1072–1077. https://doi.org/10.1246/cl.2004.1072.

[9] X. Han, D.W. Armstrong, ionic liquids in separations, *Acc. Chem. Res.* 40 (2007) 1079–1086. https://doi.org/10.1021/ar700044y.

[10] A. Stojanovic, B.K. Keppler, Ionic liquids as extracting agents for heavy metals, *Sep. Sci. Technol.* 47 (2012) 189–203. https://doi.org/10.1080/01496395.2011.620587.

[11] V. Rajadurai, B.L. Anguraj, *Ionic liquids to remove toxic metal pollution*, Springer International Publishing, 2021. https://doi.org/10.1007/s10311-020-01115-5.

[12] G. Crini, E. Lichtfouse, L.D. Wilson, N. Morin-Crini, Conventional and non-conventional adsorbents for wastewater treatment, *Environ. Chem. Lett.* 17 (2019) 195–213. https://doi.org/10.1007/s10311-018-0786-8.

[13] M. Tian, L. Fang, X. Yan, W. Xiao, K. Ho Row, Determination of heavy metal ions and organic pollutants in water samples using ionic liquids and ionic liquid-modified sorbents, *J. Anal. Methods Chem.* 2019 (2019) 19. https://doi.org/10.1155/2019/1948965

[14] M. Gharehbaghi, F. Shemirani, Ionic liquid modified silica sorbent for simultaneous separation and preconcentration of heavy metals from water and tobacco samples prior to their determination by flame atomic absorption spectrometry, *Anal. Methods.* 4 (2012) 2879–2886. https://doi.org/10.1039/c2ay25171j.

[15] N. Abdullah, N. Yusof, W.J. Lau, J. Jaafar, A.F. Ismail, Recent trends of heavy metal removal from water/wastewater by membrane technologies, *J. Ind. Eng. Chem.* 76 (2019) 17–38. https://doi.org/10.1016/J.JIEC.2019.03.029.

[16] E. Jean, D. Villemin, M. Hlaibi, L. Lebrun, Heavy metal ions extraction using new supported liquid membranes containing ionic liquid as carrier, *Sep. Purif. Technol.* 201 (2018) 1–9. https://doi.org/10.1016/J.SEPPUR.2018.02.033.

[17] Y. Ren, J. Zhang, J. Guo, F. Chen, F. Yan, Porous poly(ionic liquid) membranes as efficient and recyclable absorbents for heavy metal ions, macromol. *Rapid Commun.* 38 (2017) 1700151. https://doi.org/10.1002/MARC.201700151.

[18] A. Nezhadali, R. Mohammadi, M. Mojarrab, An overview on pollutants removal from aqueous solutions via bulk liquid membranes (BLMs): parameters that influence the effectiveness, selectivity and transport kinetic, *J. Environ. Chem. Eng.* 7 (2019). https://doi.org/10.1016/j.jece.2019.103339.

[19] M.A. Malik, M.A. Hashim, F. Nabi, Ionic liquids in supported liquid membrane technology, *Chem. Eng. J.* 171 (2011) 242–254. https://doi.org/10.1016/J.CEJ.2011.03.041.

[20] A. Kumar, A. Thakur, P.S. Panesar, Recent developments on sustainable solvents for emulsion liquid membrane processes, *J. Clean. Prod.* 240 (2019). https://doi.org/10.1016/j.jclepro.2019.118250.

[21] N.S.W. Zulkefeli, S.K. Weng, N.S. Abdul Halim, Removal of heavy metals by polymer inclusion membranes, *Curr. Pollut. Reports.* 4 (2018) 84–92. https://doi.org/10.1007/s40726-018-0091-y.

[22] R. Güell, E. Anticó, V. Salvadó, C. Fontàs, Efficient hollow fiber supported liquid membrane system for the removal and preconcentration of Cr(VI) at trace levels, *Sep. Purif. Technol.* 62 (2008) 389–393. https://doi.org/10.1016/J.SEPPUR.2008.02.015.

[23] W. Alahmad, N. Tungkijanansin, T. Kaneta, P. Varanusupakul, A colorimetric paper-based analytical device coupled with hollow fiber membrane liquid phase microextraction (HF-LPME) for highly sensitive detection of hexavalent chromium in water samples, *Talanta.* 190 (2018) 78–84. https://doi.org/10.1016/j.talanta.2018.07.056.

[24] H. Kamran Haghighi, M. Irannajad, A. Fortuny, A.M. Sastre, Selective separation of germanium(IV) from simulated industrial leachates containing heavy metals by non-dispersive ionic extraction, *Miner. Eng.* 137 (2019) 344–353. https://doi.org/10.1016/j.mineng.2019.04.021.

[25] S. Platzer, M. Kar, R. Leyma, S. Chib, A. Roller, F. Jirsa, R. Krachler, D.R. MacFarlane, W. Kandioller, B.K. Keppler, Task-specific thioglycolate ionic liquids for heavy metal extraction: synthesis, extraction efficacies and recycling properties, *J. Hazard. Mater.* 324 (2017) 241–249. https://doi.org/10.1016/j.jhazmat.2016.10.054.

[26] C.Y. Foong, M.F. Mohd Zulkifli, M.D.H. Wirzal, M.A. Bustam, L.H.M. Nor, M.S. Saad, N.S. Abd Halim, COSMO-RS prediction and experimental investigation of amino acid ionic liquid-based deep eutectic solvents for copper removal, *J. Mol. Liq.* 333 (2021). https://doi.org/10.1016/j.molliq.2021.115884.

[27] M. Regel-Rosocka, M. Wisniewski, Ionic liquids in separation of metal ions from aqueous solutions, *Appl. Ion. Liq. Sci. Technol.* (2011). https://doi.org/10.5772/23909.

[28] R. Malas, Y. Ibrahim, I. AlNashef, F. Banat, S.W. Hasan, Impregnation of polyethylene membranes with 1-butyl-3-methylimidazolium dicyanamide ionic liquid for enhanced removal of Cd^{2+}, Ni^{2+}, and Zn^{2+} from aqueous solutions, *J. Mol. Liq.* 318 (2020). https://doi.org/10.1016/j.molliq.2020.113981.

[29] L.A. Neves, J.G. Crespo, I.M. Coelhoso, Gas permeation studies in supported ionic liquid membranes, *J. Memb. Sci.* 357 (2010) 160–170. https://doi.org/10.1016/J.MEMSCI.2010.04.016.

[30] W. Zhao, G. He, F. Nie, L. Zhang, H. Feng, H. Liu, Membrane liquid loss mechanism of supported ionic liquid membrane for gas separation, *J. Memb. Sci.* 411–412 (2012) 73–80. https://doi.org/10.1016/J.MEMSCI.2012.04.016.

[31] X. Zheng, C. Ni, W. Xiao, Y. Liang, Y. Li, Ionic liquid grafted polyethersulfone nanofibrous membrane as recyclable adsorbent with simultaneous dye, heavy metal removal and antibacterial property, *Chem. Eng. J.* 428 (2022). https://doi.org/10.1016/j.cej.2021.132111.

[32] Crosslinked chitosan polyvinyl alcohol nanofibers functionalized by ionic liquid for heavy metal ions removal, *Int. J. Biol. Macromol.* 195 (2022) 132–141. https://doi.org/10.1016/j.ijbiomac.2021.12.008.

[33] F.J. Alguacil, Facilitated chromium(VI) transport across an ionic liquid membrane impregnated with Cyphos IL102, (2019). https://doi.org/10.3390/molecules24132437.

[34] G. Zante, M. Boltoeva, A. Masmoudi, R. Barillon, D. Trébouet, Supported ionic liquid and polymer inclusion membranes for metal separation, *Sep. Purif. Rev.* 51 (2022) 100–116. https://doi.org/10.1080/15422119.2020.1846564.

[35] A.P. De Los Ríos, F.J. Hernández-Fernández, L.J. Lozano, S. Sánchez-Segado, A. Ginestá-Anzola, C. Godínez, F. Tomás-Alonso, J. Quesada-Medina, On the selective separation of metal ions from hydrochloride aqueous solution by pertraction through supported ionic liquid membranes, *J. Memb. Sci.* 444 (2013) 469–481. https://doi.org/10.1016/j.memsci.2013.05.006.

[36] G. Elias, E. Marguí, S. Díez, C. Fontàs, Polymer inclusion membrane as an effective sorbent to facilitate mercury storage and detection by X-ray fluorescence in natural waters, *Anal. Chem.* 90 (2018) 4756–4763. https://doi.org/10.1021/acs.analchem.7b05430.

[37] F. Sellami, O. Kebiche-Senhadji, S. Marais, L. Colasse, K. Fatyeyeva, Enhanced removal of Cr(VI) by polymer inclusion membrane based on poly(vinylidene fluoride) and Aliquat 336, *Sep. Purif. Technol.* 248 (2020). https://doi.org/10.1016/j.seppur.2020.117038.

[38] H.I. Turgut, V. Eyupoglu, R.A. Kumbasar, I. Sisman, Alkyl chain length dependent Cr(VI) transport by polymer inclusion membrane using room temperature ionic liquids as carrier and PVDF-co-HFP as polymer matrix, *Sep. Purif. Technol.* 175 (2017) 406–417. https://doi.org/10.1016/j.seppur.2016.11.056.

[39] H. Ibrahim Turgut, V. Eyupoglu, R. Ali Kumbasar, The comprehensive investigation of the room temperature ionic liquid additives in PVC based polymer inclusion membrane for Cr(VI) transport, *J. Vinyl Addit. Technol.* 25 (2019) E107–E119. https://doi.org/10.1002/vnl.21649.

[40] X. Zheng, F. Chen, X. Zhang, H. Zhang, Y. Li, J. Li, Ionic liquid grafted polyamide 6 as porous membrane materials: enhanced water flux and heavy metal adsorption, *Appl. Surf. Sci.* 481 (2019) 1435–1441. https://doi.org/10.1016/j.apsusc.2019.03.111.

[41] Q. Shen, D.Y. Xing, F. Sun, W. Dong, F. Zhang, Designed water channels and sieving effect for heavy metal removal by a novel silica-poly(ionic liquid) nanoparticles TFN membrane, *J. Memb. Sci.* 641 (2022) 119945. https://doi.org/10.1016/J.MEMSCI.2021.119945.

[42] D. Kogelnig, A. Regelsberger, A. Stojanovic, F. Jirsa, R. Krachler, B.K. Keppler, A polymer inclusion membrane based on the ionic liquid trihexyl(tetradecyl) phosphonium chloride and PVC for solid-liquid extraction of Zn(II) from hydrochloric acid solution, *Monatshefte Fur Chemie.* 142 (2011) 769–772. https://doi.org/10.1007/s00706-011-0530-6.

13 Life Cycle Assessment of Membranes for Removing Heavy Metals from Water and Wastewater

Sisem Ektirici, Mitra Jalilzade, Adil Denizli and Deniz Türkmen

13.1 INTRODUCTION

As is known, one of the most critical factors in the selection of the process is that a solution created to eliminate any undesirable effect does not cause new problems while eliminating the effect. For this reason, the preferred method for removing heavy metals from water should be chosen in a way that causes minor damage to the water flora and the environment. In addition to this feature, it is desired that the method to be chosen should use minimum energy, have separation efficiency, have low environmental pollution, and have better final product quality. In line with these features, membrane technology has been one of the most widely used technologies for heavy metal removal from wastewater in the last 15 years. Since the development of synthetic membranes in the 1960s, membrane technologies have been used in many processes such as water treatment, food production, food safety, gas separation, air pollution control, and protein separations. By 2000, the market for membrane technologies reached $4.4 billion . In 2010 this figure increased to $10 billion, and in 2014 it reached $10.4 billion.

The favorability of the methods used for water treatment is more influenced by being environmentally friendly and cost-effective. Membrane technology has been rapidly developed and water purified due to its high separation efficiency, less pollution, and less energy consumption than conventional methods. The membrane treatment method is used for water purification due to its easy application. In this method, the separation is based not only on membrane pores' differences but also on high membrane transpressure. The common feature of these methods is the use of semipermeable membranes with different pore sizes between solvent and phase with impurities that let solvent catching impurities.

Membrane fabrication is a complex part of membrane technology, which involves various steps to produce membranes with specific characteristics. Most membranes are made of polymers. Also, ceramics and metals have been used for this purpose. Ultrafiltration, microfiltration, nanofiltration, reverse osmosis (RO), and membrane distillation are widely used methods for water treatment. The properties of membranes, such as pore size, porosity dispersion, and membranes thickness, are affected by their fabrication method and materials used. On the other hand, the fabrication method of membranes affects the membranes' properties and constituent materials.

In the life cycle assessment of an approach, all aspects of the process are considered, including raw materials, the procedure of raw materials to the product, and the product. The impact of the whole process and product from different perspectives must be studied, for example, the effect

DOI: 10.1201/9781003326281-15

of the process on global warming, shortage of fossil or power, and ecotoxicity. However, the cost impact of each method and materials for membrane fabrication must be investigated.

13.2 TYPES OF MEMBRANES ACCORDING TO THE CONSTITUENT MATERIALS AND THEIR IMPACT ON THE LIFE CYCLE

The life cycle assessment process according to ISO 14040 and 14044 (2006) includes four stages: (1) goal and scope (extraction of raw material, the procedure from raw material to product, the end of the process leads to recovery or disposing of product and by-product) (2) life cycle inventory (contains assumed and gained data) (3) environmental impact assessment (the World-H midpoint method is used to carry out the experimental impact assessment), and (4) interpretation of data (using ISO 14040 and 14044 criteria for the explanation of data (Ioannou-Ttofa *et al.*, 2016; Yadav and Samadder, 2017)). The materials used in the production of membranes must be resistant to chemicals and not be easily contaminated; they should be biodegradable and not be harmful to the environment. The membranes used in separation processes can be reviewed based on type, fabrication method, and structure. Metals, ceramics, polymers, and composites have been used in membrane fabrication.

Polymeric and composite polymeric membranes: These membranes are the most popular due to their low cost. Polyvinylidene fluoride, cellulose acetate, polypropylene, polyacrylonitrile, polysulfone, and polyethersulfone are widely used materials for polymeric membrane production (Yadav *et al.*, 2021). Different methods such as sintering, trace etching, stretching, and phase inversion have been used in membrane production, but phase inversion is the most popular in academia and industry (Azimi *et al.*, 2017). The most important part of the production of composite membranes by the phase inversion method is to dissolve the polymers in a suitable solution and prepare a homogeneous polymeric solution. In this method, dimethylacetamide, dimethylformamide, N-methyl-2-pyrrolidone, dimethyl sulfoxide, tetrahydrofuran, dimethylformamide, acetone, dioxane, triethyl phosphane, chloroform, and dichloromethane, which are highly toxic to the environment, are used as solutions. The most used solution, N-methyl-2-pyrrolidone, was especially restricted from use by the European Union in May 2021 (Subramani and Jacangelo, 2014). As a solution to this problem, eco-friendly solutions called green solutions have been used in membrane production in recent years (trimethylene glycol diacetate, methyl-5-(dimethylamino)-2-methyl-5-oxopentanoate, tributyl O-acetyl citrate, butylene carbonate, and butylene carbonate) (Cui *et al.*, 2013; Rasool, Pescarmona and Vankelecom, 2019; Galiano and Figoli, 2020). Although these solutions are eco-friendly, the production method may be toxic. Natural or bio-based polymers like fossil-based, cellulose-based, and bio-based polymers were widely used in membrane fabrication (Yadav *et al.*, 2021). Cellulose-based and fossil-based polymers are more stable than bio-based polymers thermally and chemically. The life cycle assessment of polymer membranes prepared by the phase inversion method depends on the kind of solvent, polymer, and source of power that could reduce or increase the environmental impact and cost (McManus and Taylor, 2015). Fossil-based or green solvents have fewer toxic solvents that reduce the environmental impact of membrane fabrication. Also, using electricity from a renewable source can reduce the environmental impact (Yadav *et al.*, 2021).

Ceramic membranes: These days, ceramic membranes have attracted attention due to their advantages, such as resistance to chemicals and heat, less fouling, and longer lifetime. Also, the combination of this membrane with other processes like in situ ozonation and other methods like bioreactors is preferred to the polymeric membrane because of their long-time stability, high pollutant removal capability, high flow, and low flow fouling in water and wastewater treatment. Silicon oxide, zirconium oxide, aluminum oxide, and titanium oxide are materials that are used to produce the separation layer of ceramic membranes. Alumina ceramic membranes are preferred to other ceramic membranes due to their better performance on the price scale (Asif and Zhang, 2021). Due to the pore size, ceramic membranes can be classified into microfiltration (>50 nm), ultrafiltration (2–50 nm), and nanofiltration (<2 nm) types. Microfiltration and ultrafiltration have been used to remove heavy metals from water and wastewater (Drioli and Giorno, 2010). To reduce the environmental impact of the life cycle assessment of ceramic membranes and improve water purification without physical and chemical cleaning of membrane fouling pollutants a combination of ceramic membranes and other processes like photocatalysis, ozonation, coagulation, and activated carbon adsorption have been used. Although ceramic membranes are not cost-effective for commercial use due to their high capital cost, they have comparable performance to polymeric membranes due to their lower maintenance cost and longer life than polymeric membranes. The total cost of the product life cycle for both is almost identical (Murić, Petrinić and Christensen, 2014). Membrane performance declines after use for a while due to the entrapment of colloidal particles in the membrane pores or the collapse of the pores, and, in this case, the membrane must be replaced with new ones. The lifespan of polymeric membranes (8 years) is short compared to the lifespan of ceramic membranes (20 years). Therefore, instead of focusing on the cost of investment, the cost assessment of the whole process should be considered. Although ceramic membranes are the best choice for water and wastewater treatment, membrane fouling is a disadvantage of this method. The fouling of this method can appear during intermediate, complete, standard, or cake filtration pore blocking. New prevention and control methods for fouling have been developed with optimization of process and module design improvement. The ceramic membrane's structure, surface charge, and pore size can affect the rate and amount of membrane fouling. Other processes such as oxidation, ozonation, and coagulation can be added to this method to reduce the fouling of ceramic membranes. Pre-ozonation can be more effective than in situ ozonation in protecting the membrane structure and providing a better flow. Another method used to reduce fouling is surface modification; this modification can be chemical or physical. Chemical modification is more effective in durability and can reduce the membrane life cycle's environmental and cost impacts. Ceramic membranes are more resistant to chemicals than polymeric membranes. They are suitable for water and wastewater treatment due to their better hydraulic performance (Weerasekara *et al.*, 2019; *Current Trends and Future Developments on (Bio-) Membranes*, 2022).

Liquid membrane: In recent years, a new membrane method, the liquid membrane method, has been defined and used for low concentration solutes, and this method has also been used for wastewater treatment. This method is less harmful to the environment than traditional methods. The emulsion liquid membrane exhibits greater promise relative to alternative liquid membrane technologies than bulk liquid membrane and supported liquid membrane, owing to its superior cost-effectiveness, reduced energy requirements, augmented interfacial area, and heightened mass transfer rate (Kumar, Thakur, and Panesar, 2019). Surfactant stabilizer emulsion liquid membrane technology has been used for metal ion separation, especially toxic metal ions (lead, cadmium, mercury, and nickel). This method is less stable than other methods. It is not suitable for industrial use due to the leakage of internal reagents, the coalescence of emulsion drops, membrane dissolution, and escaping of the water-soluble solute into the emulsion (Abbassian and Kargari, 2016). Additional operations can be added to increase emulsion liquid membrane stability, like using Newtonian additives and adding more surfactants. Making emulsion drops of one liquid in another liquid, forming an internal phase between liquids, and stabilizing by a surfactant are essential stages of the emulsion liquid membrane method (Kankekar, Wagh and Mahajani, 2010). An external mechanical force like a stirrer, mixer, or homogenizer is required to make an emulsion. The imposition of external mechanical force exerts influence on the cost, energy consumption, and environmental footprint within the life cycle assessment of membrane technology. Most surfactants used in this method are environmentally friendly and are obtained from renewable sources. The type of diluent in the organic phase also affects the life cycle of this method.

Favorable diluents have low toxicity, less solubility, no capability to form a new phase, moderate viscosity, extractant and surfactant adaptability, high density, and high flash point. In this method, petroleum-based solvents such as heptane, hexane, chloroform, and kerosene are used as the organic phase. These solvents affect the environmental impact of the life cycle of this method because of being toxic, nonrenewable, nonbiodegradable, and flammable, and they also affect the cost impact of the life cycle because of their expense and limited availability. Several vegetable-based organic solvents like sunflower oil, palm, rice bran oil, and palm oil were researched and applied to decrease the life cycle's environmental impact. These solvents are nontoxic, noninflammable, renewable, biodegradable, and cheaper than other solvents. Emulsion liquid membrane technology is a promising technology to treat wastewater, but the stability of emulsion and extraction efficiency affect the impact of life cycle assessment. Surfactant concentration, extractant concentration, pH of the external feed phase, the concentration of the internal phase, phase ratio, stirring speed, emulsification speed, and contact time of emulsification affect emulsion stability and life cycle impact of membranes (Kumar, Thakur and Panesar, 2019).

Ion-exchange membranes: These membranes, because of their specific ion separation, high treatment capacity, and high rate, have been used to separate and recover specific metal ions from wastewater. Nanofiber ion-exchange membranes have been used to separate and recover specific heavy metals compared to other metal ions. Although environmentally friendly ion-exchange membranes can be used to decrease the environmental impact of membranes, this method's drawback is fouling, which increases the cost impact of the process. The fouling problem of this method can be fixed by cleaning or replacing, but cleaning with a chemical causes a decrease in the performance of the membrane and also degradation of ion-exchange groups (Ran *et al.*, 2017). Traditional ion-exchange membrane technology uses functionalized monomers and organic solvents to form membranes that are toxic to the environment. To solve this problem, a solvent-free polymerization technique or liquid monomer is used to prepare these membranes. New preparation methods that are used to improve the stability of ion-exchange membranes and decrease the cost impact of these membranes are as follows: polymer material blending, pore filling, electrospinning, and in situ polymerization. The polymer blending method fixes the problems of membranes with a single polymer. Polymer blending increases membranes' selectivity, stability, and conductivity and decreases cost and swelling. This process happens in two forms: blending fluorinated and non-fluorinated polymers and functional and non-functional polymers. The limiting factors of using fluorinated polymers in this area are high price and hydrophobicity, but combining these polymers with non-fluorinated ones could improve the characteristics of these polymers. Ion functionalized polymers have high ion conductivity but low mechanical resistance, but the combination of these polymers with non-functionalized polymers improves the mechanical resistance of membranes. A combination of two or more polymers can control the properties of ion-exchange membranes, such as high ion conductivity, low water swelling, and high chemical stability (Panwar *et al.*, 2012). Pore filling and pore soaking techniques provide ion-exchange membranes with low swelling and high selectivity. Ion-exchange membranes and in situ polymerization methods decrease the environmental impact of organic solvents used in traditional production methods. This method is based on polymerizing liquid monomers instead of solid monomers in an organic solvent. Nanofiber ion-exchange monomers produced with the electrospinning method are promising because of their three-dimensional networks, high interconnected porosity, and large surface area (Jung *et al.*, 2011; Wang *et al.*, 2012, 2015; Kim *et al.*, 2013).

13.3 LIFE CYCLE ASSESSMENT OF PROMISING MEMBRANE TECHNOLOGIES

The need to treat industrial and domestic wastewater due to the increasing need for water increases the need to integrate traditional methods with new methods because of their low cost, high efficiency, high stability, and high selectivity. For this purpose, new and promising methods have been introduced: pressure-driven membrane, electrically driven membrane, membrane distillation, forward osmosis, and adsorptive membrane (El Batouti, Al-Harby and Elewa, 2021).

The pressure-driven method: This method can be in the form of low or high pressure. Ultrafiltration membranes such as low-pressure membranes because of low operation pressure and low capital cost can decrease the environmental and cost impact of the procedure, but this method is not enough on its own to extract heavy metals from water (Barakat, 2011; Yao *et al.*, 2015). Enhanced micellar ultrafiltration is one of the ultrafiltration approaches used for heavy metal extraction. This method is based on the surfactant enhancing the molecular size of the metal ions, and then the surfactant-metal ion complex forms huge micelles that can easily separate. The efficiency of this method for heavy metal removal depends on surfactant concentration, operating pressure, and medium pH and temperature. This method increases the environmental impact of the life cycle by the production of secondary pollutants (metal-surfactant complexion) (*[PDF] A Review of Studies on Micellar Enhanced Ultrafiltration for Heavy Metals Removal from Wastewater | Semantic Scholar*, no date; Schwarze, 2017). Enhanced polymer ultrafiltration is another improved kind of ultrafiltration that uses a water-soluble polymer to form a complex with heavy metal ions. The efficiency of this method is affected by polymer concentration, pH, and the kind of metal ions (Labanda, Khaidar and Llorens, 2009; Qiu, Mao and Wang, 2014). Nanofiltration membranes used as high pressure-driven membranes are capable of separating heavy metal ions because of their negatively charged surface. The quantity of heavy metal removed by this method depends on pore size and the existence of charged groups on the surface of membranes and the medium pH. This method is more efficient than ultrafiltration because the higher separation of multivalent metal ions and higher water permeability are also more efficient than RO because of lower operating pressure. As a result, this method decreases the energy impact of the water treatment procedure (Mohammad *et al.*, 2015; Pandya, 2015; Qi *et al.*, 2019).

Electrically driven membranes: Deionization via ion-exchange membranes encompasses electrodeionization, electrodialysis, and capacitive deionization processes, wherein an electrical potential difference and a charged membrane facilitate the separation of cations and anions (Handojo *et al.*, 2019). In the electrodialysis method, cation-exchange membranes (CEMs) and anion-exchange membranes (AEMs) replace the two electrodes with opposite charges. Ions attract electrodes with opposite charge when a potential difference is established between electrodes and an electrolyte solution transfer. This approach is characterized by a higher degree of simplicity compared to RO, primarily due to considerations of osmotic pressure. Nevertheless, its applicability is more apt for lower solute concentrations, a limitation attributed to heightened energy consumption (Wood *et al.*, 2010). Electrodeionization is an electrically driven method that combines electrodialysis and ion-exchange processes to decrease energy consumption (Wood *et al.*, 2010; Arar *et al.*, 2014). Membrane capacitive deionization occurs as an electrically driven membrane separates ions by electrical potential differences and adsorbs by porous electrolytes. This method has advantages like better ion separation and low energy consumption than the CEM method and disadvantages like electrode saturation (AlMarzooqi *et al.*, 2014). Electrified membranes refer to electrically driven membranes employing electrochemical reduction for the purpose of heavy metal detoxification (Duan *et al.*, 2017).

Membrane distillation: This method presents a selective barrier between two liquid (feed phase) and vapor (influent) phases driven by the difference in temperature. This method can provide pure water as well as allow heavy metal recovery. Also, because of low operating pressure, this method decreases the energy impact of the procedure (Alkhudhiri, Darwish and Hilal, 2012).

Forward osmosis: This method is driven by osmotic pressure created by the difference between feed and draw solutions. This method decreases the energy cost and environmental impact because of low energy consumption, low capital cost, and contaminants recovery. The first forward osmosis membranes are asymmetric membranes consisting of a thin selective layer and a thick support layer. Although cellulose acetate, polyamide, and polyester have been used as ingredients in the traditional kind of membranes, the composite polymer has been used as an ingredient in the modern kind. The polymeric composite decreases the cost impact because of its stability and long-time use (Qiu and Ting, 2014).

Adsorptive membranes: These membranes have specific adsorption and morphological properties that support the removal of heavy metal ions from water. Adsorptive membranes are in the form of polymeric, polymer-ceramic, electrospinning nanofiber, and nano-enhanced membranes.

These membranes must be chemically stable and reusable to decrease the procedure's cost impact. The reusability of these membranes makes this method affordable and environmentally friendly (Vo *et al.*, 2020).

13.4   ECONOMIC ANALYSIS OF MEMBRANE TECHNOLOGY

Industrial and agricultural consumption of large amounts of water require a solution to address wastewater purification. Especially in recent years, the solutions proposed for addressing the concerns because of increasing water pollution due to climate crisis, change in living standards, and the rapid progress of the pharmaceutical industry should be more effective and more economical in line with increasing demand. According to the United Nations, more than 748 million people will not have access to improved drinking water sources, and industrial water needs are projected to increase by 400% worldwide between 2000 and 2050 (UN-Water, 2015). This is a severe problem for developing countries with poor infrastructure and limited economic leverage. However, this is a global problem as it is estimated that half of the world's population will suffer from severe water shortages by 2050 (Prüss-Üstün *et al.*, 2008). In addition, the world's population is projected to increase from 7.6 billion today to 9.8 billion in 2050 (Luis and Moncayo, no date). The need for freshwater and wastewater treatment has increased dramatically. There is not enough natural freshwater to meet this tremendous need. Therefore, complacency is not a global water management option when it is necessary to avoid water shortages and conflicts.

Apart from the importance of academic studies for eliminating water pollution, there is an increasing need to assess the commercial importance of these studies on wastewater and its effective use in industry. The European Union presented a new economic package in 2015, with the demand for wastewater purification, which increases in direct proportion to the gradual increase in water consumption. In addition, the implementation of the Urban Wastewater Treatment Directive (91-271-EEC) and the Water Framework Directive of 2000 (2000/60/EC) packages on the elimination of water pollution in Europe has been an essential step for the recovery of wastewater in a high-quality way.

Although many methods are suggested to remove pesticides, bacteria, organic particles, and heavy metals from water or wastewater, the most crucial thing is that the method be low cost and highly efficient. Membrane technologies are the preferred method in wastewater treatment due to their nature-friendly characteristics, low energy consumption, and low cost (relative to the membrane technology used). The size of the membrane market is projected to grow from US $5.4 billion in 2019 to the US $8.3 billion with a Compound Annual Growth Rate (CAGR) of 8.3% in 2024. The main drivers behind the membrane market include population growth, increased awareness of wastewater reuse, and rapid industrialization. The transition from chemical to physical water treatment, strict water treatment regulations, and climate changes related to rainfall levels are also driving the membrane market. The Asia Pacific market size in 2021 was US $2.47 billion in China, India, Japan, Australia, and other Southeast Asian countries. Emerging markets such as India and China are the most promising markets, with a growth rate of 8.7% from 2022 to 2029. Europe is the second-largest region and the second-largest market after Western European countries like Germany, France, Spain, and Italy. The growth of European markets is due to growing awareness of water scarcity and government initiatives to protect freshwater sources. The lack of fresh water and the increasing demand for clean drinking water in a growing economy is driving market growth in other parts of Europe. In North America, market growth is primarily due to increasingly stringent water and wastewater treatment regulations. In Latin America, Brazil is also expected to show the highest growth rates due to the expansion of the food industry during the forecast period. The Middle East and Africa are the fastest-growing regions of the market (*Membranes Market Size, Share, Growth | Forecast Report, 2029*, no date).

RO membranes dominate the membrane market worldwide. Anticipated for the forecast period, the Reverse Osmosis Membranes Market is projected to increase from $3.10 Billion in 2022 to $6.00

Billion by 2030, reflecting a Compound Annual Growth Rate (CAGR) of 10.00%. (2023 Report By LG Chem Ltd., Toray Industries Inc. NX Filtration BV Membranium (JSC RM Nanotech) Toyobo Co Ltd. Axeon Water Technologies, Report Code: FAF-2203). Although RO first found its place in the market for separating seawater from salt, it is now used to separate bacteria, pesticides, various organic wastes, and heavy metal ions. Given the increasing adoption of RO membranes to enhance water supply, as well as for desalination and wastewater recycling, it is anticipated that RO technology will exhibit enhanced efficacy in the future. Especially in areas with water scarcity, the supply of water from a seawater desalination plant is becoming increasingly critical. However, the remedy of surface water to potable water will play an increasing number of critical functions in the forecast period.

13.5 CONCLUSION

Nowadays, membrane technology is preferred over traditional methods because of less energy consumption and more environmental protection. Different materials are used in the fabrication of membranes. Also, membrane technology is applied in different forms. Polymeric, ceramic, liquid, and ion-exchange membranes are traditional membranes used in commercial and industrial-scale water treatment. Substituting bio-based polymers and eco-friendly solvents with fossil-based polymers and hazardous solvents in the production of polymeric membranes, while also employing a liquid monomer in RO, reduces the environmental footprint of the polymeric membrane's life cycle. However, it is important to note that the production process of these polymers, which involves toxic elements, elevates the environmental impact of their life cycle. Ceramic membranes, especially ultrafiltration and microfiltration membranes, are better than polymeric membranes in terms of long-term usage, effective influencing flux, easy cleaning, less fouling, and for being environmentally friendly, but these membranes' capital cost is high. Research on ceramic membranes in aspects like cheap raw material, fouling evaluation, and efficient cleaning has increased the market of these membranes in developed countries. Ion-exchange membranes are interesting because of their high treatment capacity, for being environmentally friendly, and for their high rate, specific ion separation, and recovery. However, fouling is a disadvantage of this method and it increases the cost impact of these membranes' life cycles. Due to less solvent requirement and low energy consumption, liquid membrane technology is an attractive method for heavy metal separation. However, this method is appropriate in low concentrations of heavy metal ions.

In addition to the academic studies on the separation of heavy metals from wastewater using membranes, academic studies should be carried out at the industrial level due to the urgency and scope of the subject. For this reason, there is a need to produce membranes with low energy consumption, low cost, and high efficiency, which are easy to mass produce for use in the industry. Although the microfiltration method can find its place in the market due to its features, it is not preferred because the removal ability of the method is low in intensive conditions. In this context, forward osmosis and RO are most actively used in the industry. Since hydraulic pressure is not required for forward osmosis, the energy used in the method is low. In addition, while its environmentally friendliness makes this method selective, future studies on the development of this method are also of great importance. The RO method dominates the market, especially in China and Pacific countries, and the North American region. The method, which can separate Ni^{2+}, Cr^{6+}, and Cu^{2+} ions with an efficiency of 98%, is the most preferred membrane technology due to its low cost (Qasem, Mohammed and Lawal, 2021). In this context, RO membranes that dominate the market are produced by DuPont de Nemours Inc., General Electric Co., and Hunan Keensen Technology Co. Ltd. companies and are used to separate metal ions from wastewater worldwide (*Reverse Osmosis Membrane Market to Witness USD 5.25 Bn growth | Dominated by DuPont de Nemours Inc., General Electric Co., and Hunan Keensen Technology Co. Ltd. | Technavio*, no date).

Promising membrane-based methods such as pressure-driven and electrically driven methods remove heavy metals effectively. Also, electrolysis, electrodeionization, and distillation membrane methods are effective for heavy metal recovery from the concentrated stream. In addition to the existing methods, it can be said that the most crucial method that has been studied academically and that is promising in the future is electrodialysis. While the method includes CEM and AEM technologies, it can be a promising method in the future, as it does not require phase change, chemical consumption, or a chemical environment, and leads to high clean water recovery. The need for electrical potential and the high cost of the method is a problem that needs to be studied. Integration of ion-exchange membranes with these methods like electrodialysis, electrodeionization, and RO could increase the efficiency of this method and decrease the environmental and cost impact of these processes. These integrated methods were also investigated in lab-scale tests, and the efficiency of these processes must be investigated in pilot and commercial-scale applications.

REFERENCES

[PDF] R. Bade, Seung Hwan Lee (2011) 'A Review of Studies on Micellar Enhanced Ultrafiltration for Heavy Metals Removal from Wastewater', *Journal of Water Sustainability*, (1), pp. 85–102.

Abbassian, K. and Kargari, A. (2016) 'Modification of membrane formulation for stabilization of emulsion liquid membrane for extraction of phenol from aqueous solutions', *Journal of Environmental Chemical Engineering*, 4(4), pp. 3926–3933. doi: 10.1016/j.jece.2016.08.030.

Alkhudhiri, A., Darwish, N. and Hilal, N. (2012) 'Membrane distillation: A comprehensive review', *Desalination*, 287, pp. 2–18. doi: 10.1016/j.desal.2011.08.027.

AlMarzooqi, F. A. *et al.* (2014) 'Application of capacitive deionisation in water desalination: A review', *Desalination*, 342, pp. 3–15. doi: 10.1016/j.desal.2014.02.031.

Arar, Ö. *et al.* (2014) 'Various applications of electrodeionization (EDI) method for water treatment-A short review', *Desalination*, 342, pp. 16–22. doi: 10.1016/j.desal.2014.01.028.

Asif, M. B. and Zhang, Z. (2021) 'Ceramic membrane technology for water and wastewater treatment: A critical review of performance, full-scale applications, membrane fouling and prospects', *Chemical Engineering Journal*, 418(January), p. 129481. doi: 10.1016/j.cej.2021.129481.

Azimi, A. *et al.* (2017) 'Removal of heavy metals from industrial wastewaters: A review', *ChemBioEng Reviews*, 4(1), pp. 37–59. doi: 10.1002/cben.201600010.

Barakat, M. A. (2011) 'New trends in removing heavy metals from industrial wastewater', *Arabian Journal of Chemistry*, 4(4), pp. 361–377. doi: 10.1016/j.arabjc.2010.07.019.

Basile, A., Gensini, M., Allegrini, I., Figoli, A. (2022) 'Current Trends and Future Developments on (Bio-) Membranes'. doi: 10.1016/c2019-0-04672-9.

El Batouti, M., Al-Harby, N. F. and Elewa, M. M. (2021) 'A review on promising membrane technology approaches for heavy metal removal from water and wastewater to solve water crisis', *Water (Switzerland)*, 13(22). doi: 10.3390/w13223241.

Cui, Z. *et al.* (2013) 'Poly(vinylidene fluoride) membrane preparation with an environmental diluent via thermally induced phase separation', *Journal of Membrane Science*, 444, pp. 223–236. doi: 10.1016/j.memsci.2013.05.031.

Drioli, E. and Giorno, L. (2010) *Comprehensive Membrane Science and Engineering, Comprehensive Membrane Science and Engineering*. doi: 10.1016/c2009-1-28385-7.

Duan, W. *et al.* (2017) 'Electrochemical removal of hexavalent chromium using electrically conducting carbon nanotube/polymer composite ultrafiltration membranes', *Journal of Membrane Science*, 531(November 2016), pp. 160–171. doi: 10.1016/j.memsci.2017.02.050.

Galiano, F. and Figoli, A. (2020) 'Special issue: New trends in membrane preparation and applications?', *Molecules*, 25(5). doi: 10.3390/molecules25051132.

2023 Report By LG Chem Ltd., Toray Industries Inc. NX Filtration BV Membranium (JSC RM Nanotech) Toyobo Co Ltd. Axeon Water Technologies, Report Code: FAF-220.

Handojo, L. *et al.* (2019) 'Electro-membrane processes for organic acid recovery', *RSC Advances*, 9(14), pp. 7854–7869. doi: 10.1039/C8RA09227C.

Ioannou-Ttofa, L. *et al.* (2016) 'The environmental footprint of a membrane bioreactor treatment process through life cycle analysis', *Science of the Total Environment*, 568, pp. 306–318. doi: 10.1016/j.scitotenv.2016.06.032.

Jung, H. *et al.* (2011) 'Low fuel crossover anion exchange pore-filling membrane for solid-state alkaline fuel cells', *Journal of Membrane Science*, 373(1–2), pp. 107–111. doi: 10.1016/j.memsci.2011.02.044.

Kankekar, P. S., Wagh, S. J. and Mahajani, V. V. (2010) 'Process intensification in extraction by liquid emulsion membrane (LEM) process: A case study; enrichment of ruthenium from lean aqueous solution', *Chemical Engineering and Processing: Process Intensification*, 49(4), pp. 441–448. doi: 10.1016/j.cep.2010.02.005.

Kim, D. H. *et al.* (2013) 'Development of thin anion-exchange pore-filled membranes for high diffusion dialysis performance', *Journal of Membrane Science*, 447, pp. 80–86. doi: 10.1016/j.memsci.2013.07.017.

Kumar, A., Thakur, A. and Panesar, P. S. (2019) 'A review on emulsion liquid membrane (ELM) for the treatment of various industrial effluent streams', *Reviews in Environmental Science and Biotechnology*, 18(1), pp. 153–182. doi: 10.1007/s11157-019-09492-2.

Labanda, J., Khaidar, M. S. and Llorens, J. (2009) 'Feasibility study on the recovery of chromium(III) by polymer enhanced ultrafiltration', *Desalination*, 249(2), pp. 577–581. doi: 10.1016/j.desal.2008.06.031.

Luis, F. and Moncayo, G. (no date) 'No 主観的健康感を中心とした在宅高齢者における 健康関連指標に関する共分散構造分析 Title'.

McManus, Marcelle C., and Caroline M. Taylor. 'The changing nature of life cycle assessment.'. *Biomass and bioenergy* 82 (2015): 13–26.

Membranes Market Size, Share & Covid-19 Impact Analysis, By Technology (RO/FO, UF, NF, MF and Others), By Application (Water &Wastewater Treatment, Food &beverage, Gas Separation, and Others) and Regional Forecast, 2022-2029. Report ID : FBI102982.

Mohammad, A. W. *et al.* (2015) 'Nanofiltration membranes review: Recent advances and future prospects', *Desalination*, 356, pp. 226–254. doi: 10.1016/j.desal.2014.10.043.

Murić, A., Petrinić, I. and Christensen, M. L. (2014) 'Comparison of ceramic and polymeric ultrafiltration membranes for treating wastewater from metalworking industry', *Chemical Engineering Journal*, 255, pp. 403–410. doi: 10.1016/j.cej.2014.06.009.

Pandya, J. A. (2015) 'Nanofiltration for recovery of heavy metal from waste water', *International Journal of Environmental Science and Technology*, 3(June), pp. 29–34. doi: 10.13140/RG.2.1.5040.7524.

Panwar, V. *et al.* (2012) 'Dynamic mechanical, electrical, and actuation properties of ionic polymer metal composites using PVDF/PVP/PSSA blend membranes', *Materials Chemistry and Physics*, 135(2–3), pp. 928–937. doi: 10.1016/j.matchemphys.2012.05.081.

Prüss-Üstün, A. *et al.* (2008) 'Safer water, better health', *World Health Organization*, p. 53. Available at: http://www.who.int/quantifying_ehimpacts/publications/saferwater/en/.

Qasem, N. A. A., Mohammed, R. H. and Lawal, D. U. (2021) 'Removal of heavy metal ions from wastewater: A comprehensive and critical review', *npj Clean Water*, 4(1). doi: 10.1038/s41545-021-00127-0.

Qi, Y. *et al.* (2019) 'Polyethyleneimine-modified original positive charged nanofiltration membrane: Removal of heavy metal ions and dyes', *Separation and Purification Technology*, 222(November 2018), pp. 117–124. doi: 10.1016/j.seppur.2019.03.083.

Qiu, Y. R., Mao, L. J. and Wang, W. H. (2014) 'Removal of manganese from waste water by complexation-ultrafiltration using copolymer of maleic acid and acrylic acid', *Transactions of Nonferrous Metals Society of China (English Edition)*, 24(4), pp. 1196–1201. doi: 10.1016/S1003-6326(14)63179-4.

Qiu, G. and Ting, Y. P. (2014) 'Direct phosphorus recovery from municipal wastewater via osmotic membrane bioreactor (OMBR) for wastewater treatment', *Bioresource Technology*, 170, pp. 221–229. doi: 10.1016/j.biortech.2014.07.103.

Ran, J. *et al.* (2017) 'Ion exchange membranes: New developments and applications', *Journal of Membrane Science*, 522, pp. 267–291. doi: 10.1016/j.memsci.2016.09.033.

Rasool, M. A., Pescarmona, P. P. and Vankelecom, I. F. J. (2019) 'Applicability of organic carbonates as green solvents for membrane preparation', *ACS Sustainable Chemistry and Engineering*, 7(16), pp. 13774–13785. doi: 10.1021/acssuschemeng.9b01507.

Reverse Osmosis Membrane Market to Witness USD 5.25 Bn Growth | Dominated by DuPont de Nemours Inc., General Electric Co., and Hunan Keensen Technology Co. Ltd. | Technavio (2021).

Schwarze, M. (2017) 'Micellar-enhanced ultrafiltration (MEUF)-state of the art', *Environmental Science: Water Research and Technology*, 3(4), pp. 598–624. doi: 10.1039/c6ew00324a.

Subramani, A. and Jacangelo, J. G. (2014) 'Treatment technologies for reverse osmosis concentrate volume minimization: A review', *Separation and Purification Technology*, 122, pp. 472–489. doi: 10.1016/j.seppur.2013.12.004.

UN-Water (2015) *World Water Development Report: Water for a Sustainable World*. Available at: http://www.unesco.org/new/en/natural-sciences/environment/water/wwap/wwdr/2015-water-for-a-sustainable-world/.

Vo, T. S. *et al.* (2020) 'Heavy metal removal applications using adsorptive membranes', *Nano Convergence*, 7(1). doi: 10.1186/s40580-020-00245-4.

Wang, N. *et al.* (2015) 'Highly stable "pore-filling" tubular composite membrane by self-crosslinkable hyper-branched polymers for toluene/n-heptane separation', *Journal of Membrane Science*, 474, pp. 263–272. doi: 10.1016/j.memsci.2014.09.041.

Wang, T. *et al.* (2012) 'Preparation and properties of pore-filling membranes based on sulfonated copolyimides and porous polyimide matrix', *Polymer*, 53(15), pp. 3154–3162. doi: 10.1016/j.polymer.2012.05.049.

Weerasekara, N. A. *et al.* (2019) 'Clues to membrane fouling hidden within the microbial communities of membrane bioreactors', *Environmental Science Water Research & Technology*, 5, p. 1389. doi: 10.1039/c9ew00213h.

Wood, J. *et al.* (2010) 'Production of ultrapure water by continuous electrodeionization', *Desalination*, 250(3), pp. 973–976. doi: 10.1016/j.desal.2009.09.084.

Yadav, P. *et al.* (2021) 'Assessment of the environmental impact of polymeric membrane production', *Journal of Membrane Science*, 622(December 2020), p. 118987. doi: 10.1016/j.memsci.2020.118987.

Yadav, P. and Samadder, S. R. (2017) 'A global prospective of income distribution and its effect on life cycle assessment of municipal solid waste management: A review', *Environmental Science and Pollution Research*, 24(10), pp. 9123–9141. doi: 10.1007/s11356-017-8441-7.

Yao, Z. *et al.* (2015) 'Positively charged membrane for removing low concentration Cr(VI) in ultrafiltration process', *Journal of Water Process Engineering*, 8, pp. 99–107. doi: 10.1016/j.jwpe.2015.08.005.

14 Nanofiber Membranes for Hexavalent Chromium Removal and Reduction

*Naveed Ahmed Qambrani, Sorth Ansari,
Abdul Majeed Pirzada, Muhammad Rizwan,
Zeeshan Khatri, Rasool Bux Mahar
and Sheeraz Ahmed Memon*

14.1 INTRODUCTION

Chromium metal having a valency of six is found to be naturally occurring in highly toxic compounds (Mohamed, Nasser et al. 2017; Zhang, Shi et al. 2020). Cr(VI) is hazardous to environmental ecosystems and human beings as well. Cr(VI) is carcinogenic in nature and lethal (Gao, Sun et al. 2018; Barbosa, Souza et al. 2020). The use of Cr(VI) ions in various activities, for example, pigment production, steel manufacturing, dyeing of clothes, electroplating, leather tanning manufacturing, results in the production of wastewater with a high level of toxicity, which is harmful to the ecosystem (Celebi, Yurderi et al. 2016; Dognani, Hadi et al. 2019; Li, Chen et al. 2020a; Lu, Liu et al. 2020; Tian, Liu et al. 2020).

Chromium with valencies of three and six are basically two states of chromium oxidation (Nthumbi, Ngila et al. 2012; Zhao, Li et al. 2017). Out of these two oxidation states, Cr(III) is less toxic and is an important nutrient for living organisms but has poor solubility (Markiewicz, Komorowicz et al. 2015; Tian, Liu et al. 2020). Cr(III) exists in various forms as hydroxyl species in water, which include $CrOH^{2+}$, $Cr(OH)_4^-$, $Cr(OH)^{2+}$, $Cr_3(OH)_4^{5+}$, $Cr_2(OH)_2^{4+}$, $Cr(OH)_3$. Usually, in a highly acidic solution with a pH of less than 4, the trivalent form of chromium, which is Cr^{3+}, is predominant, and when the pH of the solution increases slightly Cr^{3+} gets transformed into $CrOH^{2+}$. At pH 4 $CrOH^{2+}$ reaches its maximum mole fraction and then it is transformed into $Cr(OH)_2^+$ and $Cr(OH)_3$. So, between the pH value of 4–7 the species of Cr(III) exists mostly in the above three forms. The $Cr_3(OH)_4^{5+}$ form of Cr(III) exists at a pH of between 6 and 8. Within a pH ranging between 5 and 14, the form of Cr(III) that is prevalent is $Cr(OH)_3$ and it gets transformed into soluble $Cr(OH)_4^-$. $Cr(OH)_4^-$ appears in a basic form with a pH value higher than 12 (Markiewicz, Komorowicz et al. 2015; Kotaś and Stasicka 2000).

Cr(VI) has opposite properties to that of Cr(III) in the sense that it has high solubility in water in a wide range of pH and is characterized by a high level of toxicity and its species are highly mobile in water (Gao, Sun et al. 2018; Shi, Ouyang et al. 2019). Cr(VI) occurs as various species in water such as CrO_4^{2-}, $HCrO_4^-$, and H_2CrO_4, as shown in Figure 14.1. Cr(VI) displays a high redox potential in an acidic solution and is highly reactive with oxygen and does not remain stable in the presence of an electron and its species is totally dependent ontheir pH value. One of the species that is highly acidic is H_2CrO_4, wherein the pH level of Cr(VI) falls to <1. When the pH of the solution is >1 then the dissociation of H_2CrO_4 into its deprotonated forms occurs. Therefore, at a pH between 1 and 6.5 the Cr(VI)-predominant form $HCrO_4^-$ is formed and CrO_4^{2-} appears in the entire

DOI: 10.1201/9781003326281-16

concentration range of the solution at a pH range > 6.5 (Markiewicz, Komorowicz et al. 2015; Kotaś and Stasicka 2000; Peng and Guo 2020).

For mitigating the effect of Cr(VI) from industrial waste and to reduce its effect on ecosystem and human health, various approaches have been investigated, which include reverse osmosis, ion exchange, electrochemical precipitation, electrolysis, solvent extraction, and adsorption (Beheshti, Irani et al. 2016; Mohamed, Nasser et al. 2017; Dai, Wu et al. 2018; Gao, Sun et al. 2018; Li, Dong et al. 2021). Out of these aforementioned approaches, the adsorption approach is considered to be an important method to remove the effect of Cr(VI) from wastewater because it is cost-effective, practical, and efficient. Adsorbents of different types, for instance, organic and inorganic, are used to enhance the removal of Cr(VI).

Recently, nanomaterials have been gaining attention for their brilliant qualities, including fine particle size, huge surface area, and surface functionality. They are derived from additives and nanoparticles, and pore sizes range from the micro to the nano scale (Ali et al. 2019).

Electrospinning is the simplest technique for producing nanofibers. Initially patented by Formhals in 1934, it had not achieved much attention until recently. This technique has been reintroduced for different purposes such as for sensors, photovoltaic cells, composites, high-performance filters, wound dressings, protective textiles, and scaffolds in tissue engineering, which mostly requires large surface areas (Miyoshi et al. 2005).

The technique of electrospinning nanofiber membranes is known for its greater capability for treating industrial water owing to its adjustable structure, high efficiency and porosity, large surface area, simplicity, and many more positive features. The positive features facilitate the fabrication of adsorbents into the electrospinning membranes (Wen, Yang et al. 2016; Zhao, Li et al. 2017; Li, Chen et al. 2020a; Li, Dong et al. 2021; Zhu, Zheng et al. 2021). The greater adsorption efficiency of Cr(VI) is reported in the presence of surface modification of the adsorbent with various groups, which include thiol, carboxyl, and amine. Nanofiber composites have shown to be effective for removing Cr(VI) from aqueous solutions (Avila, Burks et al. 2014), for instance, the amine groups in nanofibers (PVA/PEI) work as both binding and reducing sites for Cr(VI) removal (Zhang, Shi et al. 2020). Hence, reducing Cr(VI) to Cr(III) is an output yielding approach to remediate Cr(VI) (Gao, Sun et al. 2018).

Moreover, converting Cr(VI) to Cr(III) is a promising method for removing Cr(VI) from industrial wastewater. Therefore, catalytic reduction is a promising strategy and has gained popularity in recent years due to advantages such as security, selectivity, and high activity (Tian, Liu et al. 2020). Different variants of nano materials have been brought in use for reducing Cr(VI) such as zero-valent iron, Fe nanoparticles, palladium, platinum, silver, nickel, bimetallic nanoparticles, metallic sulfide, and metal free nanoparticles (Farooqi, Akram et al. 2021).

14.2 HEAVY METALS IN AQUEOUS SOLUTIONS

Heavy metals are naturally occurring elements, having fivefold higher specific gravity as compared to water, and high atomic weights. Few examples of heavy metals include arsenic (As), chromium (Cr), copper (Cu), cadmium (Cd), lead (Pb), and cobalt (Co), etc. Anthropogenic activities are the major cause of environmental contaminations from domestic, industrial, and agricultural sources (Tchounwou, Yedjou et al. 2012).

14.2.1 Chromium and Its Effects

Naturally occurring chromium originates in volcanic dust, rocks, soils, gases, and plants. It occurs in two forms: chromic and chromates. Metallic chromium(0) has no valency while chromium(II) and chromium(III) are not toxic and termed as chromous and chromic acids, respectively. Hexavalent

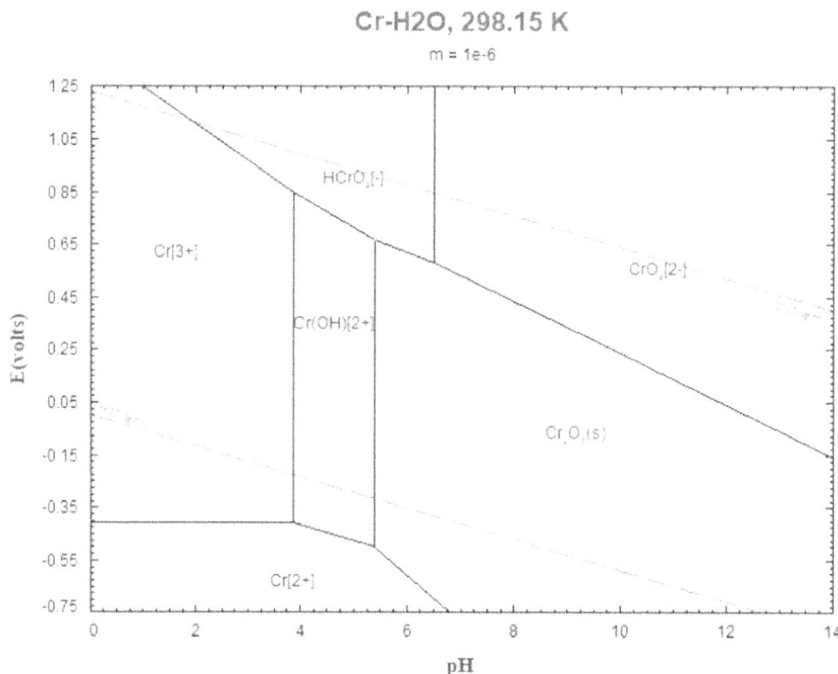

FIGURE 14.1 Eh-pH phase diagram showing speciation of chromium ions (drawn using online tool: https://www.crct.polymtl.ca/ephweb.php).

chromium(VI) as chromates is the most toxic form of chromium. The health impacts of toxic hexavalent chromium range from kidney and liver failure, lung cancer, to skin irritation (Bhaumik, Maity et al. 2011; Qusti 2014).

Chromates are the oxyanions of hexavalent chromium, which are usually industrial byproducts formed due to heating trivalent chromium in an alkaline and open environment. It is a powerful oxidizing agent in an acidic environment. Oxygen creates a strong bond with hexavalent chromium and makes chromate (CrO_4^{2-}) or dichromate ($Cr_2O_7^{2-}$) ions (Ray 2016). These two water-soluble species are extremely toxic for the tissues of the human body. The World Health Organization's (WHO) permissible limit for chromium(VI) in drinking water and wastewater is set below 50 μg/L and 500 μg/L, respectively (Hua, Yang et al. 2017).

Speciation of chromium in an aqueous environment is important because this determines the type of reaction and end product formation. At alkaline pH, chromium is usually precipitated and at acidic pH it remains soluble in water (Almeida, Cardoso et al. 2019). The speciation of chromium can be observed in Figure 14.1.

14.2.2 Heavy Metal Treatment Techniques

Several treatment technologies can be employed to remove heavy metals from water. Some have advantages over others but also have limitations (Fu and Wang 2011). The following technologies can be utilized for heavy metal removal:

14.2.2.1 Chemical Precipitation

This simple technique is quite efficient in removing heavy metals especially when the concentration is very high but is ineffective at lower concentrations. It is an economical process that generates a lot of sludge, which is difficult to dewater and handle (Ku and Jung 2001).

14.2.2.2   Ion Exchange

It is a very common technique for heavy metal removal that relies on the renewal of chemical reagents. The regeneration of resins is also responsible for secondary pollution that might cause serious problems. Also, the process is expensive and difficult to use in large-scale applications (Alyüz and Veli 2009).

14.2.2.3   Coagulation and Flocculation

It's a very common process utilized in the treatment of drinking water and wastewater. The process produces a lot of sludge that is difficult to settle and dewater. Use of chemicals for treatment is also one concern (El Samrani, Lartiges et al. 2008).

14.2.2.4   Flotation

Flotation has many advantages including good removal capacity, high overflow rates, lower detention times, and low operating cost. This technique is not suitable if the heavy metals are in soluble form and might require surface active agents as a removal aid. It also requires higher cost of operation and maintenance (Fu and Wang 2011).

14.2.2.5   Electrochemical process

It is a fast and controlled process that requires fewer chemical additions and low sludge production. High capital cost and electricity consumption restricts its widespread application (Fu and Wang 2011). Before selection of the process, it needs to be optimized for installation, cost, flexibility, reliability, selection of metal electrodes, and environmental impact, etc. (Brooks, Bahadory et al. 2010).

14.2.3   Nanotechnology

The science and technology of utilizing materials in the range of 1–100 nanometers is known as nanotechnology. It is a widespread and interdisciplinary science that has applications in domestic, industrial, transport, cosmetics, sports, medicine, and agriculture, as shown in Figure 14.2.

Nanomaterials are very popular nowadays due to their unique characteristics including electrical, optical, magnetic properties, etc. Generally, either natural nanomaterials or engineered nanomaterials (ENs) have been utilized in applications.ENs can be utilized in tires, cosmetics, electronics, sporting goods, and medicine. ENs can be carbon-based, metal-based dendrimers (highly ordered branched macromolecules) or composite-based nanomaterials (Dan 2017).

14.2.4   Electrospinning

Electrospinning is a common process to form highly porous non-woven membranes with large surface-to-volume ratio that makes these membranes very effective in various applications. Moreover, several composite materials have also been prepared with this technique to provide a variety of properties for diverse applications. It is a simple technique that makes it quite handy in industrial and academic applications (Ramakrishna, Fujihara et al. 2006).

14.2.5   Electrospinning Process

The electrospinning process depends on the voltage application, syringe pump, and fiber collection plate. The process requires a polymer solution of the desired material in the syringe that is attached to the pump. With the help of a high-voltage syringe pump, the polymeric solution is pumped out onto the collector and starts accumulating there. This forms a thin layer of nanofibers that gets accumulated with time and forms a nanomembrane layer over layer. A variation in solution ratios

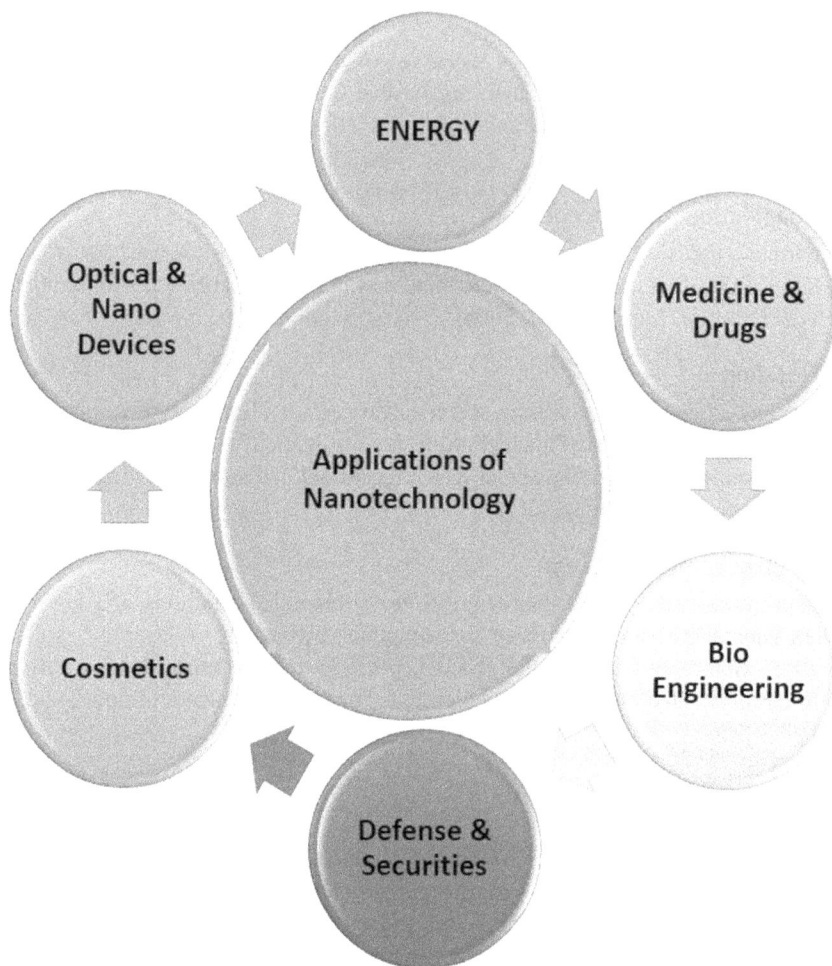

FIGURE 14.2 Application paradigms of nanotechnology.

and voltage might yield different results. So, the optimization or variation in voltage is a key step that determines the product quality as voltage creates an electrostatic repulsion force and surface tension that makes nanofiber sheets (Ramakrishna, Fujihara et al. 2006).

14.2.6 PARAMETERS AFFECTING ELECTROSPINNING

The morphological properties of nanofibers depend on the parameters which govern the process efficiency. The required morphology of nanofibers can be achieved by varying these main parameters. Three main parameters governing the electrospinning are solution, process, and ambient parameters.

14.2.6.1 Solution Parameters

14.2.6.1.1 Concentration

It is the main factor that governs the viscosity of the solution and determines the diameter of the nanofibers. Microfibers can be obtained if the concentration of the solution is very low, resulting in a small diameter of fibers and high surface tension. Higher concentration might induce the formation of beads instead of fibers. A suitable concentration will give rise to smooth fibers (Deitzel, Kleinmeyer et al. 2001).

14.2.6.1.2 Molecular Weight

Polymeric chain entanglement of the solution happens due to the molecular weight of the polymer used. Higher molecular weight results in smooth nanofibers and the formation of beads occurs when the molecular weight of the polymer is lower. The nanofiber shape totally depends on the molecular weight and concentration, which results in micro-ribbon fibers due to high molecular weight (Akduman, Perrin et al. 2014).

14.2.6.1.3 Viscosity

Viscosity determines the smooth flow of solution through the syringe and continuous nanofiber formation. It is important to optimize the viscosity of the solution as highly viscous solutions result in blockage of syringe passage and less viscous solutions do not form smooth fibers. Usually, the polymer-to-solvent ratio variation can provide the optimized viscosity of the solution. The range of viscosity depends on the type of polymer, its concentration, and the molecular weight (Ding, Kim et al. 2002).

14.2.6.1.4 Surface Tension

Surface tension has a defining role in electrospinning as a function of solvent in the polymeric solution. Several solvents have been reported including methylene chloride, dimethylformamide, and ethanol. Surface tension varies among solvents and variation in the solution ratios also provides variety in the selection of surface tension. Fiber morphology also depends on it. Usually, fixed solvent mass ratio and fixed concentration can overcome the issue of surface tension in the process (Akduman, Perrin et al. 2014).

14.2.6.1.4 Conductivity

By using different solvents and polymers the solution's conductivity can be varied. The presence of poly-electrolytes provides natural polymeric ions with the ability to increase the charge carrying capacity of polymeric jets. This results in poor nanofiber formation due to the high surface tension under an electric field. Addition of ionic salts in the polymeric solution alters the electrical conductivity of the solution. The diameter of the nanofibers is also affected by ionic salts. If the conductivity of the polymeric solution is increased, finer nanofibers can be obtained (Ding, Kim et al. 2002).

14.2.6.2 Processing Parameters

14.2.6.2.1 Voltage

The charged jets from the tip of the syringe are ejected because of the voltage applied. This makes voltage one of the key parameters of electrospinning. A lower voltage results in fibers with high diameters while a higher voltage results in finer nanofibers. The nanofiber diameter also depends on the type of polymer used. Some polymers form beads at a higher voltage, which can be adjusted by varying the syringe tip distance to collector and solution concentration (Deitzel, Kleinmeyer et al. 2001).

14.2.6.2.2 Flow Rate

Flow rate should be kept lower, as it provides enough time for the polymer to become more polarized. At higher flow rates, the solution dries quickly and forms beads with larger diameter. Smooth fibers without beads can always be achieved by low flow rates of solution (Guo, Zhou et al. 2013).

14.2.6.2.3 Collectors

Conductive substances are used as collectors to gather the charged solutions to form nanofibers. Stationary or rotatory collectors with aluminum foil wrapped around are common in electrospinning. Aluminum foil also helps in nanofiber removal from the collector. Different types of collectors have been reported such as pin type, grids, rotating rods, wire mesh, and gridded bar, etc. (Deitzel, Kleinmeyer et al. 2001).

14.2.6.2.4 Collector to Syringe Tip Distance

The distance between the tip of the syringe and collector affects the morphology and diameter of the nanofibers. The distance needs to be optimized as a short distance would not give enough time to nanofibers to dry and a long distance would form beads with a thin diameter. Drying of nanofibers is a very important parameter that needs to be considered for process optimization (Deitzel, Kleinmeyer et al. 2001).

14.2.6.3 Ambient Parameters

Ambient parameters, such as temperature and humidity, can also affect the morphology of nanofibers. Temperature is inversely proportional to viscosity; therefore, at higher temperatures less viscous solutions and at lower temperatures comparatively high viscous solutions can be obtained. Humidity also governs the viscosity of the solution, as high humidity leads to the formation of thick fibers and low humidity dries out the fibers. High humidity neutralizes the charges of the current during electrospinning (Mit-uppatham, Nithitanakul et al. 2004).

14.2.7 Adsorptive Removal of Hexavalent Chromium from Aqueous Solutions

For the removal of hexavalent chromium, an aqueous solution of zein polymer was utilized. The nanofibers obtained were ribbon-like with the variation of the zein and solvent ratio. The main removal was obtained at pH 2.0 and within 3 min. The zein nanofibers were swollen after the adsorption process (Yun, Kwak et al. 2014). Another study reported utilization of the composite material for the removal of hexavalent chromium. The chitosan/nylon-6 (CN6) and tannin/nylon-6 (TN6) composites with 100–800-nm diameter showed maximum removal at pH 3 and pH 2, respectively. An adsorption capacity of 23.9 mg/g and 62.7 mg/g was obtained for CN6 and TN6, respectively. Both composites fitted well on the Langmuir isotherm and followed pseudo second-order kinetics (Kummer, Schonhart et al. 2018). Manganese oxide nanofibers (MONFs) of 10–16-nm diameter and 94 m^2/g surface area removed hexavalent chromium at pH 2 with 14.6 mg/g adsorption capacity. MONFs also followed pseudo second-order kinetics (Qusti 2014). An electrospun nylon-66 membrane functionalized with aminopropyltriethoxysilane (APTES) containing an amine group resulted in 650 mg/g adsorption capacity of hexavalent chromium with 93% removal efficiency. The removal was best described by the Freundlich isotherm with high R^2 value (Shahram Forouz, Ravandi et al. 2016). A polyacrylonitrile/graphene oxide nanofiber decorated with zinc oxide was quite effective in hexavalent chromium removal at 690 mg/g adsorption capacity at slightly acidic pH of 6.0 (Abdel-Mottaleb, Khalil et al. 2019).

14.2.8 Chromium Removal on Different Electrospun Nanofiber Membranes

A large number of researches have indicated how to remove the toxic effects of Cr(VI) on human beings and the ecosystem as well and in this connection different variants of adsorbents have been considered for creating electrospun nanofiber membranes for removing Cr(VI) from industrial waste. Polyethylenimine and polyvinyl alcohol (PEI/PVA) are fabricated by means of electrospinning in order to remove or reduce the impact of Cr(VI). The highest adsorption limit for Cr(VI) was found to be 150 mg/g in a batch experiment. PEI/PVA display a behavior which depends upon the pH scale in the course of removing of Cr(VI) and the efficiency of removal was high from a pH of 1 to 4 and low from a pH of 4 to –9. Hence, greater adsorption removal was attained at a pH of 4. It was found that due to the existence of an amine group in electrospun nanofibers adsorption accompanied with reduction takes place (Zhang, Shi et al. 2020). The other study pointed out the production of polyacrylonitrile/iron nitrate (PAN/Fe(NO$_3$)$_3$) using an electrospinning method followed by the production of carbon/iron nanofibers by employing the carbonization process. The produced nanofibers represent a 445.3 mg/g removing ability for Cr(VI) via adsorption. The membranes made up of carbon and iron also depend upon the pH level. Therefore, the maximum removal performance was noted at a pH range of 1 and the least level of removal of Cr(VI) at a pH of 11 (Lu, Liu et al. 2020).

Another study proposes that branched functionalized magnetic iron oxide (Fe$_3$O$_4$), branched polyethylenimine (b-PEI), PAN composite, and electrospun fiber adsorbents (b-PEI-FePAN) were fabricated and also displayed the removal ability of 684.93 mg/g for mobile Cr(VI) and this aforementioned composite nanofiber membrane displayed the maximum removal of Cr(VI) at a pH scale of 2 (Zhao, Li et al. 2017). A number of researches have fabricated and designed various nanomembranes with the purpose of efficiently removing Cr(VI) from industrial waste. For instance, chitosan (CS) removes 20.5 mg/g of Cr(VI) (Li, Zhang et al. 2017), chitosan/polyethylene oxide/permutit (CS/PEO/PT) adsorbs about 208 mg/g of Cr(VI) (Wang, Wang et al. 2018), chitosan/sodium polyacrylate (CS/PAAS) adsorbs 78.92 mg/g of Cr(VI) (Jiang, Han et al. 2018), polyacrylonitrile/polypyrrole (PAN/PPy) displays a removing capacity of 74.91 mg/g of Cr(VI) (Wang, Pan et al. 2013), polydopamine/polyvinyl fluoride/polypyrrole (PDA/PVDF/PPy) removes 126.7 mg/g of Cr(VI) (Ma, Zhang et al. 2018).

Moreover, various researches were conducted by modifying the surface of nanofibers with various functional groups such as electrospun PAN nanofibers. These nanofibers were modified by using ethylenediaminetetraacetic acid and ethylenediamine, which have been used as a crosslinker and, as a result of this modification, the nanofiber surface was enhanced and it displayed an adsorption ability of 66.24 mg/g of Cr(VI) (Chaúque, Dlamini et al. 2016); other studies involved the synthesis of poly(ether sulfone) (PES) and sulfonated poly(ether sulfone) (SPES) nanofiber mats by employing the use of electrospinning and modification through the use of in situ polymerization of poly(3,4-ethylene dioxythiophene) (PEDOT) on non-woven (NW) electrospun PES and SPES mats. The adsorption capability of NW PEDOT/SPES mats was higher than that of NW PEDOT/PES mats, which is estimated to be 418.86 and 241.70 mg/g for Cr(VI). The maximum pH value for Cr(VI) came out to be 2 (Mohamadi and Abdolmaleki 2017).

Moreover, various studies have prepared organic/inorganic electrospun nanofiber composites in order to remove Cr(VI), such as polystyrene/thermoplastic polyurethane@silica (PS/TPU@SiO$_2$), which displays an adsorption capability of 57.73 mg/g for Cr(VI) (Wang, Sun et al. 2019), chitosan/graphene oxide CS/GO nanomembrane, which displays an adsorption capability of 310.4 mg/g for Cr(VI), and chitosan/polyethylene oxide/activated carbon (CS/PEO/AC), which displays an adsorption capability of 261.1 mg/g for Cr(VI) (Najafabadi, Irani et al. 2015; Shariful, Sepehr et al. 2018), The synthesis of chitosan-grafted-poly(N-vinylcaprolactam) (chitosan-g-PNVCL) nanofibers is done through the electrospinning technique. ZIF-8 metal-organic framework nanoparticles were added to the nanofibers for the purpose of adsorption of about 495.6 mg/g of Cr(VI) (Bahmani, Koushkbaghi et al. 2019). Chitosan/polyvinyl alcohol/zeolite (CS/PVA/zeolite) displays an adsorption capacity of 450 mg/g for Cr(VI) (Habiba, Siddique et al. 2017), chitosan/Fe$_3$O$_4$/oxidized multiwalled carbon nanotubes (CS/Fe$_3$O$_4$/o-MWCNTs) displays an adsorption capacity of removing of Cr(VI) of 358 mg/g (Beheshti, Irani et al. 2016), polyamide 6/Mg(OH)$_2$ displays a removing adsorption capability for Cr(VI) of 296.4 mg/g (Jia, Wang et al. 2014). Nylon-66 membranes were prepared by the process of electrospinning and then the membranes were functionalized with aminopropyltriethoxysilane, which consists of an amino group that has the ability to adsorb Cr(VI) with a capacity of 650.41 mg/g (Shahram Forouz, Ravandi et al. 2016). The adsorption removal capacities of different polymer nanofibers for Cr(VI) are provided in Table 14.1.

TABLE 14.1

Adsorption capacities of various polymer nanofibers for Cr(VI)

Polymer	Abbreviation	Adsorption (mg/g)	References
Chitosan	CS	20.5	Li, Zhang et al. (2017)
Chitosan/poly(ethylene oxide)/ permutit	CS/PEO/PT	208	Wang, Wang et al. (2018)
Chitosan/sodium polyacrylate	CS/PASS	78.92	Jiang, Han et al. (2018)
Polyethylenimine/polyvinyl alcohol	PEI/PVA	150	Zhang, Shi et al. (2020)
Polyacrylonitrile/polypyrrole	PAN/PPy	74.91	Wang, Pan et al. (2013)
Polydopamine/polyvinylidene fluoride/polypyrrole	PDA/PVDF/PPy	126.7	Ma, Zhang et al. (2018)
Polyacrylonitrile	PAN	66.24	Chaúque, Dlamini et al. (2016)
Poly(ether sulfone)/poly(3,4-ethylene dioxythiophene)	PES/PEDTO	418.86	Mohamadi and Abdolmaleki (2017)
Polyetherimide-Fe_3O_4/ polyacrylonitrile	b-PEI-Fe_3O_4/PAN	684.93	Zhao, Li et al. (2017)
Polystyrene/thermoplastic polyurethane@silica	PS/TPU@SiO_2	57.73	Wang, Sun et al. (2019)
Chitosan/graphene oxide	CS/GO	310.4	Najafabadi, Irani et al. (2015)
Chitosan/poly(ethylene oxide)/ activated carbon	CS/PEO/AC	261.1	Shariful, Sepehr et al. (2018)
Chitosan-grafted-poly(N-vinylcaprolactam)/ZIF-8	CS-PNVCL/ZIF-8	495.6	Bahmani, Koushkbaghi et al. (2019)
Chitosan/polyvinyl alcohol/zeolite	CS/PVA/zeolite	450	Habiba, Siddique et al. (2017)
Chitosan/Fe_3O_4/oxidized multiwalled carbon nanotubes	CS/Fe_3O_4/o-MWCNTs	358	Beheshti, Irani et al. (2016)
Polyamide 6/$Mg(OH)_2$	PA6/$Mg(OH)_2$	296.4	Jia, Wang et al. (2014)
Carbon/iron nanofibers (from PAN/ $Fe(NO_3)_3$)	Carbon/iron nanofibers	453.3	Lu, Liu et al. (2020)
Nylon-66	PA66	650.41	Shahram Forouz, Ravandi et al. (2016)

14.2.8.1 Catalytic Reduction of Cr(VI)

Catalytic reduction of Cr(VI) to Cr(III) has been the focus of attention for various researchers for last few years owing to several advantages, for instance, high efficiency, ecofriendliness, and economical cost as well. For carrying out the process of catalytic reduction of chromium, different types of nanoparticles were used, such as iron sulfide nanoparticles, non-metallic nanoparticles, and metallic nanoparticles, of which zero-valent iron is regarded as one of the most environmental-friendly nanoparticles for reducing chromium from Cr(VI) to Cr(III). This type of nanoparticle oxidizes into Fe^{2+}/Fe^{3+} during the reduction process of chromium from Cr(VI) to Cr(III). Iron nanoparticles have gained popularity because of being economical, the high reducing power of iron metal, and easy availability (Tian, Liu et al. 2020; Farooqi, Akram et al. 2021).

The reduction of Cr(VI) to Cr(III) takes place by the oxidization of iron nanoparticles on the carbonized surface of polyacrylonitrile/iron nitrate (PAN/$Fe(NO_3)_3$ (Lu, Liu et al. 2020). Palladium (Pd) is one of the noble metals and has a good potential for reducing Cr(VI) (Nasrollahzadeh, Bidgoli et al. 2020). Stabilizing agents are required while using palladium in the reduction process of Cr(VI) (Veerakumar and Lin 2020). Besides that, catalytic conversion can also be done by using platinum nanoparticles. The use of platinum nanoparticles is usually discouraged because of its cost (Farooqi, Akram et al. 2021). Moreover, silver nanoparticles are also employed for reducing chromium but they have low utility and stability as nano catalysts, which are important reasons for their limited use; the other reason for avoiding the use of silver is that it is not capable of catalyzing dehydrogenation of formic acid to produce hydrogen for reducing Cr(VI). Nickel (Ni) is also used for reducing chromium and it is a non-noble metal (Farooqi, Akram et al. 2021). Another technique which has also been the focus of attention for reducing Cr(VI) is N-doped carbon substance containing transition metallic nanomaterials (Zhu, Zheng et al. 2021). Instead of using monometallic nanoparticles for reducing Cr(VI), bimetallic nanoparticles are employed (Saikia, Borah et al. 2017). Iron sulfide (FeS) is used for reducing Cr(VI) and the non-metallic nanoparticles can also be used for reducing chromium from Cr(VI) to Cr(III) (Farooqi, Akram et al. 2021). $Pd@SiO_2-NH_2$ can be used for the catalytic reduction of Cr(VI) with a formic acid solution as the reducing agent (Celebi, Yurderi et al. 2016). Nanoparticles of Fe_3O_4/Pd were encapsulated and fabricated in an N-doped carbon shell (Fe_3O_4/Pd@N-C) composite and they displayed a high catalytic reduction of Cr(VI) in the presence of formic acid as the reducing agent (Tian, Liu et al. 2020).

14.2.8.2 Cr(VI) Adsorption and Reduction Mechanism

The process of adsorption usually relies on the reaction between Cr(VI) ions and functional groups including coordination chelation, electrostatic interaction, and the exchange of ions. Physical adsorption process is carried out by electrostatic contact between positively charged heavy metallic ions and negative functional groups, or between oppositely charged functional groups on the membrane surface, whereas chemical adsorption occurs via the exchange of ions between functional groups and a heavy metal. So, the adsorption of nanofibers mainly occurs because of membrane functional groups (Zhu, Zheng et al. 2021). Cr(VI) ions are adsorbed when amine groups from PVA/PEI and Fe_3O_4 particles contain a positive charge; anionic Cr(VI) ions are then adsorbed by b-PEI-FePAN through electrostatic attraction. At the same time, the reduction of Cr(VI) to Cr(III) both in the aqueous solution and on the adsorbent surface is done with the help of donating electrons from amine groups and Fe^{2+}. Conclusively, the Cr(VI) removal mechanism with the help of b-PEI-FePAN fibers is shown in Figure 14.3 (Zhao, Li et al. 2017). Electrostatic attraction between both Cr(VI) with negative charge and polyvinyl alcohol/polyethylenimine (PVA/PEI) with a positively charged nanofiber surface is the root mechanism for adsorption of Cr(VI). Amine groups in nanofibers work as both reducing and binding sites for the removal of Cr(VI) (Zhang, Shi et al. 2020).

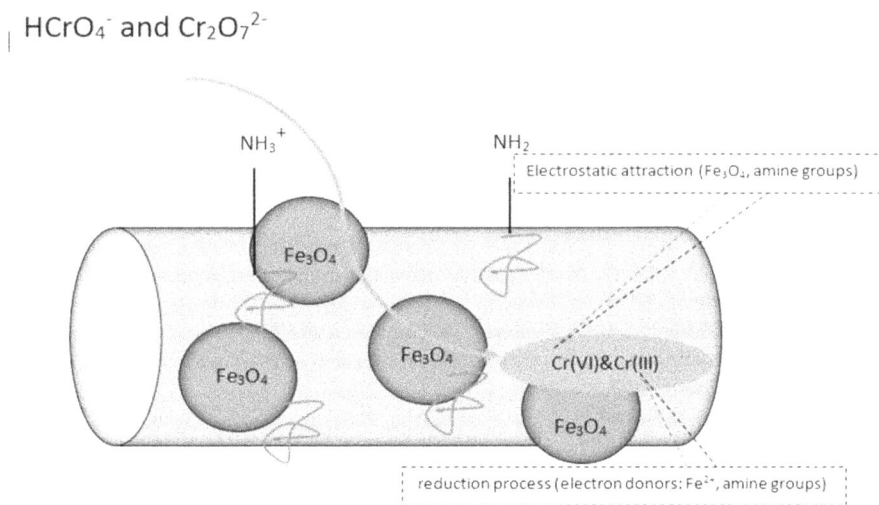

FIGURE 14.3 Cr(VI) adsorption and reduction mechanism.

14.3 CONCLUSIONS

The presence of chromium Cr(VI) in the environment is dangerous for humans and other living organisms. Anthropogenic activities are the primary cause of chromium in the environment. In this chapter, the techniques used to remove Cr(VI) from wastewater are discussed. The drawbacks in conventional Cr(VI) removal techniques necessitated the search for a better technique. Nanotechnology is an emerging technology that can remove Cr(VI) from wastewater. The addition of nanoparticles to nanofiber membranes prepared by the electrospinning method helps in the efficient removal of Cr(VI) from wastewater. The chapter also discusses the mechanism involved in removing Cr(VI) using nanofiber membranes.

REFERENCES

Abdel-Mottaleb, M., A. Khalil, T. Osman and A. Khattab (2019). "Removal of hexavalent chromium by electrospun PAN/GO decorated ZnO." *Journal of the Mechanical Behavior of Biomedical Materials* **98**: 205–212.

Akduman, C., E. Perrin, A. Kumabasar and A. Çay (2014). Effect of molecular weight on the morphology of electrospun poly(vinyl alcohol) nanofibers. XIIIth International Izmir Textile and Apparel Symposium April.

Ali, I., X. Mbianda, A. Burakov, E. Galunin, I. Burakova, E. Mkrtchyan, A. Tkachev and V. Grachev (2019). "Graphene based adsorbents for remediation of noxious pollutants from wastewater." *Environment International* **127**: 160–180.

Almeida, J. C., C. E. Cardoso, D. S. Tavares, R. Freitas, T. Trindade, C. Vale and E. Pereira (2019). "Chromium removal from contaminated waters using nanomaterials–a review." *TrAC Trends in Analytical Chemistry* **118**: 277–291.

Alyüz, B. and S. Veli (2009). "Kinetics and equilibrium studies for the removal of nickel and zinc from aqueous solutions by ion exchange resins." *Journal of Hazardous Materials* **167**(1–3): 482–488.

Avila, M., T. Burks, F. Akhtar, M. Göthelid, P. C. Lansåker, M. S. Toprak, M. Muhammed and A. Uheida (2014). "Surface functionalized nanofibers for the removal of chromium(VI) from aqueous solutions." *Chemical Engineering Journal* **245**: 201–209.

Bahmani, E., S. Koushkbaghi, M. Darabi, A. ZabihiSahebi, A. Askari and M. Irani (2019). "Fabrication of novel chitosan-g-PNVCL/ZIF-8 composite nanofibers for adsorption of Cr(VI), As(V) and phenol in a single and ternary systems." *Carbohydrate Polymers* **224**: 115148.

Barbosa, R. F., A. G. Souza, H. F. Maltez and D. S. Rosa (2020). "Chromium removal from contaminated wastewaters using biodegradable membranes containing cellulose nanostructures." *Chemical Engineering Journal* **395**: 125055.

Beheshti, H., M. Irani, L. Hosseini, A. Rahimi and M. Aliabadi (2016). "Removal of Cr(VI) from aqueous solutions using chitosan/MWCNT/Fe$_3$O$_4$ composite nanofibers-batch and column studies." *Chemical Engineering Journal* **284**: 557–564.

Bhaumik, M., A. Maity, V. Srinivasu and M. S. Onyango (2011). "Enhanced removal of Cr(VI) from aqueous solution using polypyrrole/Fe$_3$O$_4$ magnetic nanocomposite." *Journal of Hazardous Materials* **190**(1–3): 381–390.

Brooks, R. M., M. Bahadory, F. Tovia and H. Rostami (2010). "Removal of lead from contaminated water." *International Journal of Soil, Sediment and Water* **3**(2): 14.

Celebi, M., M. Yurderi, A. Bulut, M. Kaya and M. Zahmakiran (2016). "Palladium nanoparticles supported on amine-functionalized SiO$_2$ for the catalytic hexavalent chromium reduction." *Applied Catalysis B: Environmental* **180**: 53–64.

Chaúque, E. F., L. N. Dlamini, A. A. Adelodun, C. J. Greyling and J. C. Ngila (2016). "Modification of electrospun polyacrylonitrile nanofibers with EDTA for the removal of Cd and Cr ions from water effluents." *Applied Surface Science* **369**: 19–28.

Dai, S., X. Wu, J. Zhang, Y. Fu and W. Li (2018). "Coenzyme A-regulated Pd nanocatalysts for formic acid-mediated reduction of hexavalent chromium." *Chemical Engineering Journal* **351**: 959–966.

Dan, N. (2017). "Lipid-based synthetic gene carriers." In Denisa Ficai, and Alexandru Mihai Grumezescu (Eds.), *Micro and Nano Technologies, Nanostructures for Novel Therapy*, 517–538. Elsevier. https://doi.org/10.1016/B978-0-323-46142-9.00019-0

Deitzel, J. M., J. Kleinmeyer, D. Harris and N. B. Tan (2001). "The effect of processing variables on the morphology of electrospun nanofibers and textiles." *Polymer* **42**(1): 261–272.

Ding, B., H. Y. Kim, S. C. Lee, C. L. Shao, D. R. Lee, S. J. Park, G. B. Kwag and K. J. Choi (2002). "Preparation and characterization of a nanoscale poly(vinyl alcohol) fiber aggregate produced by an electrospinning method." *Journal of Polymer Science Part B: Polymer Physics* **40**(13): 1261–1268.

Dognani, G., P. Hadi, H. Ma, F. C. Cabrera, A. E. Job, D. L. Agostini and B. S. Hsiao (2019). "Effective chromium removal from water by polyaniline-coated electrospun adsorbent membrane." *Chemical Engineering Journal* **372**: 341–351.

El Samrani, A., B. Lartiges and F. Villiéras (2008). "Chemical coagulation of combined sewer overflow: heavy metal removal and treatment optimization." *Water Research* **42**(4–5): 951–960.

Farooqi, Z. H., M. W. Akram, R. Begum, W. Wu and A. Irfan (2021). "Inorganic nanoparticles for reduction of hexavalent chromium: physicochemical aspects." *Journal of Hazardous Materials* **402**: 123535.

Fu, F. and Q. Wang (2011). "Removal of heavy metal ions from wastewaters: a review." *Journal of Environmental Management* **92**(3): 407–418.

Gao, Y., W. Sun, W. Yang and Q. Li (2018). "Palladium nanoparticles supported on amine-functionalized glass fiber mat for fixed-bed reactors on the effective removal of hexavalent chromium by catalytic reduction." *Journal of Materials Science & Technology* **34**(6): 961–968.

Guo, C., L. Zhou and J. Lv (2013). "Effects of expandable graphite and modified ammonium polyphosphate on the flame-retardant and mechanical properties of wood flour-polypropylene composites." *Polymers and Polymer Composites* **21**(7): 449–456.

Habiba, U., T. A. Siddique, T. C. Joo, A. Salleh, B. C. Ang and A. M. Afifi (2017). "Synthesis of chitosan/ polyvinyl alcohol/zeolite composite for removal of methyl orange, Congo red and chromium(VI) by flocculation/adsorption." *Carbohydrate Polymers* **157**: 1568–1576.

Hua, M., B. Yang, C. Shan, W. Zhang, S. He, L. Lv and B. Pan (2017). "Simultaneous removal of As(V) and Cr(VI) from water by macroporous anion exchanger supported nanoscale hydrous ferric oxide composite." *Chemosphere* **171**: 126–133.

Jia, B.-B., J.-N. Wang, J. Wu and C.-J. Li (2014). ""Flower-like" PA6@Mg(OH)$_2$ electrospun nanofibers with Cr(VI)-removal capacity." *Chemical Engineering Journal* **254**: 98–105.

Jiang, M., T. Han, J. Wang, L. Shao, C. Qi, X. M. Zhang, C. Liu and X. Liu (2018). "Removal of heavy metal chromium using cross-linked chitosan composite nanofiber mats." *International Journal of Biological Macromolecules* **120**: 213–221.

Kotaś, J. and Z. Stasicka (2000). "Chromium occurrence in the environment and methods of its speciation." *Environmental Pollution* **107**(3): 263–283.

Ku, Y. and I.-L. Jung (2001). "Photocatalytic reduction of Cr(VI) in aqueous solutions by UV irradiation with the presence of titanium dioxide." *Water Research* **35**(1): 135–142.

Kummer, G., C. Schonhart, M. Fernandes, G. Dotto, A. Missio, D. Bertuol and E. Tanabe (2018). "Development of nanofibers composed of chitosan/nylon 6 and tannin/nylon 6 for effective adsorption of Cr(VI)." *Journal of Polymers and the Environment* **26**(10): 4073–4084.

Li, L., J. Zhang, Y. Li and C. Yang (2017). "Removal of Cr(VI) with a spiral wound chitosan nanofiber membrane module via dead-end filtration." *Journal of Membrane Science* **544**: 333–341.

Li, Q.-H., M. Dong, R. Li, Y.-Q. Cui, G.-X. Xie, X.-X. Wang and Y.-Z. Long (2021). "Enhancement of Cr(VI) removal efficiency via adsorption/photocatalysis synergy using electrospun chitosan/g-C$_3$N$_4$/TiO$_2$ nanofibers." *Carbohydrate Polymers* **253**: 117200.

Li, X., D. Chen, N. Li, Q. Xu, H. Li, J. He and J. Lu (2020a). "Efficient reduction of Cr(VI) by a BMO/Bi$_2$S$_3$ heterojunction via synergistic adsorption and photocatalysis under visible light." *Journal of Hazardous Materials* **400**: 123243.

Li, X., D. Chen, N. Li, Q. Xu, H. Li, J. He and J. Lu (2020b). "Hollow SnO$_2$ nanotubes decorated with ZnIn$_2$S$_4$ nanosheets for enhanced visible-light photocatalytic activity." *Journal of Alloys and Compounds* **843**: 155772.

Lu, Y., Z. Liu, S. W. You, L. McLoughlin, B. Bridgers, S. Hayes, X. Wang, R. Wang, Z. Guo and E. K. Wujcik (2020). "Electrospun carbon/iron nanofibers: the catalytic effects of iron and application in Cr(VI) removal." *Carbon* **166**: 227–244.

Ma, F.-f., D. Zhang, N. Zhang, T. Huang and Y. Wang (2018). "Polydopamine-assisted deposition of polypyrrole on electrospun poly(vinylidene fluoride) nanofibers for bidirectional removal of cation and anion dyes." *Chemical Engineering Journal* **354**: 432–444.

Markiewicz, B., I. Komorowicz, A. Sajnóg, M. Belter and D. Barałkiewicz (2015). "Chromium and its speciation in water samples by HPLC/ICP-MS–technique establishing metrological traceability: a review since 2000." *Talanta* **132**: 814–828.

Mit-uppatham, C., M. Nithitanakul and P. Supaphol (2004). "Ultrafine electrospun polyamide-6 fibers: effect of solution conditions on morphology and average fiber diameter." *Macromolecular Chemistry and Physics* **205**(17): 2327–2338.

Miyoshi, T., K. Toyohara and H. Minematsu (2005). "Preparation of ultrafine fibrous zein membranes via electrospinning." *Polymer International* **54**(8): 1187–1190.

Mohamadi, Z. and A. Abdolmaleki (2017). "Heavy metal remediation via poly(3,4-ethylene dioxythiophene) deposition onto neat and sulfonated nonwoven poly(ether sulfone)." *Journal of Industrial and Engineering Chemistry* **55**: 164–172.

Mohamed, A., W. Nasser, T. Osman, M. Toprak, M. Muhammed and A. Uheida (2017). "Removal of chromium(VI) from aqueous solutions using surface modified composite nanofibers." *Journal of Colloid and Interface Science* **505**: 682–691.

Najafabadi, H. H., M. Irani, L. R. Rad, A. H. Haratameh and I. Haririan (2015). "Removal of Cu^{2+}, Pb^{2+} and Cr^{6+} from aqueous solutions using a chitosan/graphene oxide composite nanofibrous adsorbent." *RSC Advances* **5**(21): 16532–16539.

Nasrollahzadeh, M., N. S. S. Bidgoli, Z. Issaabadi, Z. Ghavamifar, T. Baran and R. Luque (2020). "Hibiscus Rosasinensis L. aqueous extract-assisted valorization of lignin: preparation of magnetically reusable Pd NPs@Fe_3O_4-lignin for Cr(VI) reduction and Suzuki-Miyaura reaction in eco-friendly media." *International Journal of Biological Macromolecules* **148**: 265–275.

Nthumbi, R. M., J. C. Ngila, B. Moodley, A. Kindness and L. Petrik (2012). "Application of chitosan/polyacrylamide nanofibres for removal of chromate and phosphate in water." *Physics and Chemistry of the Earth, Parts A/B/C* **50**: 243–251.

Peng, H. and J. Guo (2020). "Removal of chromium from wastewater by membrane filtration, chemical precipitation, ion exchange, adsorption electrocoagulation, electrochemical reduction, electrodialysis, electrodeionization, photocatalysis and nanotechnology: a review." *Environmental Chemistry Letters* **18**: 2055–2068.

Qusti, A. H. (2014). "Removal of chromium(VI) from aqueous solution using manganese oxide nanofibers." *Journal of Industrial and Engineering Chemistry* **20**(5): 3394–3399.

Ramakrishna, S., K. Fujihara, W.-E. Teo, T. Yong, Z. Ma and R. Ramaseshan (2006). "Electrospun nanofibers: solving global issues." *Materials Today* **9**(3): 40–50.

Ray, R. R. (2016). "Adverse hematological effects of hexavalent chromium: an overview." *Interdisciplinary Toxicology* **9**(2): 55–65.

Saikia, H., B. J. Borah, Y. Yamada and P. Bharali (2017). "Enhanced catalytic activity of CuPd alloy nanoparticles towards reduction of nitroaromatics and hexavalent chromium." *Journal of Colloid and Interface Science* **486**: 46–57.

Shahram Forouz, F., S. Ravandi and A. Allafchian (2016). "Removal of Ag and Cr heavy metals using nanofiber membranes functionalized with aminopropyltriethoxysilane (APTES)." *Current Nanoscience* **12**(2): 266–274.

Shariful, M. I., T. Sepehr, M. Mehrali, B. C. Ang and M. A. Amalina (2018). "Adsorption capability of heavy metals by chitosan/poly (ethylene oxide)/activated carbon electrospun nanofibrous membrane." *Journal of Applied Polymer Science* **135**(7): 45851.

Shi, D., Z. Ouyang, Y. Zhao, J. Xiong and X. Shi (2019). "Catalytic reduction of hexavalent chromium using iron/palladium bimetallic nanoparticle-assembled filter paper." *Nanomaterials* **9**(8): 1183.

Tchounwou, P. B., C. G. Yedjou, A. K. Patlolla and D. J. Sutton (2012). "Heavy metal toxicity and the environment." *Molecular, Clinical and Environmental Toxicology*: 133–164.

Tian, X., M. Liu, K. Iqbal, W. Ye and Y. Chang (2020). "Facile synthesis of nitrogen-doped carbon coated Fe_3O_4/Pd nanoparticles as a high-performance catalyst for Cr(VI) reduction." *Journal of Alloys and Compounds* **826**: 154059.

Veerakumar, P. and K.-C. Lin (2020). "An overview of palladium supported on carbon-based materials: synthesis, characterization, and its catalytic activity for reduction of hexavalent chromium." *Chemosphere* **253**: 126750.

Wang, B., Z. Sun, T. Liu, Q. Wang, C. Li and X. Li (2019). "NH_2-grafting on micro/nano architecture designed PS/TPU@SiO_2 electrospun microfiber membrane for adsorption of Cr(VI)." *Desalination and Water Treatment* **154**: 82–91.

Wang, J., K. Pan, Q. He and B. Cao (2013). "Polyacrylonitrile/polypyrrole core/shell nanofiber mat for the removal of hexavalent chromium from aqueous solution." *Journal of Hazardous Materials* **244**: 121–129.

Wang, P., L. Wang, S. Dong, G. Zhang, X. Shi, C. Xiang and L. Li (2018). "Adsorption of hexavalent chromium by novel chitosan/poly(ethylene oxide)/permutit electrospun nanofibers." *New Journal of Chemistry* **42**(21): 17740–17749.

Wen, H.-F., C. Yang, D.-G. Yu, X.-Y. Li and D.-F. Zhang (2016). "Electrospun zein nanoribbons for treatment of lead-contained wastewater." *Chemical Engineering Journal* **290**: 263–272.

Yun, N. K., H. W. Kwak, M. G. Lee, S. K. Lee and K. H. Lee (2014). "Electrospun zein nanofibrous membrane for chromium(VI) adsorption." *Textile Science and Engineering* **51**(2): 57–62.

Zhang, S., Q. Shi, G. Korfiatis, C. Christodoulatos, H. Wang and X. Meng (2020). "Chromate removal by electrospun PVA/PEI nanofibers: adsorption, reduction, and effects of co-existing ions." *Chemical Engineering Journal* **387**: 124179.

Zhao, R., X. Li, Y. Li, Y. Li, B. Sun, N. Zhang, S. Chao and C. Wang (2017). "Functionalized magnetic iron oxide/polyacrylonitrile composite electrospun fibers as effective chromium(VI) adsorbents for water purification." *Journal of Colloid and Interface Science* **505**: 1018–1030.

Zhu, F., Y.-M. Zheng, B.-G. Zhang and Y.-R. Dai (2021). "A critical review on the electrospun nanofibrous membranes for the adsorption of heavy metals in water treatment." *Journal of Hazardous Materials* **401**: 123608.

15 Nanoparticles for the Removal of Heavy Metals from Wastewater

Sayed Muhammad Ata Ullah Shah Bukhari,
Liloma Shah, Sana Raza, Robina Khan and
Muhsin Jamal

15.1 INTRODUCTION

Worldwide, water is a significant naturally occurring resource, and it is vital for human beings' development and the existence of humans and living organisms. Water consumption has increased speedily because of accelerated urbanisation and industrialisation and for economical reasons water scarcity has turned out to be a significant restraint. Currently, contamination of water, particularly pollution because of heavy metals in water, turns out to be a universal ecological problem, particularly by means of household, agriculture, chemical plants, metallurgy, electroplating and mining wastewater, etc. Heavy metals might be liberated into water. For all living beings clean water is an essential requirement. Water contamination, though, has increased worldwide because of quick industrialisation (Cheraghi *et al.*, 2009). Within agriculture, the consumption and demand for clean water has increased to a large extent. Clean and fresh water consumption with a high range of contaminants in the household and industrial sectors, and further kinds of consumption are approximately 8%, 22% and 70%, respectively (Reglero *et al.*, 2009). Key classes of contaminants are ions of heavy metals and tints. Without purification, water that contains such contaminants must not be used for drinking. It is problematic to entirely treat the water when such ions of heavy metal enter it (Gybina and Prohaska 2008). Ecosystems are strongly affected by such aquatic contaminants and such pollutants are dangerous for all living beings. Hence, the elimination of such contaminants from water is important so as to avoid their detrimental impacts on the environment and on human beings. Presently, water supply faces numerous challenges. Worldwide, approximately 780 million individuals lack access to clean drinkable water (Kampa and Castanas 2008). Generally, heavy metals can be termed as a class of lanthanides, actinides and metalloid transition metals having density more than "4,000 kg/m³", and their classification is into "non-essential" and "essential metals" (Tchounwou *et al.*, 2012). Both classes are involved in a wide range of crafts and industries and are a significant portion of numerous reactions/processes (biological) (Sharma and Agrawal 2005). Severe health issues are caused when short-term and longer-term exposure to these occur even at trace concentrations. Nickel (Ni), mercury (Hg), zinc (Zn), arsenic (As), copper (Co), lead (Pb), cadmium (Cd) and chromium (Cr) are the common heavy metals found in water. Because of bio-accumulation behaviour and non-biodegradability, these metals are thought to be challenging; thus, the Environmental Protection Agency lists these metals as pollutants of priority (Dominguez-Benetton *et al.*, 2018; Lesmana *et al.*, 2009).Consumption of As, Cr, Hg, Pb and Cd beyond the standard limits could lead to severe health-related issues like gastrointestinal problems, neurological depositions, damage to the nervous system, lung cancer, raised blood pressure and bone defects and

DOI: 10.1201/9781003326281-17

numerous important illnesses. Consequently, taking into consideration the carcinogenicity and complex chemistry of heavy metals, suitable techniques are needed for extracting such metals from water sources.

Numerous electrochemical, biological, chemical and physical approaches have been studied and examined in the past for the removal of heavy metals. Conventional approaches are efficient but have numerous downsides, like unsustainability, metal specificity, low efficiency, tediousness, energy intensiveness and higher costs, thus rendering them unsuccessful for meeting ecological standards and, therefore, too hard to be implemented at the industrial level (Tahoon *et al.*, 2020). Consequently, taking into consideration the adverse impacts of such metals on the atmosphere and the wellbeing of humans, there is an urgent need for introducing approaches for heavy metal removal which are efficient, eco-friendly and cost-effective. Adsorption is actually a mass-transfer technique, in which molecules of adsorbate are attached to the adsorbent surface, through either a chemical or a physical association. In the treatment industry for water such process is the most preferred, particularly because of the regenerative capacity of adsorbents (Babel and Kurniawan, 2003). Hence, amongst the approaches studied, adsorption is thought to be the most technically feasible, safe and efficient technique due to its higher effectiveness and superficial operation (Yan *et al.*, 2014). Actually, adsorption capability alters with the type of adsorbent. Usually, adsorbents which are based on activated carbon are extensively applied for heavy metal removal but because of bio-fouling, generation of waste, incapability of recovering them from the treated water and clogging, they cannot be used at larger scales (Marsh and Rodríguez 2006; Saleem *et al.*, 2019). Consequently, there is an urgent need for efficient and novel adsorbent materials (Sarma *et al.*, 2019). Nanomaterials (NMs), which have nanoscale (varying from 1 to 100 nm) dimensions, present numerous unique biological, chemical and physical features. Such properties depend on their particular surface and structure modification (Theron *et al.*, 2008). Diverse types of nanocomposites and nanoadsorbents are widely studied and applied for treating heavy metals, inorganic substances, organic dyes and further microcontaminants (hormone active substances, biocides and customer-care products) from wastewater (Yaqoob *et al.*, 2020). Because of their nanoscale size, their features, like magnetic, optical, electrical and mechanical properties, are sufficiently different from those of conventional substances. Higher reactivity, adsorption and catalysis are shown by an extensive variety of NMs. NMs are applied in numerous areas like biology (Bujoli *et al.*, 2006), sensing (Kusior *et al.*, 2013) and catalysis (Liang *et al.*, 2012; Parmon, 2008). In wastewater treatment, NMs have received widespread attention. Stronger reactivity and adsorption capabilities are shown by NMs because of their larger surface areas and smaller sizes. Numerous types of NMs successfully remove bacteria, inorganic anions (Liu *et al.*, 2014), organic contaminants and heavy metals (Yan *et al.*, 2015). Based on several investigations, NMs seem effective in treating wastewater. Currently, the most widely investigated NMs for treating contaminated water and water comprise carbon nanotubes (CNTs), nanocomposites, zero-valent metal nanoparticles and metal oxide nanoparticles. In this chapter diverse kinds of NM-based adsorbents are discussed for eliminating heavy metals from wastewater.

15.2 WASTEWATER

It is the liquid product of municipal wastes which comprises pollutants like toxic heavy metals, inorganic soluble substances, organic substances and microbes. The presence of such pollutants alters the physical, biological and chemical properties of clean water (KAbou El-Nour *et al.*, 2010). Wastewater comprises toxic materials like trace elements, radionuclides and heavy metals and larger pathogens, comprising protozoa, bacterial and viral species. Waterborne illnesses are caused by wastewater, including lethal conditions like cholera and typhoid. In 2004, contaminated water accounted for more than 1.5 million deaths in children below five years. Treatment of wastewater is currently important due to the lethal impacts of microbes and the risks of pollution of wastewater on animal, agriculture and humans. At the government and personal level, the treatment of wastewater should be taken into consideration for preventing environmental pollution. Wastewater treatment could comprise biological, chemical and physical approaches for purification of water from numerous pollutants (Bitton, 2005). Wastewater has numerous physical properties, and comprises dyes, total solids and other substances (suspended, dissolved, volatile and fixed) (Borgohain and Mahamuni, 2002). The chemical contaminants in wastewater are categorised into gaseous, inorganic and organic chemicals. In wastewater the organic contaminants are impurities, surfactants and primary impurities. In water BOD (i.e. biological oxygen demand) and COD (i.e. chemical oxygen demand) are the best indicators of organic contaminant quality. Numerous inorganic contaminants like trace elements of phosphorus, heavy metals and nitrogen compounds are found in wastewater. Along with physical and chemical characteristics, wastewater also possesses biological characteristics. Biological contaminants are disease-causing microbes which occur in wastewater. Protozoa, viruses and bacteria are major microbes of wastewater which account for chronic and acute health effects. Diverse types of bacterial species in wastewater cause numerous waterborne illnesses like shigella, typhoid and cholera. Nanotechnology provides advanced treatment solutions for water (Daus *et al.*, 2004). Processes which are based on nanotechnology are versatile, modular and efficient, providing higher performance, lower-cost water and wastewater solutions. Furthermore, nanotechnology can be applied for restoring and cleaning unusual sources of water.

15.3   NMS USED FOR REMOVING HEAVY METALS FROM WATER

Because of properties like ion binding abilities, surface functionalities, porosity and high specific surface, NMs are extensively investigated in wastewater and water treatment applications. Actually, metal ions are also removed by them even within trace concentrations (Parvin *et al.*, 2019; Yang *et al.*, 2019). NMs are classified into silica-based, carbon-based, metal oxide and metal nanoparticles, comprising "iron oxide-based magnetic nanomaterials", zero-valent iron (ZVI) and nanocomposites, as shown in Figure 15.1. The process of adsorption of metal ions is controlled by numerous factors, like flow rate, time of contact and adsorbent concentration. For removing metal ions a major part is played by nanoparticle concentration (adsorbent). Numerous investigations suggest that the efficiency of removal increases when the dosage of the adsorbent is increased while numerous other investigations report that the efficiency of removal decreases when the dosage of the adsorbent is increased because of probable agglomeration. For instance, according to Lei *et al.* when the concentration of the "Dopamine-Modified Magnetic Nano-Adsorbent" was raised from "10 to 50 mg" the capability of adsorption of Cd^{2+} ions was decreased (Lei *et al.*, 2019). The possible cause stated was adsorbent agglomeration at higher amounts, which hinders the process of adsorption. Actually, in other investigations it was reported that when the level of the gas "industry-based adsorbent" was

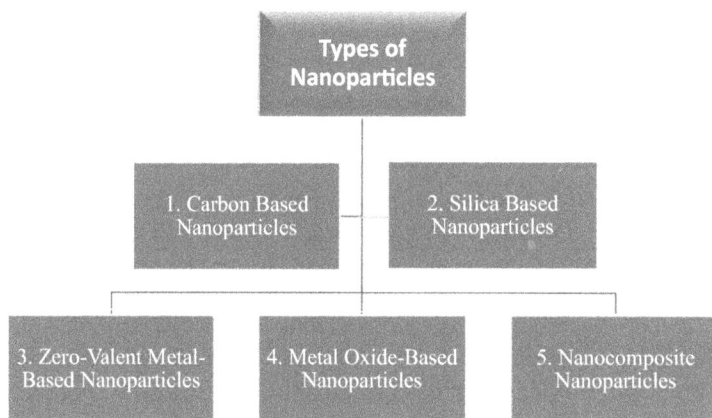

FIGURE 15.1   Commonly used types of nanoparticles.

raised from "0.25–1.25 g/100 ML" the efficiency of removal of Cd^{2+} ions was increased from 83% to 89% (Kumar and Kumar 2019). It was noticed that no agglomeration occurs in the gas phase; consequently, adsorption merely relies on the adsorbent concentration added. On the basis of such investigations, it can be concluded that the removal process is affected by adsorbent concentration, but it relies on adsorbent chemical properties and its nature, implying that either decrease or increase in removal could happen. However, lower and high nanoparticle volumes with equivalent concentration lead to similar results.

15.4   USE OF CARBON-BASED NMS FOR HEAVY METAL REMOVAL

The following NMs are utilised for removing heavy metals.

15.4.1   CARBON-BASED NMs

NMs which are carbon-based, like graphene oxide (GO), graphene, activated carbon, fullerenes and CNTs are extensively utilised in the diagnosis of disease, delivery of drugs, purification of water, electronics, sensors and energy storage (Pyrzyńska and Bystrzejewski, 2010). Furthermore, exclusive properties are shown by them which likewise enable the removal of inorganic and organic contaminants, and this makes them a suitable alternate for wastewater treatment. Hence, they are regarded as promising adsorbents for metallic contaminants (El-Sayed, 2020; Smith et al., 2015).

15.4.1.1   Fullerenes

Along with the defects and lower capability of aggregation, fullerenes are NMs which are valuable for metal adsorption from wastewater due to the adsorbate's penetration into spaces amongst the nanoclusters of carbon (Baby et al., 2019; Lucena et al., 2011). Alekseeva et al. carried out comparative investigations of nanocomposites of polystyrene and fullerenes for eliminating Cu^{2+} and fullerenes presented the best results (Alekseeva et al., 2016; Baby et al., 2019).

15.4.1.2   Carbon Nanotubes

CNTs have been widely studied in previous years. Many extraordinary properties, including thermal, mechanical, vibrational, electronic and optical properties, are shown by them (Popov 2004). Several reports exist on their uses for elimination of heavy metals from contaminated water (Gupta et al., 2016). Classification of carbon-based nanotubes is actually done into "single-walled CNTs (SWCNTs)" and "multi-walled CNTs (MWCNTs)" (Martel et al., 1998; Yu et al., 2018). In treating wastewater which contains heavy metals, numerous properties are demonstrated by CNTs particularly comprising fast kinetics of adsorption, higher capability of adsorption and larger specific surface area (Lu et al., 2016). CNTs displayed the best adsorption effects for Cr(VI), Tl(I), Cu(II), Pb(II) and Mn(VII), etc. (Kabbashi et al., 2009; Pu et al., 2013; Tang et al., 2012; Tuzen and Soylak, 2007; Yadav and Srivastava, 2017). CNTs have active adsorption sites which generally contain external groove sites, internal sites, interstitial channels and an exterior surface. For improving the efficiency of adsorption of CNTs towards different heavy metals, certain functional groups, like -OH, -NH$_2$, -COOH, etc., are usually introduced on to the CNTs' surface via endohedral filling, heat treatment or chemical modification (Kumar et al., 2014). For instance, it was stated that oxidants like NaOCl, H_2SO_4, HNO_3 and $KMnO_4$ might enhance the capability of adsorption unusually by modifying the CNTs' surface (El-Sheikh et al., 2011). Mohamed et al. demonstrated Hg(II) removal by applying a "functionalized-CNTs absorbent" (Alomar et al., 2017).

SWCNTs-COOH are very efficient in absorbing molecules as compared to those SWCNTs which are bare and unmodified (Baby et al., 2019). SWCNTs-COOH showed capabilities of adsorption of 55.89, 77.00 and 96.02 (mg/g) towards Cd^{2+}, Cu^{2+} and Pb^{2+} ions. When compared with SWCNTs which were functionalised, the SWCNTs which were not functionalised absorbed Cd^{2+} (24.07 mg/g),

Cu^{2+} (24.29 mg/g) and Pb^{2+} (33.55 mg/g). Investigators Alijani *et al.* established a nanocomposite by applying "magnetite cobalt sulphide" and SWCNTs for Hg removal; the resultant nanocomposites attained 90% mercury adsorption (Alijani and Shariatinia, 2018; Baby *et al.*, 2019). Mubarak *et al.* stated that magnetic biochar and functionalised CNTs worked best as adsorbents for Zn(II) removal. By applying magnetic biochar and functionalised CNTs the ions of Zn were removed at 75% and 99%, respectively (Mubarak *et al.*, 2013; Ouni *et al.*, 2019). Hayati *et al.* stated that CNTs that are modified with four generations of "polyamidoamine dendrimer (PAMAM/CNT)" are extremely efficient at adsorption of Pb(II) and Cu(II) (Hayati *et al.*, 2016). Applying oxidised MWCNTs, Lashedeen *et al.* assessed capabilities of adsorption for removing Ni(II), Cr(VI), Pb(II), Cd(II) and Cu(II) (Lasheen *et al.*, 2015). "Novel-modified magnetic multiwalled carbon nanotubes (Fe_3O_4/O-MWCNT)" were developed by Jiang and collaborators by co-precipitation and by evaporation based on the acid purification technique (Alimohammady *et al.*, 2017). For adsorbing Cd(II) from aqueous solutions, a modified form of MWCNTs functionalised via 3-aminopyrazole (MWCNTsf) was developed. MWCNTs, MWCNTsCOOH and MWCNTs-f sowed displayed 35.7%, 65.9% and 83.7% removal capabilities towards Cd(II), respectively (Alimohammady *et al.*, 2017; Baby *et al.*, 2019). Yang *et al.* used oxidised multi-walled CNTs for investigating adsorption of Ni^{2+} and noticed that the capability of CNTs increases when pH is varied from 0% to 99.9% (Cho *et al.*, 2010). For treatment of water, the chemical associations amongst ions of heavy metals and functional groups of MWCNT are accountable for their adsorptive capabilities. Along with capabilities of desorbing Cd^{2+}, Pb^{2+} and Cr^{6+} from water, oxidised multi-walled CNTs displayed higher efficiency and capacities of adsorption (Farghali *et al.*, 2017; Robati, 2013). Given the adsorptive capacity of heavy metal ions, multi-walled CNTs made of composite materials are used for purifying water. When MWCNTs-MnO_2-Fe_2O_3, MWCNTs Al_2O_3, MWCNTs-Fe_2O_3, MWCNTs-ZrO_2 and MWCNTs-Fe_3O_4 nanocomposites were used they effectively removed ions of heavy metals like Pb^{2+}, Cu^{2+}, Ni^{2+}, Cr^{6+} and As^{3+} from water (Luo *et al.*, 2013; Ntim and Mitra 2012; Tang *et al.*, 2012; Yang *et al.*, 2009). As compared with unoxidised MWCNTs, functionalised MWCNTs showed 20% more adsorption efficiency.

15.4.1.3 Graphene/GO

It is one more carbon-based material which has a higher capacity for removing heavy metals from contaminated water (Ali *et al.*, 2019; Amin *et al.*, 2014; Chowdhury and Balasubramanian 2014). The arrangement of carbon atoms in it is in a honeycomb lattice or in a hexagonal manner and it is a 2D material. Graphene can be classified into two main kinds: "reduced graphene oxide (RGO)" and GO. These indicate capabilities of remediating numerous environmental contaminants (Woo *et al.*, 2018). Moreover, numerous functional groups, like epoxide, carboxyl and so on, on the RGO and GO surfaces act as the active regions for removing ions of metals from water (Azizighannad and Mitra, 2018; Smith *et al.*, 2019). Functional groups like -COOH, -OH and -CH(O)CH- possess higher densities of negative charge together with hydrophilic features, which encourage their associations with metal ions which have positive charge so as to enable extraction (Xu and Wang 2017). It is reported that materials which are graphene- and GO/RGO-based possess the best capabilities of extracting metals ions from polluted water (Xu and Wang 2017). For instance, few-layered nanosheets of GO were studied for eliminating Co^{2+} and Cd^{2+} from polluted water using the batch mode (Zhao *et al.*, 2011). According to Tabish findings, researchers may be able to employ the porous graphene as an adsorbent to remove heavy metal ions and other impurities from water. As^{3+} could be efficiently removed from water by using porous graphene and the removal efficiency was 80%. In spite of recycling and the ability to be regenerated, the material maintains its efficiency in water treatment (Tabish *et al.*, 2018). In situ co-precipitating of GO from Fe3O4 and its application in water decontamination resulted in a partially reduced GO nanocomposite. Ions of Pb^{2+} are efficiently removed from aqueous solutions when such constructed nanocomposites are applied, and the adsorption capability is 373.14 mg/g (Guo *et al.*, 2018). Numerous metal ions can be removed from aqueous solutions by applying reduced GO and the functionalisation is done with "4-sulphophenylazo (rGOs)", as directed by Zhang *et al.* Maximum adsorption capacities for Cr^{3+}, Cd^{2+}, Ni^{2+}, Cu^{2+} and Pb^{2+} are 191, 267, 66, 59 and 689 (mg/g), respectively (Zhang *et al.*, 2018). Investigators from China manufactured a nanocomposite of zinc oxide from "tea polyphenols and reduced graphene oxide (TPG-ZnO)". By using it ions of heavy metals were removed and it also demonstrated anti-bacterial action (Zheng *et al.*, 2018). In larger amounts of aqueous solutions, nanosheets of GO, made from graphite by applying the modified Hummers technique, efficiently removed Co^{2+} and Cd^{2+} (Thines *et al.*, 2017; Zhao *et al.*, 2011). A temperature of about 303 K and a pH of 6.0 provided the maximum sorption capabilities for Co^{2+} and Cd^{2+} on nanosheets of GO (Thines *et al.*, 2017). According to Deng *et al.*, by applying a simpler process of electrolysis functionalised graphene (GNS PF6) was established and this was made by electrolysis with "potassium hexafluorophosphate" as the electrolyte. In order to remove ions of Cd^{2+} and Pb^{2+} from water, functionalised graphene ("GNS PF6") was made, with adsorption capabilities of 73.42 (mg/g) and 406.6 (mg/g) for Cd and Pb, respectively (Deng *et al.*, 2010; Thines *et al.*, 2017). Dawid Pakulski *et al.* investigated that with a temperature of 25°C and a pH of 6, maximum capacity of adsorption of extremely oxidised/terpyridine hybrids for Co(II), Zn(II) and Ni(II) was 336, 421 and 462 mg/g, respectively, the maximum reported within literature for GO-based sorbents and pure GO (Pakulski *et al.*, 2021). Jie Lu *et al.* assessed strength and adsorption of "GO/paper" by applying a solution of heavy metal ions and industrial wastes from a bleaching plant of the paper industry. Kinetics and isotherms of adsorption were also assessed by them. The highest capabilities of adsorption of GO/paper for Cd^{2+}, Ni^{2+} and Pb^{2+} are 31.35, 29.04 and 75.41 mg/g, respectively. S. Kabiri *et al.* characterised and made nanosheets of graphene with porous grains of silica (diatomaceous earth) and "thiol functionalized graphene" composites for elimination of Hg^{2+} (Kabiri *et al.*, 2016).

15.4.1.4 Carbon Florets

Numerous heavy metal ions (Hg^{2+}, Cd^{2+}, Cr^{6+} and As^{3+}) are efficiently removed via 3D "dendritic mesoporous nanostructured carbon florets (NCFs)" with higher precise surface area and an effortlessly available open-ended assembly of pores (1.23 cm³/g). For pH values of 2 to 13 the adsorption of water was effective and wide-ranging with NCF. It has chemical stability and a hydrophilic surface.

15.4.2 Silica-Based NMs

NMs which are silica-based are one more type of significant NM for removing metals ions because of best surface features and non-toxicity. The surface of nanosilica can be modified by groups such as -SH, -NH₂, etc. For instance, Kotsyuda *et al.* developed nanospheres of silica that were biofunctionalised via phenyl and 3-aminopropyl groups and studied their efficiency of removal towards cationic thiazine dye and Cu(II) (Kotsyuda *et al.*, 2017). The results indicate that silica nanospheres which were functionalised had greater capabilities of adsorption towards methylene blue and Cu(II) as compared to nano silica which was amino-functionalised. Najafi *et al.* studied removal efficiencies for Ni(II), Pb(II) and Cd(II) by using three silica-based NMs, comprising NH₂-SNHS (i.e. amino-functionalised silica nano hollow sphere), SNHS (i.e. non-functionalised silica nano hollow sphere) and NH₂-SG (i.e. amino-functionalised silica gel) (Najafi *et al.*, 2012). On applying NH₂-SNHS, the maximum capabilities of adsorption for Ni(II), Cd(II) and Pb(II) were 31.29, 40.73 and 96.79 mg/g, respectively. Further, for surface adjustment, silica has been widely used to make nanocomposites, amongst which magnetic silica constituents have gained a lot of interest. Pogorilyi *et al.* through Stöber reaction effectively coated particles of magnetite with silica layers and displayed their application capacity on an industrial scale (Pogorilyi *et al.*, 2014). Mahmoud *et al.* carried out "nanopolyaniline and crosslinked nanopolyaniline" immobilisation onto nano silica for creating nanocomposites "Sil-Phy-NPANI and Sil-Phy-CrossNPANI" (Mahmoud *et al.*, 2016). Through the batch method the capacities of adsorption of Sil-Phy-CrossNPANI and Sil-Phy-NPANI for Hg(II), Pb(II), Cu(II) and Cd(II) were associated. The maximum capabilities of adsorption of "Sil-Phy-NPANI" for Pb(II), Hg(II), Cd(II) and Cu(II) were 900, 600, 800 and 1,700 (μmol/g), respectively, whereas capabilities of adsorption of "Sil-Phy-CrossNPANI" for these four ions were 1,450, 1,350, 1,050 and 1,650 μmol/g, respectively. Such investigation showed that "Sil-Phy-CrossNPANI" might act as an effective adsorbent for Cd(II), Pb(II) and Hg(II).

15.4.3   Zero-Valent Metal-Based Nanoparticles

In current decades NMs which are zero-valent metal-based demonstrate their capabilities in the remediation and treatment of water. For instance, NMs of Ag are utilised for disinfection of wastewater because of their anti-microbial property (Srinivasan *et al.*, 2013). Zero-valent zinc which was nanosized showed best capability of degradation towards dioxins (Bokare *et al.*, 2013). For treating ions of heavy metals, ZVI is discussed in depth here. Furthermore, numerous other noble metals which are nanosized are likewise discussed.

15.4.3.1   Zero-Valent Iron

ZVI which is a nanosized compound that consists of "ferric oxide" and an Fe(0) coating (O'Carroll *et al.*, 2013). It received a lot of interest as an innovative adsorbent for treating numerous types of metals, like Cu(II), Cd(II), Cr(VI), Ni(II) and Hg(II), etc. (Liu *et al.*, 2015; Seyedi *et al.*, 2017). Principally, reducing capability is provided by Fe(0) whereas the ferric oxide shell offers the regions of electrostatic and reactive contact with heavy metal ions. Furthermore, the dimensions of nZVI particles are manageable and, on the surface, plentiful reactive regions are available (Cundy *et al.*, 2008). Properties like larger precise surface area and reducing capability contribute towards the potential of nanosized ZVI in the removal of metals from wastewater. The process by which nanosized ZVI removes the diverse ions of heavy metals might alter consistent with the standard potentials of heavy metals (Huang *et al.*, 2013). For instance, Pb(II) has E0, which is somewhat more positive as compared to Fe(II), so the mechanism of removal primarily included sorption and reduction, which can be expressed as follows:

$$\text{Sorption: FeOOH} + M^{2+} \rightarrow \equiv \text{FeOOM}^+ + H^+ \tag{15.1}$$

$$\text{Reduction: Fe}^0 + M^{2+} \rightarrow \equiv \text{Fe}^{2+} + M^0 \tag{15.2}$$

nZVI has numerous advantages for removal of heavy metals; however, its limitations and short-comings cannot be ignored. In aqueous solutions nZVI is oxidised with water and oxygen, which hinders or slows the process of reduction of heavy metals (Tratnyek *et al.*, 2009). Moreover, nZVI is effortlessly aggregated, which lowers the movement and surface areas of the reaction (O'Carroll *et al.*, 2013). Also, it is hard to separate nZVI from wastewater. For improving the activities of nZVI, numerous types of modification approaches are established, like doping nZVI with other metals (Pt, Ni, Cu and Pd, etc.) or surface chemical modifications (Fu *et al.*, 2014). Huang *et al.* made an innovative nZVI-modified material. This was done via combination of sodium dodecyl sulphate (SDS) with nZVI. SDS has the best capabilities of dispersion and migration and it is an anionic surfactant (Huang *et al.*, 2015). In the batch adsorption experiment, the extreme capability of removal of such innovative nZVI for Cr(VI) was "253.68 mg/g". This indicated that it is the best adsorbent, having lower aggregation and enhanced capability of adsorption. Diverse factors like initial concentration, time of contact, pH and dosage were likewise studied and under optimal circumstances a maximum efficiency of removal, i.e. 98.919%, was attained. When Au-doped nZVI was used, higher level Cd(II) removal was achieved. Such result indicated that Au-doped nZVI might be applied for treating wastewater which contains nitrate and Cd(II). Along with the two modification techniques discussed above, nanocomposites which are nZVI-based are likewise garnering interest. For instance, Zarime *et al.* established a novel nZVI-based nanocomposite by using a lower-price bentonite for treating Ni(II), Zn(II), Cd(II), Co(II), Pb(II) and Cu(II) in water (Zarime *et al.*, 2018). When bentonite is introduced to nZVI it can hinder nZVI particles from aggregation and it offers more sites for adsorption of heavy metals to the particles of nZVI. As compared to bentonite alone, when a nanocomposite of bentonite-nZVI was used it showed a high capability of removal towards such heavy metals. It is not just applied at the laboratory scale; there also exist studies on using nZVI for treating groundwater in situ (Zhang 2003).

15.4.3.2   Silver (Ag) Nanoparticles

In contrast to nZVI, studies on other metallic NMs that might be applied for heavy metal removal are not adequate. There have been numerous studies on the association between Hg(II) and Ag nanoparticles (Fan *et al.*, 2009; Morris *et al.*, 2002). The reactivity is not high for bulk silver and Hg(II). A higher reactivity is displayed by nanoparticles of Ag because with decrease in particle size the reduction potential of Ag is lowered (Pradhan *et al.*, 2002). Sumesh *et al.* developed an innovative silver nanoparticle-based adsorbent by coordinating Ag with "mercaptosuccinic acid (MSA)" (Sumesh *et al.*, 2011). Two diverse substances were examined and made by altering the Ag to MSA ratio. The results showed that as compared to common adsorbents Ag@MSA (1:6) has a high capability of removal towards Hg(II). Moreover, according to the investigators the cost for Hg(II) removal by using Ag@MSA was reasonable. As such, Ag@MSA might be used as a suitable substitute for removing Hg(II). Lisha et al. examined the removal efficiency of gold nanoparticles towards Hg(II).

15.4.3.3   Gold (Au) Nanoparticles

Lisha et al. examined the removal efficiency of gold nanoparticles towards Hg(II, such nanoparticles were supported on AI (Lisha and Pradeep, 2009). Both column and batch tests were conducted. For reducing Hg(II) to Hg(0), $NaBH_4$ was employed. The results indicated that the removal capability of Au nanoparticles towards Hg stretched upto 4.065 g/g. And such removal capacity was more than with common adsorbents. The cost of employing such type of Au nanoparticles for treating Hg(II) was lower and used Au nanoparticlescould be efficiently recovered again. It was demonstrated that Au nanoparticles supported on aluminium might be used for treating wastewater. In one study, Ojea-Jiménez *et al.* established Au nanoparticles which were citrate-coated, and they were employed for removing Hg(II) from water (Ojea-Jiménez *et al.*, 2012). Ions of citrate serve like a weak reducing agent that reduces "Hg(II) to Hg(0)", and consequently prevents $NaBH_4$ being used (Sneed *et al.*, 1961).

15.4.4   Metal Oxide-Based Nanoadsorbents

Metal oxide nanoparticles offer specific affinity, higher surface area and higher capability of removal and are the best nanoadsorbent for treating wastewater (Hua *et al.*, 2012; Yang *et al.*, 2019). Their size may vary from 1 to 100 nm and they're nanosized. Zirconium oxides, magnesium oxides, aluminium oxides, titanium oxide, iron oxides, nickel oxides, manganese oxide and zinc oxide are included in metal oxides (Hua *et al.*, 2012; Taman *et al.*, 2015; Yang *et al.*, 2019). Such nanosized metal oxides are employed to eliminate heavy metals from wastewater. Complexation influences the mechanism of metal oxide adsorption among metal oxides and dissolved metals. This is a two-step process. In the first step the ions of metal are adsorbed on an exterior surface and in the second step they are diffused along the micropore via rate-limiting intraparticle diffusion (Wang *et al.*, 2020).

15.4.4.1   Nanosized Cerium Oxides

Amongst cerium oxides, the most advantageous and common metal oxide is ceria, which is used in industrial practices, including for luminescence, in adsorbents, gas sensors, fuel cells, polishing materials and for UV shielding and blocking materials (Brosha *et al.*, 2002; Wang *et al.*, 2007; Yabe and Sato, 2003). The adsorptive characteristics of ceria depend on the surface areas, shapes, sizes and morphologies. The use of nanoscales additionally encourages newer features for nanoscale ceria, like photovoltaic response (Corma *et al.*, 2004), phase transformation (Wang *et al.*, 2001), blue shifting absorption spectra (Tsunekawa *et al.*, 2000), new catalytic activity and lattice expansion (Tsunekawa *et al.*, 2000). Nanoparticles of ceria were made via oxidation of "Ce^{3+} to Ce^{4+}". During this process hexamethylenetetramine (HMT) was used and this was done under alkaline conditions (Zhang *et al.*, 2004). In such a procedure, the stabilisation of nanocrystals of CeO_2 in the solution is done through HMT via dual electrical layer creation, which prevents nanoparticle aggregation. Throughout the Cr(VI) adsorption process on nanoparticles of ceria, no Cr(VI) was noticed in the solid phase. Apart from nanoparticles, the fabrication of ceria is effectively done in other forms, like hollow structures (Strandwitz and Stucky, 2009), 3D flower-like structures (Zhong *et al.*, 2007), nanopolyhedrons (Si *et al.*, 2005), nanotubes (Tang *et al.*, 2002), nanowires (Yu *et al.*, 2005) and nanorods (Liu *et al.*, 2009). Cao *et al.* (2010) offered a "template-free microwave-supported hydrothermal" technique for preparation of hollow nanospheres of ceria. When comparison was done with those techniques which were based on templates such as hard-template techniques in which solid templates are applied (Bian *et al.*, 2009) or softer template techniques in which organic surfactants are used (Yang *et al.*, 1998), it was found that those processes which are template-free are more economical. Spatial dispersion is efficiently enhanced by a hollow internal space, which not only results in a high surface area but it also increases molecular transportation to active regions. Hollow ceria nanospheres are formed of CeO2 nanocrystals and range in size from 14 nm to 260 nm. Ions of heavy metals are best adsorbed by such hollow nanospheres of ceria, for instance, for Pb(II) the adsorption rate was 9.2 (mg/g) and for Cr(VI) it was 15.4 (mg/g).

15.4.4.2 Copper Oxide

A study showed that when copper oxide is used as a nanoadsorbent it effectively removes heavy metals including Cd and Fe and its adsorption capacity is 131.33 and 94.34 (mg/g), respectively (Taman *et al.*, 2015). Furthermore, according to Hassan *et al.* (2017) copper oxide possesses a higher capacity of removing other heavy metals like nickel and Cd and its adsorption capabilities are 322.50 mg/g and 64.935 mg/g, respectively.

15.4.4.3 Titanium Oxide

Titanium oxide is found to remove heavy metals in aqueous solutions (Gebru and Das, 2017; Lu *et al.*, 2016). Gebru and Das (2017) stated that when "electrospun cellulose acetate/titanium oxide" was used as a nanoadsorbent it effectively removed copper and lead and its effectiveness of removal was 98.9% and 99.7% and its capability of adsorption was 23 mg/g and 25 mg/g, respectively (Gebru and Das, 2017). Li *et al.* (2014) reported that titanium oxide was employed for removing chromium. Freundlich isotherm was followed by such process of adsorption. The maximum capacity of adsorption was 117.4 (mg/g) (Li *et al.*, 2014). Consequently, it was found that titanium oxide can be used as an effective nanoadsorbent in the treatment of wastewater.

15.4.4.4 Zinc Oxide

Furthermore, zinc oxide is a NM which is manufactured and employed extensively for eliminating heavy metals due to its properties like unusual capability of removal, reasonable costs and higher surface area (Ghiloufi *et al.*, 2016; Khan *et al.*, 2019; Le *et al.*, 2019). Ghiloufi *et al.* (2016) investigated the adsorption capabilities of doped gallium zinc oxide nano powders for Cr and Cd elimination from aqueous solutions. As compared to bare zinc oxide, doped gallium zinc oxide at 1 wt% showed higher removal of heavy metal ions. Also, when gallium was introduced in nanoparticles of zinc oxide it enhanced heavy metal removal and their uptake. Zinc oxide green synthesis likewise showed that at pH 5 the capacity of removal of lead was higher at approximately 93%. Furthermore, such study showed satisfactory adsorption capacities (Azizi *et al.*, 2017). Le *et al.* (2019) likewise stated that numerous metals ions like Ag(I), Pb(II) and Cu(II) could be efficiently removed by using zinc oxide and its capacity of adsorption was above 85%.

15.4.4.5   Iron Oxide

Nanoparticles of iron oxide are widely investigated and researched as a nanoadsorbent for removing heavy metals because of their magnetic properties, easy method of isolation, best capacities of adsorption, smaller size and higher surface area (Dave and Chopda, 2014; Nizamuddin *et al.*, 2019; Vélez *et al.*, 2016). Baalousha (2009) investigated iron oxide features of agglomeration with respect to diverse factors which included pH and concentration of particle. When magnetic γ-Fe_2O_3-biochar was used it effectively adsorbed arsenic and 3.147 mg/g was its highest capability of adsorption (Zhang *et al.*, 2013). Lin and Chen (2014) likewise stated that "carbonized Fe_3O_4/phenol-formaldehyde resins" were able to adsorb arsenic and they showed higher adsorption for it, which was 216.9 mg/g. Additionally, mercury was removed effectively from water when nanoparticles of iron oxide like γ-Fe_2O_3 and Fe_3O_4 were used and the effectiveness of removal was 87% (Vélez *et al.*, 2016). However, nanoadsorbents which are based on metal oxides like iron oxides, zinc oxides and titanium oxides possess numerous drawbacks when they are utilised in a suspension particularly for treating wastewater (Lu *et al.*, 2016; Nizamuddin *et al.*, 2019). Because of the complex process of production of metal oxides and their smaller size it is extremely hard to recover them from treated water and likewise their cost of production hampers their use (Lu *et al.*, 2016; Nizamuddin *et al.*, 2019). Hence, according to investigators functionalisation of nanoadsorbents which are metal oxide-based might overcome such drawbacks and could increase adsorption capabilities. For instance, copolymers addition or ligands addition like mercaptobutyric acid, L-glutathione and ethylenedi-amine tetraacetic acid may enhance the adsorption capability of such metal oxides (Ge *et al.*, 2012; Lu *et al.*, 2016).

15.4.4.6   Nanosized Magnesium Oxides

Several investigations concentrated on the synthesis of nanosized magnesium oxides, which were of numerous structures, like nanocubes (Stankic *et al.*, 2005), 3D entities (Klug and Dravid, 2002), nanowires (Tang *et al.*, 2002; Yin *et al.*, 2002), fishbone fractal nanostructures induced by Co (Zhu *et al.*, 2001), nanorods (Mo *et al.*, 2005), nanotubes and nanobelts. Gao *et al.* (2008) investigated a technique for fabrication of magnesium oxides of diverse structures and studied their impact on the capacities of adsorption to contaminants. Nanosized magnesium oxides might likewise be utilised for removing heavy metals.

15.4.5   Nanocomposite Adsorbents

There are disadvantages associated with each nanoadsorbent. Consequently, for overcoming all such issues the fabrication of innovative nanocomposites is an efficient approach for wastewater and water treatment. Modification of nanoadsorbents is a novel approach wherein nanoparticles are incorporated with carbon/metals/polymers for producing numerous kinds of nanocomposites. Currently, numerous kinds of nanocomposites are established that are magnetic nanocomposites, both organic polymer and inorganic polymer (Yang *et al.*, 2019). The capacity of adsorption can be enhanced by such nanocomposites and these offer precise associations with the desired pollutants; hence, the best adsorption capability and higher capacity for removing heavy metals from wastewater could be accomplished (Yang *et al.*, 2019; Zhao *et al.*, 2018). Nanocomposites are advantageous in numerous ways: they are vulnerable to harsh chemical surroundings and higher temperature, have lower consumption of energy, best mechanical features and lower costs. Furthermore, adsorbents like hybrid nanocomposites likewise offer astonishing benefits in term of magnetic features and physiochemical stabilities for applications which are related to the treatment of wastewater (Nizamuddin *et al.*, 2019). Numerous studies stated that nanocomposite adsorbents possess longevity, and they are re-usable. Furthermore, they have properties of regeneration, which is the crucial factor and makes them cost-effective in metal elimination from polluted water (Ahmaruzzaman, 2019; Hadi-Najafabadi *et al.*, 2015; Mahmoudi *et al.*, 2019; Nasir *et al.*, 2019; Razzaz *et al.*, 2016).

15.4.5.1 Organic Polymer-Supported Nanocomposites

Numerous characteristics are displayed by polymeric hosts such as ecological soundness, easy regeneration, tunable functional groups and best mechanical strength (Zhao *et al.*, 2018). There exist two kinds of nanocomposites which are polymer-supported: biopolymer-supported and synthetic organic polymer-supported nanocomposites (Lu and Astruc 2018). In order to fabricate the polymer, two methods are employed, i.e. in situ synthesis and direct compounding (Zhang *et al.*, 2016). Organic polymers which are synthetic such as polyaniline (PAN), polystyrene (PS), polyaniline, etc. (Rajakumar *et al.*, 2014), are extensively investigated for nanocomposite fabrication for heavy metal treatment. For instance, Afshar *et al.* carried out fabrication of magnetic nanocomposites, i.e. "polypyrrole-polyaniline/Fe_3O_4", and studied its removal capability of Pb(II) in aqueous solutions (Afshar *et al.*, 2016). At pH = 8–10 once the Pb(II) level was 20 (mg/L) the nanocomposite effectively removed it and the effectiveness of removal was 100%. In addition to organic polymers, biopolymers like cellulose, chitosan, alginate, etc., are widely employed as nanocomposite supports. The most common biopolymer is cellulose. It has hydroxyl groups on its glucose ring, which offers plentiful association regions for heavy metal ions. Consequently, it serves as the best nanoadsorbent (Cai *et al.*, 2017). Suman *et al.* established NC-AgNPs (i.e. nanocellulose-Ag nanoparticles). They were utilised for removing microbes, heavy metals and dyes in water via column adsorption. The results indicated that they could remove Cr(III) with 98.30% and Pb(II) with 99.48% efficiency. One more biodegradable and eco-friendly substance is chitosan; it possesses good capacities for heavy metal removal because of the existence of -OH and $-NH_2$ in its structures. Saad *et al.* established a "ZnO/chitosan core-shell nanocomposite (ZOCS)" that has some benefits: its biological toxicity is lesser and it is cost-effective; also, its capacities of removal for Cd(II), Cu(II) and Pb(II) were studied (Saad *et al.*, 2018). Results from batch adsorption presented that the highest proficiencies of adsorption for Cu(II), Cd(II) and Pb(II) were 117.6, 135.1 and 476.1 (mg/g), respectively, and such nanocomposites might be employed continually as they have the best capability of adsorption. Apart from this one more biodegradable biopolymer, which is also biocompatible and non-lethal, is alginate. It is extracted from brown seaweed (Esmat *et al.*, 2017). Gokila *et al.* carried out chitosan/alginate nanocomposite fabrication for Cr(VI) removal from wastewater (Gokila *et al.*, 2017). In a batch adsorption experiment its highest capability of adsorption was 108.8 mg/g.

15.4.5.2 Magnetic Nanocomposites

Because of their easy capability of separation, considerable attention has been received by magnetic nanocomposites. These are frequently based on iron oxides and magnetic iron. Their fabrication might be done in three ways: (1) alteration of surface of magnetic iron/iron oxide nanoparticles by using functional groups like NH_2, -SH, etc., (2) iron/iron oxide nanoparticle encapsulation with other substances, like polypyrrole, MnO_2, polyrhodanine, polyethylenimine and humic acid, etc., to form a structure similar to core-shell (Kim *et al.*, 2013; Lü *et al.*, 2018; Mirrezaie *et al.*, 2014; Song *et al.*, 2011); (3) nanoparticles of iron/iron oxide coating with certain porous substance like CNTs, GO, etc. (Elmi *et al.*, 2017). Recently, Huang *et al.* made an innovative magnetic composite. This magnetic composite has amino-decorated metal-organic frameworks which serve as the shell and a nanoscale $Fe_3O_4@SiO_2$ which acts as the core (Huang *et al.*, 2018). Methylene blue and Pb(II) were effectively removed by an amino-decorated metal-organic framework. Ge *et al.* carried out fabrication of nanocomposite of "Fe@MgO" (Ge *et al.*, 2018). It showed best capabilities of removal towards methyl orange and Pb(II). Its capacity of removal was 6947.9 and 1476.4 mg/g, respectively. It was indicated that it could be used for the treatment of wastewater. Magnetic nanocomposites have higher capabilities of removing heavy metals because they possess the property of easy separation. Diverse NMs applied for removing heavy metals are provided in Table 15.1.

TABLE 15.1
Different NMs used for removal of heavy metals

S/no.	Adsorbent	Target metal	Adsorption capacity (mg/g)/ efficiency (%)
	Types of CNTs applied for removing diverse ions of metals		
1	SWCNT	Hg^{2+}	4.16%
2	Oxidised MWCNT	Cu^{2+}	78%
3	MWCNT-COOH-functionalised nanotube	Pb^{2+}	99.1%
4	SWCNTs-polysulphone nanocomposite-based membrane	Pb^{2+}, As^{3+}	94.2%, 87.6%
5	MSWCNT-CoS	Hg^{2+}	166.6%
6	Acidified MWCNTs	Pb^{2+}, Cu^{2+}, Ni^{2+}	93%, 78%, 83%
7	MWCNTs	Cd^{2+}	94.2% (pH = 7) 100% (pH = 10) 10.7% (pH = 2)
8	Oxidised SWCNTs	Ni^{2+}	47.86 38.46
9	Polyethyleneimine crosslinked GO	Cr(VI)	436.3
10	CNTs-GO	Pb(II), Cu(II)	350.87 mg/g 318.47 mg/g
11	Functionalised MWCNTs	Pb^{2+}	93%
12	MWCNTs-iron oxide composite	CO^{2+}	0.15 mmol/g
13	Thiol-functionalised SWCNTs (SWCNTs-SH)	Hg^{2+}	131.58
14	MWNCTs-hydroxyapatite	Co^{+}	16.26
15	Iron oxide-coated MWCNTs (Fe-MWCNTs)	As(III)	1.723
16	Functionalised GO-embedded calcium alginate (GOCA) beads	Pb(II), Hg(II) and Cd(II)	602, 374 and 181
17	$MnFe_2O_4$/GO	As(V), As(III) and Pb(II)	207, 146 and 673
18	RGO/NiO	Cr(VI)	198
19	Oxidised MWCNTs	Ni^{2+}	83%
20	Oxidised MWCNTs	Cu^{2+}	78%
21	GO-alpha cyclodextrin-polypyrrole	Cr^{6+}	66.67%
22	GOCA beads	Hg^{2+}, Cd^{2+}, Pb^{2+}	37.4%, 18.1%, 60.2%
	Silica-based materials for the adsorption of heavy metals		
1	Amino-functionalised and pure silica nano hollow sphere (NH_2-SNHS, SHNS) and silica gel (NH_2-SG)	Ni^{2+}	0.84% (SHNS), 2.59% (NH_2-SG), and 3.13% (NH_2-SNHS) mg/g
2	Amino-functionalised mesoporous silica	Cr^{6+}	8.205%
3	Functionalised silica with -SH	Hg^{2+}	50.5%
4	Amino-functionalised silica gel in tea polyphenol extracts	Cu^{2+}	99.59%
5	Amino-functionalised and "pure silica nano hollow sphere (NH_2-SNHS, SHNS) and silica gel (NH_2-SG)"	Pb^{2+}	26.85% (SHNS), 54.35% (NH_2-SG), and 96.78% (NH_2-SNHS)
6	Organically functionalised silica gel	Cu^{2+}	1.99%
	Nanosized metal oxides for heavy metal removal from water		
1	Hematite (α-Fe_2O_3)	Cu(II)	84.46 mg/g
2	ZnO	Pb(II)	6.7 mg/g
3	CeO_2	Cr(VI), Pb(II)	15.4 mg/g (Cr), 9.2 mg/g (Pb)
	Goethite (α-FeOOH)	Cu(II)	100% removal
4	Hematite-magnetite hybrid	Cd^{2+}	99.84%

5	Hematite-magnetite hybrid	Cd^{2+}	99.84%
6	Hematite-magnetite hybrid	Pb^{2+}	97.67%
Zero-valent metal-based NMs for heavy metal removal from water			
1	nZVI with SDS	Cr^{6+}	253.68 mg/g
2	nZVI and Au-doped nZVI nanoparticles	Cd^{2+} and nitrates	40 mg/g to 188 mg/g and 3%
3	Zero-valent iron	As(III)	3.5 mg/g
Nanocomposite NMs for heavy metal removal from water			
1	CNT-coated poly-amidoamine dendrimer (PAMAM) nanocomposite (CNT/PAMAM)	Zn(II), Co(II) and As(III)	470, 494 and 432 mg/g
2	Hydroxyapatite/zeolite nanocomposite (HAp/NaP)	Cd(II) and Pb(II)	40.16 and 55.55 mg/g
3	Polypyrrole-polyaniline/Fe_3O_4 magnetic nanocomposite	Pb(II)	100%
4	Chitosan/alginate nanocomposite	Cr(VI)	108.8 mg/g

Carbon nanotubes (CNTs); single-walled carbon nanotubes (SWCNTs); multi-walled carbon nanotubes (MWCNTs); reduced graphene oxide (RGO); graphene oxide (GO).

Adapted and modified from Kumar, S., 2021. Carbon based nanomaterial for removal of heavy metals from wastewater: a review. *International Journal of Environmental Analytical Chemistry*, pp. 1–18; Hua, M., Zhang, S., Pan, B., Zhang, W., Lv, L. and Zhang, Q., 2012. Heavy metal removal from water/wastewater by nanosized metal oxides: a review. *Journal of Hazardous Materials*, *211*, pp. 317–331; Yang, J., Hou, B., Wang, J., Tian, B., Bi, J., Wang, N., Li, X. and Huang, X., 2019. Nanomaterials for the removal of heavy metals from wastewater. *Nanomaterials*, *9*(3), p. 424; Ibrahim, H., Sazali, N., Salleh, W.N.W., Hasrul, N., Ngadiman, A., Fadil, N.A. and Harun, Z., 2021. Outlook on the carbon-based materials for heavy metal removal; Dave, P.N. and Chopda, L.V., 2014. Application of iron oxide nanomaterials for the removal of heavy metals. *Journal of Nanotechnology*, *2014*; Nik Abdul Ghani, N.R., Jami, M.S. and Alam, M.Z., 2021. The role of nanoadsorbents and nanocomposite adsorbents in the removal of heavy metals from wastewater: a review and prospect. *Pollution*, *7*(1), pp. 153–179; Kumar, R., Rauwel, P. and Rauwel, E., 2021. Nanoadsorbants for the removal of heavy metals from contaminated water: current scenario and future directions. *Processes*, *9*(8), p. 1379).

15.5 CONCLUSION

In this chapter, the most widely investigated NMs like nanocomposites, CNTs, metal oxide nanoparticles (iron oxides, ZnO and TiO_2) and nanoparticles which are zero-valent metal-based (Zn, Fe and Ag) are emphasised. Furthermore, their uses in the treatment of wastewater and water are deliberated. NMs seem to be more effective for the treatment of wastewater and water. Industrialisation, quick urbanisation, climatic alterations and growth of populations has posed certain risks such as water pollution because of which numerous pollutants are released into water like dyes, heavy metals, etc. So there is an urgent need for introducing novel approaches so as to treat such polluted water and to remove contaminants like heavy metals which pose health risks and to reduce water scarcity. Nanotechnology seems to be effective in this regard. So far, an extensive variety of nanoadsorbents have been effectively used for adsorbing heavy metals from wastewater. They could effectively remove heavy metals. However, studies on the use of nanoadsorbents for treating wastewater are not sufficient. Consequently, further studies need to be conducted on the utilisation of nanoadsorbents with regard to their synthesis, reusability and capacity of removal and suitable approaches are needed for improving the workability and effectiveness of nanoadsorbents in treating wastewater.

REFERENCES

Afshar, A., Sadjadi, S.A.S., Mollahosseini, A. and Eskandarian, M.R., 2016. Polypyrrole-polyaniline/Fe_3O_4 magnetic nanocomposite for the removal of Pb(II) from aqueous solution. *Korean Journal of Chemical Engineering*, *33*(2), pp. 669–677.

Ahmaruzzaman, M., 2019. Nano-materials: novel and promising adsorbents for water treatment. *Asian Journal of Water, Environment and Pollution*, *16*(3), pp. 43–53.

Alekseeva, O.V., Bagrovskaya, N.A. and Noskov, A.V., 2016. Sorption of heavy metal ions by fullerene and polystyrene/fullerene film compositions. *Protection of Metals and Physical Chemistry of Surfaces*, *52*(3), pp. 443–447.

Ali, I., Mbianda, X.Y., Burakov, A., Galunin, E., Burakova, I., Mkrtchyan, E., Tkachev, A. and Grachev, V., 2019. Graphene based adsorbents for remediation of noxious pollutants from wastewater. *Environment International*, *127*, pp. 160–180.

Alijani, H. and Shariatinia, Z., 2018. Synthesis of high growth rate SWCNTs and their magnetite cobalt sulfide nanohybrid as super-adsorbent for mercury removal. *Chemical Engineering Research and Design*, *129*, pp. 132–149.

Alimohammady, M., Jahangiri, M., Kiani, F. and Tahermansouri, H., 2017. A new modified MWCNTs with 3-aminopyrazole as a nanoadsorbent for Cd(II) removal from aqueous solutions. *Journal of Environmental Chemical Engineering*, *5*(4), pp. 3405–3417.

AlOmar, M.K., Alsaadi, M.A., Hayyan, M., Akib, S., Ibrahim, M. and Hashim, M.A., 2017. Allyl triphenyl phosphonium bromide based DES-functionalized carbon nanotubes for the removal of mercury from water. *Chemosphere*, *167*, pp. 44–52.

Amin, M.T., Alazba, A.A. and Manzoor, U., 2014. A review of removal of pollutants from water/wastewater using different types of nanomaterials. *Advances in Materials Science and Engineering*, *2014*, 1–24.

Azizi, S., Mahdavi Shahri, M. and Mohamad, R., 2017. Green synthesis of zinc oxide nanoparticles for enhanced adsorption of lead ions from aqueous solutions: equilibrium, kinetic and thermodynamic studies. *Molecules*, *22*(6), p. 831.

Azizighannad, S. and Mitra, S., 2018. Stepwise reduction of graphene oxide (GO) and its effects on chemical and colloidal properties. *Scientific Reports*, *8*(1), pp. 1–8.

Baalousha, M., 2009. Aggregation and disaggregation of iron oxide nanoparticles: influence of particle concentration, pH and natural organic matter. *Science of the Total Environment*, *407*(6), pp. 2093–2101.

Babel, S. and Kurniawan, T.A., 2003. Low-cost adsorbents for heavy metals uptake from contaminated water: a review. *Journal of Hazardous Materials*, *97*(1–3), pp. 219–243.

Baby, R., Saifullah, B. and Hussein, M.Z., 2019. Carbon nanomaterials for the treatment of heavy metal-contaminated water and environmental remediation. *Nanoscale Research Letters*, *14*(1), pp. 1–17.

Bian, S.W., Ma, Z., Zhang, L.S., Niu, F. and Song, W.G., 2009. Silica nanotubes with mesoporous walls and various internal morphologies using hard/soft dual templates. *Chemical Communications*, (10), pp. 1261–1263.

Bitton, G., 2005. Anaerobic digestion of wastewater and biosolids. *Wastewater Microbiology*, *3*, pp. 345–371.

Bokare, V., Jung, J.L., Chang, Y.Y. and Chang, Y.S., 2013. Reductive dechlorination of octachlorodibenzo-p-dioxin by nanosized zero-valent zinc: modeling of rate kinetics and congener profile. *Journal of Hazardous Materials*, *250*, pp. 397–402.

Borgohain, K. and Mahamuni, S., 2002. Formation of single-phase CuO quantum particles. *Journal of Materials Research*, *17*(5), pp. 1220–1223.

Brosha, E.L., Mukundan, R., Brown, D.R., Garzon, F.H. and Visser, J.H., 2002. Development of ceramic mixed potential sensors for automotive applications. *Solid State Ionics*, *148*(1–2), pp. 61–69.

Bujoli, B., Roussière, H., Montavon, G., Laïb, S., Janvier, P., Alonso, B., Fayon, F., Petit, M., Massiot, D., Bouler, J.M. and Guicheux, J., 2006. Novel phosphate–phosphonate hybrid nanomaterials applied to biology. *Progress in Solid State Chemistry*, *34*(2–4), pp. 257–266.

Cai, J., Lei, M., Zhang, Q., He, J.R., Chen, T., Liu, S., Fu, S.H., Li, T.T., Liu, G. and Fei, P., 2017. Electrospun composite nanofiber mats of cellulose@organically modified montmorillonite for heavy metal ion removal: design, characterization, evaluation of absorption performance. *Composites Part A: Applied Science and Manufacturing*, *92*, pp. 10–16.

Cao, C.Y., Cui, Z.M., Chen, C.Q., Song, W.G. and Cai, W., 2010. Ceria hollow nanospheres produced by a template-free microwave-assisted hydrothermal method for heavy metal ion removal and catalysis. *The Journal of Physical Chemistry C*, *114*(21), pp. 9865–9870.

Cheraghi, M., Lorestani, B. and Yousefi, N., 2009. Effect of waste water on heavy metal accumulation in Hamedan Province vegetables. *International Journal of Botany*, *5*(2), pp. 109–193.

Cho, H.H., Wepasnick, K., Smith, B.A., Bangash, F.K., Fairbrother, D.H. and Ball, W.P., 2010. Sorption of aqueous Zn[II] and Cd[II] by multiwall carbon nanotubes: the relative roles of oxygen-containing functional groups and graphenic carbon. *Langmuir*, *26*(2), pp. 967–981.

Chowdhury, S. and Balasubramanian, R., 2014. Recent advances in the use of graphene-family nanoadsorbents for removal of toxic pollutants from wastewater. *Advances in Colloid and Interface Science*, *204*, pp. 35–56.

Corma, A., Atienzar, P., Garcia, H. and Chane-Ching, J.Y., 2004. Hierarchically mesostructured doped CeO_2 with potential for solar-cell use. *Nature Materials*, *3*(6), pp. 394–397.

Cundy, A.B., Hopkinson, L. and Whitby, R.L., 2008. Use of iron-based technologies in contaminated land and groundwater remediation: a review. *Science of the Total Environment*, *400*(1–3), pp. 42–51.

Daus, B., Wennrich, R. and Weiss, H., 2004. Sorption materials for arsenic removal from water: a comparative study. *Water Research*, *38*(12), pp. 2948–2954.

Dave, P.N. and Chopda, L.V., 2014. Application of iron oxide nanomaterials for the removal of heavy metals. *Journal of Nanotechnology*, *2014*.

Deng, X., Lü, L., Li, H. and Luo, F., 2010. The adsorption properties of Pb(II) and Cd(II) on functionalized graphene prepared by electrolysis method. *Journal of Hazardous Materials*, *183*(1–3), pp. 923–930.

Dominguez-Benetton, X., Varia, J.C., Pozo, G., Modin, O., Ter Heijne, A., Fransaer, J. and Rabaey, K., 2018. Metal recovery by microbial electro-metallurgy. *Progress in Materials Science*, *94*, pp. 435–461.

Elmi, F., Hosseini, T., Taleshi, M.S. and Taleshi, F., 2017. Kinetic and thermodynamic investigation into the lead adsorption process from wastewater through magnetic nanocomposite Fe_3O_4/CNT. *Nanotechnology for Environmental Engineering*, *2*(1), pp. 1–13.

El-Sayed, M.E., 2020. Nanoadsorbents for water and wastewater remediation. *Science of the Total Environment*, *739*, p. 139903.

El-Sheikh, A.H., Al-Degs, Y.S., Al-As' ad, R.M. and Sweileh, J.A., 2011. Effect of oxidation and geometrical dimensions of carbon nanotubes on Hg(II) sorption and preconcentration from real waters. *Desalination*, *270*(1–3), pp. 214–220.

Esmat, M., Farghali, A.A., Khedr, M.H. and El-Sherbiny, I.M., 2017. Alginate-based nanocomposites for efficient removal of heavy metal ions. *International Journal of Biological Macromolecules*, *102*, pp. 272–283.

Fan, Y., Liu, Z. and Zhan, J., 2009. Synthesis of starch-stabilized Ag nanoparticles and Hg^{2+} recognition in aqueous media. *Nanoscale Research Letters*, *4*(10), pp. 1230–1235.

Farghali, A.A., Abdel Tawab, H.A., Abdel Moaty, S.A. and Khaled, R., 2017. Functionalization of acidified multi-walled carbon nanotubes for removal of heavy metals in aqueous solutions. *Journal of Nanostructure in Chemistry*, *7*(2), pp. 101–111.

Fu, F., Dionysiou, D.D. and Liu, H., 2014. The use of zero-valent iron for groundwater remediation and wastewater treatment: a review. *Journal of Hazardous Materials*, *267*, pp. 194–205.

Gao, C., Zhang, W., Li, H., Lang, L. and Xu, Z., 2008. Controllable fabrication of mesoporous MgO with various morphologies and their absorption performance for toxic pollutants in water. *Crystal Growth and Design*, *8*(10), pp. 3785–3790.

Ge, F., Li, M.M., Ye, H. and Zhao, B.X., 2012. Effective removal of heavy metal ions Cd^{2+}, Zn^{2+}, Pb^{2+}, Cu^{2+} from aqueous solution by polymer-modified magnetic nanoparticles. *Journal of Hazardous Materials*, *211*, pp. 366–372.

Ge, L., Wang, W., Peng, Z., Tan, F., Wang, X., Chen, J. and Qiao, X., 2018. Facile fabrication of Fe@MgO magnetic nanocomposites for efficient removal of heavy metal ion and dye from water. *Powder Technology*, *326*, pp. 393–401.

Gebru, K.A. and Das, C., 2017. Removal of Pb(II) and Cu(II) ions from wastewater using composite electrospun cellulose acetate/titanium oxide (TiO_2) adsorbent. *Journal of Water Process Engineering*, *16*, pp. 1–13.

Ghiloufi, I., El Ghoul, J., Modwi, A. and El Mir, L., 2016. Ga-doped ZnO for adsorption of heavy metals from aqueous solution. *Materials Science in Semiconductor Processing*, *42*, pp. 102–106.

Gokila, S., Gomathi, T., Sudha, P.N. and Anil, S., 2017. Removal of the heavy metal ion chromiuim(VI) using chitosan and alginate nanocomposites. *International Journal of Biological Macromolecules*, *104*, pp. 1459–1468.

Guo, T., Bulin, C., Li, B., Zhao, Z., Yu, H., Sun, H., Ge, X., Xing, R. and Zhang, B., 2018. Efficient removal of aqueous Pb(II) using partially reduced graphene oxide-Fe_3O_4. *Adsorption Science & Technology*, *36*(3–4), pp. 1031–1048.

Gupta, V.K., Moradi, O., Tyagi, I., Agarwal, S., Sadegh, H., Shahryari-Ghoshekandi, R., Makhlouf, A.S.H., Goodarzi, M. and Garshasbi, A., 2016. Study on the removal of heavy metal ions from industry waste by carbon nanotubes: effect of the surface modification: a review. *Critical Reviews in Environmental Science and Technology*, *46*(2), pp. 93–118.

Gybina, A.A. and Prohaska, J.R., 2008. Copper deficiency results in AMP-activated protein kinase activation and acetylCoA carboxylase phosphorylation in rat cerebellum. *Brain Research*, *1204*, pp. 69–76.

Hadi-Najafabadi, H., Irani, M., Roshanfekr Rad, L., Heydari Haratameh, A. and Haririan, I., 2015. Removal of Cu^{2+}, Pb^{2+} and Cr^{6+} from aqueous solutions using a chitosan/graphene oxide composite nanofibrous adsorbent. *RSC Advances*, 5(21), pp. 16532–16539.

Hassan, K.H., Jarullah, A.A. and Saadi, S.K., 2017. Synthesis of copper oxide nanoparticle as an adsorbent for removal of Cd(II) and Ni(II) ions from binary system. *International Journal of Applied Environmental Sciences*, 12(11), pp. 1841–1861.

Hayati, B., Maleki, A., Najafi, F., Daraei, H., Gharibi, F. and McKay, G., 2016. Synthesis and characterization of PAMAM/CNT nanocomposite as a super-capacity adsorbent for heavy metal (Ni^{2+}, Zn^{2+}, As^{3+}, Co^{2+}) removal from wastewater. *Journal of Molecular Liquids*, 224, pp. 1032–1040.

Hua, M., Zhang, S., Pan, B., Zhang, W., Lv, L. and Zhang, Q., 2012. Heavy metal removal from water/wastewater by nanosized metal oxides: a review. *Journal of Hazardous Materials*, 211, pp. 317–331.

Huang, D.L., Chen, G.M., Zeng, G.M., Xu, P., Yan, M., Lai, C., Zhang, C., Li, N.J., Cheng, M., He, X.X. and He, Y., 2015. Synthesis and application of modified zero-valent iron nanoparticles for removal of hexavalent chromium from wastewater. *Water, Air, & Soil Pollution*, 226(11), pp. 1–14.

Huang, L., He, M., Chen, B. and Hu, B., 2018. Magnetic Zr-MOFs nanocomposites for rapid removal of heavy metal ions and dyes from water. *Chemosphere*, 199, pp. 435–444.

Huang, P., Ye, Z., Xie, W., Chen, Q., Li, J., Xu, Z. and Yao, M., 2013. Rapid magnetic removal of aqueous heavy metals and their relevant mechanisms using nanoscale zero valent iron (nZVI) particles. *Water Research*, 47(12), pp. 4050–4058.

Kabbashi, N.A., Atieh, M.A., Al-Mamun, A., Mirghami, M.E., Alam, M.D.Z. and Yahya, N., 2009. Kinetic adsorption of application of carbon nanotubes for Pb(II) removal from aqueous solution. *Journal of Environmental Sciences*, 21(4), pp. 539–544.

Kabiri, S., Tran, D.N., Cole, M.A. and Losic, D., 2016. Functionalized three-dimensional (3D) graphene composite for high efficiency removal of mercury. *Environmental Science: Water Research & Technology*, 2(2), pp. 390–402.

KAbou El-Nour, K.M., Eftaiha, A.A., Al-Warthan, A. and Ammar, R.A., 2010. Synthesis and applications of silver nanoparticles. *Arabian Journal of Chemistry*, 3(3), pp. 135–140.

Kampa, M. and Castanas, E., 2008. Human health effects of air pollution. *Environmental Pollution*, 151(2), pp. 362–367.

Khan, I., Saeed, K. and Khan, I., 2019. Nanoparticles: properties, applications and toxicities. *Arabian Journal of Chemistry*, 12(7), pp. 908–931.

Kim, E.J., Lee, C.S., Chang, Y.Y. and Chang, Y.S., 2013. Hierarchically structured manganese oxide-coated magnetic nanocomposites for the efficient removal of heavy metal ions from aqueous systems. *ACS Applied Materials & Interfaces*, 5(19), pp. 9628–9634.

Klug, K.L. and Dravid, V.P., 2002. Observation of two-and three-dimensional magnesium oxide nanostructures formed by thermal treatment of magnesium diboride powder. *Applied Physics Letters*, 81(9), pp. 1687–1689.

Kotsyuda, S.S., Tomina, V.V., Zub, Y.L., Furtat, I.M., Lebed, A.P., Vaclavikova, M. and Melnyk, I.V., 2017. Bifunctional silica nanospheres with 3-aminopropyl and phenyl groups. Synthesis approach and prospects of their applications. *Applied Surface Science*, 420, pp. 782–791.

Kumar, R., Khan, M.A. and Haq, N., 2014. Application of carbon nanotubes in heavy metals remediation. *Critical Reviews in Environmental Science and Technology*, 44(9), pp. 1000–1035.

Kumar, P. and Kumar, P., 2019. Removal of cadmium (Cd-II) from aqueous solution using gas industry-based adsorbent. *SN Applied Sciences*, 1(4), pp. 1–8.

Kusior, A., Klich-Kafel, J., Trenczek-Zajac, A., Swierczek, K., Radecka, M. and Zakrzewska, K., 2013. TiO_2–SnO_2 nanomaterials for gas sensing and photocatalysis. *Journal of the European Ceramic Society*, 33(12), pp. 2285–2290.

Lasheen, M.R., El-Sherif, I.Y., Sabry, D.Y., El-Wakeel, S.T. and El-Shahat, M.F., 2015. Removal of heavy metals from aqueous solution by multiwalled carbon nanotubes: equilibrium, isotherms, and kinetics. *Desalination and Water Treatment*, 53(13), pp. 3521–3530.

Le, A.T., Pung, S.Y., Sreekantan, S. and Matsuda, A., 2019. Mechanisms of removal of heavy metal ions by ZnO particles. *Heliyon*, 5(4), p. e01440.

Lei, T., Li, S.J., Jiang, F., Ren, Z.X., Wang, L.L., Yang, X.J., Tang, L.H. and Wang, S.X., 2019. Adsorption of cadmium ions from an aqueous solution on a highly stable dopamine-modified magnetic nano-adsorbent. *Nanoscale Research Letters*, 14(1), pp. 1–17.

Lesmana, S.O., Febriana, N., Soetaredjo, F.E., Sunarso, J. and Ismadji, S., 2009. Studies on potential applications of biomass for the separation of heavy metals from water and wastewater. *Biochemical Engineering Journal*, 44(1), pp. 19–41.

Li, L., Duan, H., Wang, X. and Luo, C., 2014. Adsorption property of Cr(VI) on magnetic mesoporous titanium dioxide–graphene oxide core–shell microspheres. *New Journal of Chemistry*, 38(12), pp. 6008–6016.

Liang, X.J., Kumar, A., Shi, D. and Cui, D., 2012. Nanostructures for medicine and pharmaceuticals. *Journal of Nanomaterials*, 2012.

Lin, Y.F. and Chen, J.L., 2014. Magnetic mesoporous Fe/carbon aerogel structures with enhanced arsenic removal efficiency. *Journal of Colloid and Interface Science*, 420, pp. 74–79.

Lisha, K.P. and Pradeep, T., 2009. Towards a practical solution for removing inorganic mercury from drinking water using gold nanoparticles. *Gold Bulletin*, 42(2), pp. 144–152.

Liu, F., Yang, J., Zuo, J., Ma, D., Gan, L., Xie, B., Wang, P. and Yang, B., 2014. Graphene-supported nanoscale zero-valent iron: removal of phosphorus from aqueous solution and mechanistic study. *Journal of Environmental Sciences*, 26(8), pp. 1751–1762.

Liu, T., Wang, Z.L. and Sun, Y., 2015. Manipulating the morphology of nanoscale zero-valent iron on pumice for removal of heavy metals from wastewater. *Chemical Engineering Journal*, 263, pp. 55–61.

Liu, X., Zhou, K., Wang, L., Wang, B. and Li, Y., 2009. Oxygen vacancy clusters promoting reducibility and activity of ceria nanorods. *Journal of the American Chemical Society*, 131(9), pp. 3140–3141.

Lu, F. and Astruc, D., 2018. Nanomaterials for removal of toxic elements from water. *Coordination Chemistry Reviews*, 356, pp. 147–164.

Lu, H., Wang, J., Stoller, M., Wang, T., Bao, Y. and Hao, H., 2016. An overview of nanomaterials for water and wastewater treatment. *Advances in Materials Science and Engineering*, 2016.

Lü, T., Qi, D., Zhang, D., Lü, Y. and Zhao, H., 2018. A facile method for emulsified oil-water separation by using polyethylenimine-coated magnetic nanoparticles. *Journal of Nanoparticle Research*, 20(4), pp. 1–9.

Lucena, R., Simonet, B.M., Cárdenas, S. and Valcárcel, M., 2011. Potential of nanoparticles in sample preparation. *Journal of Chromatography A*, 1218(4), pp. 620–637.

Luo, C., Tian, Z., Yang, B., Zhang, L. and Yan, S., 2013. Manganese dioxide/iron oxide/acid oxidized multi-walled carbon nanotube magnetic nanocomposite for enhanced hexavalent chromium removal. *Chemical Engineering Journal*, 234, pp. 256–265.

Mahmoud, M.E., Fekry, N.A. and El-Latif, M.M., 2016. Nanocomposites of nanosilica-immobilized-nanopolyaniline and crosslinked nanopolyaniline for removal of heavy metals. *Chemical Engineering Journal*, 304, pp. 679–691.

Mahmoudi, E., Ng, L.Y., Ang, W.L., Chung, Y.T., Rohani, R. and Mohammad, A.W., 2019. Enhancing morphology and separation performance of polyamide 6,6 membranes by minimal incorporation of silver decorated graphene oxide nanoparticles. *Scientific Reports*, 9(1), pp. 1–16.

Marsh, H. and Rodríguez, F.R., 2006. Activation processes (chemical). *Activated Carbon*, 6, 322–365.

Martel, R., Schmidt, T., Shea, H.R., Hertel, T. and Avouris, P., 1998. Single-and multi-wall carbon nanotube field-effect transistors. *Applied Physics Letters*, 73(17), pp. 2447–2449.

Mirrezaie, N., Nikazar, M. and Hasan Zadeh, M., 2014. Synthesis of magnetic nanocomposite Fe_3O_4 coated polypyrrole (PPy) for chromium(VI) removal. In *Advanced Materials Research* (Vol. 829, pp. 649–653). Trans Tech Publications Ltd.

Mo, M., Yu, J.C., Zhang, L. and Li, S.K., 2005. Self-assembly of ZnO nanorods and nanosheets into hollow microhemispheres and microspheres. *Advanced Materials*, 17(6), pp. 756–760.

Morris, T., Copeland, H., McLinden, E., Wilson, S. and Szulczewski, G., 2002. The effects of mercury adsorption on the optical response of size-selected gold and silver nanoparticles. *Langmuir*, 18(20), pp. 7261–7264.

Mubarak, N.M., Alicia, R.F., Abdullah, E.C., Sahu, J.N., Haslija, A.A. and Tan, J., 2013. Statistical optimization and kinetic studies on removal of Zn^{2+} using functionalized carbon nanotubes and magnetic biochar. *Journal of Environmental Chemical Engineering*, 1(3), pp. 486–495.

Najafi, M., Yousefi, Y. and Rafati, A.A., 2012. Synthesis, characterization and adsorption studies of several heavy metal ions on amino-functionalized silica nano hollow sphere and silica gel. *Separation and Purification Technology*, 85, pp. 193–205.

Nasir, A.M., Goh, P.S., Abdullah, M.S., Ng, B.C. and Ismail, A.F., 2019. Adsorptive nanocomposite membranes for heavy metal remediation: recent progresses and challenges. *Chemosphere*, 232, pp. 96–112.

Nizamuddin, S., Siddiqui, M.T.H., Mubarak, N.M., Baloch, H.A., Abdullah, E.C., Mazari, S.A., Griffin, G.J., Srinivasan, M.P. and Tanksale, A., 2019. Iron oxide nanomaterials for the removal of heavy metals and dyes from wastewater. *Nanoscale Materials in Water Purification*, (Vol. 1, pp. 447–472).

Ntim, S.A. and Mitra, S., 2012. Adsorption of arsenic on multiwall carbon nanotube–zirconia nanohybrid for potential drinking water purification. *Journal of Colloid and Interface Science*, 375(1), pp. 154–159.

O'Carroll, D., Sleep, B., Krol, M., Boparai, H. and Kocur, C., 2013. Nanoscale zero valent iron and bimetallic particles for contaminated site remediation. *Advances in Water Resources*, *51*, pp. 104–122.

Ojea-Jiménez, I., López, X., Arbiol, J. and Puntes, V., 2012. Citrate-coated gold nanoparticles as smart scavengers for mercury(II) removal from polluted waters. *ACS Nano*, *6*(3), pp. 2253–2260.

Ouni, L., Ramazani, A. and Taghavi Fardood, S., 2019. An overview of carbon nanotubes role in heavy metals removal from wastewater. *Frontiers of Chemical Science and Engineering*, *13*(2), pp. 274–295.

Pakulski, D., Gorczyński, A., Marcinkowski, D., Czepa, W., Chudziak, T., Witomska, S., Nishina, Y., Patroniak, V., Ciesielski, A. and Samorì, P., 2021. High-sorption terpyridine–graphene oxide hybrid for the efficient removal of heavy metal ions from wastewater. *Nanoscale*, *13*(23), pp. 10490–10499.

Parmon, V., 2008. Nanomaterials in catalysis. *Materials Research Innovations*, *12*(2), pp. 60–61.

Parvin, F., Rikta, S.Y. and Tareq, S.M., 2019. Application of nanomaterials for the removal of heavy metal from wastewater. In *Nanotechnology in Water and Wastewater Treatment* (pp. 137–157). Elsevier.

Pogorilyi, R.P., Melnyk, I.V., Zub, Y.L., Carlson, S., Daniel, G., Svedlindh, P., Seisenbaeva, G.A. and Kessler, V.G., 2014. New product from old reaction: uniform magnetite nanoparticles from iron-mediated synthesis of alkali iodides and their protection from leaching in acidic media. *Rsc Advances*, *4*(43), pp. 22606–22612.

Popov, V.N., 2004. Carbon nanotubes: properties and application. *Materials Science and Engineering: R: Reports*, *43*(3), pp. 61–102.

Pradhan, N., Pal, A. and Pal, T., 2002. Silver nanoparticle catalyzed reduction of aromatic nitro compounds. *Colloids and Surfaces A: Physicochemical and Engineering Aspects*, *196*(2–3), pp. 247–257.

Pu, Y., Yang, X., Zheng, H., Wang, D., Su, Y. and He, J., 2013. Adsorption and desorption of thallium(I) on multiwalled carbon nanotubes. *Chemical Engineering Journal*, *219*, pp. 403–410.

Pyrzyńska, K. and Bystrzejewski, M., 2010. Comparative study of heavy metal ions sorption onto activated carbon, carbon nanotubes, and carbon-encapsulated magnetic nanoparticles. *Colloids and Surfaces A: Physicochemical and Engineering Aspects*, *362*(1–3), pp. 102–109.

Rajakumar, K., Kirupha, S.D., Sivanesan, S. and Sai, R.L., 2014. Effective removal of heavy metal ions using Mn_2O_3 doped polyaniline nanocomposite. *Journal of Nanoscience and Nanotechnology*, *14*(4), pp. 2937–2946.

Razzaz, A., Ghorban, S., Hosayni, L., Irani, M. and Aliabadi, M., 2016. Chitosan nanofibers functionalized by TiO_2 nanoparticles for the removal of heavy metal ions. *Journal of the Taiwan Institute of Chemical Engineers*, *58*, pp. 333–343.

Reglero, M.M., Taggart, M.A., Monsalve-Gonzalez, L. and Mateo, R., 2009. Heavy metal exposure in large game from a lead mining area: effects on oxidative stress and fatty acid composition in liver. *Environmental Pollution*, *157*(4), pp. 1388–1395.

Robati, D., 2013. Pseudo-second-order kinetic equations for modeling adsorption systems for removal of lead ions using multi-walled carbon nanotube. *Journal of Nanostructure in Chemistry*, *3*(1), pp. 1–6.

Saad, A.H.A., Azzam, A.M., El-Wakeel, S.T., Mostafa, B.B. and Abd El-latif, M.B., 2018. Removal of toxic metal ions from wastewater using ZnO@chitosan core-shell nanocomposite. *Environmental Nanotechnology, Monitoring & Management*, *9*, pp. 67–75.

Saleem, J., Shahid, U.B., Hijab, M., Mackey, H. and McKay, G., 2019. Production and applications of activated carbons as adsorbents from olive stones. *Biomass Conversion and Biorefinery*, *9*(4), pp. 775–802.

Sarma, G.K., Sen Gupta, S. and Bhattacharyya, K.G., 2019. Nanomaterials as versatile adsorbents for heavy metal ions in water: a review. *Environmental Science and Pollution Research*, *26*(7), pp. 6245–6278.

Seyedi, S.M., Rabiee, H., Shahabadi, S.M.S. and Borghei, S.M., 2017. Synthesis of zero-valent iron nanoparticles via electrical wire explosion for efficient removal of heavy metals. *CLEAN–Soil, Air, Water*, *45*(3), p. 1600139.

Sharma, R.K. and Agrawal, M., 2005. Biological effects of heavy metals: an overview. *Journal of Environmental Biology*, *26*(2), pp. 301–313.

Si, R., Zhang, Y.W., You, L.P. and Yan, C.H., 2005. Rare-earth oxide nanopolyhedra, nanoplates, and nanodisks. *Angewandte Chemie International Edition*, *44*(21), pp. 3256–3260.

Smith, S.C. and Rodrigues, D.F., 2015. Carbon-based nanomaterials for removal of chemical and biological contaminants from water: a review of mechanisms and applications. *Carbon*, *91*, pp. 122–143.

Smith, A.T., LaChance, A.M., Zeng, S., Liu, B. and Sun, L., 2019. Synthesis, properties, and applications of graphene oxide/reduced graphene oxide and their nanocomposites. *Nano Materials Science*, *1*(1), pp. 31–47.

Sneed, M.C., 1961. *Comprehensive Inorganic Chemistry: Brasted, RC Sulfur, Selenium, Tellurium, Polonium, and Oxygen* (Vol. 8). Van Nostrand.

Song, J., Kong, H. and Jang, J., 2011. Adsorption of heavy metal ions from aqueous solution by polyrhodanine-encapsulated magnetic nanoparticles. *Journal of Colloid and Interface Science*, *359*(2), pp. 505–511.

Srinivasan, N.R., Shankar, P.A. and Bandyopadhyaya, R., 2013. Plasma treated activated carbon impregnated with silver nanoparticles for improved antibacterial effect in water disinfection. *Carbon*, *57*, pp. 1–10.

Stankic, S., Müller, M., Diwald, O., Sterrer, M., Knözinger, E. and Bernardi, J., 2005. Size-dependent optical properties of MgO nanocubes. *Angewandte Chemie International Edition*, *44*(31), pp. 4917–4920.

Strandwitz, N.C. and Stucky, G.D., 2009. Hollow microporous cerium oxide spheres templated by colloidal silica. *Chemistry of Materials*, *21*(19), pp. 4577–4582.

Sumesh, E., Bootharaju, M.S. and Pradeep, T., 2011. A practical silver nanoparticle-based adsorbent for the removal of Hg^{2+} from water. *Journal of Hazardous Materials*, *189*(1–2), pp. 450–457.

Tabish, T.A., Memon, F.A., Gomez, D.E., Horsell, D.W. and Zhang, S., 2018. A facile synthesis of porous graphene for efficient water and wastewater treatment. *Scientific Reports*, *8*(1), pp. 1–14.

Tahoon, M.A., Siddeeg, S.M., Salem Alsaiari, N., Mnif, W. and Ben Rebah, F., 2020. Effective heavy metals removal from water using nanomaterials: a review. *Processes*, *8*(6), p. 645.

Taman, R., Ossman, M.E., Mansour, M.S. and Farag, H.A., 2015. Metal oxide nano-particles as an adsorbent for removal of heavy metals. *Journal of Advanced Chemical Engineering*, *5*(3), pp. 1–8.

Tang, C., Bando, Y. and Sato, T., 2002. Oxide-assisted catalytic growth of MgO nanowires with uniform diameter distribution. *The Journal of Physical Chemistry B*, *106*(30), pp. 7449–7452.

Tang, W.W., Zeng, G.M., Gong, J.L., Liu, Y., Wang, X.Y., Liu, Y.Y., Liu, Z.F., Chen, L., Zhang, X.R. and Tu, D.Z., 2012. Simultaneous adsorption of atrazine and Cu(II) from wastewater by magnetic multi-walled carbon nanotube. *Chemical Engineering Journal*, *211*, pp. 470–478.

Tchounwou, P.B., Yedjou, C.G., Patlolla, A.K. and Sutton, D.J., 2012. Heavy metal toxicity and the environment. *Molecular, Clinical and Environmental Toxicology*, pp. 133–164.

Theron, J., Walker, J.A. and Cloete, T.E., 2008. Nanotechnology and water treatment: applications and emerging opportunities. *Critical Reviews in Microbiology*, *34*(1), pp. 43–69.

Thines, R.K., Mubarak, N.M., Nizamuddin, S., Sahu, J.N., Abdullah, E.C. and Ganesan, P., 2017. Application potential of carbon nanomaterials in water and wastewater treatment: a review. *Journal of the Taiwan Institute of Chemical Engineers*, *72*, pp. 116–133.

Tratnyek, P.G., Sarathy, V., Nurmi, J., Baer, D.R., Amonette, J.E., Chun, C.L., Penn, R.L. and Reardon, E.J., 2009. *Aging of Iron Nanoparticles in Water: Effects on Structure and Reactivity* (No. PNNL-SA-61980). Pacific Northwest National Laboratory (PNNL), Environmental Molecular Sciences Laboratory (EMSL).

Tsunekawa, S., Ishikawa, K., Li, Z.Q., Kawazoe, Y. and Kasuya, A., 2000. Origin of anomalous lattice expansion in oxide nanoparticles. *Physical Review Letters*, *85*(16), p. 3440.

Tuzen, M. and Soylak, M., 2007. Multiwalled carbon nanotubes for speciation of chromium in environmental samples. *Journal of Hazardous materials*, *147*(1–2), pp. 219–225.

Vélez, E., Campillo, G.E., Morales, G., Hincapié, C., Osorio, J., Arnache, O., Uribe, J.I. and Jaramillo, F., 2016, February. Mercury removal in wastewater by iron oxide nanoparticles. *Journal of Physics: Conference Series*, *687*(1), p. 012050.

Wang, Z., Saxena, S.K., Pischedda, V., Liermann, H.P. and Zha, C.S., 2001. In situ X-ray diffraction study of the pressure-induced phase transformation in nanocrystalline CeO_2. *Physical Review B*, *64*(1), p. 012102.

Wang, L., Shi, C., Pan, L., Zhang, X. and Zou, J.J., 2020. Rational design, synthesis, adsorption principles and applications of metal oxide adsorbents: a review. *Nanoscale*, *12*(8), pp. 4790–4815.

Wang, L., Zhang, K., Song, Z. and Feng, S., 2007. Ceria concentration effect on chemical mechanical polishing of optical glass. *Applied Surface Science*, *253*(11), pp. 4951–4954.

Woo, Y.C., Kim, S.H., Shon, H.K. and Tijing, L.D., 2018. Introduction: membrane desalination today, past, and future.

Xu, L. and Wang, J., 2017. The application of graphene-based materials for the removal of heavy metals and radionuclides from water and wastewater. *Critical Reviews in Environmental Science and Technology*, *47*(12), pp. 1042–1105.

Yabe, S. and Sato, T., 2003. Cerium oxide for sunscreen cosmetics. *Journal of Solid State Chemistry*, *171*(1–2), pp. 7–11.

Yadav, D.K. and Srivastava, S., 2017. Carbon nanotubes as adsorbent to remove heavy metal ion (Mn^{7+}) in wastewater treatment. *Materials Today: Proceedings*, *4*(2), pp. 4089–4094.

Yan, J., Han, L., Gao, W., Xue, S. and Chen, M., 2015. Biochar supported nanoscale zerovalent iron composite used as persulfate activator for removing trichloroethylene. *Bioresource Technology*, *175*, pp. 269–274.

Yan, H., Li, H., Tao, X., Li, K., Yang, H., Li, A., Xiao, S. and Cheng, R., 2014. Rapid removal and separation of iron(II) and manganese(II) from micropolluted water using magnetic graphene oxide. *ACS Applied Materials & Interfaces*, *6*(12), pp. 9871–9880.

Yang, J., Hou, B., Wang, J., Tian, B., Bi, J., Wang, N., Li, X. and Huang, X., 2019. Nanomaterials for the removal of heavy metals from wastewater. *Nanomaterials*, 9(3), p. 424.

Yang, S., Li, J., Shao, D., Hu, J. and Wang, X., 2009. Adsorption of Ni(II) on oxidized multi-walled carbon nanotubes: effect of contact time, pH, foreign ions and PAA. *Journal of Hazardous Materials*, 166(1), pp. 109–116.

Yang, P., Zhao, D., Margolese, D.I., Chmelka, B.F. and Stucky, G.D., 1998. Generalized syntheses of large-pore mesoporous metal oxides with semicrystalline frameworks. *Nature*, 396(6707), pp. 152–155.

Yaqoob, A.A., Parveen, T., Umar, K. and Mohamad Ibrahim, M.N., 2020. Role of nanomaterials in the treatment of wastewater: a review. *Water*, 12(2), p. 495.

Yin, Y., Zhang, G. and Xia, Y., 2002. Synthesis and characterization of MgO nanowires through a vapor-phase precursor method. *Advanced Functional Materials*, 12(4), pp. 293–298.

Yu, T., Joo, J., Park, Y.I. and Hyeon, T., 2005. Large-scale nonhydrolytic sol–gel synthesis of uniform-sized ceria nanocrystals with spherical, wire, and tadpole shapes. *Angewandte Chemie International Edition*, 44(45), pp. 7411–7414.

Yu, G., Lu, Y., Guo, J., Patel, M., Bafana, A., Wang, X., Qiu, B., Jeffryes, C., Wei, S., Guo, Z. and Wujcik, E.K., 2018. Carbon nanotubes, graphene, and their derivatives for heavy metal removal. *Advanced Composites and Hybrid Materials*, 1(1), pp. 56–78.

Zarime, N.A., Yaacob, W.Z.W. and Jamil, H., 2018, April. Removal of heavy metals using bentonite supported nano-zero valent iron particles. *AIP Conference Proceedings, 1940*(1), p. 020029.

Zhang, W.X., 2003. Nanoscale iron particles for environmental remediation: an overview. *Journal of Nanoparticle Research*, 5(3), pp. 323–332.

Zhang, C.Z., Chen, B., Bai, Y. and Xie, J., 2018. A new functionalized reduced graphene oxide adsorbent for removing heavy metal ions in water via coordination and ion exchange. *Separation Science and Technology*, 53(18), pp. 2896–2905.

Zhang, M., Gao, B., Varnoosfaderani, S., Hebard, A., Yao, Y. and Inyang, M., 2013. Preparation and characterization of a novel magnetic biochar for arsenic removal. *Bioresource Technology*, 130, pp. 457–462.

Zhang, F., Jin, Q. and Chan, S.W., 2004. Ceria nanoparticles: size, size distribution, and shape. *Journal of Applied Physics*, 95(8), pp. 4319–4326.

Zhang, Y., Wu, B., Xu, H., Liu, H., Wang, M., He, Y. and Pan, B., 2016. Nanomaterials-enabled water and wastewater treatment. *NanoImpact*, 3, pp. 22–39.

Zhao, G., Huang, X., Tang, Z., Huang, Q., Niu, F. and Wang, X., 2018. Polymer-based nanocomposites for heavy metal ions removal from aqueous solution: a review. *Polymer Chemistry*, 9(26), pp. 3562–3582.

Zhao, G., Li, J., Ren, X., Chen, C. and Wang, X., 2011. Few-layered graphene oxide nanosheets as superior sorbents for heavy metal ion pollution management. *Environmental Science & Technology*, 45(24), pp. 10454–10462.

Zheng, S., Hao, L., Zhang, L., Wang, K., Zheng, W., Wang, X., Zhou, X., Li, W. and Zhang, L., 2018. Tea polyphenols functionalized and reduced graphene oxide-ZnO composites for selective Pb^{2+} removal and enhanced antibacterial activity. *Journal of Biomedical Nanotechnology*, 14(7), pp. 1263–1276.

Zhong, L.S., Hu, J.S., Cao, A.M., Liu, Q., Song, W.G. and Wan, L.J., 2007. 3D flowerlike ceria micro/nano-composite structure and its application for water treatment and CO removal. *Chemistry of Materials*, 19(7), pp. 1648–1655.

Zhu, Y.Q., Hsu, W.K., Zhou, W.Z., Terrones, M., Kroto, H.W. and Walton, D.R.M., 2001. Selective Co-catalysed growth of novel MgO fishbone fractal nanostructures. *Chemical Physics Letters*, 347(4–6), pp. 337–343.

16 Polymeric Membranes for Water and Wastewater Treatment

Maria Wasim, Aneela Sabir, Muhammad Shafiq and Rafi Ullah Khan

16.1 INTRODUCTION

For a prolonged period of time, various type of contaminants (comprising heavy metals) have been removed from wastewater by utilizing numerous techniques including adsorption [1], chemical precipitation [2], and microbial-based decomposition [3]. Almost all of the contaminants pose a serious threat to the ecosystem and to human beings. Regrettably, the above-mentioned techniques exhibit various demerits and limitations that include the procreation of a generous quantity of lethal sludge with liquefied surplus, excessive time consumption, inadequate efficacy, enormous usage of raw material, i.e. resins and solvent media.

In recent years, due to shortage of fresh water supply along with progressive enhancements with regard to pure water contamination, the treatment of wastewater with the help of membrane separation processes has gained worldwide interest from scientists. Membrane separation technologies using various kinds of membranes have shown great potential for the treatment and purification of organic contaminants [4], gas and solvents [5] and heavy metals [6] from wastewater. Owing to their high efficacy, ease of operation and structural flexibility they can be used for separation as well as adsorption purposes. Traditional pressure-driven membrane processes that include ultrafiltration (UF), microfiltration (MF), reverse osmosis (RO) and nanofiltration (NF) have been used for the removal of extensive amounts of organic macro-sized contaminants; on the contrary, NF and RO have shown significant efficacy in withdrawing micro-sized contaminants [7].

Conversely, a few membrane-based technologies like membrane distillation (MD) have shown substantial capability for the eradication of heavy metal ions that may include uranium fluoride and arsenic [8]. Arsenic is bland and unscented in nature and is considered a highly lethal metal for human beings that come in contact with it either from air or from the food supply. The existence of this compound in the crust of the earth is in the range of 2,000 and 5,000 µg/kg (ppm) [9].

Most importantly, membranes are perm-selective barriers that allow the eradication of lethal metal ions from wastewater streams. Up till now, various studies have been undertaken to develop membranes for this specific application with relevant features like perm selectivity and physio-chemical properties with regard to the selectivity of metal ions. In order to attain this target, numerous techniques have been employed like sintering, electrospinning, track etching, phase inversion, stretching and interfacial polymerization [10, 11]. Numerous inorganic and organic materials have been used for the formation of membranes. Till now, polymers have been extensively employed in organic-based materials for the synthesis of membranes, while inorganic-based membranes have been synthesized employing various kind of materials like glass, metals and ceramics [12].

Predominantly, ceramic-based membranes exhibits high mechanical, thermal and chemical stability in contrast to polymeric membranes. Moreover, surface charge and hydrophilicity of ceramic-based membranes are higher in comparison to polymeric membranes. Furthermore, ceramic-based

DOI: 10.1201/9781003326281-18

membranes can be utilized under severe conditions of oxidation, temperature or pH [13]. Though polymeric materials facilitate flexibility in design, the combination of materials offers multiple benefits and such materials have been widely employed for eradication of heavy metals.

16.2 POLYMERIC-BASED MEMBRANES FOR THE ERADICATION FOR HEAVY METALS

Presently, for the treatment of wastewater, polymeric-based membranes are extensively used Owing to the properties of pore formation technique like cost effectiveness and flexible nature [14] it has gained tremendous interest in membrane-based filtration technology like UF, NF, MF, RO and electrodialysis. Currently, most of the research has been conducted to investigate the use of polymeric membranes for water treatment applicability. These membranes are found to be the most capable material for the eradication of various types of contaminants that may include inorganic contaminants, suspended solids and organic compounds [15].

16.2.1 SYNTHETIC POLYMERS

Polymeric membranes are widely prepared either with natural or with synthetic-based polymers. They are employed to fabricate membrane (selective barrier) among the two neighboring phases that fine-tunes the transport phenomena. Overall, perm selectivity depends on to the characteristics (such as size shape and chemical nature) of the species to be transported. Including its physical and chemical properties mainly its porous behavior. In practical use, synthetic polymers satisfy the physiochemical properties for application in membrane filtration processes.

Odais and Moussa [16] determined the performance of RO and NF process for the eradication of cadmium and copper metals ions from industrial effluents employing polyamide spiral wound membranes. In the RO process, data indicate that selectivity of cadmium and copper can be up to 99% and 98%, respectively, while for the NF process, approximately 90% of copper metal ions were eradicated. Furthermore, data showed that such membranes are able to eradicate more than one type of metal ions. Such membranes can decrease the concentration of ion up till 3 ppm from 500 ppm, which shows a selectivity efficiency of 99.4%. One more widely used polymeric membrane material is polyethersulfone (PES), which has gained interest because of its inert chemical nature, high mechanical and thermal stability and wide-ranging pH tolerance. However, such polymers exhibit elevated biofouling phenomena when employed for aqueous-based filtration. For that purpose, numerous studies have been conducted to modify the properties of the surface of these hydrophobic membranes, thereby increasing their selectivity performance. The simplest method to alter the properties of the surface of a membrane is by immersing it into a polyelectrolyte-based solution. The adsorption of the polyelectrolyte solution onto the surface of membranes substantially increases the perm selectivity with regard to the removal of heavy metal ions owing to the presence of a cheating agent in their chemical structure. In this respect, Mokhtar and his coworkers [17] altered PES-based membranes made of polyelectrolyte multilayers composed of poly allylamine hydrochloride with poly styrene sulfonate. The synthesized membranes were further used to eradicate metals like zinc, nickel and copper from a widespread concentration of solution (aqueous) of metals (50–1,200 ppm). An analysis of membrane performance indicates that separation methodology was effective for a sole metal or for a mixture of metals, with a selectivity of 90%. Using the same methodology for alteration of membranes of poly acrylonitrile (PAN), for instance, Qin and his coworkers [18] fabricated (+) charged membranes via polyelectrolyte deposition. Scientists have employed poly acrylonitrile membranes functionalized with poly sodium 4-styrene sulfonate and poly ethyleneimine to form a layered assembly to efficaciously eradicate cadmium, nickel, copper and zinc ions. The data revealed that the selectivity of nickel and cadmium increases with the number of layers present but, on the other hand, flux decreases. Further use of polymeric layered structure can be for the fabrication of hollow fibers membranes. A significant advantage of a multi-layered structure is that a low cost material can be applied for the substrate. On the other hand, high-performance polymers can be employed as a selective layer. For instance, Zhu *et al.* [19] synthesized improved selective membranes having twin-layered NF membranes for the removal of heavy metal ions. Scientists have employed poly benzimidazole (PBI), which acts as a selective layer, while the support layer is a blend of poly vinylpyrrolidone (PVP) and PES employed as a substrate. Owing to exceptional charge properties and elevated chemical resistivity of PBI, the prepared twin-layered membrane displayed a remarkable salt selectivity that is because of the Donnan exclusion effect increment and low adsorption on to the surface of PBI of heavy metal ions.

16.2.2 NATURAL POLYMERS

Polymers obtained from biological sources have been widely employed for the eradication of heavy metal ions from the ecosystem. The most important polysaccharides employed for this application are cellulose, chitosan [20], carrageenan [21] and alginate [22].

16.2.2.1 Chitosan

It is acquired from chitin that has been widely employed for the fabrication of filtration membranes owing to its high rejection, stable nature and remarkable film forming capability. Moreover, chitosan is a biopolymer polysaccharide that has a high number of hydroxyl and amino groups. The presence of both these groups increases the functionality of amino acids and therefore they have attracted interest for water treatment. Chitosan is a partially deacetylated polymer, or it can be termed as a poly amino saccharide having a structure similar to glucose, connected with poly (1 → 4)-2-amino-2 deoxy-D-glucose units. The molecular structure of chitosan is depicted in Figure 16.1.

It has been developed to be employed in composite membranes. Various methodologies for various applications can be employed for their fabrication. According to Liu and his coworkers [23], the usage of nanostructured materials like carbon nanotubes in chitosan media may affect the performance property of nanocomposite chitosan-based membranes in the adsorption process. Salehi and his coworkers [24] have fabricated chitosan-based composites that can applied for the removal of copper ions from water. In this, multiwalled carbon nanotubes were functionalized to bear an amino group on it, which are incorporated in the polymeric media of chitosan/PVA. The presence of Multiwalled carbon nanotubes (MWCNTs) enhances the permeability owing to the presence of nanochannels. Moreover, it facilitates the adsorption of copper ions. A few other benefits are the reusability of these membranes by employing eluent-ethylenediaminetetraacetic acid (EDTA-Na) and the removal of copper from adsorptive membranes, which ultimately lead to promising thermodynamics: an endothermic, spontaneous and entropy-driven adsorption process.

Chitosan has been used as a coated material on the substrate in order to achieve the characteristics of both the active and the supporting layer [24]. Chitosan along with its derivatives are a favorable

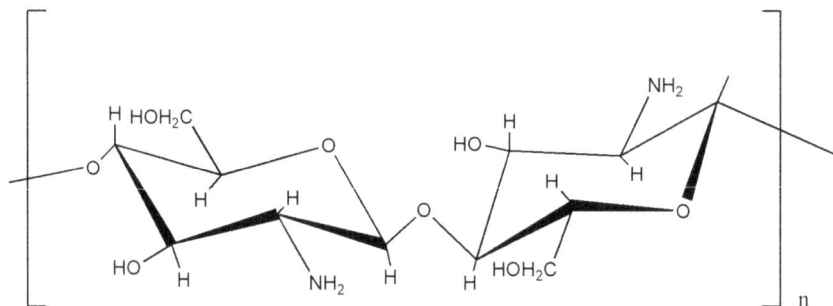

FIGURE 16.1 Molecular structure of chitosan.

substitute for the introduction of amine and hydroxyl groups to a variety of polymers used for the preparation of membranes. It may include poly sulfone, poly ethersulfone (PES) and polyvinylidene fluoride (PVDF). The charge present on the surface of hydrophobic polymers can be altered by the introduction of chitosan, a hydrophilic polymer. This approach suggests the possible alteration in the liability of these biopolymers in prompting the positive charge on polymeric membranes [25].

16.2.2.2 Cellulose

It is a linearly arranged homopolymer type of a β-D,1,4-glucose unit connected to the glycosidic bond (Figure 16.2). It is one of the most plentiful natural materials [26]. According to Roi and his coworkers [27], cellulose is produced by the photosynthesis process. Cellulose is composed of microfibers that are nanosized in diameter and owing to its distinctive feature, it is employed as a material for membrane fabrication. Four kinds of polymorphs of cellulose exist that are termed as cellulose I, II, III and IV. Out of these four, cellulose I exists naturally and is the most crystalline kind that is present in two variant forms. The main difference in these forms is the magnitude of hydrogen bonding present in the chains that lead to the alteration in the filling order of the lattice.

Ulewicz *et al.* [28] utilized a cellulose triacetate-based membrane modified with alkylimidazole mainly for the eradication of zinc and manganese ions from water. In this cellulose triacetate acts as a substrate while alkylimidazole and plasticizer act as ion carriers in the solvent. The results indicated that zinc was eradicated from the aqueous solution of zinc and manganese sulfates. An increase in the concentration of the carrier affects the perm selectivity. Selectivity up to 92.5% was recorded for zinc ions.

According to Ritchie and his coworkers [29], the adsorption capacity of membranes can be increased by adding nanoparticles on to the surface of a membrane. Based on this, Ibrahim and his coworkers [30] fabricated alpha zirconium phosphate nanoparticles on to the surface of cellulose membranes for eradication of heavy metals from polluted water.

16.2.2.3 Carrageenan

It is an anionic biopolymer mainly obtained from red seaweed; it has excellent solubility in water, and is a remarkable gelling agent employed particularly in the food industry [31]. Various studies showed that the carrageenan-based aerogel acts as an excellent adsorbent material for the removal of heavy metals from contaminated water [32]. The modification of carrageenan is to alter its physical and chemical properties [33]. The characteristics of gel like gelation time, transition temperature (from sol to gel), syneresis and hardness totally depend on the arrangement of the carrageenan structure, which is correlated to the molecular arrangement of the polysaccharide and its accumulation [34] (Figure 16.3).

Numerous technologies have been proposed for the fabrication of carrageenan-based membranes like polyelectrolyte-based [35] and hybrid membranes [36]. Conversely, only some of these have been utilized for the eradication of heavy metals from contaminated water. Prasannan and his coworkers [21] described a simplified strategy to synthesize super oleophobic membranes by

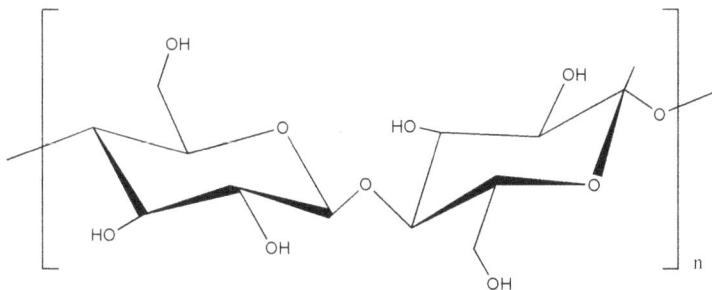

FIGURE 16.2 Molecular structure of cellulose.

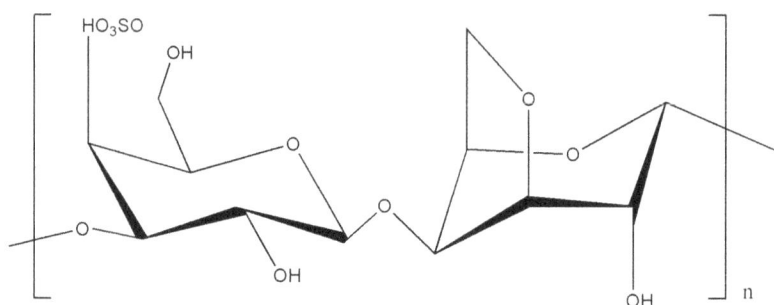

FIGURE 16.3　Molecular structure of κ-carrageenan.

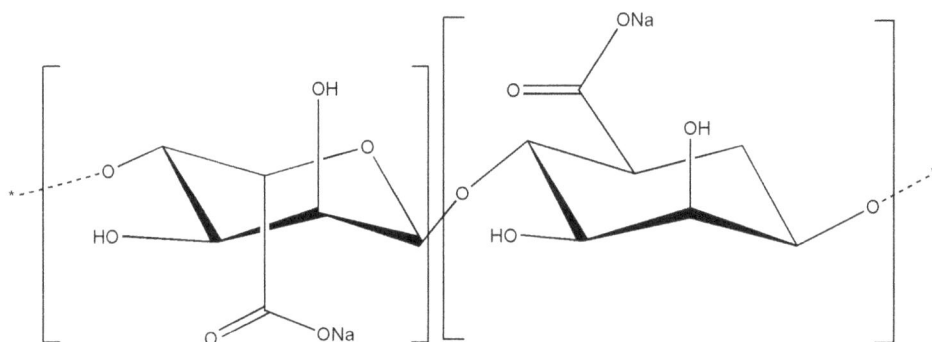

FIGURE 16.4　Molecular structure of sodium alginate.

hydrolyzation of poly acrylonitrile with carrageenan and nanoclay. The results indicate that carrageenan and clay-modified membranes show remarkable adsorption for heavy metal ions by a simple separation process.

16.2.2.4　Alginate

Alginate is a natural polymer derived from microorganisms and brown algae. It is mainly derived in the sodium alginate form, and consists of ample carboxyl and hydroxyl functionalities scattered throughout the polymeric main chain. When it is prepared as a solution in water it forms a high-viscosity gel that can precipitate out in accordance with the ionic force applied or based on the pH of the solution. They are extensively employed in pharmaceutical, medical and food-based industries and because of their properties related to stabilization, they act as thickeners, fixators, etc. [37] (Figure 16.4).

Sodium alginate consists of a huge number of functional groups that include carboxyl and hydroxyl functionalities distributed in the main chain, which can undergo functionalization via crosslinking, esterification and etherification [37, 38]. These functionalities can undergo a chemical reaction with heavy metal ions via a complex reaction or by ion exchange. Generally, sulfonic group functionalized sodium alginates have been developed as nanocomposites or hydrogels for their applicability in the treatment of contaminated water. Hence, it may be employed as a component for adsorption in order to eradicate heavy metals [39, 40].

Alginate membranes have filtration efficacy for organic contaminants as well as for heavy metal ions can be a suitable contender [41]. Alginate membranes for NF of organic solvents were developed by Aburabie and his coworkers [22] via the crosslinking of sodium alginate in a solution of calcium chloride. The study suggests that alginate-based membranes can be suitable for green NF of an organic solvent. Dudek *et al.* [42] developed a composite-based membrane of alginate that comprises an organic filler it was studied that the presence of inorganic and organic materials in membranes are promising candidate pervaporation dehydration of ethanol and water mixture having the increased separation factor. Chen and his coworkers [43] fabricated ionic membranes with alginate and with poly vinylalcohol acting as an adsorbent, which showed high adsorption capacity of chromium. Moreover, they also exhibited remarkable selectivity for chromium. Beni *et al.* [44] produced a membrane reactor made of a membrane of *Sargassum glaucescens* brown algae nanoparticles. Data revealed that the nanoparticles of alginate eradicate cobalt and nickel. Moreover, Googerdchian and his coworkers [45] synthesized nanoparticles of hydroxyapatite by a mechanical method and later mixed the particles with sodium alginate to fabricate granular particles. Hydroxyapatite/sodium alginate membranes used for the adsorption of lead showed high efficacy.

16.3   NANOMATERIALS USED FOR HEAVY METAL REMOVAL

Various kinds of nanomaterials are used to eradicate heavy metal ions from contaminated water. Graphene oxide (GO) [46, 47], Fe_3O_4 nanoparticles [48, 49], bentonite clay [50] and zeolite [51] are commonly employed. Addition of nanoparticles into the membranes can effectively increase their performance in terms of permeability and the selectivity of membranes. Novel types of nanomaterials termed as titanate nanotubes were employed for the eradication of arsenic metal ions from an aqueous solution. A nanomaterial composed of polyethersulfone and titanate nanotubes was used to form a thin membrane. Owing to the lower value of point of zero change, an enormous quantity of hydroxyl functional groups present on the surface were attributed by the high surface area [52]. Mixed matrix membranes homogenized with nanomaterials having multiple functionalities can increase the permeability of water and have antifouling characteristics [53, 54]. The compatibility of nanomaterials and polymers can be an issue for the development of high-performance membranes [55]. Different types of fillers or nanoparticles including TiO_2, iron oxide, ZnO, SiO_2, carbon nanotubes and GO are added for the development of nanocomposite-based membranes [47]. The homogenously distributed nanoparticles in membrane media are reduced by poor dispersion capability and elevated property to aggregate that can influence the performance of membrane. Poor dispersion and agglomeration can cause the formation of microdefects that impact the properties of membranes [56].

GO has a high surface area and distinctive properties and has therefore gained the interest of researchers for use as a nanofiller. GO composed of hydroxyl, epoxy and carboxyl groups are the functionalized nanosheets of carbon, which can be employed as a modifier or as basic material for membrane formation [57]. Zeolite is a microporous crystalline inorganic material that has a well-ordered pore structure and distinctive ion exchange and sorption characteristics. Due to these they are extensively employed for water treatment processes that may include the removal of heavy metals and radioactive materials and for adsorption and water softening [58]. Due to its intrinsic adsorptive characteristics, chemical structure and versatile nature in the presence of oxygen-containing functionalities on the basal plane of GO, the selectivity of heavy metal ions is increased [59]. The modification of the surface of GO nanoparticles by polymerization for making a suitable monomer like polyaniline can result in better adsorption [47]. Bentonite clay offers superior benefits with regard to the isolation or the fractionation of heavy metal ions from contaminated water. The presence of a negative charge on clay facilitates the hydrophilic nature and ion exchange ability and hence the adsorption of heavy metal ions [50]. Ferric oxide offers such distinctive properties like biodegradation, biocompatibility, thermal and chemical stability and magnetic characteristics and is employed to eradicate selective metal ions from aqueous media [48].

16.4   CURRENT ADVANCES IN NANOCOMPOSITE MEMBRANES FOR HEAVY METAL ERADICATION

Nowadays, numerous materials have been suggested as a filling material for the fabrication of nanocomposite-based membranes. GO is considered as the most favorable for the fabrication of such membranes for the selectivity of toxic compounds and organic molecules from contaminated water

[48]. The presence of GO positively affects performance with regard to the selectivity of molecules and ions. The presence of an interconnected nanosized capillary network with connected interlayer spaces, in combination with the gaps or spaces in between the edges of non-interconnected adjacent GO nanosheets and the holes of GO nanosheets [60], helps in the formation of a channel to pass ions or molecules via GO-based membranes in aqueous media, thus providing high selectivity. Without any doubt, numerous parameters including ion charge, molecules size and their interaction (like cation-π interaction between GO sheets and ions, electrostatic and metal coordination) can affect the selectivity of GO. Undoubtedly, such parameters have made GO more valuable as a nanomaterial for the eradication of pharmaceutical waste from polluted water [61]. Moreover, the insertion of GO can bring about some changes in the characteristics of polymeric-based membranes such as a change from a hydrophobic to a hydrophilic nature, which is achieved by enhancing the rate of permeation [62]. Chang and his coworkers [63] have demonstrated the combined influence of both PVP and GO on the performance property of a PVDF ultrafiltration membrane. The data depicted that hydrophilicity and antifouling property were improved by the incorporation of PVP and GO. Researchers further reported that the upgradation might be due to the creation of hydrogen bonds between GO and PVP. Given the advantages GO offers for use in polymeric-based membranes, researchers are now keen to modify its structure to further improve its properties. On the basis of data findings, a modification or alteration in the positive charge is required for better removal of heavy metal ions [64]. According to this, Xu *et al.* [65] modified GO via covalent functionalization with 3-aminopropyltriethoxysilane (APTES). The organosilane-modified GO was then added to PVDF membranes; the resulting membranes illustrated remarkable high hydrophilic character, high water flux and rejection of protein in contrast to pure PVDF- and GO-incorporated PVDF membranes.

Lately, numerous studies [66–68] developed a crosslinking network of a composite GO along with isophorone diisocyanate, which was then coated on a PVDF membrane for microfiltration usage. Fundamentally, the crosslinking method helps to enhance eradication of dyes (greater than 96%) and heavy metal ions (lead, copper, cadmium and chromium between 40% and 70%) in contrast to a GO/PVDF membrane. It is noteworthy that composite-based membranes exhibit a high rate of permeability under low external pressure. The addition of a functionalized magnetic GO/metformin hybrid was also evaluated and it showed a substantial enhancement in water flux due to the change in roughness and hydrophilicity of the surface of PES NF membranes. Moreover, the selectivity of copper and dyes increased remarkably in the presence of hydrophilic functionalities on the surface of a hybrid membrane. The 0.5 wt% of hybrid in an NF membrane showed 92% copper ion rejection [69]. Generally, GO-incorporated NF membranes can be employed for various water treatment applications that may include eradication of natural organic matter, decolorization and water softening [70].

16.5　CONCLUSION AND FUTURE TRENDS

An overview of both synthetic and natural polymeric membranes for the separation or eradication of various kinds of toxic heavy metal ions (such as cadmium, copper, lead, nickel, manganese, chromium, etc.) for meeting the current demand for fresh water is provided in this chapter. Notably, numerous researches have provided convincing evidence that polymer or nanocomposite membranes are able to eradicate micropollutants. Nanocomposite-based membranes have demonstrated remarkable capability for the eradication of heavy metal ions, which was totally dependent on to the appropriate selection of inorganic filler. Therefore, the adsorption capacity of the filler and its sieving capacity should be taken into account for fabricating nanocomposite-based membranes for the eradication of heavy metals. Coming to the viability of the process, nanocomposite-based membranes also offer adequate features to conduct the separation process efficiently with a high rate of permeability, which is a vital feature in terms of productivity. Finally, it is expected that the research society will be on the lookout for novel inorganic materials that not only overshadow the few demerits like rate of retention and permeability of polymeric membranes but also provide physical and chemical characteristics that include mechanical, thermal and chemical stability.

REFERENCES

1. Yang, B.S.S.a.R.T., Ultrasound enhanced adsorption and desorption of phenol on activated carbon and polymeric resin. *Industrial and Enginnering Chemistry Research*, 2001. **40**(22): pp. 4912–4918.
2. Hu, X.-j., et al., Adsorption of chromium(VI) by ethylenediamine-modified cross-linked magnetic chitosan resin: Isotherms, kinetics and thermodynamics. *Journal of Hazardous Materials*, 2011. **185**(1): pp. 306–314.
3. Yang, K., et al., Microbial diversity in combined UAF–UBAF system with novel sludge and coal cinder ceramic fillers for tetracycline wastewater treatment. *Chemical Engineering Journal*, 2016. **285**: pp. 319–330.
4. Wasim, M., et al., Preparation and characterization of composite membrane via layer by layer assembly for desalination. *Applied Surface Science*, 2017. **396**: pp. 259–268.
5. Pendergast, M.M. and E.M.V. Hoek, A review of water treatment membrane nanotechnologies. *Energy & Environmental Science*, 2011. **4**(6): pp. 1946–1971.
6. Asim, S., et al., The effect of nanocrystalline cellulose/gum arabic conjugates in crosslinked membrane for antibacterial, chlorine resistance and boron removal performance. *Journal of Hazardous Materials*, 2018. **343**: pp. 68–77.
7. Rajesha, B.J., et al., Effective composite membranes of cellulose acetate for removal of benzophenone-3. *Journal of Water Process Engineering*, 2019. **30**: p. 100419.
8. Criscuoli, A. and A. Figoli, Pressure-driven and thermally-driven membrane operations for the treatment of arsenic-contaminated waters: A comparison. *Journal of Hazardous Materials*, 2018. **370**: pp. 147–155.
9. Figoli, A., A. Criscuoli, and J. Hoinkis, Review of membrane processes for arsenic removal from drinking water: Challenges for safe water production. In *The Global Arsenic Problem*, 2010. pp. 131–145.
10. Lalia, B.S., et al., A review on membrane fabrication: Structure, properties and performance relationship. *Desalination*, 2013. **326**: pp. 77–95.
11. Peydayesh, M., T. Mohammadi, and O. Bakhtiari, Effective treatment of dye wastewater via positively charged TETA-MWCNT/PES hybrid nanofiltration membranes. *Separation and Purification Technology*, 2018. **194**: pp. 488–502.
12. Ulbricht, M., Advanced functional polymer membranes. *Polymer*, 2006. **47**(7): pp. 2217–2262.
13. Ng, L.Y., et al., Polymeric membranes incorporated with metal/metal oxide nanoparticles: A comprehensive review. *Desalination*, 2013. **308**: pp. 15–33.
14. Wasim, M., A. Sabir, and R.U. Khan, Membranes with tunable graphene morphology prepared via Stöber method for high rejection of azo dyes. *Journal of Environmental Chemical Engineering*, 2021. **9**(5): p. 106069.
15. Grylewicz, A. and S. Mozia, Polymeric mixed-matrix membranes modified with halloysite nanotubes for water and wastewater treatment: A review. *Separation and Purification Technology*, 2021. **256**: p. 117827.
16. Qdais, H.A. and H. Moussa, Removal of heavy metals from wastewater by membrane processes: A comparative study. *Desalination*, 2004. **164**(2): pp. 105–110.
17. Mokhter, M.A., et al., Preparation of polyelectrolyte-modified membranes for heavy metal ions removal. *Environmental Technology*, 2017. **38**(19): pp. 2476–2485.
18. Qin, Z., et al., Synthesis of positively charged polyelectrolyte multilayer membranes for removal of divalent metal ions. *Journal of Materials Research*, 2013. **28**(11): pp. 1449–1457.
19. Zhu, W.-P., et al., Dual-layer polybenzimidazole/polyethersulfone (PBI/PES) nanofiltration (NF) hollow fiber membranes for heavy metals removal from wastewater. *Journal of Membrane Science*, 2014. **456**: pp. 117–127.
20. Khulbe, K.C. and T. Matsuura, Removal of heavy metals and pollutants by membrane adsorption techniques. *Applied Water Science*, 2018. **8**(1): p. 19.
21. Prasannan, A., et al., Robust underwater superoleophobic membranes with bio-inspired carrageenan/laponite multilayers for the effective removal of emulsions, metal ions, and organic dyes from wastewater. *Chemical Engineering Journal*, 2020. **391**: p. 123585.
22. Aburabie, J.H., T. Puspasari, and K.-V. Peinemann, Alginate-based membranes: Paving the way for green organic solvent nanofiltration. *Journal of Membrane Science*, 2020. **596**: p. 117615.
23. Liu, Y.-L., W.-H. Chen, and Y.-H. Chang, Preparation and properties of chitosan/carbon nanotube nanocomposites using poly(styrene sulfonic acid)-modified CNTs. *Carbohydrate Polymers*, 2009. **76**(2): pp. 232–238.
24. Salehi, E., P. Daraei, and A. Arabi Shamsabadi, A review on chitosan-based adsorptive membranes. *Carbohydrate Polymers*, 2016. **152**: pp. 419–432.

25. Ji, Y.-L., et al., Fabrication of chitosan/PDMCHEA blend positively charged membranes with improved mechanical properties and high nanofiltration performances. *Desalination*, 2014. **357**: pp. 8–15.

26. Choi, H.S., et al., Development of Co-hemin MOF/chitosan composite based biosensor for rapid detection of lactose. *Journal of the Taiwan Institute of Chemical Engineers*, 2020. **113**: pp. 1–7.

27. Rol, F., et al., Recent advances in surface-modified cellulose nanofibrils. *Progress in Polymer Science*, 2019. **88**: pp. 241–264.

28. Radzyminska-Lenarcik, E. and M. Ulewicz, The application of polymer inclusion membranes based on CTA with 1-alkylimidazole for the separation of zinc(II) and manganese(II) ions from aqueous solutions. *Polymers*, 2019. **11**(2): p. 242.

29. Ritchie, S., et al., Surface modification of silica- and cellulose-based microfiltration membranes with functional polyamino acids for heavy metal sorption. Langmuir, 1999. **15**: pp. 6346–6357.

30. Ibrahim, Y., et al., Synthesis of super hydrophilic cellulose-alpha zirconium phosphate ion exchange membrane via surface coating for the removal of heavy metals from wastewater. *Science of the Total Environment*, 2019. **690**: pp. 167–180.

31. Alnaief, M., R. Obaidat, and H. Mashaqbeh, Effect of processing parameters on preparation of carrageenan aerogel microparticles. *Carbohydrate Polymers*, 2018. **180**: pp. 264–275.

32. Mola Ali Abasiyan, S., F. Dashbolaghi, and G.R. Mahdavinia, Chitosan cross-linked with κ-carrageenan to remove cadmium from water and soil systems. *Environmental Science and Pollution Research*, 2019. **26**(25): pp. 26254–26264.

33. Campo, V.L., et al., Carrageenans: Biological properties, chemical modifications and structural analysis – A review. *Carbohydrate Polymers*, 2009. **77**(2): pp. 167–180.

34. Piculell, L., S. Nilsson, and P. Ström, On the specificity of the binding of cations to carrageenans: Counterion N.M.R. spectroscopy in mixed carrageenan systems. *Carbohydrate Research*, 1989. **188**: pp. 121–135.

35. Liew, J., et al., Synthesis and characterization of modified κ-carrageenan for enhanced proton conductivity as polymer electrolyte membrane. *PLOS ONE*, 2017. **12**: p. e0185313.

36. Nogueira, L.F.B., et al., Formation of carrageenan-CaCO$_3$ bioactive membranes. *Materials Science and Engineering: C*, 2016. **58**: pp. 1–6.

37. Wang, M., et al., Efficient removal of heavy metal ions in wastewater by using a novel alginate-EDTA hybrid aerogel. *Applied Sciences*, 2019. **9**(3): p. 547.

38. Wang, B., et al., Alginate-based composites for environmental applications: A critical review. *Critical Reviews in Environmental Science and Technology*, 2018. **49**: pp. 1–39.

39. Wang, Z., et al., Bioavailability of wilforlide A in mice and its concentration determination using an HPLC-APCI-MS/MS method. *Journal of Chromatography B*, 2018. **1090**: pp. 65–72.

40. Lee, K.Y. and D.J. Mooney, Alginate: Properties and biomedical applications. *Progress in Polymer Science*, 2012. **37**(1): pp. 106–126.

41. Fatin-Rouge, N., et al., Removal of some divalent cations from water by membrane-filtration assisted with alginate. *Water Research*, 2006. **40**(6): pp. 1303–1309.

42. Dudek, G. and R. Turczyn, New type of alginate/chitosan microparticle membranes for highly efficient pervaporative dehydration of ethanol. *RSC Advances*, 2018. **8**: pp. 39567–39578.

43. Chen, J.H., et al., Cr(III) ionic imprinted polyvinyl alcohol/sodium alginate (PVA/SA) porous composite membranes for selective adsorption of Cr(III) ions. *Chemical Engineering Journal*, 2010. **165**(2): pp. 465–473.

44. Esmaeili, A. and A. Aghababai Beni, Novel membrane reactor design for heavy-metal removal by alginate nanoparticles. *Journal of Industrial and Engineering Chemistry*, 2015. **26**: pp. 122–128.

45. Googerdchian, F., A. Moheb, and R. Emadi, Lead sorption properties of nanohydroxyapatite–alginate composite adsorbents. *Chemical Engineering Journal*, 2012. **200–202**: pp. 471–479.

46. Kochameshki, M.G., et al., Grafting of diallyldimethylammonium chloride on graphene oxide by RAFT polymerization for modification of nanocomposite polysulfone membranes using in water treatment. *Chemical Engineering Journal*, 2017. **309**: pp. 206–221.

47. Ghaemi, N., S. Zereshki, and S. Heidari, Removal of lead ions from water using PES-based nanocomposite membrane incorporated with polyaniline modified GO nanoparticles: Performance optimization by central composite design. *Process Safety and Environmental Protection*, 2017. **111**: pp. 475–490.

48. Gholami, A., et al., Preparation and characterization of polyvinyl chloride based nanocomposite nanofiltration-membrane modified by iron oxide nanoparticles for lead removal from water. *Journal of Industrial and Engineering Chemistry*, 2014. **20**(4): pp. 1517–1522.

49. Daraei, P., et al., Novel polyethersulfone nanocomposite membrane prepared by PANI/Fe$_3$O$_4$ nanoparticles with enhanced performance for Cu(II) removal from water. *Journal of Membrane Science*, 2012. **415–416**: pp. 250–259.

50. Hebbar, R.S., A.M. Isloor, and A.F. Ismail, Preparation and evaluation of heavy metal rejection properties of polyetherimide/porous activated bentonite clay nanocomposite membrane. *RSC Advances*, 2014. **4**(88): pp. 47240–47248.

51. Mukhopadhyay, M., et al., Removal of arsenic from aqueous media using zeolite/chitosan nanocomposite membrane. *Separation Science and Technology*, 2018. **54**: pp. 1–7.

52. Gohari, R.J., et al., Arsenate removal from contaminated water by a highly adsorptive nanocomposite ultrafiltration membrane. *New Journal of Chemistry*, 2015. **39**(11): pp. 8263–8272.

53. Zinadini, S., et al., Preparation of a novel antifouling mixed matrix PES membrane by embedding graphene oxide nanoplates. *Journal of Membrane Science*, 2014. **453**: pp. 292–301.

54. Zhu, J., et al., Fabrication of a novel "loose" nanofiltration membrane by facile blending with Chitosan–Montmorillonite nanosheets for dyes purification. *Chemical Engineering Journal*, 2015. **265**: pp. 184–193.

55. Wang, C., et al., Covalent organic framework modified polyamide nanofiltration membrane with enhanced performance for desalination. *Journal of Membrane Science*, 2017. **523**: pp. 273–281.

56. Liu, F., M.R.M. Abed, and K. Li, Preparation and characterization of poly(vinylidene fluoride) (PVDF) based ultrafiltration membranes using nano γ-Al_2O_3. *Journal of Membrane Science*, 2011. **366**(1): pp. 97–103.

57. Zhang, Z.-B., et al., Layer-by-layer assembly of graphene oxide on polypropylene macroporous membranes via click chemistry to improve antibacterial and antifouling performance. *Applied Surface Science*, 2015. **332**: pp. 300–307.

58. Li, Z., et al., Removal of arsenic from water using Fe-exchanged natural zeolite. *Journal of Hazardous Materials*, 2011. **187**(1): pp. 318–323.

59. Khan, A.A.P., et al., Conventional surfactant-doped poly (o-anisidine)/GO nanocomposites for benzaldehyde chemical sensor development. *Journal of Sol-Gel Science and Technology*, 2016. **77**(2): pp. 361–370.

60. He, L., et al., Promoted water transport across graphene oxide–poly(amide) thin film composite membranes and their antibacterial activity. *Desalination*, 2015. **365**: pp. 126–135.

61. Sophia, C., et al., Application of graphene based materials for adsorption of pharmaceutical traces from water and wastewater - A review. *Desalination and Water Treatment*, 2016. **57**: pp. 1–14.

62. Xia, S., et al., Preparation of graphene oxide modified polyamide thin film composite membranes with improved hydrophilicity for natural organic matter removal. *Chemical Engineering Journal*, 2015. **280**: pp. 720–727.

63. Chang, X., et al., Exploring the synergetic effects of graphene oxide (GO) and polyvinylpyrrodione (PVP) on poly(vinylylidenefluoride) (PVDF) ultrafiltration membrane performance. *Applied Surface Science*, 2014. **316**: pp. 537–548.

64. Yu Zhang, S.Z., and Tai-Shung Chung, Nanometric graphene oxide framework membranes with enhanced heavy metal removal via nanofiltration. *Environmental Science and Technology*, 2015. **49**(16): pp. 10235–10242.

65. Xu, Z., et al., Organosilane-functionalized graphene oxide for enhanced antifouling and mechanical properties of polyvinylidene fluoride ultrafiltration membranes. *Journal of Membrane Science*, 2014. **458**: pp. 1–13.

66. Zhang, C., et al., Graphene oxide quantum dots incorporated into a thin film nanocomposite membrane with high flux and antifouling properties for low-pressure nanofiltration. *ACS Applied Materials & Interfaces*, 2017. **9**(12): pp. 11082–11094.

67. Zhang, A., et al., In situ formation of copper nanoparticles in carboxylated chitosan layer: Preparation and characterization of surface modified TFC membrane with protein fouling resistance and long-lasting antibacterial properties. *Separation and Purification Technology*, 2017. **176**: pp. 164–172.

68. Zhang, P., et al., Cross-linking to prepare composite graphene oxide-framework membranes with high-flux for dyes and heavy metal ions removal. *Chemical Engineering Journal*, 2017. **322**: pp. 657–666.

69. Abdi, G., et al., Removal of dye and heavy metal ion using a novel synthetic polyethersulfone nanofiltration membrane modified by magnetic graphene oxide/metformin hybrid. *Journal of Membrane Science*, 2018. **552**: pp. 326–335.

70. Wei, Y., et al., Multilayered graphene oxide membranes for water treatment: A review. *Carbon*, 2018. **139**: pp. 964–981.

17 Role of Bacterial Cell Transmembranes as a Biopolymer to Remediate Mercury in Industrial Wastewater

Aatif Amin, Mohsin Gulzar Barq and Sunbul Rasheed

17.1 INTRODUCTION

Mercury pollution has become the foremost problem because of its pervasive use in industries and unprocessed disposal of wastes containing heavy metals (Budnik et al. 2019). Industrial effluents released from sewage sludge, oil spillages, animal wastes, chemical wastes, eroded sediments, littering, fertilizers, herbicides, pesticides, agrochemical activities and industries contain heavy metals and cause severe problems in the environment (Liu et al. 2021). Lethal heavy metals commonly used include cadmium (Cd), copper (Cu), arsenic (As), chromium (Cr), lead (Pb), mercury (Hg), and nickel (Ni) (Vareda et al. 2019). Mercury is the third most toxic and harmful global pollutant released in the environment through a multitude of natural sources such as volcanic eruptions, emissions from water bodies or soil and human activities such as combustion of fossil fuels, mining, coal burning, industrial processes and waste incinerators, etc. Mercury is redistributed to the environment through physical, chemical and biological processes, and, once released, in water, soil and the atmosphere. Such processes result in an imminent surge of mercury in both the terrestrial and aquatic environment (Frossard et al. 2018).

On top of that, increasing concentration of mercury in water, canals and rivers is due to the disposal of waste from different sectors such as industry, household, agriculture and municipalities, etc. (Akhtar et al. 2021). These sectors have seriously contributed toward a decline in both the quality and the quantity of fresh water because they consume about one-third of portable water and the contaminants released by them comprise different natural and chemical pollutants (Kirby et al. 2019).

Industrial waste water is one of the significant reasons for irreversible damage to the environment. Inappropriate treatment and direct discharge of these perilous effluents in the sewerage drains ultimately contaminates the groundwater as well as other significant water bodies, harming aquatic and terrestrial life (Sivaranjanee et al. 2021). Under-treated waste water can likewise cause other potential ecological contamination in air, land surface, soil, and so forth. Disposal of industrial waste water used in irrigating crops can cause severe damage to the quality of the crops produced and can also reach the food chain (Sandeep et al. 2019). The consumption of such Hg-contaminated plants as food by humans and animals poses serious ill effects on human health. So, it is important to remediate heavy metals from industrial wastewater prior to disseminating them into water bodies by developing novel approaches to bioremediate toxic forms of mercury (Hg^{2+}) in the agriculture system.

310

DOI: 10.1201/9781003326281-19

17.2   OXIDATION STATES OF MERCURY

Mercury exists in three oxidation forms such as metallic or elemental (Hg^0), mercuric ion (Hg^{2+}) and mercurous ion (Hg^{2+}) (Rani et al. 2021). The oxidation states of mercury decide the intensity of toxicity. Elemental mercury is extracted from an ore known as cinnabar. It has low solubility and is comparatively less toxic than the other oxidation forms (O'Connor et al. 2019). It can be broken down into droplets and it easily vaporizes at ambient conditions. It is a lethal neurotoxin and affects the central nervous system (CNS). In vitro elemental mercury can be transformed into a highly toxic state by the action of peroxidase and catalase enzyme. Mercuric ions are highly reactive in nature. Due to this property, they can bond strongly with the nitrogen atom of nuclear material of cell and cysteine residues of protein. Organomercurials possess a role in biomagnification and bioaccumulation of mercury in tissues of living organisms (Donadt et al. 2021). Inorganic mercury compounds are formed by the combination of mercury with other group of elements such as halogens, oxygen and sulfur (Amin et al. 2015). This state is less toxic and affects the gastrointestinal tract as well as the CNS. All these states of mercury are allocated different parts in human body as shown in Table 17.1.

17.3   GLOBAL STATUS OF MERCURY TOXICITY

Mercury is a toxic pollutant that has been added in the form of manures, fertilizers, lime and sludge to agricultural fields and it suppresses plant growth as a result of cellular organelle damage and the breakdown of membranes, which become genotoxic, altering the process of photosynthesis, respiration, protein synthesis and carbohydrate metabolism (Eagles-Smith et al. 2018). Living organisms require heavy metals in trace amounts but the consumption of such Hg-contaminated plants as food poses serious ill effects on human health by bioaccumulation (Raj et al. 2019).

Detoxification of mercury from industrial wastewater is more challenging than that of organic pollutants, as they can be converted into water and carbon dioxide, but mercury cannot be mineralized and can only be converted to a less toxic form to reduce its bioavailability (Naguib et al. 2019). Besides, mercury obstructs the biodegradation of organic and inorganic pollutants making co-contaminated industrial water difficult to detoxify. Also, its toxicity to human beings, plants and animals is becoming a major medical/health concern (Sakamoto et al. 2018).

17.4   CHEMICAL AND PHYSICAL METHODS FOR REMEDIATION OF MERCURY

The real problem of this decade is contamination of the ecosystem because of migration of heavy metal contaminants into noncontaminated areas in the form of leachates through soil and sewage sludge. These harmful contaminants are removed from the environment by various chemical and physical methods (Teng et al. 2020).

TABLE 17.1

Oxidation states of mercury and their allocation in different parts of the human body

Oxidation states	Examples	Allocation in humans	References
Volatile mercury	Dimethyl mercury $(CH_3)_2Hg$ Elemental mercury (Hg^0)	Accumulation in gray matter of cerebral hemispheres	Navarro-Sempere et al. (2021)
Inorganic mercury	Mercuric ions (Hg^{2+}) Hg halogens ($HgBr_2$) Hg hydroxide (HgOH) Hg oxide (HgO)	Dissociates to bivalent mercury in liver. Bioaccumulation in liver and kidneys	Pamphlett et al. (2018)
Organic mercury	Methyl mercury (CH_3Hg^+) Methyl mercury chloride (CH_3HgCl)	Accumulation in kidneys, liver and affects the gastrointestinal tract	Svoboda et al. (2021)

Chemical methods for mercury remediation cause modification in the chemical properties of contaminants present in polluted areas to lessen their hazardous effect (Eckley et al. 2020). Chemical methods show effectiveness at a field scale but they form by-products which are harmful for living organisms. Chemical treatment techniques for contaminated soil and water include ion exchange, precipitation, nanoremediation, chemical leaching and extraction (Eckley et al. 2020).

Physical methods need a broad range of spectrum to detoxify mercury in densely polluted areas. This method is very expensive as compared to other techniques because it requires additional processing to remove all the contaminants (Teng et al. 2020). Removal of mercury by physical separation methods is based on particle size distribution of contaminants. Physical treatment of mercury includes activated carbon desorption, electroremediation, heat treatment, vitrification technology and soil replacement, etc. (Wang et al. 2019).

These physio-chemical technologies used for soil remediation reduce the land available for irrigation, as they eliminate all natural activities (Verma et al. 2019). Such conventional methods are expensive and far from optimum. For example, mercury-contaminated soil is remediated by onsite management or digging and subsequent disposal to a landfill site. This method of remediation shifts heavy metal contaminants elsewhere in an adjacent environment (Mansoor et al. 2021).

An alternative method to excavation and disposal to landfill is washing of contaminated soil. This method is very expensive and yields a residue rich in heavy metals, which need further treatment. Though physio-chemical technologies have various advantages, they possess certain disadvantages due to which they are still not applicable as enduring solutions for the removal of heavy metals from industrial wastes (Chen et al. 2018b).

17.5   BIOREMEDIATION OF HEAVY METALS

It is an urgent need to develop novel approaches to bioremediate toxic forms of mercury (Hg^{2+}) in the agriculture system as these lands cover large areas, which become unsuitable for sustainable agriculture (Sharma et al. 2021). Detoxification by microorganisms is economically achievable because it completely removes organic compounds such as petroleum hydrocarbons and heavy metal ions from polluted sites (Hou et al. 2020).

Toxic chemicals are biologically treated with controlled microbial processes under ambient conditions accompanied with effective management of indigenous microbial communities to confirm successful in vitro biodetoxification activity (Manzoor et al. 2020). Heavy metals are toxic for bacterial and fungal communities but some microorganisms are resistant to heavy metals because they have ability to remediate them from the ecosystem or transform them to a completely benign or less toxic state (Medfu Tarekegn et al. 2020). Microbial tolerance toward contaminants mainly mercury is completely significant for biodetoxification of polluted sites as the process needs microorganisms to come into contact with the contaminants to achieve bioremediation (Chen et al. 2018a).

Mercury detoxification by bacteria is highly dependent on the nature of the environment. But it is the most ecofriendly and cost-effective method to remediate highly toxic mercury to a less toxic state. Use of mercury-resistant bacteria is an attractive biodetoxification alternative with promising ability to treat industrial wastewater and polluted soil (Christakis et al. 2021). The biogeochemical cycle present in the soil ecosystem is directly or indirectly maintained by microorganisms because they recycle essential minerals and metallic ions such as phosphorous, nitrogen, sulfur, copper, aluminum and iron, which play a role in sustainability of life (Kappler et al. 2021).

Mercury-resistant bacteria are involved in biodetoxification by adsorption, precipitation, redox reaction, complex formation and methylation, as shown in Figure 17.1. Moreover, the chemical reactions that occur between bacteria and metal ions are widely classified into various processes such as metal-siderophore interaction, intracellular interaction, extracellular immobilization, cell wall-linked metals, volatilization and modification of metal ions (Zango Usman et al. 2020).

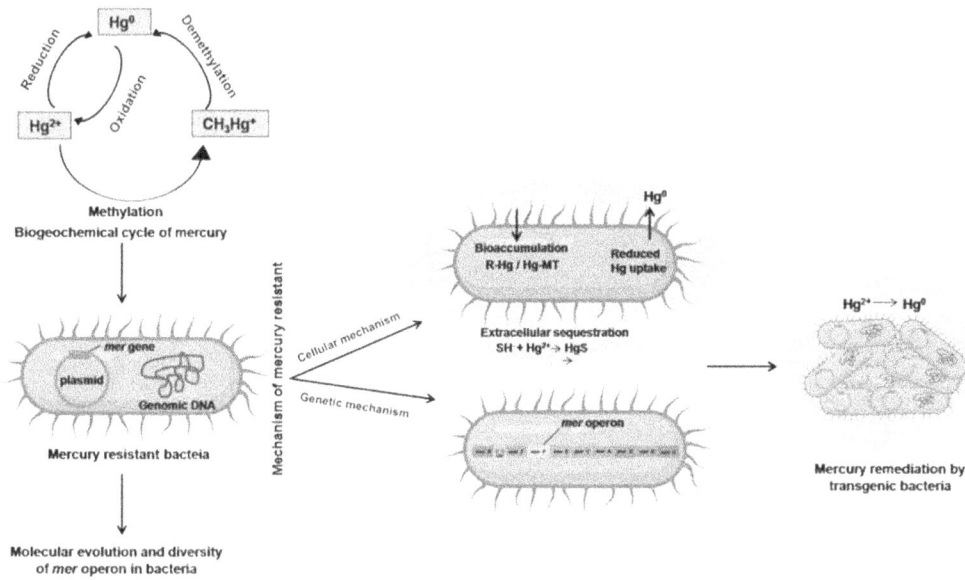

FIGURE 17.1 Mechanism of mercury resistance in bacteria and their role in biodetoxification of mercury (Priyadarshanee et al. 2022).

17.6 MECHANISM OF MERCURY BIODETOXIFICATION

Biodetoxification of mercury-contaminated soils, deposits and groundwater remains an uncertain challenge. Bacteria have developed several astounding arrays of resistance systems to overcome the environment contaminated by heavy metals (Al-Ansari 2022). For this purpose, structural surfaces of bacteria are very significant to study the interactions with heavy metal ions. Based on the composition of the cell wall, bacteria can be classified as Gram positive or Gram negative. Gram positive bacterial cell walls are composed of 90% peptidoglycan with a minute amount of teichoic acid whereas Gram negative cell walls possess an additional layer of lipopolysaccharides (LPS), phospholipids and a small layer of peptidoglycans. These structures attain negative charge, which helps them to interact with metal ions present in the environment (Lake et al. 2019).

Nature offers a rich toolbox for the transformation of mercury. Detoxification of highly toxic mercuric ion is based on a resistance system known as the *mer* operon (Norambuena et al. 2020). This system consists of a set of clustered genes that encodes essential proteins such as mercuric reductase MerA, which is a homodimeric flavin-dependent disulfide oxido-reductase enzyme, and it encodes for organomercury lyase (MerB), a periplasmic protein MerP and some transmembrane protein (MerT, MerE, MerC, MerG and MerF) that help in the transportation of Hg^{2+} to the cytoplasm where MerA and regulatory proteins (MerR, MerD) cause its reduction (Zheng et al. 2018). The MerR protein acts as a transcriptional activator or repressor in the presence and absence of mercury, respectively, thereby regulating the overall expression of *mer* operon, as shown in Figure 17.2. Moreover, there is an additional protein MerD that acts in downregulation operon expression and is encoded by some proteobacteria (Kumari et al. 2020).

The *mer* operon is a system for sustained biological biodetoxification of mercury and the management of mercury-polluted sites (Balan et al. 2018). Mercury-resistant bacteria are supplemented to increase the amount in mercury-contaminated sites where they actively convert MeHg, Hg^{2+} and Hg^0 (Zhao et al. 2020). Moreover, bacteria that use the *mer* operon system to resist mercury have been successfully operated on a technical scale to efficiently remediate mercury from chlor-alkali bioreactor wastewater, producing water fit for disposal to industrial waste streams (Manirethan et al. 2020). The wide understanding of functions of the *mer* operon has enabled the bioengineering

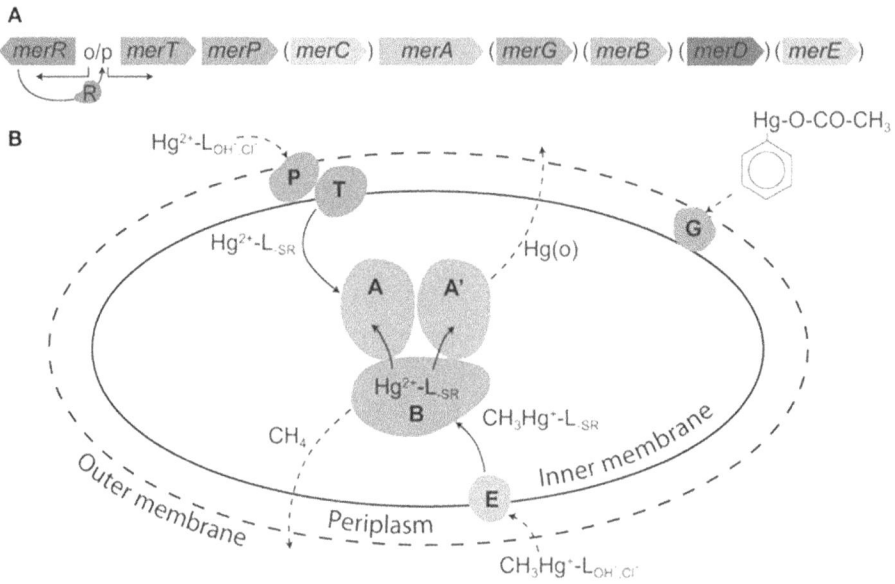

FIGURE 17.2   The *mer* system. (A) A generic bacterial Hg-detoxification system containing genes in parentheses showing those that are present in specific but not the majority of operons. (B) The cellular *mer*-encoded mercury remediation mechanisms (Boyd et al. 2012).

of cells, molecules and bacteria effective in transforming highly toxic mercury to a less toxic state (Saravanan et al. 2022). These tools have been engineered in different arrangements into bacteria in nature through horizontal gene transfer.

17.7   TRANSMEMBRANES AS BIOPOLYMERS FOR MERCURY BIOREMEDIATION

Membrane proteins are categorized as proteins that can either permanently anchor to be part of the membrane of cells or organelles (integral membrane proteins) or only temporarily bind to the lipid bilayer or to other integral proteins (peripheral member proteins). For integral membrane proteins, two categories of protein were classified (Wu 2019). One is transmembrane proteins, which span across the membrane, and the other one is integral monotonic proteins which merely associate with one side of the biological membrane. The *mer* operon consists of several transmembrane genes such as *mer*T, *mer*E, *mer*P, *mer*E, *mer*G, *mer*F, and a novel *mer*H that act as polymers and encode for different proteins that play a role in the detection, transportation and reduction of toxic mercury ions (Ghimire et al. 2019). MerT is the most significant transmembrane protein that aids in the uptake of alkylmercury and inorganic mercury and transports it to the cytoplasm of the cell. MerT has three transmembrane sites with cysteine residues positioned at the first transmembrane site Cys-Cys (Shettihalli et al. 2021).

The *mer*P gene encodes for a periplasmic protein MerP that contains two cysteine residues that replace nucleophiles (Cl^-) bound to Hg^{2+}; in this way, Hg^{2+} might be linked to the cysteine residues of MerP and transported to cysteine residues on the merT protein found in the cytoplasmic membrane and then transferred to the other pair of the merT's cytosolic cysteine residues (Hwang et al. 2020). Mercury enters the cytoplasmic membrane where cytosolic thiols (cysteines or glutathione) compete with cytosolic cysteine of *mer*T to bind with Hg^{2+} and prepare for *mer*A activity (Thomas et al. 2020).

Mercury sensitivity takes place by the deletion of *mer*T and *mer*P genes from Tn501 transposons whereas supersensitivity of mercury occurs by the expression of Tn501 *mer*T and *mer*P in the absence of mercuric reductase (MerA) (Naguib et al. 2018). Furthermore, concentration of mercury necessary for induction of *mer*A-lacZ transcriptional fusions increases by the mutation in *mer*P and *mer*T genes. The *mer*P gene plays a significant role in mercury resistance because when it is absent, transportation of organic and inorganic mercury increases in the presence of *mer*C, *mer*E, *mer*F and *mer*T transporters. Broad-spectrum mercury transporter proteins such as biopolymers like MerE, MerT, MerC and MerF transport inorganic and alkylmercury into the cells (Priyadarshanee et al. 2022).

An inner membrane cytoplasmic protein is encoded by *mer*C that uptakes organic and inorganic alkylmercury across the cytoplasmic membrane until it reaches the active site of mercuric reductase. In some bacterial strains, *mer*C can acts alone as the main transporter of mercury when *mer*T and *mer*P are absent whereas in the presence of these genes there is no effect of *mer*C on the resistance and transportation of mercury (Rani et al. 2021). In some studies, it was stated that *mer*C is more effective in establishing a successful mercurial biodetoxification system because it is considered more efficient in the transportation of mercury.

The *mer*F gene is involved in the transportation of both inorganic and organic mercury across the cytoplasmic membrane (Amin 2018). A study conducted by Schué revealed that *mer*H encodes a membrane protein that uses two cysteine residues in the transportation of mercury across the inner membrane. A metallo-chaperone, homolog to the C-terminal domain, known as TRASH, plays a role in sensing and trafficking heavy metals, which explains the role of *mer*H in trafficking of mercury to the MerR transcription factor (Naguib et al. 2018). Increased intracellular retention of mercury clarifies the low rate of *mer* operon induction and the need for *mer*H in metal trafficking. Transportation of organic and inorganic mercury compounds across the membrane is carried out by *mer*E and *mer*G. Organomercurials are lipid-soluble compounds and have the ability to cross the plasma membrane by simple diffusion and be transported inside by *mer*G or *mer*E genes (Barman et al. 2021).

17.8 STRUCTURAL STUDIES OF TRANSMEMBRANES BY NMR SPECTROSCOPY

Nuclear magnetic resonance (NMR) is a phenomenon which determines the biochemical processes such as molecular structure, affinity, transportation, reactivity and 3D structures of proteins (Herrling et al. 2019). It is possible to find an individual signal for every atom of a protein. These signals are not only allocated to a particular site, but also characterized by frequencies that offer distance and orientation constrictions as input for structural prediction. NMR spectroscopy is a well-established and broadly employed technique that has been considered as a potential tool for the determination of globular protein structures that reorient in the solution. In the mercury detoxification system, the toxic atoms are carried to MerT from MerP, and mercury is transported to mercuric reductase (MerA) (Amin et al. 2021).

Besides this transport mechanism, there are many pairs of cysteine residues present that bind to Hg^{2+} in a linear bi-coordinate pattern. The motifs, particularly CC-, -CXC-, -CXXC-, and -CXXXXXC-, are present in MerT/MerF, MerE, MerP and MerC, respectively (Priyadarshanee et al. 2022). Mutagenesis of the cysteine residues shows that the first -CC- pair in MerT and MerF and a single cysteine residue in MerP, Cl7, are significant.

NMR spectroscopy shows that MerF has two membrane-spanning segments. It performs a function in the transportation of toxic mercury ions in the cell by holding two adjacent pairs of cysteine residues that play an important role in transporting mercury across the cytoplasmic membrane. Independently, MerF is enough for transporting mercury across the cell membrane (Chill et al. 2019).

NMR studies of MerP present that both cysteine 14 and 17 of reduced MerP degrade and get exposed in the absence and presence of Hg^{2+}, respectively. The oxidized structure closely resembles the Hg^{2+} bound form and extensive structural modifications are observed as compared to the reduced form. Differential effects in the interaction of these cysteine residues with MerT is prominent because Cys17 is degraded in the reduced form of protein and its position is unclear in the Cyt14Ser mutation, which can halt the acceptance of Hg^{2+} by wild-type MerT (Wu 2019).

NMR spectroscopy demonstrates that MerP binds to toxic mercury in a linear and bi-coordinate manner. However, current NMR studies of mutant peptides established in the immediate neighborhood of these two cysteines indicated substantial flexibility in this region and the simultaneous acquisition of novel metal specificities. The wild-type peptide has ca. two- to four-fold higher affinity for mercury as compared to other thiophilic transition metals (Wang et al. 2021). It is assumed that MerP passes the bound mercuric ion to the cysteine pair in the first transmembrane helix of MerT (MerC or MerF) via a three-coordinate intermediate.

17.9   MERCURY-RESISTANT BACTERIA AS BIOFERTILIZERS

The use of microorganisms in the detoxification of mercury is an advanced technique for reducing mercury because it is efficient, ecofriendly and cost-effective. Remediation of mercury pollutants by microorganisms is broadly adopted. Microorganisms exhibit several characteristics such as the production of hydrogen sulfide (H_2S), which converts toxic mercuric ions Hg^{2+} to less toxic ions Hg^0 by precipitating Hg^{2+} into HgS and detoxifying the harmful effects of inorganic and organic compounds of mercury (Huang et al. 2019). Microorganisms resistant to heavy metals play a significant role in agronomical processes. Plant growth-promoting rhizobacteria not only prevent plants from the adverse effects of heavy metals but induce soil fertility and increase crop yield by the supplementation of vital nutrients and growth promoters (Amin 2018). These substantial characteristics possessed by microorganisms make them a preferable choice for use as biofertilizers. Mercury-resistant bacteria are used as biofertilizers, which are a greener substitute for chemical fertilizers and are used for the detoxification of heavy metals and greenhouse gases alleviating in the arable lands. Therefore, microorganisms are used as biofertilizers to lessen mercury pollution due to the dual benefit of plant growth promotion and mercury resistance by producing hydrogen sulfide (Guo et al. 2020).

Owing to its property to enhance the health of the plant by producing plant growth hormone, potassium and phosphate solubilization and nitrogen fixation and mercury resistance, *Azotobacter* can be the preferable option to be utilized as a biofertilizer because it is cost-effective and environmentally friendly and provides a sustainable crop yield (Guo et al. 2020; Soniari et al. 2019). Considering all the advantages of *Azotobacter,* it may prove to be a potential candidate for improving crop yield in future endeavors. Therefore, it is important to conduct more research to improve the isolation, screening and characterization techniques of heavy metal-resistant and plant growth-promoting bacterial species. Moreover, to provide all the benefits from a biofertilizer, it is necessary to find out compatible bacterial species that will make a valuable association with the specific plant genotype.

17.10   CONCLUSION

Membrane and transmembrane biopolymers such as MerT, MerF and MerE of *mer* operon harboring Hg-resistant bacteria play a key role in the biodetoxification of toxic forms of mercury. Such membranes transfer toxic Hg(II) from outside to the inside of the bacterial cell and then transfer it to the main enzymatic protein, mercuric reductase (MerA), which reduces Hg(II) to the elemental form of mercury Hg(0). Elemental mercury, Hg(0), is then finally moved out of the cell. Based on the fundamental role in the applications of such membranes, the use of Hg-resistant bacteria harboring the *mer* operon is an ecofriendly and cost-effective approach. Such bacteria make beneficial associations with plant roots and are used not only for biodetoxification of mercury to clean up the environment but also as a biofertilizer to replace chemical fertilizers.

REFERENCES

Akhtar, N., Syakir Ishak, M. I., Bhawani, S. A. & Umar, K. 2021. Various natural and anthropogenic factors responsible for water quality degradation: A review. *Water*, vol. 13, pp. 26–60.

Al-Ansari, M. M. 2022. Biodetoxification mercury by using a marine bacterium *Marinomonas* sp. RS3 and its *mer*A gene expression under mercury stress. *Environmental Research*, vol. 205, pp. 112–452.

Amin, A. 2018. *Molecular Characterization of Microorganisms to Detoxify Mercury Pollutants and their Role in Enhancing Plant Growth*. PhD thesis, University of the Punjab, Lahore. http://prr.hec.gov.pk/jspui/bitstream/123456789/10147/1/Aatif%20Amin_Microbio%20%26%20Molecular%20Genetics_2018_UoPunjab_PRR.pdf.

Amin, A. & Latif, Z. 2015. Phytotoxicity of Hg and its detoxification through microorganisms in soil. *Advancements in Life Sciences*, vol. 2, pp. 98–105.

Amin, A., Naveed, M., Munawar, U., Sarwar, A. & Latif, Z. 2021. Characterization of mercury-resistant rhizobacteria for plant growth promotion: An in vitro and in silico approach. *Current Microbiology*, vol. 78, pp. 3968–3979.

Balan, B. M., Shini, S., Krishnan, K. P. & Mohan, M. 2018. Mercury tolerance and biosorption in bacteria isolated from Ny-Ålesund, Svalbard, Arctic. *Journal of Basic Microbiology*, vol. 58, pp. 286–295.

Barman, D. & Jha, D. K. 2021. Metallotolerant microorganisms and microbe-assisted phytoremediation for a sustainable clean environment. *Microbes in Microbial Communities*, vol. 73, pp. 307–336.

Boyd, E. & Barkay, T. 2012. The mercury resistance operon: From an origin in a geothermal environment to an efficient detoxification machine. *Frontiers in Microbiology*, vol. 3, pp. 349.

Budnik, L. T. & Casteleyn, L. 2019. Mercury pollution in modern times and its socio-medical consequences. *Science of The Total Environment*, vol. 654, pp. 720–734.

Chen, J., Dong, J., Chang, J., Guo, T., Yang, Q., Jia, W. & Shen, S. 2018a. Characterization of an Hg(II)-volatilizing *Pseudomonas* sp. strain, DC-B1, and its potential for soil remediation when combined with biochar amendment. *Ecotoxicology and Environmental Safety*, vol. 163, pp. 172–179.

Chen, S., Lin, W., Chien, C., Tsang, D. C. & Kao, C. 2018b. Development of a two-stage biotransformation system for mercury-contaminated soil remediation. *Chemosphere*, vol. 200, pp. 266–273.

Chill, J. H., Qasim, A., Sher, I. & Gross, R. 2019. NMR perspectives of the KcsA potassium channel in the membrane environment. *Israel Journal of Chemistry*, vol. 59, pp. 1001–1013.

Christakis, C. A., Barkay, T. & Boyd, E. S. 2021. Expanded diversity and phylogeny of mer genes broadens mercury resistance paradigms and reveals an origin for MerA among thermophilic archaea. *Frontiers in Microbiology*, vol. 12, pp. 13–20.

Donadt, C., Cooke, C. A., Graydon, J. A. & Poesch, M. S. 2021. Mercury bioaccumulation in stream fish from an agriculturally-dominated watershed. *Chemosphere*, vol. 262, pp. 128–259.

Eagles-Smith, C. A., Silbergeld, E. K., Basu, N., Bustamante, P., Diaz-Barriga, F., Hopkins, W. A., Kidd, K. A. & Nyland, J. F. 2018. Modulators of mercury risk to wildlife and humans in the context of rapid global change. *Ambio*, vol. 47, pp. 170–197.

Eckley, C. S., Gilmour, C. C., Janssen, S., Luxton, T. P., Randall, P. M., Whalin, L. & Austin, C. 2020. The assessment and remediation of mercury contaminated sites: A review of current approaches. *Science of The Total Environment*, vol. 707, pp. 136031–136048.

Frossard, A., Donhauser, J., Mestrot, A., Gygax, S., Bååth, E. & Frey, B. 2018. Long-and short-term effects of mercury pollution on the soil microbiome. *Soil Biology and Biochemistry*, vol. 120, pp. 191–199.

Ghimire, P. S., Tripathee, L., Zhang, Q., Guo, J., Ram, K., Huang, J., Sharma, C. M. & Kang, S. 2019. Microbial mercury methylation in the cryosphere: Progress and prospects. *Science of The Total Environment*, vol. 697, pp. 134–150.

Guo, J., Muhammad, H., Lv, X., Wei, T., Ren, X., Jia, H., Atif, S. & Hua, L. 2020. Prospects and applications of plant growth promoting rhizobacteria to mitigate soil metal contamination: A review. *Chemosphere*, vol. 246, pp. 125–823.

Herrling, M. P., Lackner, S., Nirschl, H., Horn, H. & Guthausen, G. 2019. Recent NMR/MRI studies of biofilm structures and dynamics. *Annual Reports on NMR Spectroscopy*, vol. 97, pp. 163–213.

Hou, D., O'connor, D., Igalavithana, A. D., Alessi, D. S., Luo, J., Tsang, D. C., Sparks, D. L., Yamauchi, Y., Rinklebe, J. & Ok, Y. S. 2020. Metal contamination and bioremediation of agricultural soils for food safety and sustainability. *Nature Reviews Earth & Environment*, vol. 1, pp. 366–381.

Huang, Z., Wei, Z., Xiao, X., Tang, M., Li, B., Ming, S. & Cheng, X. 2019. Bio-oxidation of elemental mercury into mercury sulfide and humic acid-bound mercury by sulfate reduction for Hg^0 removal in flue gas. *Environmental Science & Technology*, vol. 53, pp. 12923–12934.

Hwang, H., Hazel, A., Lian, P., Smith, J. C., Gumbart, J. C. & Parks, J. M. 2020. A minimal membrane metal transport system: Dynamics and energetics of mer proteins. *Journal of Computational Chemistry*, vol. 41, pp. 528–537.

Kappler, A., Bryce, C., Mansor, M., Lueder, U., Byrne, J. M. & Swanner, E. D. 2021. An evolving view on biogeochemical cycling of iron. *Nature Reviews Microbiology*, vol. 19, pp. 360–374.

Kirby, M. A., Nagel, C. L., Rosa, G., Zambrano, L. D., Musafiri, S., Ngirabega, J. D. D., Thomas, E. A. & Clasen, T. 2019. Effects of a large-scale distribution of water filters and natural draft rocket-style cookstoves on diarrhea and acute respiratory infection: A cluster-randomized controlled trial in Western Province, Rwanda. *PLoS Medicine*, vol. 16, pp. 100–281.

Kumari, S., Jamwal, R., Mishra, N. & Singh, D. K. 2020. Recent developments in environmental mercury bioremediation and its toxicity: A review. *Environmental Nanotechnology, Monitoring & Management*, vol. 13, pp. 100–283.

Lake, R. J., Yang, Z., Zhang, J. & Lu, Y. 2019. DNAzymes as activity-based sensors for metal ions: recent applications, demonstrated advantages, current challenges, and future directions. *Accounts of Chemical Research*, vol. 52, pp. 3275–3286.

Liu, S., Wang, X., Guo, G. & Yan, Z. 2021. Status and environmental management of soil mercury pollution in China: A review. *Journal of Environmental Management*, vol. 277, pp. 111–442.

Manirethan, V., Gupta, N., Balakrishnan, R. M. & Raval, K. 2020. Batch and continuous studies on the removal of heavy metals from aqueous solution using biosynthesised melanin-coated PVDF membranes. *Environmental Science and Pollution Research*, vol. 27, pp. 24723–24737.

Mansoor, S., Kour, N., Manhas, S., Zahid, S., Wani, O. A., Sharma, V., Wijaya, L., Alyemeni, M. N., Alsahli, A. A. & El-Serehy, H. A. 2021. Biochar as a tool for effective management of drought and heavy metal toxicity. *Chemosphere*, vol. 271, pp. 129–458.

Manzoor, M. M., Goyal, P., Gupta, A. P. & Gupta, S. 2020. Heavy metal soil contamination and bioremediation. *Bioremediation and Biotechnology*, vol. 2, pp. 221.

Medfu Tarekegn, M., Zewdu Salilih, F. & Ishetu, A. I. 2020. Microbes used as a tool for bioremediation of heavy metal from the environment. *Cogent Food & Agriculture*, vol. 6, pp. 178–317.

Naguib, M. M., El-Gendy, A. O. & Khairalla, A. S. 2018. Microbial diversity of operon genes and their potential rules in mercury bioremediation and resistance. *The Open Biotechnology Journal*, vol. 12, pp. 1–5.

Naguib, M. M., Khairalla, A. S., El-Gendy, A. O. & Elkhatib, W. F. 2019. Isolation and characterization of mercury-resistant bacteria from wastewater sources in Egypt. *Canadian Journal of Microbiology*, vol. 65, pp. 308–321.

Navarro-Sempere, A., Segovia, Y., Rodrigues, A. S., Garcia, P. V., Camarinho, R. & García, M. 2021. First record on mercury accumulation in mice brain living in active volcanic environments: A cytochemical approach. *Environmental Geochemistry and Health*, vol. 43, pp. 171–183.

Norambuena, J., Miller, M., Boyd, J. M. & Barkay, T. 2020. Expression and regulation of the mer operon in *Thermus thermophilus*. *Environmental Microbiology*, vol. 22, pp. 1619–1634.

O'connor, D., Hou, D., Ok, Y. S., Mulder, J., Duan, L., Wu, Q., Wang, S., Tack, F. M. & Rinklebe, J. 2019. Mercury speciation, transformation, and transportation in soils, atmospheric flux, and implications for risk management: A critical review. *Environment International*, vol. 126, pp. 747–761.

Pamphlett, R. & Jew, S. K. 2018. Inorganic mercury in human astrocytes, oligodendrocytes, corticomotoneurons and the locus ceruleus: implications for multiple sclerosis, neurodegenerative disorders and gliomas. *Biometals*, vol. 31, pp. 807–819.

Priyadarshanee, M., Chatterjee, S., Rath, S., Dash, H. R. & Das, S. 2022. Cellular and genetic mechanism of bacterial mercury resistance and their role in biogeochemistry and bioremediation. *Journal of Hazardous Materials*, vol. 423, pp. 126–985.

Raj, D. & Maiti, S. K. 2019. Sources, toxicity, and remediation of mercury: An essence review. *Environmental Monitoring and Assessment*, vol. 191, pp. 1–22.

Rani, L., Srivastav, A. L. & Kaushal, J. 2021. Bioremediation: An effective approach of mercury removal from the aqueous solutions. *Chemosphere*, vol. 280, pp. 130–654.

Sakamoto, M., Nakamura, M. & Murata, K. 2018. Mercury as a global pollutant and mercury exposure assessment and health effects. *Nihon eiseigaku zasshi. Japanese Journal of Hygiene*, vol. 73, pp. 258–264.

Sandeep, G., Vijayalatha, K. & Anitha, T. 2019. Heavy metals and its impact in vegetable crops. *International Journal of Chemical Studies*, vol. 7, pp. 1612–1621.

Saravanan, A., Kumar, P. S., Ramesh, B. & Srinivasan, S. 2022. Removal of toxic heavy metals using genetically engineered microbes: Molecular tools, risk assessment and management strategies. *Chemosphere*, vol. 11, pp. 134–341.

Sharma, P., Pandey, A. K., Kim, S.-H., Singh, S. P., Chaturvedi, P. & Varjani, S. 2021. Critical review on microbial community during in-situ bioremediation of heavy metals from industrial wastewater. *Environmental Technology & Innovation*, vol. 24, pp. 101–826.

Shettihalli, A. K., Palanirajan, S. K. & Gummadi, S. N. 2021. Are cysteine residues of human phospholipid scramblase 1 essential for Pb^{2+} and Hg^{2+} binding-induced scrambling of phospholipids? *European Biophysics Journal*, vol. 50, pp. 745–757.

Sivaranjanee, R. & Kumar, P. S. 2021. A review on remedial measures for effective separation of emerging contaminants from wastewater. *Environmental Technology & Innovation*, vol. 23, pp. 101–741.

Soniari, N. N. & Atmaja, I. W. D. 2019. Isolation and identification of *Azotobacter* of some type of land use in Jegu villages. *International Journal of Biosciences and Biotechnology*, vol. 6, pp. 106–113.

Svoboda, M., Bureš, J., Drápal, J., Geboliszová, K., Haruštiaková, D., Nepejchalová, L., Skočovská, M. & Svobodová, Z. 2021. A survey of mercury content in pig tissues carried out in the Czech Republic during years 2015–2019. *Acta Veterinaria Brno*, vol. 90, pp. 287–293.

Teng, D., Mao, K., Ali, W., Xu, G., Huang, G., Niazi, N. K., Feng, X. & Zhang, H. 2020. Describing the toxicity and sources and the remediation technologies for mercury-contaminated soil. *RSC Advances*, vol. 10, pp. 23221–23232.

Thomas, S. A., Mishra, B. & Myneni, S. C. 2020. Cellular mercury coordination environment, and not cell surface ligands, influence bacterial methylmercury production. *Environmental Science & Technology*, vol. 54, pp. 3960–3968.

Vareda, J. P., Valente, A. J. & Durães, L. 2019. Assessment of heavy metal pollution from anthropogenic activities and remediation strategies: A review. *Journal of Environmental Management*, vol. 246, pp. 101–118.

Verma, S. & Kuila, A. 2019. Bioremediation of heavy metals by microbial process. *Environmental Technology & Innovation*, vol. 14, pp. 100–369.

Wang, J., Xing, Y., Xie, Y., Meng, Y., Xia, J. & Feng, X. 2019. The use of calcium carbonate-enriched clay minerals and diammonium phosphate as novel immobilization agents for mercury remediation: Spectral investigations and field applications. *Science of The Total Environment*, vol. 646, pp. 1615–1623.

Wang, Y., Selvamani, V., Yoo, I.-K., Kim, T. W. & Hong, S. H. 2021. A novel strategy for the microbial removal of heavy metals: Cell-surface display of peptides. *Biotechnology and Bioprocess Engineering*, vol. 26, pp. 1–9.

Wu, J. 2019. *Oriented Sample Solid-State NMR of the Mercury Detoxification Membrane Protein MerFt*. Master of Science, University of California, San Diego. https://escholarship.org/uc/item/7wb7k3vq.

Zango Usman, U., Mukesh, Y., Vandana, S., Sharma, J., Sanjay, P., Sidhartha, D. & Sharma Anil, K. 2020. Microbial bioremediation of heavy metals: Emerging trends and recent advances. *Research Journal of Biotechnology*, vol. 15, p. 1.

Zhao, J., Liang, X., Zhu, N., Wang, L., Li, Y., Li, Y.-F., Zheng, L., Zhang, Z., Gao, Y. & Chai, Z. 2020. Immobilization of mercury by nano-elemental selenium and the underlying mechanisms in hydroponic-cultured garlic plant. *Environmental Science: Nano*, vol. 7, pp. 1115–1125.

Zheng, R., Wu, S., Ma, N. & Sun, C. 2018. Genetic and physiological adaptations of marine bacterium *Pseudomonas stutzeri* 273 to mercury stress. *Frontiers in Microbiology*, vol. 9, pp. 6–82.

18 A Review of the Effect of Ionic/Zwitterionic Materials on Membrane Performance for Heavy Metal Removal Application

Syed Ibrahim Gnani Peer Mohamed and Arun M. Isloor

1.1 INTRODUCTION

The continuous progress in the field of science and technology led to many developments in industrial production, which positively benefited modern society to improve its living standard and economic growth. The developments in industries such as batteries, coatings, fuels, steel, aeronautics and photographic films necessitated the usage of a high quantity of various metals. In spite of their positive attributes, heavy metals are responsible for severe environmental pollution, which negatively impacts humans and animals in the world [1].

In general, "heavy metals" refer to metals and metalloids with higher density and atomic weight. In early 1817, Gmelin categorized 25 metals as heavy metals with a density ranging from 5.31 to 22.0 g/cm^3 [2]. Given the environmental impact, the most frequently identified and extremely examined "heavy metals" include lead (Pb^{2+}), nickel (Ni^{2+}), copper (Cu^{2+}), manganese (Mn^{2+}), zinc (Zn^{2+}), mercury (Hg^{2+}), iron (Fe^{3+}), arsenic (As^{3+}, As^{5+}) and chromium (Cr^{6+}), which are difficult to degrade. Consequently, these metals accumulate in living organisms and most of these heavy metals are extremely toxic or carcinogenic [3, 4]. Specifically, the excess amount of zinc in the human body causes vomiting, anemia, nausea, skin irritation and stomach cramps. Lead can damage the central nervous system, the kidneys, the reproductive system and the liver. Chromium causes lung carcinoma and skin irritation. Copper ingestion in the human body induces vomiting, convulsions and stomach cramps. Nickel and cadmium are human carcinogens and affect the lungs and kidneys. Similarly, mercury damages the central nervous system and is a known neurotoxin [5, 6].

The main source of heavy metal pollution is due to anthropogenic sources such as mining, metal refining, chemicals production in industries, fertilizer and pesticide spraying and natural sources such as volcanic eruptions and earthquakes. Thus, if these metal ions are not treated properly, they mix with surface water and groundwater and enter the human and animal body, causing severe health effects [3]. Therefore, to conserve public safety and the environment, the effective removal of heavy metal ions from water sources is very important and highly challenging.

In the past few decades, tremendous efforts have been taken to mitigate heavy metal pollution. In literature, methods such as adsorption, chemical precipitation, ion exchange, electrochemical treatment, electrodialysis and membrane separation are reported to remove heavy metal ions from the contaminated water. However, most of the proposed processes suffer from several disadvantages such as high cost, low separation efficiency, poor selectivity, not being appropriate for continuous

DOI: 10.1201/9781003326281-20

separation processes and so on. Membrane technology has proven to be an environmentally friendly, highly efficient and less energy-consuming technology.

Polymeric membranes can be classified into four types based on pore size: microfiltration (MF), ultrafiltration (UF), nanofiltration (NF) and reverse osmosis (RO). MF membranes exhibit poor removal efficiency for heavy metal ions due to the presence of larger pore sizes. UF membranes have slightly lower pore size than MF membranes. Still, UF membranes exhibit lower heavy metal ion removal efficiency. Thus, surface charge modification of the UF membrane, polymer-enhanced UF and incorporation of adsorbents in the UF membrane matrix to make adsorptive removal membranes have been developed [7]. NF membranes exhibit pore size in between UF and RO membranes. As the pore size of NF membranes is smaller than that of UF membranes, they presents higher heavy metal ion removal efficiency than UF and MF membranes. RO membranes exhibit superior metal ion removal efficiency than all other membranes. However, the higher operating pressure and lower water flux restrict its usage in heavy metal ion removal applications [8]. As shown in Figure 18.1, the increased number of publications per year indicates that membrane-based separation technology has great potential in the mitigation of heavy metal pollution.

In recent days, many different types of membrane materials have been developed and utilized for fabricating heavy metal removal membranes. However, the bottleneck associated with membrane separation is fouling, which deteriorates membrane performance by blocking the membrane pores and leads to poor water permeability and high maintenance costs. Thus, it is very challenging to control the fouling of polymeric membranes. In addition, heavy metal separation membranes work on the principles of Donnan and size exclusion to remove the heavy metal ions present in wastewater. However, this is possible with some compromise in the permeability to maintain the size of the pores and surface charge. Thus, it is also desired to have some metal-binding coordination sites on the membrane surface to improve their performance.

Research on the incorporation of ionic liquids (ILs) in the membrane matrix is ongoing. ILs are green reagents and are non-volatile. These are molten salt, which exists as a liquid at room temperature. In general, ILs consist of cationic moieties such as pyridinium, quaternary ammonium and imidazolium and corresponding counter anions such as Cl^-, Br^-, hexafluorophosphate (PF_6) and bis(trifluoromethylsulfonyl) imide (TFSI). The availability of a variety of polymerizable cationic moieties and the corresponding counter anions provides a feasible option for further chemical and surface properties optimization [9]. ILs also help in increasing heavy metal ion adsorption by increasing electrostatic interaction.

Zwitterionic polymers or nanoparticles are novel groups of hydrophilic materials which possess both positive and negative charges in the core moiety. The presence of these zwitterionic functional

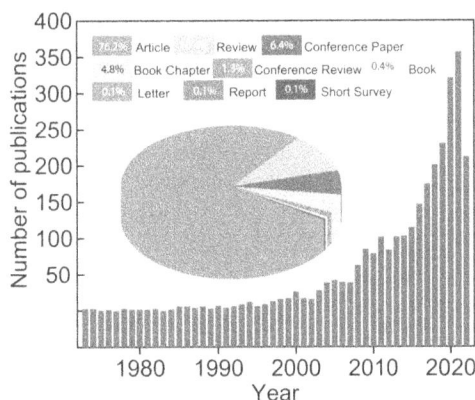

FIGURE 18.1 The number of publications per year based on the keyword "heavy metal removal membrane" from the Scopus database as of June 2022. The inset shows the document types.

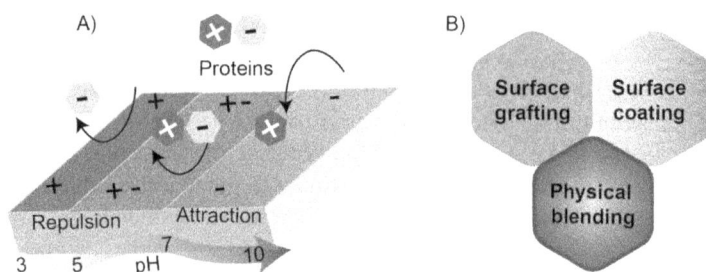

FIGURE 18.2 (A) The antifouling mechanism of zwitterionic materials as a function of pH and (B) schematic of methods of zwitterionic functionalization of polymeric membranes.

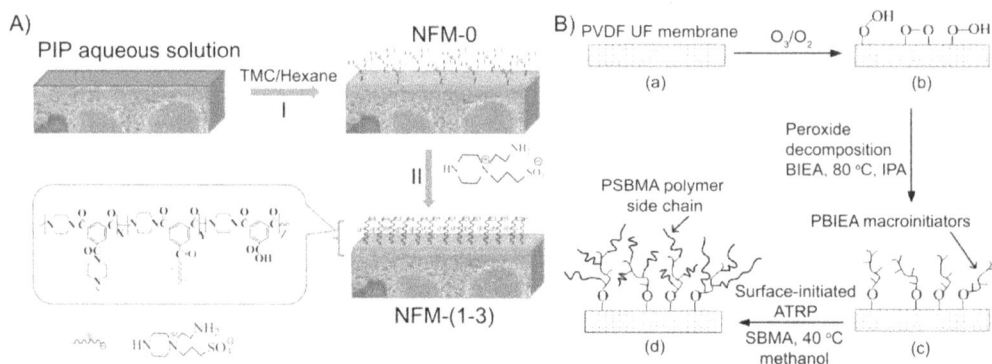

FIGURE 18.3 (A) The surface grafting-to and [17] (B) grafting-from approach for the synthesis of zwitterionic polymeric membranes [18].

groups helps in forming a hydration layer on the membrane surface and offers electrostatic repulsion at neutral pH, which avoids foulant adsorption [10] (Figure 18.2A). In addition, it also acts as a ligand for coordinating with different metal ions [7]. Zwitterionic materials found their application in many fields such as drug delivery [12], antifouling coating [13], biomedicals [14], biosensors [15], membranes [16] and so on.

There are different types of approaches that have been proposed by researchers to fabricate the zwitterionic functionalization of polymeric membranes. As presented in Figure 18.2B, there are three types of approaches such as surface grafting, surface coating and physical blending that are well-known in the literature. In surface grafting, grafting-to and grafting-from are famous techniques. In the case of grafting-to, the end-functionalized polymer chain directly binds to the membrane surface via covalent bonding. For instance, the fabrication of zwitterionic material via ring-opening of 1.3 propane sultone with amine is reported (Figure 18.3A) [17]. However, in the grafting-from technique, the zwitterionic material is formed on the membrane surface during the grafting process. The surface grating-from technique was demonstrated to modify the polyvinylidene fluoride (PVDF) UF membrane via atom transfer radical polymerization (ATRP) (Figure 18.3B) [18]. The surface coating-to method includes the post-modification of membranes with zwitterionic polymers. However, in the surface coating-from technique, the zwitterionic materials are formed in situ [19]. Finally, the zwitterionic membranes are prepared via physical blending. This technique is more popular and easier than surface grafting and coating. It avoids the usage of specific instruments and reduces the preparation time and cost. In this chapter, the effect of ionic/zwitterionic materials on the performance enhancement of polymeric membranes for heavy metal removal application is discussed.

1.2 ADSORPTIVE REMOVAL/UF MEMBRANES

Traditionally, methods such as ion exchange [20], chemical precipitation [21], electrochemical treatment [22] and flotation [23] are established for treating heavy metal-contaminated water. However, difficulties, for example, management of a large amount of sludge, low efficiency, usage of costlier resin and its regeneration, make these processes complicated [3]. Recently, membrane-based separation processes have attracted many researchers' interest and have become an attractive candidate for heavy metal removal applications [24].

Membrane-based separation processes such as RO, NF and UF are well-established for effective water treatment [24, 25]. Among these processes, UF membranes play a vital role in water treatment as they operate at low pressure and can be easily fabricated [24]. However, UF membranes are severely hampered by fouling when treated with wastewater. Thus, the development of fouling-resistant and high-flux UF membranes is challenging. In recent literature, several novel methodologies have been put forward to fabricate ionic/zwitterionic UF and adsorptive removal membranes. The presence of ionic/zwitterionic material on the membrane surface increases heavy metal removal via Donnan exclusion or adsorption. In addition, it enhances the antifouling property of UF membranes.

Usage of polymer beads for the adsorption of heavy metals is limited due to the reduced active binding sites and specific surface area. Thus, research is ongoing on the fabrication of an adsorptive membrane via electrospinning for heavy metal removal applications. It improves the active binding sites and specific surface area. In the electrospinning process, an electric field is applied to the tip of the needle; as a result, it produces the polymer solution jet from the tip of the needle. The as-produced polymer liquid jet travels toward the collector to neutralize its charge. During this process, the solvent molecules evaporate from the polymer solution and lead to the formation of the nanofiber on the collector [26].

Functionalization of ILs on the chitosan/polyvinyl alcohol (PVA) nanofiber membrane to improve adsorption capacity was reported [27]. Adsorption analysis indicated that the chitosan/PVA blend nanofiber functionalized with IL exhibited higher removal of 82.5% for the Pb^{2+} ion. However, the chitosan/PVA blend nanofiber showed only 18% of Pb^{2+} ion removal. The increased adsorption capacity of chitosan/PVA/IL nanofiber was attributed to the presence of IL, which increased the electrostatic attraction between the heavy metal ion and the Cl^- ion present in the IL. Furthermore, the effect of pH on heavy metal adsorption was also studied. The adsorption capacity was increased by increasing the pH. The maximum adsorption capacity of 39 mg/g was observed at pH 9. However, adsorption was low at a pH below 5. Low adsorption capacity below a pH of 5 was attributed to the increased electrostatic repulsion between the heavy metal ion and the adsorbent. In addition, the increased concentration of H^+ and H_3O^+ at a pH below 5 also competed over Pb^{2+} ion adsorption. The adsorption isotherm analysis indicated that the Freundlich isotherm fits well with the current study, which demonstrates that adsorption follows the multilayer adsorption pathway.

Ren et al. demonstrated the usage of poly(ionic liquid) (PIL) membranes for heavy metal ion removal [28]. Here, imidazolium-based PILs were prepared along with poly(ethylene glycol) diacrylate (PEGDA) via photocrosslinking. The effect of water uptake and swelling ratio on metal ion adsorption was also evaluated. It was noted that the increased water uptake increases the dissociation of metal ions and can lead to an increase in the diffusion of metal ions. However, when hydrophilic PIL was converted into hydrophobic PIL by exchanging with a hydrophobic counter anion, metal ion adsorption efficiency was increased. At the same time, the higher concentration of the hydrophobic part in the PIL led to a decrease in metal ion adsorption due to a reduction in the coordination efficiency. Furthermore, the microstructure of the membrane also played an important role in determining heavy metal ion adsorption efficacy. The presence of a microstructure in the polymer could accommodate more amount of metal ions in the cavity. Finally, the recyclability of this polymer was studied by acid treatment. The PIL membrane could exhibit consistent adsorption even after ten cycles.

Recently, zwitterionic modification of polyvinyl chloride (PVC) with L-cysteine was demonstrated for arsenic(V) removal [29]. The as-synthesized zwitterionically functionalized PVC was blended with polysulfone. L-cysteine exhibited an isoelectric point at pH 5.1. Thus, a zwitterionically modified PVC/polysulfone blend membrane could show a neutral charge. As arsenic (V) presents as $HAsO_4^{2-}$ at the neutral pH, the positive ion (amine group) in the zwitterionic functional group attracts, at the same time, to maintain the electrical neutrality the negative ion (acid group) rejects the $HAsO_4^{2-}$. Thus, the composite membrane could exhibit 73% of arsenic(V) rejection via the Donnan exclusion effect. In the same study, the surface charge was further modified with the addition of a small amount of TiO_2. The added TiO_2 increased the surface negative charge and led to arsenic(V) rejection of 85%. In addition, the zwitterionic membrane also showed better antifouling performance when bovine serum albumin (BSA) was used as a model foulant. The blend membrane could exhibit a flux recovery ratio (FRR) of up to 81.8%.

Li et al. reported the functionalization of glass fiber with zwitterionic ornithine methacrylamide via radical polymerization (Figure 18.4) [11]. The as-prepared membrane exhibited superior heavy metal removal ability. Especially, the membrane presented superior removal ability for heavy metals such as Cu^{2+} and Ni^{2+} ions. The membrane exhibited adsorptive removal of around 36.3 and 6.9 mg/g for Cu^{2+} and Ni^{2+} ions, respectively. The reason for the improved adsorptive removal of Cu^{2+} and Ni^{2+} ions was due to the chelation effect offered by the zwitterionic polymer. It was also mentioned that among the two heavy metal ions tested, the Cu^{2+} ion was adsorbed more on the zwitterionic polymer-coated glass fiber membrane. The higher adsorption of Cu^{2+} ions was due to effective chelation between the Cu^{2+} ions and zwitterionic polymer.

Although the UF membrane could exhibit higher flux than NF and RO membranes, the larger pore size of the UF membrane leads to a reduction in heavy metal removal ability with time. Thus, polymer-enhanced UF was proposed for the effective removal of heavy metal ions present in wastewater. In this context, a zwitterionic glycine betaine/polysulfone blend membrane was prepared by Moideen et al. [30]. The zwitterionic blend membrane presented 80.2 and 71.4% of Pb^{2+} and Cd^{2+} ions removal efficiency via polymer-enhanced UF. Here, the polymer polyethyleneimine (PEI) was used as a complexing agent. The heavy metal ions form a complex with PEI; consequently, the size of the metal complex increases, increasing heavy metal removal. The reason for the slightly lower percentage of Ni^{2+} when compared to Pb^{2+} was attributed to the stability of the Pb^{2+}-PEI complex. The blend membrane could also exhibit FRR better due to the presence of zwitterionic glycine

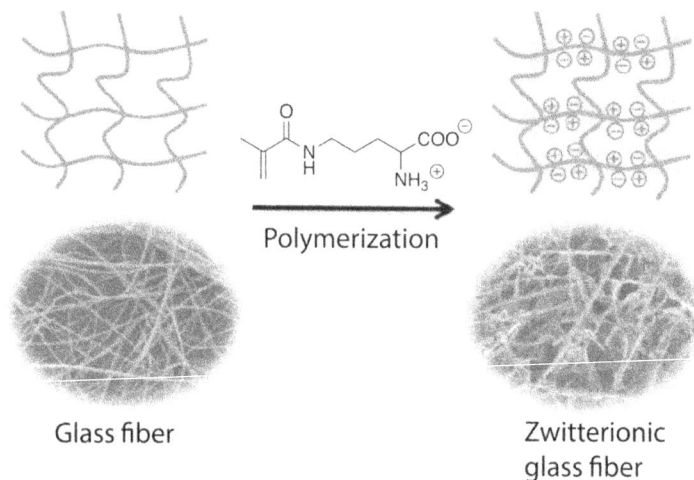

Glass fiber → Polymerization → Zwitterionic glass fiber

FIGURE 18.4 Synthetic route for the functionalization of glass fiber membrane with zwitterionic ornithine methacrylamide via radical polymerization [33].

betaine in the polysulfone membrane matrix. From an environmental point of view, these polymer-enhanced UF membranes increase the cost and the processing time of the UF membrane. Therefore, it is highly desirable to have technology with higher heavy metal removal and without the need for any additional coordinating polymer for the heavy metal ions.

1.3 NF MEMBRANES

NF membranes are effective pre-treatment membranes for RO. The smaller pore size of NF membranes compared to UF membranes offers selective separation of large varieties of divalent ions and small organic molecules with the molecular weight cut-off (MWCO) in the range of 200–1,000 Da [31]. NF membranes mainly work on the Donnan exclusion principle and size exclusion comes into the picture during the NF of uncharged species. The NF membrane has versatile applications such as heavy metal removal, dye separation, desalination, separation of antibiotics and so on [32]. The main bottleneck in the NF membrane is the fouling during the operation. Many efforts have been undertaken to mitigate NF membrane fouling; the incorporation of zwitterionic materials in the NF membrane exhibited superior antifouling performance and flux.

NF membranes are extensively used for the treatment of wastewater contaminated with heavy metal ions. NF membranes operate at relatively low pressure and are cost-effective compared to RO [33]. To improve the fouling resistance of membranes, different zwitterionic materials such as carboxybetaine, sulfobetaine and phosphobetaine and ILs containing pyridinium, quaternary ammonium and imidazolium moieties have been reported in the literature [34, 35]. The membranes functionalized with an ionic/zwitterionic surface bestow improved antifouling performance by forming a hydration layer on the membrane surface. The as-formed hydration layer avoids the attachment of foulant molecules on the membrane surface.

Zwitterionically functionalized octa glycidyloxypropyl-silsesquioxane nanoparticles incorporated in a NF membrane were reported (Figure 18.5) [36]. Here, the octa glycidyloxypropyl-silsesquioxane (glycidyl POSS) nanoparticles were functionalized with the zwitterionic molecule L-cysteine (Figure 18.5). The L-cysteine was attached to glycidyl-polyhedral oligomeric silsesquioxane (POSS) via epoxide ring-opening. The as-synthesized zwitterionic nanoparticles were blended with polyetherimide (PEI) in different ratios. The mean pore size of the membrane was increased with an increase in the concentration of functionalized nanoparticles. However, the porosity of the membrane was reduced with the higher loading of nanoparticles, which was attributed to the agglomeration of the nanoparticles at the higher concentration. The nanocomposite membrane exhibited maximum Cr^{2+} removal of 84%. However, the Cr^{2+} removal efficiency of the nanocomposite membrane started decreasing with an increased concentration of functionalized nanoparticles. The lower removal efficiency of the nanocomposite membrane with a higher concentration of zwitterionic POSS nanoparticles was ascribed to the increased pore size of the nanocomposite membrane. The nanocomposite membrane also exhibited an improved FRR. The improved FRR of the membrane was due to the improvement in surface hydrophilicity by the as-added zwitterionic nanoparticles.

The effective removal of heavy metals can be accomplished by thin-film composite (TFC) membranes. These are state-of-the-art NF membranes prepared via interfacial polymerization (IP). The TFC NF membrane consists of a porous UF or MF support (generally polysulfone) and selective polyamide (PA) layer. The selective PA layer controls the permeability and solute rejection of the TFC membrane. In literature, many techniques such as support modification, usage of different monomers for IP, post-treatment, adjusting temperature and time of IP and so on have been proposed to fine-tune the PA of TFC membranes.

Zhang et al. demonstrated the post-modification of a TFC membrane with phosphonium IL [37]. In this work, tetrakis (hydroxymethyl) phosphonium chloride was grafted into the PA layer after IP (Figure 18.6). The as-added ionic liquid significantly influenced membrane properties such as morphology, surface zeta potential and hydrophilicity. The post-modified TFC membrane in the

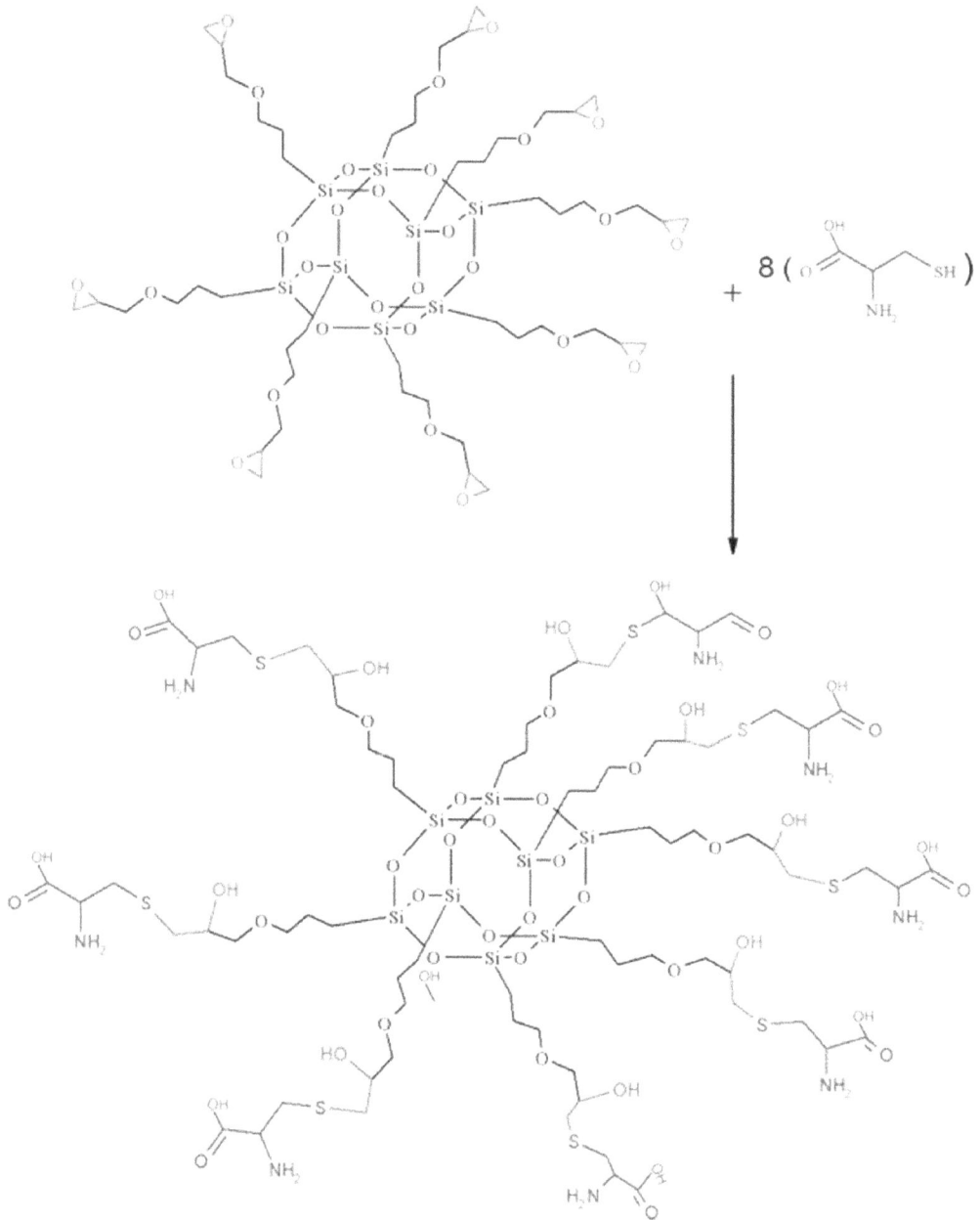

FIGURE 18.5 Zwitterionic functionalization of octa glycidyloxypropyl-silsesquioxane nanoparticles with L-cysteine [40].

presence of dimethylamino pyridine (DMAP) as a base exhibited an increase of pure water permeability from 9.4 to 17.3 L/m^2 h bar. The increased water permeability was attributed to the improved hydrophilicity by the addition of ILs. In addition, the post-modification time also influenced membrane performance. The increased post-modification time decreased the water permeability due to the formation of a denser modified PA layer. This modified TFC NF membrane was employed to remove heavy metal ions such as Pb^{2+}, Zn^{2+}, Cu^{2+}, Cd^{2+} and Ni^{2+}. Among these metal ions tested, the NF membrane exhibited slightly lower rejection for Pb^{2+} ions. The lower rejection of this membrane

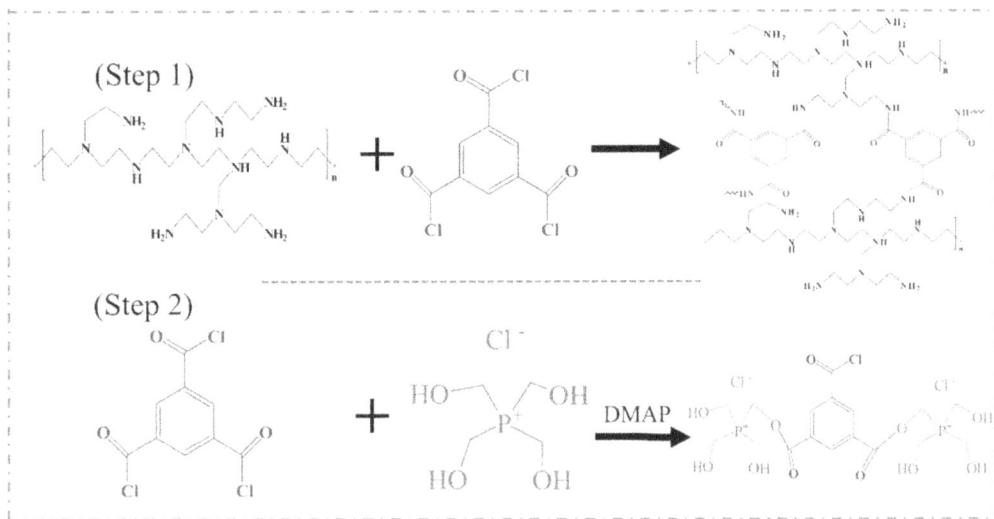

FIGURE 18.6 Fabrication of ionic liquid functionalized TFC membrane [41].

was attributed to the smallest hydrated radius of Pb^{2+} ions. Furthermore, the IL-functionalized TFC NF membrane exhibited antibacterial activity by destabilizing the *Escherichia coli* cell wall.

The fabrication of a zwitterionic TFC membrane was demonstrated to remove heavy metals such as selenite and arsenate ions [38]. In this study, phosphobetaine zwitterionic copolymer was integrated into the TFC membrane PA layer via IP. When the zwitterionic copolymer loading was 50 wt%, the pore size was reduced to 0.63 nm. However, the pore size started increasing on increasing the zwitterionic copolymer content above 50 wt%. The increase in pore size was attributed to the decreased crosslinking degree in the PA layer. The zwitterionically modified TFC membrane also exhibited enhanced water permeability when compared to the pristine TFC membrane without compromise in the solute rejection. However, copolymer loading above 50 wt% exhibited deterioration in the salt rejection. Thus, the TFC membrane prepared with 50 wt% of the zwitterionic copolymer was tested for selenium and arsenate removal application. The modified membrane exhibited higher removal of 98.2% and 99.8% for selenite and arsenate, respectively. The enhanced rejection behavior of the zwitterionic TFC membrane compared to the pristine TFC membrane was attributed to the as-formed hydration layer and reduced pore size. The presence of both positive and negative charges on the hydration layer assisted in increasing the repulsion of both charged ions. The higher removal of arsenate was due to the higher hydrated radius compared to selenite.

Recently, the development of thin-film nanocomposite (TFN) membranes for heavy metal removal applications has been emerging. Although the TFC membrane could exhibit improved heavy metal removal efficiency, the water permeability is reasonably low as compared to TFN membranes. The TFN membrane was first proposed by Hoek et al. by incorporating zeolite nanoparticles in the PA layer [39]. TFN membranes are almost like TFC membranes; however, TFN membranes have nanomaterials in the PA layer. It was reported that the presence of nanomaterials in the PA layer manipulates the surface chemistry, charge and hydrophilicity and creates nanovoids. The incorporation of nanomaterials on the PA layer during IP can be done in two ways: by dispersing the nanomaterials either in the aqueous phase or in the organic phase. Most researchers have claimed that the incorporation of nanomaterials in the organic phase will alleviate the loss of nanomaterials during IP [40]. At the same time, the type of methodology also depends on the surface hydrophilicity of the nanomaterials. Hydrophilic nanomaterials are preferably dispersed in the aqueous phase and hydrophobic nanomaterials are generally dispersed in the organic phase during IP. Thus, the functionalization of nanomaterials to make them suitable for the organic phase is also another ongoing research area. Therefore, the development of TFN membranes with improved water permeability without compromise in heavy metal removal is challenging.

The development of poly(ionic liquid) (PIL) functionalized SiO_2 nanomaterials incorporated TFN membranes was reported by Shen et al. [41]. SiO_2 was modified with 1-allyl-3-methyl imidazolium hexafluorophosphate via ATRP. Figure 18.7 presents the synthetic methodology for the ionic modification of SiO_2. The functionalized SiO_2 exhibited an increased positive charge with an isoelectric point (IEP) of 5.2 compared to pristine SiO_2 due to the presence of ILs. The enhanced surface charge also contributed to the enhanced colloidal stability of the modified SiO_2. The ionic SiO_2 was added into the aqueous phase during the IP. The water flux of this TFN membrane was tested with varying concentrations of pristine SiO_2 and ionic SiO_2 in the PA layer. The TFN membrane prepared with 0.3% pristine SiO_2 exhibited improved water flux and solute rejection. However, the performance started decreasing when the concentration of pristine SiO_2 was above 0.3% in the PA layer due to partial agglomeration. At the same time, the TFN membrane prepared with 0.5% ionic SiO_2 could exhibit higher flux and solute rejection. The improved loading of ionic SiO_2 was attributed to its colloidal stability. Furthermore, the TFN membrane could present higher heavy metal removal compared to the pristine PA membrane. The TFN membrane prepared with 0.5% ionic SiO_2 demonstrated higher rejection of 90.2, 88.3 and 71.9% to Cu^{2+}, Ni^{2+} and Cd^{2+} ions, respectively. The slightly lower rejection of the Cd^{2+} ion, when compared to other metal ions, was attributed to its higher diffusivity (0.87×10^{-9} m^2 s^{-1}). Overall, the TFN membrane could show 15% higher heavy metal removal ability than the pristine PA membrane. The higher heavy removal ability of this TFN membrane was ascribed to the increased positive charge of the TFN membrane, which improved the electrostatic repulsion. The effect of the presence of other interfering ions in the feed was also studied. The results indicated that the TFN membrane could still exhibit stable performance toward heavy metal rejection.

The incorporation of zwitterionic nanoparticles in the selective PA layer of the TFN NF membrane was described [42]. In this work, the polymeric nanoparticles were prepared by distillation precipitation polymerization (DPP) and zwitterionic modification was performed via thiol-ene click chemistry (Figure 18.8). The as-synthesized zwitterionic nanomaterial was incorporated into the selective PA layer by dispersing in the organic phase. The surface zeta potential of the zwitterionic TFN membrane was less negative due to the presence of zwitterionic nanoparticles. The water permeability was increased with the increase of nanomaterial loading; however, the solute rejection started decreasing with higher loading of the nanomaterials. The decrease in the solute rejection was attributed to the agglomeration of nanomaterials at the higher concentration, which created more defects in the PA layer. The membrane exhibited water permeability of 11.4 L/m^2 h bar and an MWCO of 743 Da. As reported, the zwitterionic membrane could remove 99.4% and 95.6% of Pb^{2+}

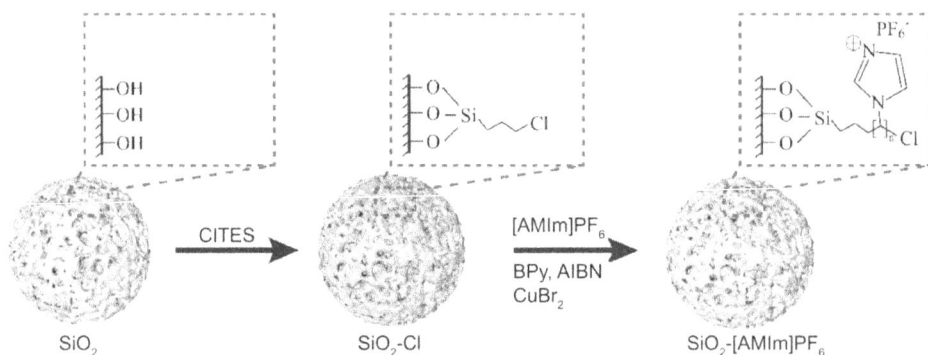

FIGURE 18.7 Synthetic methodology for the preparation of ionic liquid functionalized SiO_2 [45].

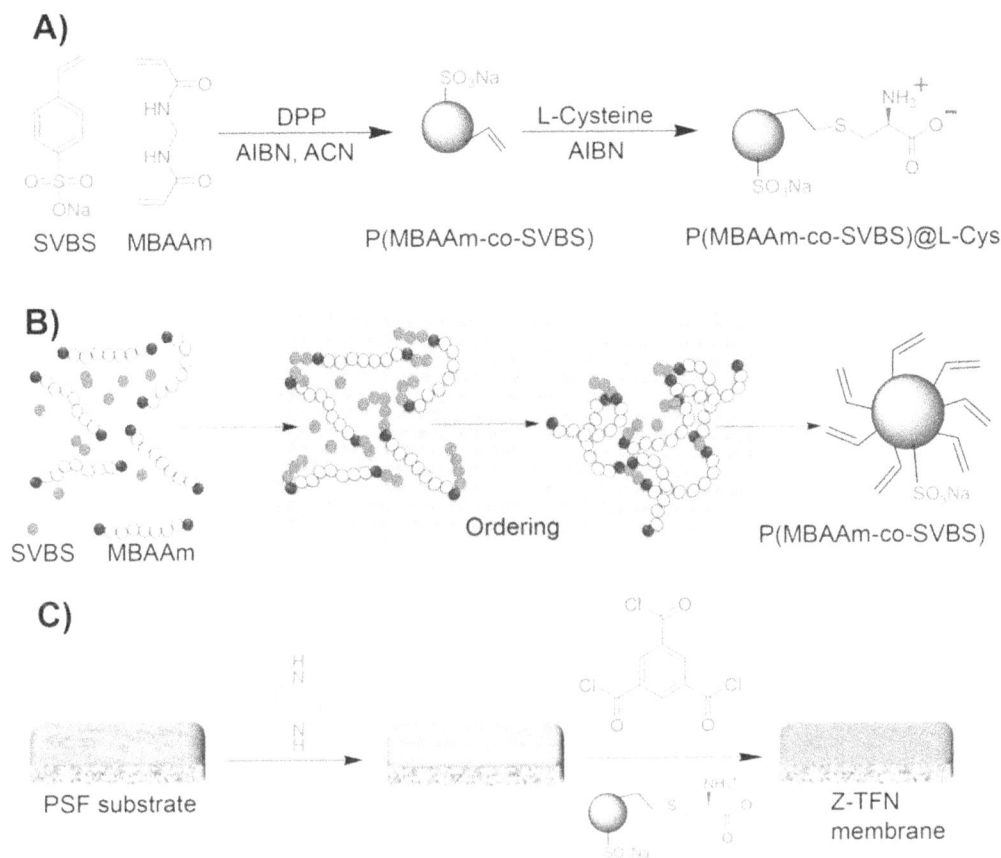

FIGURE 18.8 Synthesis of zwitterionic nanoparticles and their formation mechanism and schematic of fabrication of zwitterionic TFN membrane [46].

and Cd^{2+} ions, respectively. The preferable mechanism for heavy metal removal of this membrane was Donnan exclusion. The TFN membrane could also exhibit FRR of 83.1% with a 25% of reduction in flux when tested with BSA as a model foulant.

1.4   CONCLUSIONS AND FUTURE TRENDS

In summary, the usage of ionic/zwitterionic materials incorporated in polymeric membranes for heavy metal removal applications is gaining more visibility among researchers across the globe. These ionic/zwitterionic materials have the massive potential to reduce the fouling on the membrane surface via hydration layer formation. These materials also enhance heavy metal removal by improving electrostatic interaction. In addition, the optimal loading of ionic/zwitterionic materials into the polymeric membrane matrix increases the water permeability without compromising the solute rejection. Much research work has been done to understand the effect of ionic/zwitterionic materials on the performance enhancement of polymeric membranes; still, there is a lot of room for development, especially, the development of a cost-effective synthetic route for the synthesis of ionic/zwitterionic materials and their chemistry for the modification of polymeric membranes. Another bottleneck is the optimization of pore size in UF membranes for heavy metal removal applications. Although UF membranes have high water permeability, heavy metal rejection is low due to the large pore size when compared to NF membranes. Therefore, more efforts should be undertaken to make these ionic/zwitterionic modified polymeric membranes attractive candidates for heavy metal removal applications.

REFERENCES

[1] W. Liang, G. Wang, C. Peng, J. Tan, J. Wan, P. Sun, Q. Li, X. Ji, Q. Zhang, Y. Wu, Recent advances of carbon-based nano zero valent iron for heavy metals remediation in soil and water: A critical review, *J. Hazard. Mater.* 426 (2022) 127993.

[2] S.J. Hawkes, What is a "heavy metal"?, *J. Chem. Educ.* 74 (1997) 1374.

[3] X. Feng, R. Long, L. Wang, C. Liu, Z. Bai, X. Liu, A review on heavy metal ions adsorption from water by layered double hydroxide and its composites, *Sep. Purif. Technol.* 284 (2022) 120099.

[4] S. Ibrahim, M. Mohammadi Ghaleni, A.M. Isloor, M. Bavarian, S. Nejati, Poly(homopiperazine–amide) thin-film composite membrane for nanofiltration of heavy metal ions, *ACS Omega* 5 (2020) 28749–28759.

[5] F. Fu, Q. Wang, Removal of heavy metal ions from wastewaters: A review, *J. Environ. Manage.* 92 (2011) 407–418. https://doi.org/10.1016/j.jenvman.2010.11.011.

[6] N. Abdullah, N. Yusof, W. Lau, J. Jaafar, A. Ismail, Recent trends of heavy metal removal from water/wastewater by membrane technologies, *J. Ind. Eng. Chem.* 76 (2019) 17–38.

[7] I.K. Moideen, A.M. Isloor, A.A. Qaiser, A.F. Ismail, M.S. Abdullah, Separation of heavy metal and protein from wastewater by sulfonated polyphenylsulfone ultrafiltration membrane process prepared by glycine betaine enriched coagulation bath, *Korean J. Chem. Eng.* 35 (2018) 1281–1289. https://doi.org/10.1007/s11814-018-0018-8.

[8] M.C. Nayak, A.M. Isloor, B. Lakshmi, H.M. Marwani, I. Khan, Polyphenylsulfone/multiwalled carbon nanotubes mixed ultrafiltration membranes: Fabrication, characterization and removal of heavy metals Pb^{2+}, Hg^{2+}, and Cd^{2+} from aqueous solutions, *Arab. J. Chem.* 13 (2020) 4661–4672.

[9] A. Berthod, M. Ruiz-Angel, S. Carda-Broch, Ionic liquids in separation techniques, *J. Chromatogr. A* 1184 (2008) 6–18.

[10] G. P. S. Ibrahim, A.M. Isloor, A.F. Ismail, R. Farnood, One-step synthesis of zwitterionic graphene oxide nanohybrid: Application to polysulfone tight ultrafiltration hollow fiber membrane, *Sci. Rep.* 10 (2020) 6880. https://doi.org/10.1038/s41598-020-63356-2.

[11] W. Li, K. Chu, L. Liu, Multipurpose zwitterionic polymer-coated glass fiber filter for effective separation of oil–water mixtures and emulsions and removal of heavy metals, *ACS Appl. Polym. Mater.* 3 (2021) 1276–1284.

[12] M. Harijan, M. Singh, Zwitterionic polymers in drug delivery: A review, *J. Mol. Recognit.* 35 (2022) e2944.

[13] B. He, Y. Du, B. Wang, X. Zhao, S. Liu, Q. Ye, F. Zhou, Self-healing polydimethylsiloxane antifouling coatings based on zwitterionic polyethylenimine-functionalized gallium nanodroplets, *Chem. Eng. J.* 427 (2022) 131019.

[14] M. Zhang, P. Yu, J. Xie, J. Li, Recent advances of zwitterionic-based topological polymers for biomedical applications, *J. Mater. Chem. B* 10(14) (2022) 2338–2356.

[15] Q. Wang, Z.-H. Ren, W.-M. Zhao, L. Wang, X. Yan, A. Zhu, F. Qiu, K.-K. Zhang, Research advances on surface plasmon resonance biosensors, *Nanoscale* 14 (2022) 564–591. https://doi.org/10.1039/D1NR05400G.

[16] G.P.S. Ibrahim, A.M. Isloor, Inamuddin, A.M. Asiri, N. Ismail, A.F. Ismail, G.M. Ashraf, Novel, one-step synthesis of zwitterionic polymer nanoparticles via distillation-precipitation polymerization and its application for dye removal membrane, *Sci. Rep.* 7 (2017) 15889. https://doi.org/10.1038/s41598-017-16131-9.

[17] Y.-F. Mi, Q. Zhao, Y.-L. Ji, Q.-F. An, C.-J. Gao, A novel route for surface zwitterionic functionalization of polyamide nanofiltration membranes with improved performance, *J. Membr. Sci.* 490 (2015) 311–320. https://doi.org/10.1016/j.memsci.2015.04.072.

[18] Y.-C. Chiang, Y. Chang, A. Higuchi, W.-Y. Chen, R.-C. Ruaan, Sulfobetaine-grafted poly(vinylidene fluoride) ultrafiltration membranes exhibit excellent antifouling property, *J. Membr. Sci.* 339 (2009) 151–159. https://doi.org/10.1016/j.memsci.2009.04.044.

[19] R. Zhang, Y. Liu, M. He, Y. Su, X. Zhao, M. Elimelech, Z. Jiang, Antifouling membranes for sustainable water purification: Strategies and mechanisms, *Chem. Soc. Rev.* 45 (2016) 5888–5924. https://doi.org/10.1039/C5CS00579E.

[20] S.T. Hussain, S.A.K. Ali, Removal of heavy metal by ion exchange using bentonite clay, *J. Ecol. Eng.* 22(1) (2021) 104–111.

[21] J. Kim, S. Yoon, M. Choi, K.J. Min, K.Y. Park, K. Chon, S. Bae, Metal ion recovery from electrodialysis-concentrated plating wastewater via pilot-scale sequential electrowinning/chemical precipitation, *J. Clean. Prod.* 330 (2022) 129879.

[22] Y. Song, S. Gao, X. Yuan, R. Sun, R. Wang, Two-compartment membrane electrochemical remediation of heavy metals from an aged electroplating-contaminated soil: a comparative study of anodic and cathodic processes, *J. Hazard. Mater.* 423 (2022) 127235.

[23] G. Pooja, P.S. Kumar, S. Indraganti, Recent advancements in the removal/recovery of toxic metals from aquatic system using flotation techniques, *Chemosphere* 287 (2022) 132231.

[24] G.S. Ibrahim, A.M. Isloor, A.M. Asiri, A. Ismail, R. Kumar, M.I. Ahamed, Performance intensification of the polysulfone ultrafiltration membrane by blending with copolymer encompassing novel derivative of poly(styrene-co-maleic anhydride) for heavy metal removal from wastewater, *Chem. Eng. J.* 353 (2018) 425–435.

[25] G.S. Ibrahim, A.M. Isloor, E. Yuliwati, A.F. Ismail, Carbon-based nanocomposite membranes for water and wastewater purification, in: Woei-Jye Lau; Ahmad Fauzi Ismail; Arun Isloor; Amir Al-Ahmed (Eds.), *Advanced Nanomaterials for Membrane Synthesis and Its Applications*, Elsevier, 2019: pp. 23–44.

[26] K. Ohkawa, D. Cha, H. Kim, A. Nishida, H. Yamamoto, Electrospinning of chitosan, *Macromol. Rapid Commun.* 25 (2004) 1600–1605.

[27] N. Rosli, W.Z.N. Yahya, M.D.H. Wirzal, Crosslinked chitosan/poly(vinyl alcohol) nanofibers functionalized by ionic liquid for heavy metal ions removal, *Int. J. Biol. Macromol.* 195 (2022) 132–141.

[28] Y. Ren, J. Zhang, J. Guo, F. Chen, F. Yan, Porous poly(ionic liquid) membranes as efficient and recyclable absorbents for heavy metal ions, Macromol. *Rapid Commun.* 38 (2017) 1700151.

[29] V. Nayak, R.G. Balakrishna, M. Padaki, K. Soontarapa, Zwitterionic ultrafiltration membranes for As(V) rejection, *Chem. Eng. J.* 308 (2017) 347–358.

[30] I. Moideen K, A.M. Isloor, B. Garudachari, A. Ismail, The effect of glycine betaine additive on the PPSU/PSF ultrafiltration membrane performance, *Desalination Water Treat.* 57 (2016) 24788–24798.

[31] S.I. Gnani Peer Mohamed, S. Nejati, M. Bavarian, All-polymeric thin-film nanocomposite membrane for organic solvent nanofiltration, *ACS Appl. Polym. Mater.* 3 (2021) 6040–6044.

[32] L.W. Jye, A.F. Ismail, *Nanofiltration Membranes: Synthesis, Characterization, and Applications*, Crc Press, 2016.

[33] G.S. Ibrahim, A.M. Isloor, R. Farnood, Fundamentals and basics of reverse osmosis, in: Angelo Basile, Alfredo Cassano, Navin K. Rastogi (Eds.), *Current Trends and Future Developments on (Bio-) Membranes*, Elsevier, 2020: pp. 141–163.

[34] C.Y. Foong, M.D.H. Wirzal, M.A. Bustam, A review on nanofibers membrane with amino-based ionic liquid for heavy metal removal, *J. Mol. Liq.* 297 (2020) 111793.

[35] G. Ibrahim, A. Isloor, A. Asiri, R. Farnood, Tuning the surface properties of Fe_3O_4 by zwitterionic sulfobetaine: application to antifouling and dye removal membrane, *Int. J. Environ. Sci. Technol.* 17 (2020) 4047–4060.

[36] S. Bandehali, F. Parvizian, A. Moghadassi, S.M. Hosseini, High water permeable PEI nanofiltration membrane modified by L-cysteine functionalized POSS nanoparticles with promoted antifouling/separation performance, *Sep. Purif. Technol.* 237 (2020) 116361.

[37] X. Zhang, J. Zheng, P. Jin, D. Xu, S. Yuan, R. Zhao, S. Depuydt, Y. Gao, Z.-L. Xu, B.V. der Bruggen, A PEI/TMC membrane modified with an ionic liquid with enhanced permeability and antibacterial properties for the removal of heavy metal ions, *J. Hazard. Mater.* 435 (2022) 129010. https://doi.org/10.1016/j.jhazmat.2022.129010.

[38] Y. He, J. Liu, G. Han, T.-S. Chung, Novel thin-film composite nanofiltration membranes consisting of a zwitterionic co-polymer for selenium and arsenic removal, *J. Membr. Sci.* 555 (2018) 299–306.

[39] B.-H. Jeong, E.M. Hoek, Y. Yan, A. Subramani, X. Huang, G. Hurwitz, A.K. Ghosh, A. Jawor, Interfacial polymerization of thin film nanocomposites: A new concept for reverse osmosis membranes, *J. Membr. Sci.* 294 (2007) 1–7.

[40] G. Lai, W. Lau, S. Gray, T. Matsuura, R.J. Gohari, M. Subramanian, S. Lai, C. Ong, A. Ismail, D. Emazadah, A practical approach to synthesize polyamide thin film nanocomposite (TFN) membranes with improved separation properties for water/wastewater treatment, *J. Mater. Chem. A* 4 (2016) 4134–4144.

[41] Q. Shen, D.Y. Xing, F. Sun, W. Dong, F. Zhang, Designed water channels and sieving effect for heavy metal removal by a novel silica-poly(ionic liquid) nanoparticles TFN membrane, *J. Membr. Sci.* 641 (2022) 119945.

[42] G. Syed Ibrahim, A.M. Isloor, M. Bavarian, S. Nejati, Integration of zwitterionic polymer nanoparticles in interfacial polymerization for ion separation, *ACS Appl. Polym. Mater.* 2 (2020) 1508–1517.

Index

Note: **Bold** page numbers refer to tables; *italic* page numbers refer to figures.

Taylor & Francis Group
an **informa** business

Taylor & Francis eBooks

www.taylorfrancis.com

A single destination for eBooks from Taylor & Francis
with increased functionality and an improved user
experience to meet the needs of our customers.

90,000+ eBooks of award-winning academic content in
Humanities, Social Science, Science, Technology, Engineering,
and Medical written by a global network of editors and authors.

TAYLOR & FRANCIS EBOOKS OFFERS:

A streamlined
experience for
our library
customers

A single point
of discovery
for all of our
eBook content

Improved
search and
discovery of
content at both
book and
chapter level

REQUEST A FREE TRIAL
support@taylorfrancis.com

Routledge
Taylor & Francis Group

CRC Press
Taylor & Francis Group

For Product Safety Concerns and Information please contact our EU
representative GPSR@taylorandfrancis.com
Taylor & Francis Verlag GmbH, Kaufingerstraße 24, 80331 München, Germany

www.ingramcontent.com/pod-product-compliance
Lightning Source LLC
Chambersburg PA
CBHW080904220326
41598CB00034B/5474

* 9 7 8 1 0 3 2 3 5 3 0 6 7 *